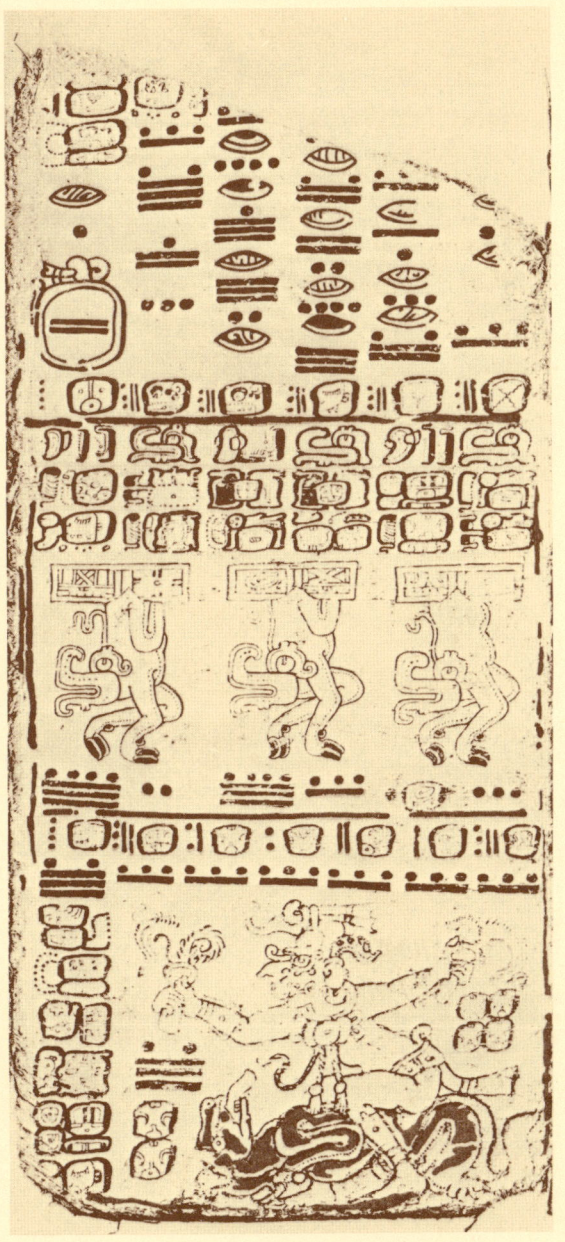

Reproduced here are three pages from the Codex Dresdensis, dated ca. 1200–1250 A.D. On the left is page 24 of Chapter 4, which consists of a table of the planetary motions (584-day synodical revolutions) of Venus. The Maya Venus tables are so accurate that only one day of error will be accumulated in 6000 years. The symbol for Venus as the morning star appears several times on this page. It has the appearance of a four-pointed star with a dot-centered circle in each lobe (see third column from left, about one-fourth distance from top of page).

In the center is page 53 from Chapter 5, which gives tables for moon phases and the dates on which solar and lunar eclipses might be visible. This page contains some dire predictions based on solar and lunar occultations. Thus, the upper part of page 53 shows the death god Cizin seated on his throne of bones. The moon hovers between light and shadow in the third glyph above his head. The picture below is of the moon goddess, hanging from the celestial band by a cord, her eye closed in death, her hands limp. The oval shell symbols (two appear on this page) represented the number zero in Maya calculations.

The right-hand frame, page 45, is probably the initial page of a missing almanac. The two columns in the upper left specify a date in time based on an accepted starting point in Maya history. The left column states that starting point. The second column tells the reader how many days must be added. A multiplication table (multiples of 364) occupies the remaining four columns. The year length was 364 days in the Maya calendar of the 1200s.

The middle and lower parts refer to another almanac or almanacs, probably related to weather and crops, and probably divinatory. The center shows a monster with upturned snout, cloven hoofs, and spots falling from the sky. The presence of axe glyphs above signifies impact from the sky, i.e., thunderbolts. Thus, the monster is identified as a "lightning beast." In the lower figure a god with torches in each hand sits on a deer dying of thirst, signifying drought. The presence of numbers refers to the time cycles of wet and dry weather.

From J.E.S. Thompson, *A Commentary on the Dresden Codex* (Philadelphia: American Philosophical Society, 1972), pp. 24, 45, and 53. (Courtesy of the New York Public Library.)

Handbook of Physical and Mechanical Testing of Paper and Paperboard

VOLUME 2

An aged god reading a Maya codex—a screenwise-folded book. Detail from a carved vase of pre-Columbian age.

Handbook of Physical and Mechanical Testing of Paper and Paperboard

VOLUME 2

Edited by
RICHARD E. MARK
Empire State Paper Research Institute
State University of New York
College of Environmental Science and Forestry
Syracuse, New York

Associate Editor
KOJI MURAKAMI
Faculty of Agriculture
Kyoto University
Kyoto, Japan

MARCEL DEKKER, INC.   New York and Basel

Cover illustration: Scanning electron micrograph of the top side of a 15 g/m^2 machine-made paper; the furnish is bleached softwood kraft beaten to 285 mℓ Canadian standard freeness. Machine direction is horizontal. The basis weight of this paper is approximately equal to that of the handmade paper depicted on the cover of Vol. 1. Also, the magnification of this picture, about 80 ×, is the same as that on the cover of Vol. 1, illustrating the greater coarseness and degree of collapse of these softwood fibers compared with papermaking bark fibers. Paper sample supplied by the Swedish Forest Products Research Laboratory. Photomicrograph by Dr. Ikuo Furukawa of Tottori University, Tottori, Japan, while on leave at the Empire State Paper Research Institute.

Library of Congress Cataloging in Publication Data
(Revised for volume 2)
Main entry under title:

Handbook of physical and mechanical testing of paper and paperboard.

Includes indexes.
1. Paper testing. 2. Paperboard testing.
I. Mark, Richard E.
TS1109.H32 1983   676'.2'0287   83-1925
ISBN 0-8247-1871-2 (v. 1)
ISBN 0-8247-7052-8 (v. 2)

COPYRIGHT © 1984 by MARCEL DEKKER, INC. ALL RIGHTS RESERVED

Neither this book nor any part may be reproduced or transmitted in any form or by any means, electronic or mechanical, including photocopying, microfilming, and recording, or by any information storage and retrieval system, without permission in writing from the publisher.

MARCEL DEKKER, INC.
270 Madison Avenue, New York, New York 10016

Current printing (last digit):
10 9 8 7 6 5 4 3 2 1

PRINTED IN THE UNITED STATES OF AMERICA

During the final preparation days of this book, all of us associated with this work were saddened to learn of the untimely death of our esteemed colleague, Dr. Fred Shafizadeh.

We are proud to dedicate this volume to his memory.

PREFACE

We have tried to assemble here as many of the modern aspects of properties testing in the paper and paperboard field as possible. One of our objectives is to enable those concerned with planning, specifying, and evaluating the physical and mechanical testing of these materials to take advantage of the many advances and improvements that have taken place in recent years. We also feel that it is important and useful to codify the excellent pioneering work that has provided the essential base for these advances, which has been included to the maximum extent possible.

As with their associates in other aspects of the broad and dynamic paper industry, the pioneers in paper testing have come from many diverse areas of the world; the authors who have contributed to these volumes live on four continents and represent a spectrum of industrial, government, and educational institutions, bringing their own special insights to the areas covered in their respective chapters. It goes without saying that the very nature of the subject matter covered here has wide-ranging applicability; our intention has been to treat the topics in our various chapters in a holistic manner.

We wish to give some justly deserved praise to the contributors to this handbook and their affiliated organizations—academic, industrial, and governmental. Creation of a multiauthored book of this type requires more than professional expertise. It also requires a great deal of diligence, perseverance, and willingness to accept criticism, as well as laborious revision and adjustment to the coverage of related topics in other chapters. In short, it is hard work that calls for dedication to purpose. An editor fortunate enough to have become associated with this group of contributors is lucky indeed, for they have been exceptional in all these aspects, as has Associate Editor Koji Murakami. The organizations of which each contributor is a part must also be praised, for without their cooperation and support, even the most dedicated contributor would find it difficult if not impossible to do all that is required. Finally, we are all deeply indebted to the management and staff of Marcel Dekker, Inc. who have combined rigorous publishing professionalism with genuinely compassionate accommodation to the severe personal traumas that accompanied the preparation of these volumes. Thank you, all.

Richard E. Mark

ACKNOWLEDGMENTS

The history of pre-Columbian paper and books in the New World is little known. Through the generous contributions of several leading companies in the paper industry, it has been possible to include color illustrations of several aspects of paper and papermaking in early Mesoamerica. We are indebted to these donors: Abitibi-Price, Inc., Crown Zellerbach Corporation, International Paper Company, Mead Corporation Foundation, St. Regis Paper Company, Scott Paper Company, Westvaco Corporation, and Weyerhaeuser Company.

CONTRIBUTORS

Gary A. Baum, Ph.D. Paper Physics Section, The Institute of Paper Chemistry, Appleton, Wisconsin

Jens Borch, Ph.D.* Paper Technology Center, Xerox Corporation, Webster, New York

Charles J. Green, Jr., B.S. Paper Technology Center, Xerox Corporation, Webster, New York

Holger Hollmark, Ph.D. Paper Technology Department, Swedish Forest Products Research Laboratory, Stockholm, Sweden

Rikizo Imamura, D.Eng.† Department of Wood Science and Technology, Faculty of Agriculture, Kyoto University, Kyoto, Japan

Takashi Kadoya, D.Eng. Department of Forest Products, The University of Tokyo, Tokyo, Japan

M. Bruce Lyne, Ph.D. Surface Physics Section, Pulp and Paper Research Institute of Canada, Pointe Claire, Quebec, Canada

Richard E. Mark, D.For. Empire State Paper Research Institute, Department of Paper Science and Engineering, State University of New York College of Environmental Science and Forestry, Syracuse, New York

Shinji Matsuda, B.Eng. Technical Research Laboratory, Tomoegawa Paper Company, Shizuoka-shi, Japan

---

*Current affiliation: Printer Materials Technology Group, General Products Division, IBM Corporation, Tucson, Arizona.
†Dr. Imamura is currently Professor Emeritus of Kyoto University.

Koji Murakami, D.Ag. Department of Wood Science and Technology, Faculty of Agriculture, Kyoto University, Kyoto, Japan

Saburo Nakagawa, B.Ag. Scientific Research Laboratory—Mochimune Mill, Tomoegawa Paper Company, Shizuoka-shi, Japan

Vance C. Setterholm, B.Sc. Criteria for Fiber Product Design, U.S. Forest Products Laboratory, Forest Service, U.S. Department of Agriculture, Madison, Wisconsin

Fred Shafizadeh, Ph.D., D.Sc.[†] Wood Chemistry Laboratory, Department of Chemistry, University of Montana, Missoula, Montana

Tetsu Uesaka, D.Ag.[*] Empire State Paper Research Institute, State University of New York College of Environmental Science and Forestry, Syracuse, New York

Makoto Usuda, D.Ag. Department of Forest Products, The University of Tokyo, Tokyo, Japan

---

[*]Current affiliation: Oji Paper Co. Ltd., Central Research Laboratory, Tokyo, Japan.
[†]Now deceased.

CONTENTS

Preface *v*
Acknowledgments *vi*
Contributors *vii*
Contents of Volume 1 *xi*
Paper, Books, and Paper Testing *xiii*

INTERACTIONS WITH LIGHT

16. Optical and Appearance Properties — 1
    Jens Borch

INTERACTIONS WITH LIQUIDS AND GASES

17. Porosity and Gas Permeability — 57
    Koji Murakami and Rikizo Imamura

18. Wetting and the Penetration of Aqueous Liquids — 103
    M. Bruce Lyne

19. The Penetration of Nonaqueous Liquids — 123
    Takashi Kadoya and Makoto Usuda

20. Absorbency of Tissue and Toweling — 143
    Holger Hollmark

ELECTRICAL AND THERMAL INTERACTIONS

21. Electrical Properties : I. Theory — 171
    Gary A. Baum

22. Electrical Properties : II. Practical Considerations and Methods of Measurement of Electrical Properties — 201
    Shinji Matsuda

23. Thermal Properties — 241
    Saburo Nakagawa and Fred Shafizadeh

OTHER PHYSICAL PARAMETERS

24. Structure and Structural Anisotropy — 283
    Richard E. Mark

25. Determination of Fiber-Fiber Bond Properties — 379
    Tetsu Uesaka

26. Dimensional Property Measurements — 403
    Vance C. Setterholm

27. Curl, Expansivity, and Dimensional Stability — 415
    Charles J. Green, Jr.

28. Fiber Structure — 445
    Richard E. Mark

Appendix   485
Subject Index   489

CONTENTS OF VOLUME 1

INTRODUCTION TO PAPER TESTING

1. Retrospect and Prospect of Physical and Mechanical Testing of Paper and Paperboard

   Alfred H. Nissan

THEORY AND TEST FOR MECHANICAL PARAMETERS

2. Models for Describing the Elastic, Viscoelastic, and Inelastic Mechanical Behavior of Paper and Board

   Richard W. Perkins, Jr.

3. Specimen Design for Mechanical Testing of Paper and Paperboard

   Tetsu Uesaka

4. Observations on Load-Deformation Testing

   Vance C. Setterholm and Dennis E. Gunderson

5. Failure Phenomena

   Jay A. Johnson, Keith A. Bennett, and Henry M. Montrey

6. The Measurement of Viscoelastic Behavior for the Characterization of Time-, Temperature-, and Humidity-Dependent Properties

   Petter Kolseth and Alf de Ruvo

7. Bending Stiffness, with Special Reference to Paperboard
   Christer Fellers and Leif A. Carlsson

8. Edgewise Compression Strength of Paper
   Christer Fellers

9. Corrugated Fiberboard
   John W. Koning, Jr.

10. Mechanical Properties of Fibers
    Richard E. Mark and Peter P. Gillis

11. Mechanical Properties of Tissue
    Holger Hollmark

GENERAL INSTRUMENTATION

12. Conditioned Test Atmospheres
    John L. de Yong

13. Data Acquisition and Processing by Mini- or Microcomputers
    Jack M. Mendel and Carley V. Davis

14. Equipping the Paper and Board Testing Laboratory
    Helen R. Schuierer

15. Interlaboratory Reference Systems
    Helen R. Schuierer

    Appendix: Standards Index
    Subject Index

PAPER, BOOKS, AND PAPER TESTING: THE ORIGINS OF PAPER (AND BOOKS) IN THE NEW WORLD

Much has been written about the "humanity" of paper: its influence on civilization as its geographical and cultural distribution has widened; the increasing diversity in its use; and its central role in providing a convenient vehicle for the acquisition, storage, and dissemination of both tangible goods (as in packaging), and these same functions applied to intangible areas, such as communication and the infinitely diverse applications of human knowledge.

It is altogether in keeping with this humanistic, universal view of paper and civilization that we have acknowledged in our jacket design and elsewhere in these volumes, the role that paper played in the waxing and waning of human culture in several ancient societies not often associated with paper or papermaking.

The origin of paper as invented by T'sai Lun in China is a well-known story. The art of making paper also arose independently in Mesoamerica sometime before A.D. 660, and perhaps many centuries before that. As with their counterparts in the orient, Maya, Toltec, Aztec, and Zapotec papermakers utilized the bark fibers of trees of the Moraceae family. They devised their own techniques for fiber separation, washing, beating, felting, couching, sizing, drying, hot pressing, coating, and converting. They developed pigments, dyes, inks, and glues. Some of their writing was done in codices—handwritten books—fitted with covers of leather, jaguar skin, or wood, the latter often studded with decorative stones. From the few codices that survived the conquest, one senses that Mesoamerica may have been on the verge of printing when the conquistadores arrived, for the Mayas and Aztecs were already using wood, clay, and metal stamps for decorative stampings of ceramics and weavings.

Paper was of tremendous importance in pre-Columbian Mesoamerica. It was used in many rituals. It was a substantial article of commerce, and served as a vehicle to sustain and enlarge commerce generally. The Maya originators of paper built libraries—guarded stone buildings—to house their records, documents, and sacred books, which were consulted before decisions were made in matters ranging from crop planting to war. The Aztecs, who improved on the techniques of papermaking that the Maya had started, employed paper for a much broader range of records keeping, including land surveys, engineering plans, tax rolls, and tribute lists, and they had even begun to use the medium for communication when the ships of Cortés arrived at their shores.

How important a role was assigned to paper in the governance of the Aztec domain can be deduced from the tribute lists that were housed in the libraries at Tenochtitlan. Areas of Mesoamerica that were under Aztec control were required to contribute substantial quantities of foodstuffs, spices, fabrics, blankets, skins, garments, shields, incense, jade, metals, firewood, hewn timbers, and other products of value, both natural and manufactured. The tribute lists were, of course, on paper. A handful of these tribute lists, such as those in the Codex Mendoza, have been preserved. From the lists in that codex we learn that two areas of present-day southern Mexico were required to contribute an amount of paper that seems impressive when we recall that each step of the operation, from the cutting of the trees to the carrying of the bundles of paper to the Aztec capital, was done by manual labor. The tribute towns had to deliver 16,000 *resmas* twice yearly—an annual contribution amounting to 480,000 sheets.

The Aztecs did not allow any paper or other tributes to be toted to their storehouses without prior approval of the ruler's inspectors. In order to carry out their responsibilities, the inspectors had standards of quality to verify; evidently some form of quality control testing was conducted in the papermaking towns, probably by both fabricators and inspectors.

Paper also played a role in communication in the empire that Cortés encountered in 1519—a situation he used as an aid, after minor skirmishing, to bring about the empire's downfall. The shrewd Spaniard noted that Indian artists were sketching scenes of the ships, the military deployment of the soldiers and horses, and the equipment that he had landed. What was being painted on the pads, he reasoned, would be the messages delivered to the capital. He then arranged for a mounted drill and demonstration firing of his cannon, which was duly recorded by the artists. As Cortés expected, the renditions of the scene on the beach caused great consternation in Tenochtitlan, for the Aztecs had no knowledge of horsemanship, gunpowder, or firearms, and could not be sure if they were dealing with men or gods. To act on either assumption was fraught with uncertainty and danger; while Moctezuma II temporized and, with lavish gifts, tried to induce the Europeans to leave, Cortés learned where all the disaffections and weaknesses, civil and military, lay in the restive, tribute-burdened satrapies.

The rest is history. The Aztec and all the other Indian nations of Mesoamerica were conquered militarily. Following that conquest came a spiritual assalt on their customs and traditions, and a progressive disintegration of these once-proud societies ensued. It does not seem to be an overstatement to say that the loss of their paper records played a significant role in that disintegration. The chroniclers of the time duly recorded the extreme anguish of the Indians as the contents of their libraries were burned.

So complete was the destruction of the paper documents of the Maya ordered by Bishop Diego de Landa in 1561, that only four of the Maya codices have survived, and each of these relics has damaged and/or missing pages. Destruction of the Aztec libraries was not as total, and about 500 Aztec books were collected by a handful of individuals who recognized their value, monetary or otherwise. However, the man responsible for collecting most of them, the Chevalier Lorenzo Boturini, was shortly imprisoned on religious grounds. Partly "torn, pillaged, and dispersed" and partly stored with little care in a damp location for many years thereafter, the remains of Boturini's collection were finally auctioned off in 1804. At that time they were examined by the eminent scientist Alexander von Humboldt, who declared them to be so deteriorated that "there exists at present only an eighth part of the hieroglyphic manuscripts taken from the Italian traveler."

After succeeding vicissitudes and misadventures, there remain today less than 20 of the Aztec codices. Not a single Toltec or Zapotec book is known to have survived.

And so the accumulated centuries of writings by the Mesoamerican peoples—the paper record—has largely been lost and forgotten. Gone are their maps and hieroglyphic charts, their tribute lists and tax rolls, their herbals, their land surveys, the plans for their canals, roads, buildings, and other engineering and architectural works, the records of their wars and migrations, their histories, genealogies, charts of the constellations, calendars and almanacs, and other known writings on agriculture and crops, fishing, soils, astronomy and astrology, mythology, disease and medicine, mathematics, commerce, songs and chants, cosmogony (including the legendary *History of Heaven and Earth* of the Toltecs) and especially prophecy, religion, and sacred themes. In the front inside cover, we have reproduced a few pages of the most famous relict Maya book, the Dresden Codex.

Although giving recognition to the Amerindian paper-making societies constitutes only a minor (but we hope interesting and thought-provoking) part of these volumes, we have made one use of the number system of the Maya in the chapter texts. Their vigesimal system (see rear inside cover design) lends itself rather ideally to the problem of setting off itemizations (for example, lists of principles to follow in test design) from the rest of the textual material in a given section or subsection. We think it fulfills a real need in the publishing field, especially as regards technical books. Any comments on this usage will be welcomed by the editor.

Richard E. Mark

INTERACTIONS WITH LIGHT

Emblem of Amacoztitlan

The "Town of the Yellow Paper" was situated on a river, as indicated by blue water and droplets in its symbol. The name is thought to derive from amatl (paper), coztic (yellow), and tlan (place), although it is interesting that Amacuztic was the Aztec name for *Ficus petiolaris*, a Mesoamerican papermaking fiber tree. Two white teeth surmounted by a red gum were a form of Aztec representation of "a place where something is found." This town, number 301 on the Aztec tribute list, was required every 6 months to deliver 8000 reams of cut sheets approximately 35 × 46 cm in size, unlike other towns that supplied white paper in rolls.

16
OPTICAL AND APPEARANCE PROPERTIES

JENS BORCH*

Paper Technology Center
Xerox Corporation
Webster, New York

I.	Introduction	2
II.	Principles of Measurement	2
	A. Light-Sheet Interactions	2
	B. Routine Reflectance Methods	4
	C. Colorimetry	6
	D. Fluorescence and Whiteness	14
	E. Gloss and Surface Finish	16
	F. Light Transmittance Methods	20
III.	Instrumentation Characteristics	23
	A. Instrument Types	23
	B. Design Principles	26
	C. Calibration	30
IV.	Interpretation of Optical Measurements	32
	A. Kubelka-Munk Theory	32
	B. Particulate Scattering Approaches	39
	C. Interpretation of Gloss	40
	D. Formation Theory	42
	References	44

---

*Current affiliation: General Products Division, IBM Corporation, Tucson, Arizona.

I. INTRODUCTION

In all applications for which paper is used to print on or write on, the appearance of the paper contributes to the image quality that is obtainable. Thus, while the paper base must be suitable for the reproduction process that is used, the appearance characteristics will ultimately determine whether the copy is acceptable to the customer. Color is created in paper and paperboard for decorative or functional reasons. Gloss or shininess is often produced to make a product more effective in sales.

Various testing methods have been developed to measure the appearance characteristics of paper and paperboard. In order to be able to comprehend the principles behind optical testing and to be able to apply the measurement techniques effectively it is necessary to understand the manner in which paper interacts with light. This chapter examines the principles behind optical and appearance measurements of paper and paperboard. It also describes the proper choice of instrumentation and the optical theories that are available for the interpretation of optical measurements.

II. PRINCIPLES OF MEASUREMENT

A. Light-Sheet Interactions

The Physics of Light Visible light is radiation in the wavelength range 400 to 700 nm, which is perceived by the human eye at the wavelength-dependent luminosity (lightness) response shown in Fig. 1. Light interacts with matter according to the physical laws valid for the electromagnetic spectrum extending from short gamma and X rays to long radio waves. In paper and board, matter consists mainly of cellulosic fibers, pigment particles, and air voids of certain size and shape. Fibers and fiber fragments are generally considerably larger than the wavelength range of visible light. Pigment particles and air voids are comparable to or even smaller than this size range.

From a fundamental viewpoint, the ratio of particle size to wavelength is extremely important in determining how the individual particles interact

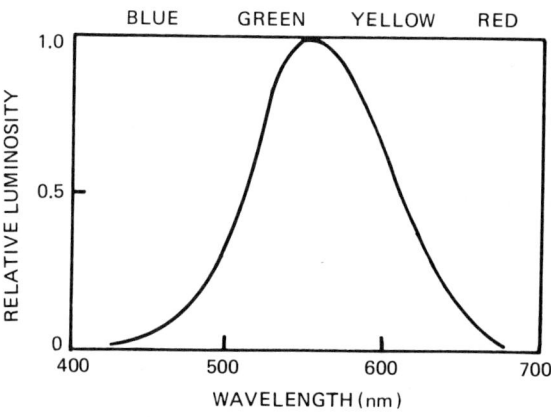

Fig. 1 Luminosity response of the human eye in the visible spectrum.

light. The larger particles that usually predominate in any sheet structure will surface refract and reflect light according to the laws of Fresnel; the smaller particles interact with light in the more complex manner described by Rayleigh and Mie [79,156]. In both cases, the *refractive index* and the *absorption coefficient* of the particle determine the energy scattered and absorbed by the particle.

Reflectance, Transmittance, and Absorptance Due to the multitude, complexity, and close vicinity of light-interacting particles in a sheet of paper, it is customary to consider the effect rather than the exact nature of light-sheet interactions in paper physics. The radiation is either reflected from, transmitted through, or absorbed in the sheet structure, as shown in Fig. 2. *Reflectance* is the ratio of reflected light intensity to incident light intensity. *Transmittance and absorptance* are similar definitions for transmitted and absorbed light intensities. Of the three light fluxes, both reflected and transmitted light are susceptible to photometric measurement. However, as will be shown, except for rather specific purposes—such as the determination of paper formation, paper transparency, on-line opacity, and research applications—the description of appearance properties is mainly obtained through reflectance measurements.

Routine reflectance measurements, such as those described in ensuing pages, will seldom show the exact manner in which light interacts with paper constituents. For example, a sheet made of fully bleached pulp fibers diffuses the incident light to such a degree that details of particle scattering are lost. The reflected light is spatially uniform and comparable to the light distribution from reflectance standards, such as magnesium oxide, barium sulfate, and the "perfect" diffuser that reflects light of equal intensity in any direction [74]. Paper is *opaque*, in contrast to *transparent*—which describes, for example, cellulosic sheets like parchment paper and cellophane sheets where the gross fiber structure is not present. Similarly, the paper sheets are white and do not visually show the light absorption characteristics of transparent cellulose, which make cellulosic films appear yellowish.

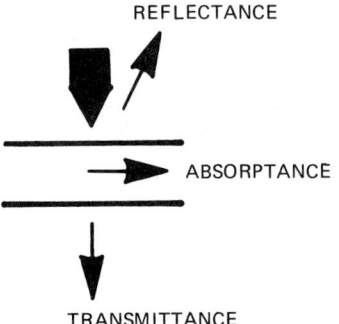

Fig. 2 The effects of light-sheet interactions.

Table 1 Standard Methods for Opacity and Brightness

	Opacity	Brightness
International Organization for Standardization	ISO 2471	ISO 2470
United States	TAPPI T 425 T-519	TAPPI T 452
	ASTM D 589	ASTM D 985
Japan	JIS P-8138	JIS P-8123
Canada	CPPA E.2	CPPA E.1
Scandinavia	SCAN P8	SCAN P3

B. Routine Reflectance Methods

Opacity and Brightness Instrumental assessment of visual paper sheet appearance has been performed for more than fifty years. Routine measurements are usually those of *opacity* and *brightness*, for which national standard methods have been issued by national pulp and paper associations or materials testing organizations, as, for example, the American Society for Testing and Materials (ASTM) (Table 1). In the area of paper optics, testing methods are greatly influenced by procedures devised by the International Commission on Illumination (CIE) [74]. Lately, the International Organization for Standardization (ISO) has undertaken the task of integrating different procedures into universally acceptable methods.

Unfortunately, even the most basic reflectance measurements such as opacity and brightness still to some degree reflect varying measurement conditions and instrumentation design [31,60,68,73,101,131,140,141,159,160,169]. Therefore, the meaning of paper reflectance terms such as *reflectance*, *brightness*, *opacity*, and so on, is nonspecific unless definitions are provided through standard methods or other procedures. These terms should rely only on spectral and geometric light characteristics and should not be governed by instrument design.

Light Geometry The International Commission on Illumination (CIE) specifies four geometries for light reflectance measurements (Fig. 3). Two of these are commonly used in the paper industry, namely, *diffuse illumination-near normal viewing* and *45° illumination-normal viewing*. Diffuse illumination-near normal viewing (diffuse-0° in Fig. 3) is the geometry agreed upon for routine paper reflectance measurements by the International Organization for Standardization (ISO 2469, and ISO 2470-2471 in Table 1). However, 45° illumination-normal viewing (45°-0° in Fig. 3) is also common, especially in the standards of the Technical Association of the Pulp and Paper Industry (TAPPI) in the United States (TAPPI T-452). It is important to realize that these geometries will not produce identical reflectance values for any kind of sheet structure. A change from normal to diffuse incident light makes both transmitted and reflected light distributions more diffuse. Light reversal—that is, interchange of detector and light sources (45°-0° ↔ 0°-45°

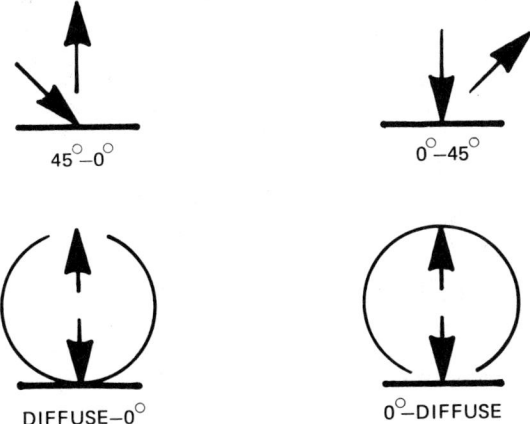

Fig. 3 CIE-specified light geometries for reflectance measurements.

and diffuse-0° ↔ 0°-diffuse in Fig. 3)—produces identical results only when the irradiated sample is a perfect diffuser, a requirement that is approximated but never fulfilled for paper and paperboard because of the size and orientation of the fibers in the paper surface [11,47]. Similarly, a collimated incident light beam (as in the 45°-0° and 0°-45° geometries) introduces the possibility of directionality in the reflected light [140]. That is, surface formation and anisotropy might produce fluctuations in measured light when the sheet is rotated under fixed incident light.

The feasibility of further treatment of reflectance and transmittance measurements by theories like that of Kubelka and Munk depends upon the manner in which the data are obtained. Kubelka-Munk calculations require diffuse incident light. The stepwise calculation of, for example, coating optics of heterogeneous layers shown on pp. 34-36 does not require specific incident radiation geometry as long as transmitted and reflected light fluxes are measured correctly.

ISO Standardization The issue of ISO paper standards represents an effort to create international agreement on optical paper measurements [19]. The optical standards ISO 2469-2471 are based upon the following concepts and terminology:

Opacity (Paper Backing) (R_0/R_∞) This is the ratio between the amount of light reflected from a single sheet with a black (total light absorbing) backing, R_0 (luminance factor), to the amount of light reflected by the same sheet of paper backed by an opaque pile of the same paper, R_∞ (intrinsic luminance factor). Using the diffuse reflectance method, spectral and geometric light characteristics should be as specified in ISO 2471.

Opacity (White Backing) (R_0/R_W) In contrast to opacity (paper backing), opacity (white backing) requires only one sheet of paper. It is the ratio between the amount of light reflected from a single sheet with a black backing, R_0, to the amount of light reflected by the same sheet of paper backed by a standard white, R_W, where the white standard has an absolute reflectance of W. This measurement has often been termed *contrast ratio* when performed according to TAPPI T-425 (W = 0.89).

Directional Blue Reflectance Factor (R_∞) This "brightness" measurement is the reflectance of an opaque pile of sheets relative to a perfect reflecting diffuser (CIE 45-20-195) at an effective wavelength of 457 nm and as determined with an instrument employing illumination at 45° and normal viewing (TAPPI T-452). In practice, the measurement of R_∞ means that there is no measurable change in reflectance when the thickness of the sheet structure being measured for reflectance is doubled.

Diffuse Blue Reflectance Factor (ISO Brightness) (R_∞) The diffuse blue reflectance factor is similar to the directional blue reflectance factor. This measurement is obtained as specified in ISO 2470 (diffuse illumination-near normal viewing). All reflectance measurements should be given in percentage units.

Due to the difference in light geometry, the directional blue reflectance factor is lower than the diffuse blue reflectance factor [32]. The directional blue reflectance factor can be increased to the approximate level of ISO Brightness by measuring it relative to magnesium oxide (TAPPI T-452) rather than relative to the perfect diffuser [18]. Measurement of ISO Brightness requires the elimination of gloss components in the reflected light flux [19, 145].

There is no doubt that the development of ISO standards in the 1960s created uncertainty as to how optical paper properties should be specified [60,140,159]. Paper technologists who had been using perfectly reliable but different characterization methods were not about to change their procedures. This reluctance was intensified in cases where a change in method produced lower values, as, for example, when the perfect diffuser rather than magnesium oxide was applied as the reference standard. Stenius et al. [145] have derived approximate relationships between ISO Brightness and SCAN Brightness, both of which are based upon diffuse incident illumination. Using previously published data by Dearth et al. [32], Budde and Chapman [18] have demonstrated that ISO brightness values are similar to standard brightness values obtained with the IPC brightness tester according to TAPPI T-452. Exact relationships are not feasible since IPC brightness utilizes 45°-0° light geometry and ISO brightness uses diffuse-0° geometry.

Japanese standards (JIS P-8138 and P-8123 in Table 1) reflect U.S. practice (TAPPI T-452 and T-425), with the exception that the specimen size for brightness measurement is larger.

C. Colorimetry

Spectral Reflectance The majority of optical measurements that are routinely performed on paper and paperboard are reflectance measurements like those described above. That is, a limited number of well-defined reflectance values are used to describe the sheet appearance. In contrast, *colorimetry* is the quantitative measurement of spectral reflectance variation through the whole visible spectrum.

The need for more complete paper optics characteristics is most pronounced when paper is required to exhibit specific spectral reflectance properties similar to those that are routinely demanded, for example, for paints in the paint industry. White writing and printing papers are required to show a pleasing appearance as defined by certain color characteristics that cannot be given by standard reflectance values solely. Similarly, paper and paperboard for packaging are often required to exhibit quite specific color characteristics. These requirements, in addition to instrumentation that has become more reliable and easy to operate, have created an increased use of color measurements both in the mill and in the laboratory.

The CIE Standard System The basic operation in colorimetry is the determination of the *tristimulus values* X, Y, and Z in the CIE standard system [36,96,144]. The tristimulus values are calculated through the following equations:

$$X = k \int_\lambda P_\lambda x(\lambda) \, d\lambda \tag{1}$$

$$Y = k \int_\lambda P_\lambda y(\lambda) \, d\lambda \tag{2}$$

$$Z = k \int_\lambda P_\lambda z(\lambda) \, d\lambda \tag{3}$$

where
  $P_\lambda \, d\lambda$ = spectral power function
  k = a normalizing factor
  $x(\lambda), y(\lambda), z(\lambda)$ = color matching functions for a "standard" observer

These functions are shown in Fig. 4 [74,78,172,173]. The color matching functions for the standard observer are based upon the color matches of a human eye of normal color vision. Therefore, the $y(\lambda)$ response is the luminosity response shown in Fig. 1 that defines *lightness* (color intensity). The additional $x(\lambda)$ and $z(\lambda)$ responses produce the amount of the necessary two extra primary colors required to produce for the standard observer the color of the spectral stimulus of unit radiance. The concept of primary colors, color stimuli, and their effect on the human eye (color vision) have been discussed in more detail by Judd and Wyszecki [78].

The standard observer was adapted in 1931 by CIE based upon experiments where observers obtained color matches at a 2° viewing angle. A supplementary observer was introduced in 1964 for a viewing angle of 4° and up [78,173]. The latter is characterized by a slightly higher $z(\lambda)$ peak and reflects visual color matching of fields of large angular subtense.

The *object color* is now derived by replacing $P_\lambda \, d\lambda$ with $R_\lambda H_\lambda \, d\lambda$ where R_λ is the spectral reflectance of the object and $H_\lambda \, d\lambda$ is the spectral distribution of the illuminant irradiating the object. The equation system is normalized by specifying

$$k = \frac{100}{H_\lambda y(\lambda) \, d\lambda} \tag{4}$$

but other more appropriate constants may be chosen [78]. The equation system is fully defined by the spectral object reflectance R_λ and the spectral

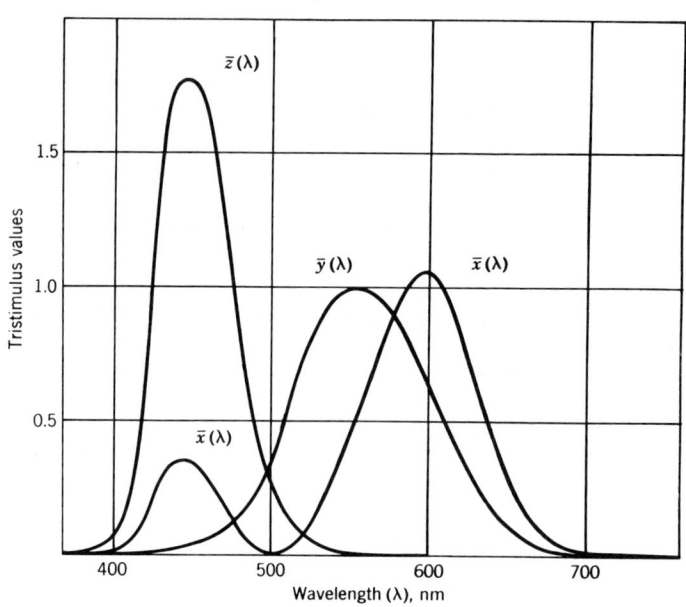

Fig. 4 Color matching functions for 1931 CIE standard observer. (From Ref. 78.)

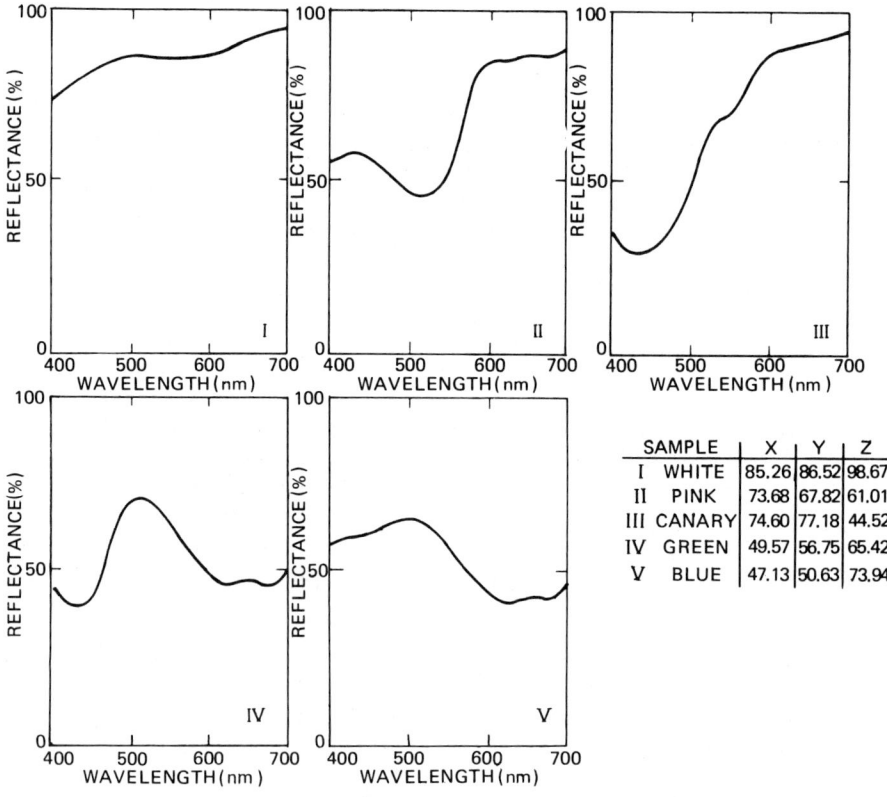

Fig. 5 Spectral object reflectance variations (Eastman white reflectance standard) for white and colored papers (Xerox 4024 DP and DP Colors).

distribution H_λ of the light illuminating the paper. The appropriate summations are often performed by minicomputers attached directly to the photometric instrumentation (Chap. 0) [9,123]. Figure 5 shows examples of spectral object reflectance variations for white and colored bond papers measured relative to a barium sulfate standard. The X, Y, Z coordinates were calculated for the spectral distribution of the CIE C illuminant described below.

Chromaticity Coordinates The description of color measurements in *chromaticity coordinates* or *trichromatic coefficients* is merely an expression of X, Y, and Z values in the following form:

$$x = \frac{X}{X + Y + Z} \qquad (5)$$

$$y = \frac{Y}{X + Y + Z} \qquad (6)$$

$$z = \frac{Z}{X + Y + Z} \qquad (7)$$

and where y may be plotted against x as shown in Fig. 6 for the sheets analyzed in Fig. 5.

Hue, the attribute of color perception by means of which the paper is judged to be red, yellow, and so on, is described quantitatively as the *dominant wavelength*, that is, the wavelength of light that if added to the illumination will match a given color. *Saturation* is the quality of color sensation by which the paper shows different purities of any one dominant wavelength. It is measured as *excitation purity*—the percent departure from a neutral gray of the same lightness. Both dominant wavelength and saturation purity can be determined using the chromaticity diagram (TAPPI T 527) [78] or by means of calculation aids [96a,143a].

Illuminants In 1931 CIE defined standard illuminants to represent incandescent light, sunlight, and daylight (illuminants A, B, and C. in Fig. 7) [78]. Their positions in the chromaticity diagram are given in Fig. 8. The variation in X-Y coordinates in which the illuminants are situated in Fig. 8 defines their *color temperature*, that is, the temperature to which a black body must be heated to produce the same color stimuli.

For paper color, the energy function H_λ used to define tristimulus coordinates is generally that of illuminant C which has also been specified for ISO reflectance measurement. However, none of the illuminants given above are adequate to simulate the spectral distributions of different daylight conditions. CIE has therefore added an additional series known as D illuminants, based on spectrophotometric measurements at different locations (Fig. 9) [7, 78]. None of these can be reproduced in detail by artificial sources [40]. Nonetheless, light sources simulating illuminant D_{65} (overcast north sky) produce more adequate appearance characteristics for paper containing fluorescent brighteners, since the ultraviolet emitted energy is enhanced over that of illuminant C in the wavelength range below 400 nm (Fig. 9 compared to Fig. 7). Other D illuminants are useful when it is necessary to assess color under a variety of daylight conditions (TAPPI T 508).

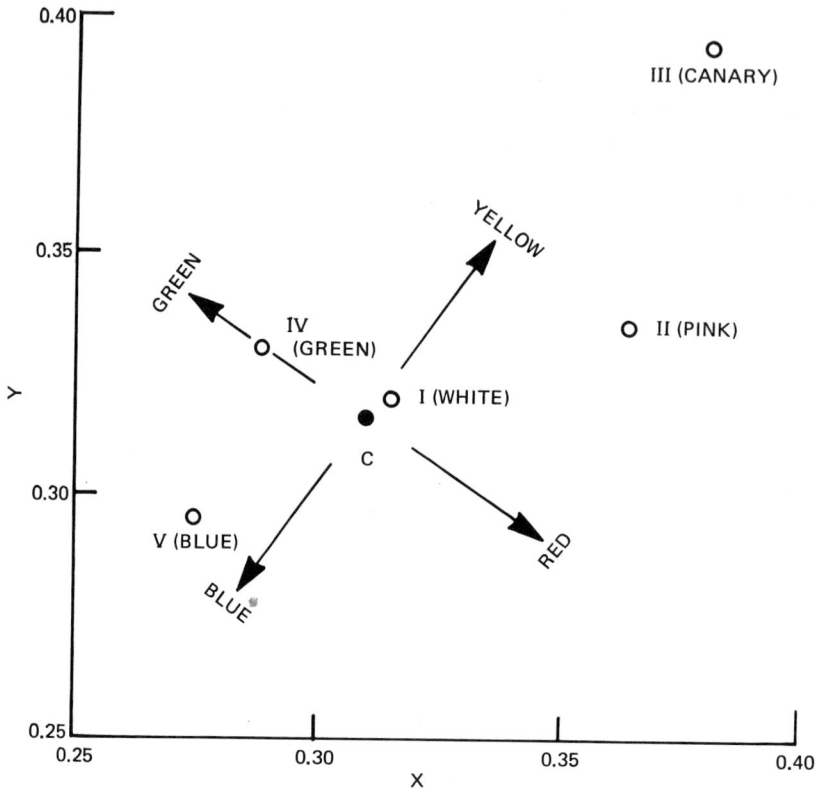

Fig. 6 Chromaticity coordinates for illuminant C (Xerox 4024 DP and DP Colors).

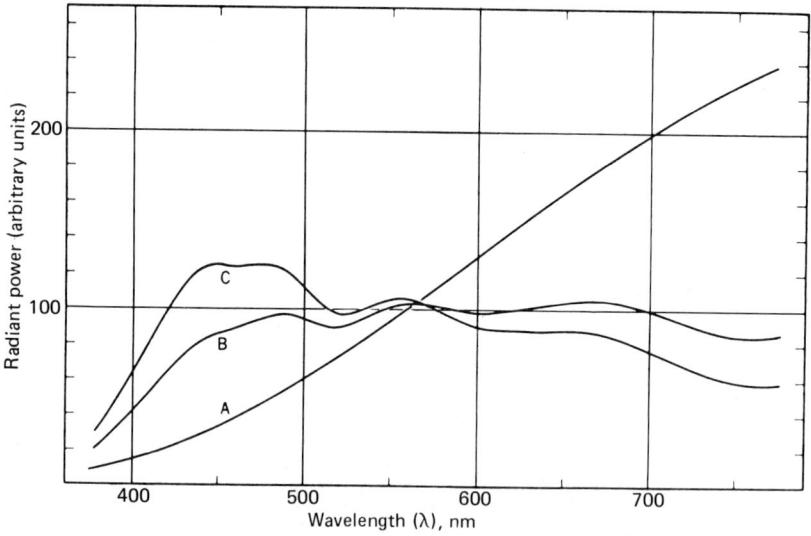

Fig. 7 Relative spectral power distributions of CIE illuminants A, B, and C. (From Ref. 78.)

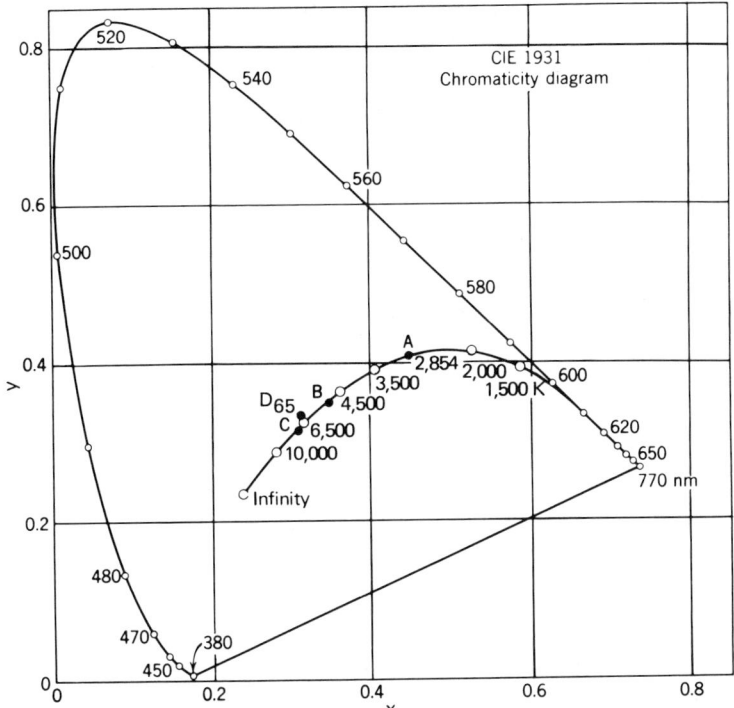

Fig. 8 CIE 1931 chromaticity diagram. (From Ref. 78).

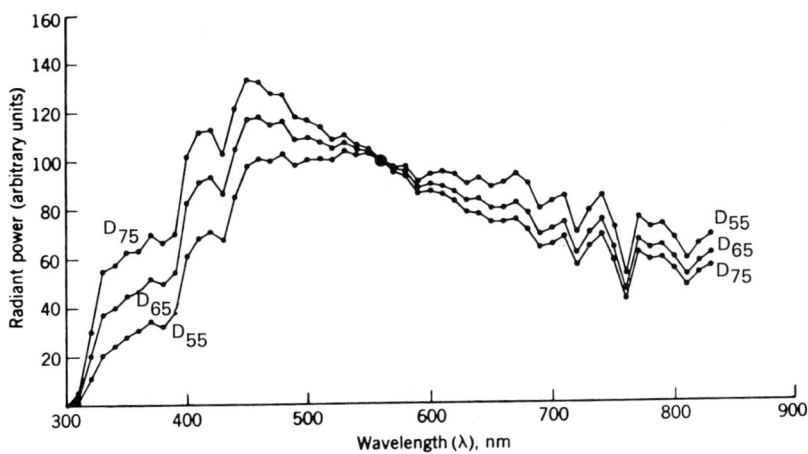

Fig. 9 Relative spectral power distributions of CIE D illuminants. (From Ref. 78.)

Table 2 Chromaticity Coordinates and Color Temperatures of CIE Illuminants

	CIE Coordinates				
	1931		1964		Temperature
Illuminant	x	y	x'	y'	(°K)
A	0.4476	0.4074	0.4512	0.4059	2856
B	0.3484	0.3516	0.3498	0.3527	4874
C	0.3101	0.3162	0.3104	0.3191	6774
D_{55}	0.3324	0.3475	0.3341	0.3487	5503
D_{65}	0.3127	0.3290	0.3138	0.3310	6504
D_{75}	0.2990	0.3150	0.2996	0.3173	7504

Table 2 lists current CIE illuminants, their chromaticity coordinates, and their color temperatures. The 1964 CIE coordinates reflect the use of the 1964 chromaticity diagram (viewing fields larger than 2°).

In practice, a range of illuminants is generally created by a limited number of lamps using different filters. This can greatly expand the range of visual color impressions that a single sheet of paper may create. *Metameric* matches occur when the same X, Y, Z coordinates can be calculated for different P_λ variations. This is possible for either the same R_λ at different H_λ (same object color illuminated by different illuminants) or the same H_λ for different R_λ (different object colors under the same illumination). Metamerism has been discussed in detail by Judd and Wyszecki [78].

Visual Color Matching Visual color matching is still widespread in the paper industry (TAPPI T 508 and T TAPPI 515). The human eye is quite sensitive to changes in hue. Lightness affects the eye less. The term *strength* is used to indicate color difference between two samples by specifying the concentration difference of the dyes in the color formula [147]. Strength differences in single dye concentrations create color differences withing single color areas (pink in red, canary in yellow, etc.).

For general color matching, the Munsell system is the most widely used system that does not rely on instrumental measurements (ASTM D 1535) [106]. A number of manufacturers produce color charts that simulate the Munsell standards and that can be visually matched to paper products under well-defined illumination conditions. Huey [61,62] has reviewed the practices for color matching in industries where color is assessed visually.

Color Scales Even more versatile are the colorimeters that produce numerical color characteristics, which are easier to relate to visual color perception than tristimulus values. In the paper industry, opponent color systems like

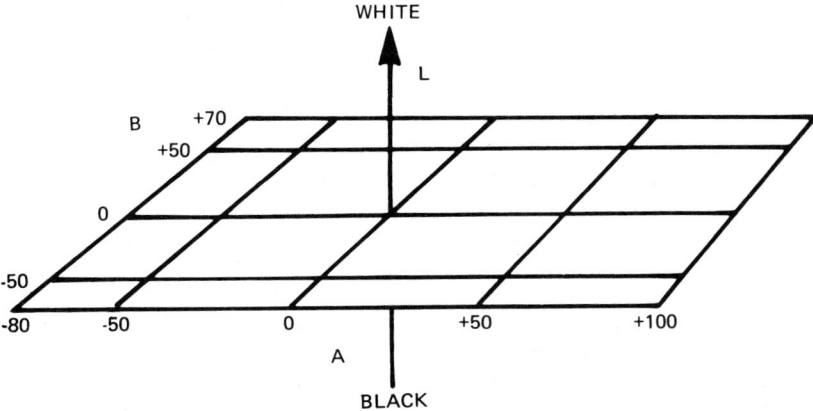

Fig. 10 L, a, b color space.

the L, a, b color space are now widely used for color characterization (Fig. 10) [65,71]. Two color scales have been proposed by CIE [120a]. The L*, a*, b* color space is suitable for small color differences [84]. The L*, u*, v* formula applies to larger color differences or when these result from additively mixed lights [84,120a]. Both Judd and Wyszecki [78] and Hunter [70] have described the application of various color spaces, and calculator programs are available for color space calculations [96b,173a].

It should be realized that any color space can be mathematically derived from the more universally accepted X, Y, Z values under the illumination conditions that pertain to the color space [25]. For example, L, a, b coordinates (illuminant C) are derived in the following manner [71]:

$$L = 10.0Y^{1/2} \tag{8}$$

$$a = 17.5Y^{-1/2}(1.02X - Y) \tag{9}$$

$$b = 7.0Y^{-1/2}(Y - 0.847X) \tag{10}$$

The preference of L, a, b to X, Y, Z relies on the user's easier identification of visual shades. Thus, positive a values indicate red, negative indicate green. Similarly, b shows yellow when positive and blue when negative. The zero level of both a and b indicates grayness that increases to white when L increases. Color spaces have been graduated to provide meaningful scales for the visual perception of color changes [100].

Color Difference A change in color from one point to another in a color space like that shown in Fig. 10 is simply described as the distance ΔE between the points:

$$\Delta E = (\Delta L^2 + \Delta a^2 + \Delta b^2)^{1/2} \tag{11}$$

where ΔL, Δa, and Δb are differences in coordinates [3]. Similarly, the Friele-MacAdams-Chickering color difference formula (FMC II color difference) recommended by CIE expresses color differences in terms of a single number [22]. A computer program is available for the calculation of FMC II color differences in terms of X, Y, Z coordinates for the two colors constituting the color pair to be evaluated [12].

A color difference expressed by a single number will never adequately describe the three-dimensional shift in color coordinates. However, color differences like those described above are useful when the performance characteristics of colorimeters are evaluated and compared with each other [9,10,99]. Small color differences are especially applicable to the measurement of metamerism (metamerism index) and to the description of changes in whiteness.

Hunter has presented a comprehensive review of color scales and color difference systems [70]. The development of color scales and their relation to visual color perception by the Optical Society of America has been described by Nickerson [108]. Wyszecki [173] has given concise definitions of CIE colorimetry procedures. The American Society for Testing and Materials has issued a detailed method for instrumental evaluation of color differences (ASTM D 2244).

D. Fluorescence and Whiteness

Paper Whiteness *Fluorescence* is the ability of a substance to absorb light at one wavelength and to emit it at a higher wavelength. In the paper industry, optical brighteners are often added to improve the appearance properties of printing and writing papers [142].

Figure 11 shows the expanded area of the chromaticity diagram where natural whites and fluorescent whites are perceived under CIE C illumination

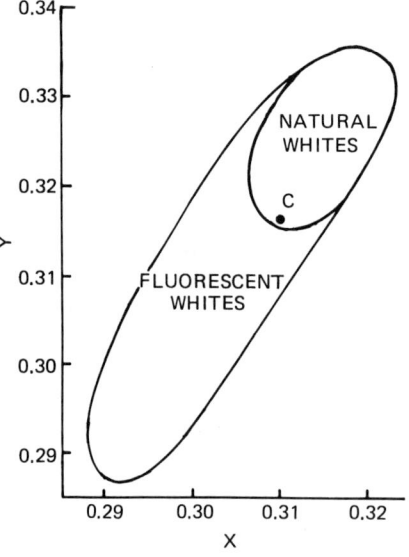

Fig. 11 Chromaticity areas of natural and fluorescent whites. (From Ref. 154).

[154]. Due to the nature of the optical brighteners added, the fluorescent whites extend toward the blue end of the color space, thereby greatly expanding the chromaticity domain of whites. Fluorescence and whiteness are possibly the colorimetric paper appearance properties that are most difficult to characterize. Anyone who has been involved in the shade evaluation of white bond papers will realize the difficulties involved in getting a consensus on whiteness among a number of sheet candidates showing slightly different hue [40b]. They can all be accurately characterized through their tristimulus values, yet the human eye does not perceive color as an absolute physical sensation. Each individual sheet is subject to the color of the sheets to which it is being compared and to the nature of the illuminants. Corte [29] has discussed the psychological uncertainty in whiteness measurements and has recommended the design of instrumentation that simulates psychological sensation in color and other appearance measuring techniques. This is not merely a challenge of accurately defining spectral illumination distributions, geometries, and sample positions, but will in some instances require the generation of additional physical stimuli to achieve meaningful responses.

Fluorescence Spectroscopy Ideally, a complete spectral characterization of a fluorescent material requires that spectral reflectances be obtained at each individual wavelength as discussed by Judd and Wyszecki (Fig. 12A) [78]. The incoming intensity $(I_0)_\lambda$ generates reflected and fluorescent radiation over a wavelength band $\Sigma\lambda$ that is wider than the single wavelength of irradiation. Consequently, it is necessary to scan reflected radiation spectrophotometrically and separate it into the different spectral components $[(I_R)_{\Sigma\lambda} = (I_R)_{\lambda_1} + (I_R)_{\lambda_2} + \cdots$ in Fig. 12A]. The procedure must be repeated for all wavelengths smaller than that for which maximum fluorescence is emitted. This is time consuming and generally not feasible using conventional spectrophotometric instrumentation. Instead, paper fluorescence measurements are obtained using reversed optics, as shown in Fig. 12B [31,158]. The spectrum is obtained by scanning reflected instead of incoming light with the monochromator. A comparison with the spectrum obtained for nonreversed optics (Fig. 12A) permits the determination of nonfluorescent reflected light by locating the lowest reflectance at each wavelength [59].

Numerical methods for describing whiteness and fluorescence are closely linked to the design of the instrumentation that is required to measure them. For example, the addition of fluorescent dyestuff to bleached pulp can be quantified simply by insertion of an ultraviolet filter between light source and paper sample in the Elrepho brightness meter [48]. Suitable combinations of colorimetric measurements can be applied when the fluorescent com-

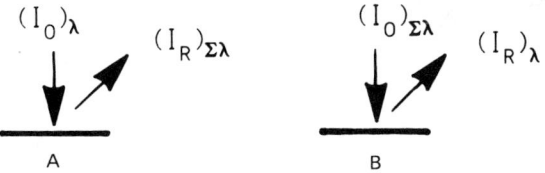

Fig. 12 "Forward" (A) contra "reversed " (B) light geometry in fluorescence spectrophotometry.

ponent is added by use of the proper light source and filters [92,120]. Two methods of whiteness evaluation of paper have been described by Grum and Patek [45]. The simpler of these is based on relative radiance by selecting four wavelengths and summing spectral reflectances after a statistical weighting procedure [45,149a]. Similarly, methods have been given for predicting fluorescence measurements under changing illumination without experimental measurements [44].

Vaeck [154] and Ganz [39] have reviewed the determination of whiteness for industries where this appearance characteristic is important. Ganz [40a] has proposed the following formulas for whiteness:

$$W = Y - 800(x - x_0) - 1700(y - y_0) \quad \text{(neutral)}$$

$$W = Y - 1700(x - x_0) - 900(y - y_0) \quad \text{(green)}$$

$$W = Y + 800(x - x_0) - 3000(y - y_0) \quad \text{(red)}$$

where W quantifies whiteness, Y is the luminosity-dependent tristimulus factor (luminance factor), x, y, z are chromaticity coordinates for the sample and x_0, y_0, z_0 are those of the perfect diffuser (see p. 30). The appropriate formula is chosen on hue preference (neutral, green or red in the whiteness formulas above). Methods for such measurements on paper (TAPPI Useful Method 548, JIS Z-8902) are less established, and changes in both procedures and instrumentation are likely to occur in the future.

E. Gloss and Surface Finish

Specular Reflectance *Gloss*, like color, is psychophysical [67]. The physical effect is one for which reflected light intensity is directional (Fig. 13). Gloss, or directional (specular) light reflectance, is created in the uppermost surface layer of the sheet structure (thickness less than half the wavelength of the incident light). Consequently, gloss is strongly dependent upon the

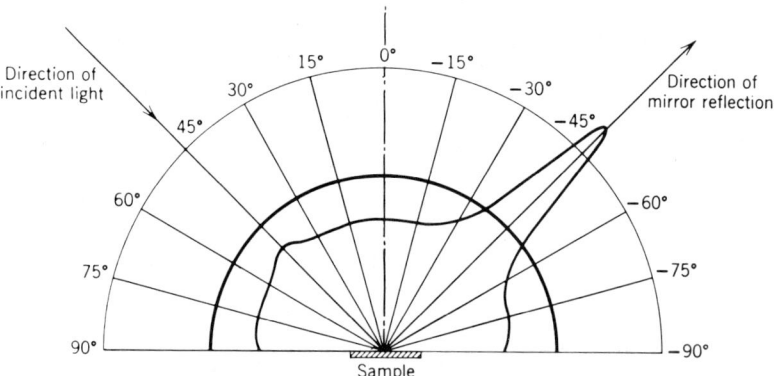

Fig. 13 Directional light reflectance (gloss). Reflected light intensity is enhanced in the direction of mirror reflection (−45°). (From Ref. 78.)

Table 3 Types of Gloss and Their Measurement

Kind of glossiness	Gloss-factor measurement	
Specular	Incident flux ϕ_1 (at 60°) = reflected flux from perfect mirror; ϕ_V (at −60°) = reflected flux from sample; Gloss factor = ϕ_V/ϕ_1	
Sheen	Incident flux ϕ_1 (at 85°) = reflected flux from perfect mirror; ϕ_V (at −85°) = reflected flux from sample; Gloss factor = ϕ_V/ϕ_1	
Contrast	Incident flux ϕ_1 (at 60°); ϕ_{V1} (at −60°) = reflected flux from sample; ϕ_{V2} (at 0°) = reflected flux from sample; Gloss factor = ϕ_{V1}/ϕ_{V2}	
Distinctness of image	Incident flux ϕ_1 (at i); ϕ_V (at −j) = reflected flux from sample; Angle of view (−j) differs by a few minutes of arc from angle of mirror reflection (−j); Gloss factor = rate of change of ϕ_V with angle of incidence (i)	
Absence of bloom	Incident flux ϕ_1 (at i); ϕ_{V2} (at −j) = reflected flux from sample; Angle of view (−j) differs from the angle of mirror reflection (−i) by a few degrees; Gloss factor = ϕ_{V1}/ϕ_{V2}	

Source: From Ref. 78.

smoothness of the paper surface, as discussed by Van den Akker and Sears [163] and demonstrated experimentally for clay-coated sheets by Gate et al. [42].

Since gloss is directional, it is attractive to consider its physical basis as one where diffuse reflectance is being enhanced by Fresnel reflection from the sheet surface [53]. Fresnel reflected light flux is dependent upon the fractive index ratio n at the interface and the angle of incidence i, according to the following formula:

$$\frac{I_R}{I_0} = \frac{1}{2}\left[\frac{\sin^2(i-r)}{\sin^2(i+r)} + \frac{\tan^2(i-r)}{\tan^2(i+r)}\right] \quad (12)$$

where
I_R = intensity of reflected beam
I_0 = intensity of incoming beam

Further, i and r (angle of refraction) are related such that [64]:

$$r = \sin^{-1}\left(\frac{\sin i}{n}\right) \quad (13)$$

In contrast, a perfect reflecting diffuser provides reflected light flux that is independent of the angle of reflection. A combination of Fresnel reflected and diffusely reflected light will yield light distributions of the type shown in Fig. 13, with an increase in the specular component at higher angles of incidence due to the nature of the equations shown above.

Paper Gloss Perception Since gloss is directional and is seldom measured without diffusely scattered light components, it is necessary to achieve a measuring geometry that is meaningful for the surface that is being analyzed. Hunter distinguishes between five different kinds of surface gloss (Table 3) [64,78]. The gloss perception that generally is applicable to paper and paperboard is of a low-gloss nature (shininess at grazing angles) that requires a relatively large receptor angle (from 75° in TAPPI T 480 to 85° in ASTM D 523) [152]. Papers of higher gloss as, for example, waxed and cast-coated sheets, are measured at a much smaller receptor angle (20° in TAPPI T 653). Figure 14 describes numerical gloss values obtained using different geometries for perceived visual gloss ratings. The measurement geometries shown in Fig. 14 and described above are those that are commonly used in the United States; measurement practice elsewhere may not reflect U.S. procedures [90]. A series of Japanese gloss studies have attempted to clarify the meaning and measurement of gloss and its effect on paper color (see pp. 40-41).

Contrast gloss or *luster* (Table 3) is a measure of the ratio of specularly reflected light to diffusely reflected light. Relying on the principle of polarization, the first commercially available glossmeter, the Ingersoll Glarimeter, was designed to measure contrast gloss [72]. Initially polarized light does not become depolarized when Fresnel reflected. In contrast, the diffusely reflected light cannot maintain polarization due to the multiple light scattering events beneath the sheet surface. The same effect has been utilized in instrumentation design suitable for determining the effect of the

Optical and Appearance Properties 19

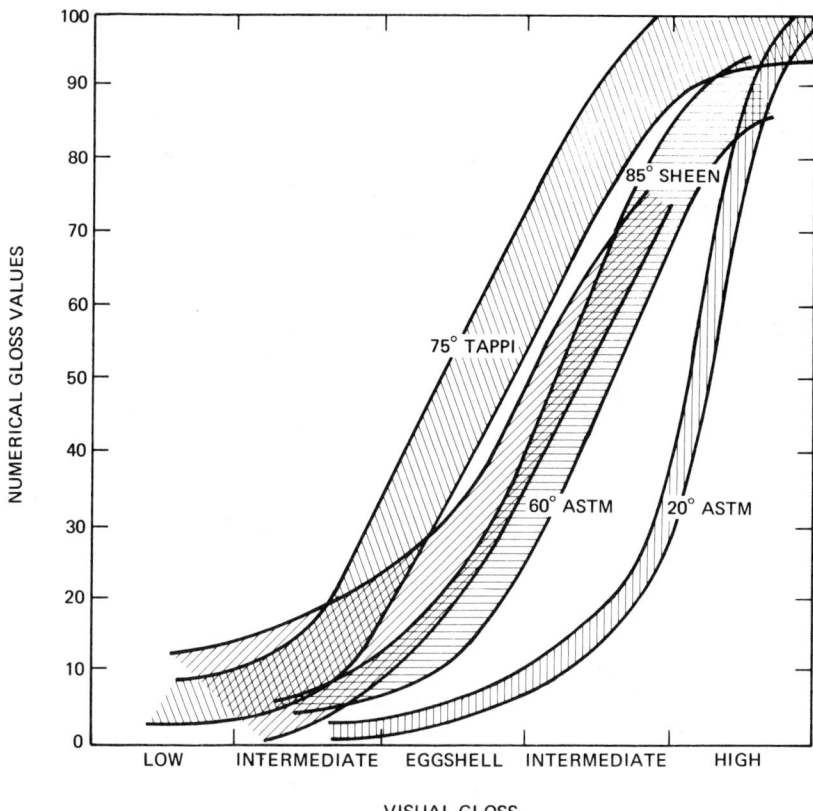

Fig. 14 Numerical gloss values for visual gloss perceptions at different measuring angles. (From Ref. 70.)

paper surface in high quality printing applications at high print gloss levels [17,87,119].

A novel contrast gloss method has been described by Nishiwaki [110]. Based upon instrumentation design by Fukushima [53], reflected light distributions are measured on a goniophotometer where the sample is rotated and the illuminator and receptor axes are fixed. Diffusely and specularly reflected light components are evaluated by plotting the logarithmic intensity against the square of the rotation angle. *Distinctness-of-image gloss* is obtained by comparing specular reflectance to the reflectance obtained when the angle is changed slightly from that of the specular direction (Table 3) [70, 107].

Surface Unevenness and Finish All routine reflectance measurements require an illuminated paper surface area that is sufficiently large to eliminate serious fluctuations in reflectance values from different spots on the same sheet structure. In contrast, if the area measured for reflectance is limited in size, reflectance will vary due to surface unevenness that shows up in scanning. This effect can simulate the appearance phenomenon of *mottle* (optical unevenness) of both printed and nonprinted sheets [119]. The reflectometer measures the reflectance of a small paper area situated within a larger one that is similar in size to that commonly measured by routine instrumentation. The reflectance of a large area is simultaneously measured and compared with the reflectance from the smaller, spotlike area that will fluctuate due to optical unevenness [88,89]. *Speckle* is the nonuniform appearance of printed matter that is caused by nonprinted depressions in the paper surface. Poulter [119] has reviewed the measurement of print unevenness. A mathematical model that accounts for both mottle and speckle of solid prints has been proposed by Wahren and Bryntse [164]. As will be shown in the following, improved methods for characterizing both printed and nonprinted paper structure on a micro scale rely on instrumentation that can be designed specifically for the measurement of reflectance and transmittance from areas of micro dimensions. Data analysis methods utilizing optical character recognition (OCR) systems and novel printing methods such as nonimpact inkjet devices impose specific demands on micro-scale paper surface structure and appearance. Consequently, equipment and techniques for measuring and analyzing variations in micro-scale reflectance are gaining increasing use [41,77,117,174].

The appearance of surface finish can be as important as the more general reflectance characteristics. There exists here a papermaking terminology that is nonspecific and based upon visual perception and finishing techniques. Uncoated printing papers may be *machine finished* or increased in smoothness to levels of *English, writing*, and *glazed finish*. A grainy, rough texture is *antique finish*, which, with increased fineness, becomes *eggshell* and *vellum finish*. *Cockle finished* papers are desirable in business paper manufacture. There are no specific guidelines for paper finish terms, and well-defined instrumental methods for the measurement of these appearance characteristics are lacking. Emerton et al. [35] have presented a selection of photomicrographs of printing paper surfaces.

F. Light Transmittance Methods

Transmittance Contra Reflectance The number and variety of reflectometers that are in use today demonstrate the extensive use of reflectance measurements for routine control of the appearance properties of paper and paperboard. Nevertheless, this does not mean that in principle optical appearance characteristics can be analyzed only using reflectometry. When photometric instrumentation for optical paper characterization was still in its infancy, transmittance methods were shown to give reliable measurement of some paper appearance properties such as contrast ratio and printing opacity (opacity, paper backing) [95]. Harrison [52] has presented an extensive review of early opacity measurement methods.

Properly done, the measurement of the transmitted light flux can present advantages to the measurement of reflected flux. Reflectance and transmittance are complementary, always adding up to 100% minus absorbed light (Fig. 2). Therefore, variations in transmittance will often be considerably larger than variations in reflectance when considered on a relative basis. Together with a properly applied light scattering theory like that of Kubelka and Munk, transmittance measurements can be taken advantage of in measurement of sheet opacity, both in the laboratory and in the mill (on-line appearance control). Lathrop [86] and Van den Akker [160] have discussed theoretical aspects of transmittance measurements. Springer [135] and Lodzinski [91] have demonstrated the on-line application of appearance control using light transmittance. Tsuchida et al. [153] have applied transmittance measurements on relatively large paper areas to formation analysis.

Unfortunately, light transmitted through an opaque paper sheet does not lend itself to measurement as easily as does light reflected from the same structure. The spatial distribution of transmitted light is less well defined and more sensitive to incoming light geometry, especially for collimated incoming light flux [73]. Changes in the formation of the sheet sometimes introduce obscure changes in transmitted light distributions, as discussed below in relation to formation testing (Fig. 16) [115]. Therefore, transmittance characterization is limited to instances where good correlation between measurement and sheet properties can be established [135] or for specialized purposes (transparency and formation). In general, reflectance methods using well-designed instrumentation are more desirable for appearance characterization, including sheet opacity.

Transparency *Transparency ratio* is the ratio of near regular transmittance to near hemispherical transmittance.

This measurement characterizes glassine papers, for which transmitted light intensity distribution becomes directional, similar to that for gloss reflected light shown in Fig. 13. The regular transmittance is the light transmittance normal to the sheet surface (Fig. 15). The hemispherical light transmittance is a measure of the total light transmittance obtained by placing a light-collecting sphere behind the sample (Fig. 15) (TAPPI T 522). The measurement reflects the clarity with which an object placed at a distance from the sheet is visually perceived through the sheet structure [70].

Formation "Formation is the manner in which the fibers are distributed, disposed, and intermixed to constitute the paper " [115]. It affects paper appearance and influences the strength properties of paper as well (Chap. 11). For example, semitransparent paper products will show a cloudy or blotchy appearance if the sheet constituents are not sufficiently evenly distributed in the paper structure.

Fig. 15 Measurement of regularly and diffusely (hemispherically) transmitted light.

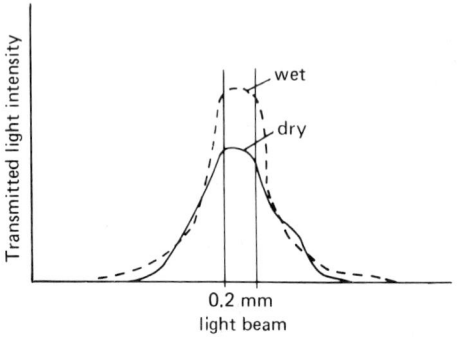

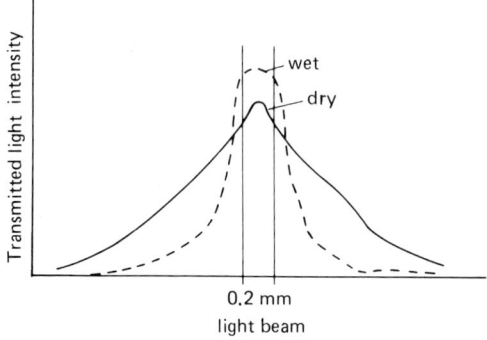

Fig. 16 Transmitted light intensity distributions for sheets of (*upper*) unbleached and (*lower*) bleached pulps. (From Ref. 115.)

The psychological aspect of paper formation is difficult to characterize satisfactorily [50]. Consequently, instrumentation for formation measurements vary in design principle and are often of an experimental nature [115]. Most methods depend on the scanning of paper sheets with visible light and the detection of light transmitted through microdomains of the sheet structure. The scanning area is similar in size to that of the smaller area used for the measurement of optical unevenness by reflectance comparison as described earlier in this chapter (0.1 to 1 mm beam diameter). The light intensity is monitored continuously and either is plotted as a function of paper position, thereby providing a "map" of paper basis weight variation [20], or is treated electronically to give a measure of paper floc frequency [113]. Since the radiation beam size is so small, the distribution of light transmitted through the paper becomes very sensitive both to the nature of the sheet structure and to the sheet thickness. Figure 16 shows transmitted light intensities obtained for sheets of bleached and unbleached pulps [115]. In spite of lower basis weight, the light is diffused more by the bleached sample. Also, reducing the Fresnel index ratio by wetting the fibers affects the bleached structure more than the light absorbing structure. To some degree, the influence of composition and scattering in the sheet structure can be eliminated by using beta-ray transmission in a scanning or stationary mode. In the stationary mode, beta radiographs are recorded on X-ray films and these are then scanned in the formation tester by means of visible light [56,114].

III. INSTRUMENTATION CHARACTERISTICS

A. Instrument Types

Classification Photometers for paper appearance measurements are generally of the following types: *spectrophotometers, colorimeters, glossmeters* and *goniophotometers*, and *microdensitometers*. These are not specific to the paper industry, and design and cost often reflect other applications. Wendlandt [167] and Frei and MacNeil [38] have described a number of specific instruments available for light transmittance and reflectance measurements. Wurzburg [171] has discussed instrumentation applied specifically to paper color measurements. The Institute of Paper Chemistry [32–34] and the Rensselaer Color Measurement Laboratory at Rensselaer Polytechnic Institute [8–11,99,123] have analyzed the performance characteristics of several instruments used for paper and paperboard reflectance characterization. Wyszecki [173] has listed instruments applicable to colorimetry.

The Spectrophotometer Figure 17 shows a typical spectrophotometer. A double prism provides monochromatic light that is split into two beams alternately irradiating the sample and a surface for which the reflectance is known (the standard). Reversible optics allow the recording of fluorescence in the reversed mode (Fig. 12B). The complete recording of light reflectance through the visible range can be quite time consuming when measure-

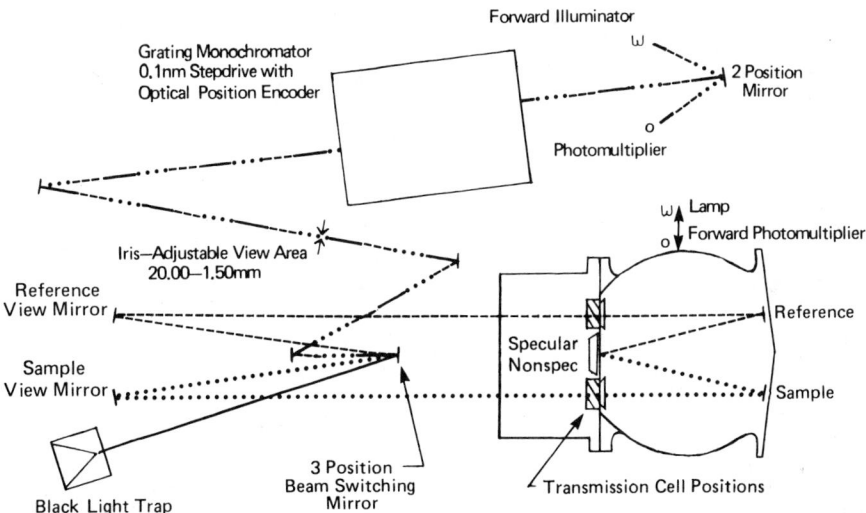

Fig. 17 Diano Match-Scan Color Spectrophotometer. (Courtesy of Diano Corp., Woburn, Massachusetts.)

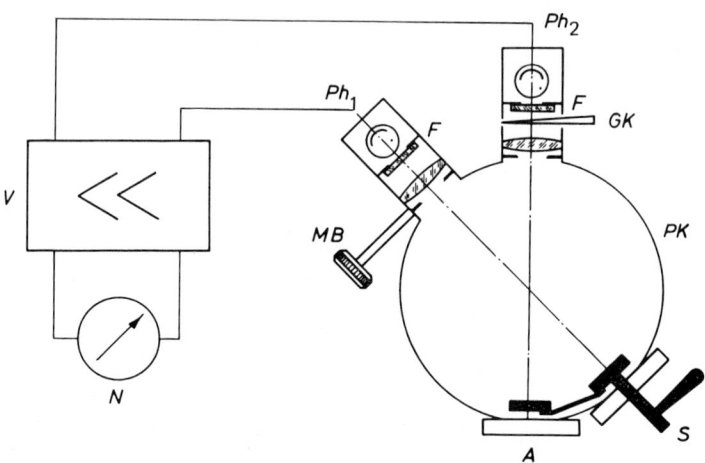

Measuring principle of the ELREPHO

Ph_1	photocell	A	sample
Ph_2	photocell	MB	measuring diaphragm
F	filter	N	balancing instrument
GK	neutral wedge	PK	photometer sphere
S	swing-in standard	V	amplifier

Fig. 18 Zeiss Elrepho Colorimeter. (Courtesy of Carl Zeiss, Inc., New York, New York.)

ments are obtained for a large number of wavelengths. Therefore, routine reflectance measurements are usually performed on the more handy colorimeters (Fig. 18) or abridged spectrophotometers where the monochromator is replaced by narrow-band filters. However, because of its spectral characteristics, the spectrophotometer usually surpasses the filter colorimeter with regard to color measurement accuracy, and it is often the standard type with which less sophisticated instrumentation is compared [8,9].

The Colorimeter In the filter colorimeter, different spectral characteristics are achieved through the placement of filters in front of the photodetectors, as shown in Fig. 18 [32]. This arrangement permits more freedom with regard to illumination and vastly speeds up the time required to obtain routine reflectance measurements. A great number of instruments have been designed that rely both on 45° illumination-normal viewing and on diffuse illumination-near normal viewing as, for example, in Fig. 18. Diffuse illumination requires the elimination of gloss and incident light on the sample, which come directly from the light source. Traps and baffles are used to remove these.

The Goniophotometer The main characteristic of the goniophotometer is the continuous angular adjustment of the photodetector system relative to fixed or adjustable radiation geometries, as seen in Fig. 19 [69]. Here, both sample holder and lamp arm are rotatable such that any combination of incident

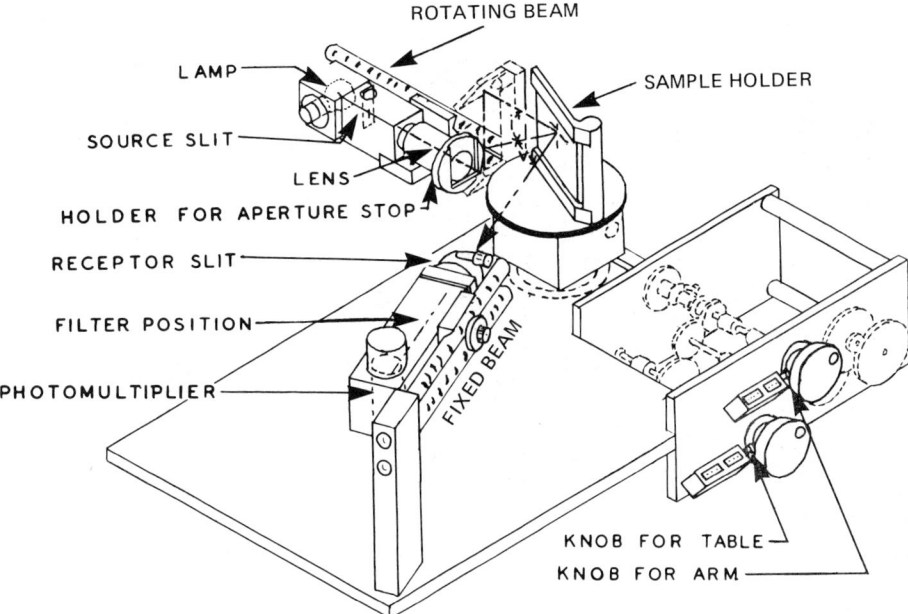

Fig. 19 Hunter goniophotometer. (From Ref. 69.)

and reflected light angles can be created in the horizontal plane. Other configurations are possible for both reflected and transmitted light geometries, additional azimuthal adjustments, and measurement of spectral light distributions (spectrogoniophotometer) [69,70]. The more complex systems are precision instruments that demand considerable attention to optical alignment and measurement procedures.

Photometers, like glossmeters that require light measurement at fixed specific angles, sometimes permit adjustment to account for various gloss types [55,107]. In this case, they are therefore goniophotometers, where freedom in angular measurement is more restricted. Both glossmeters and goniophotometers require the accurate definition of both incident and reflected light distributions, especially if high angular resolution is required. Due to their complex design and the time-consuming measurement procedure, goniophotometers such as that shown in Fig. 19 are mostly applied in the research laboratory [170]. Routine measurements of directional light are obtained mainly using fixed geometries (glossmeters).

The Microdensitometer   The microdensitometer measures reflected or transmitted light from areas of the sheet structure that are sufficiently small for surface unevenness or formation testing. The STFI formation tester shown in Fig. 20 is a research design, but it eliminates a number of difficulties encountered with similar, commercially available testers [115]. The incoming light is diffused by illumination from inside a Teflon cylinder, and light transmitted through the sample can be further diffused by placing a diffusing mat over the test specimen. Reflectance attachments allow the collection of

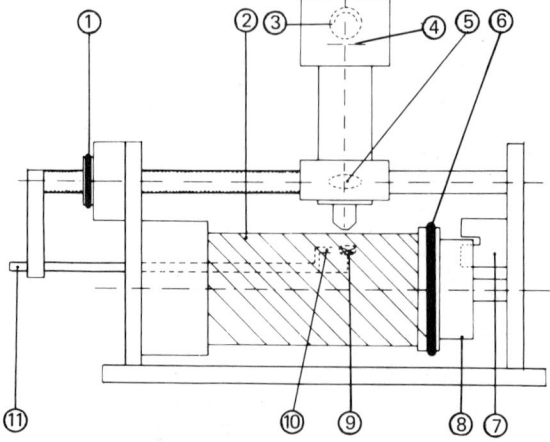

1. Driving belt and screw for axial motion of measuring equipment and lamp
2. Teflon cylinder
3. Photomultiplier
4. Aperture
5. Lens
6. Driving belt for rotating the cylinder
7. Track—hold unit
8. Translucent ring
9. Lamp
10. Cooling air nozzle
11. Cooling air intake

Fig. 20 STFI formation tester. (From Ref. 115.)

reflected light from scanned microareas, a capability also found in instrumentation specifically designed for measurement of surface unevenness.

The microdensitometer is unique among paper appearance testing instruments in that reflectance or transmittance values are collected and compared from different parts of the same sheet structure. As seen in Fig. 20, this is accomplished by the automatic revolution of the cylinder and axial motion of the photodetector system.

B. Design Principles

Optical Geometry The issue of proper reflectometer design is a controversial one [31,140,158,159]. It has rightly been pointed out that the stability of the diffuse reflecting coating in the instruments employing diffuse illumination is often difficult to maintain. Such instruments therefore introduce an extra variability that can be avoided by using instrumentation employing 45° illumination-normal viewing [31,159]. As discussed above, different design geometries will not provide identical measurement of any reflectance type for any paper product, and the choice of design depends upon the use of the measurements (mill comparisons contra Kubelka-Munk applications, etc.). For color measurements that require sheet stacking (R_∞ values), instrumentation geometry becomes less important and measurement agreement

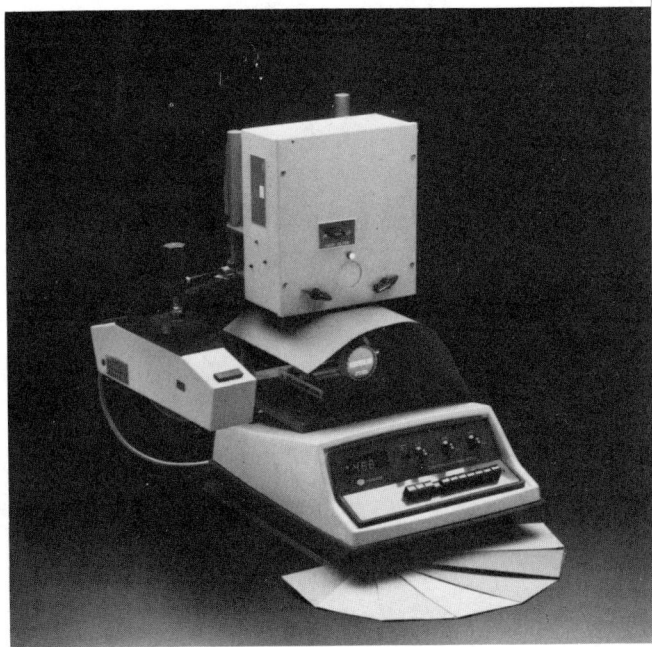

Fig. 21 Hunter D25-2 Colorimeter with additional gloss sensor. (Courtesy of Hunterlab Associates Laboratory, Inc., Fairfax, Virginia.)

between different instrument types has in some cases been shown to be better than for instruments of similar types [10].

Glossmeters measure directional reflectance and do not, therefore, employ diffusing spheres. However, since all measured light will in practice show some angular distribution (Fig. 13), the maximum light acceptance angle and light distribution of the photodetector system will determine instrument response [51,66]. Colorimeters of modular design allow reflectance measurements under different viewing and illumination geometries and sometimes also permit gloss measurements (Fig. 21).

Spectral Characteristics The meaning of any light measurement relies strongly on how well its spectral characteristics are generated by light source, photodetector, and possible filters. In particular, the increasing use of fluorescent brighteners in fine paper manufacture has made it imperative to employ in reflectance instrumentation light sources that contain a suitable amount of ultraviolet radiation [39,92,140]. Older instruments that illuminate mainly in the visible range of the spectrum are less suitable than similar types that emit sufficient ultraviolet light—as, for example, the xenon lamp type. The switch to UV lamp types has increased the problem of maintaining instrument stability and has not eliminated spectral differences when the performance characteristics of different instruments are compared. Figure 22 shows the chromaticity coordinates obtained for the same fluorescent sample using five different instruments, all equipped with xenon lamps

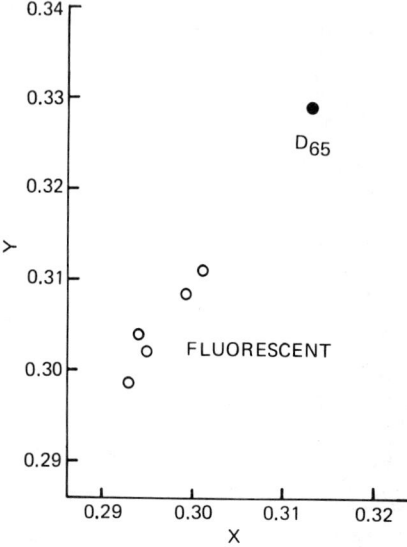

Fig. 22 Chromaticity coordinates of fluorescent sample measured with different instrument types. (From Ref. 39.)

[39]. The values vary significantly due to differences in spectral fluorescence response. TAPPI has published technical information sheets on the effects of different light sources when papers, including those containing fluorescent agents are evaluated (TIS 017-1). A method using white metameric sample pairs has been devised for testing light source and natural daylight for relative UV content [3].

Light sources, photodetectors, and possibly filters and diffuse sphere coatings are all limited in regard to long-term stability. To some degree, instrument drift is counteracted by the use of microprocessor technology in the more modern instrument types. The microprocessor controls the instrument, allows the rapid collection of reflectance data at predetermined wavelengths, and also contains storage capability for reference and calibration data [98]. Digital output microcomputers automatically derive specific colorimetric data when programmed for specific calculations. Billmeyer [6,8] has compared the performance of several instrumentation types and has stressed the proper interpretation of colorimetric measurements. A number of reviews describe modern instrumentation and color measuring practices in Europe and North America [8,134].

On-Line Control   Radiation sources are ideally suited for on-line control since radiation of suitable intensity will not interfere with the paper structure. For example, the monitoring of moisture and caliper using beta rays, IR radiation, and microwave techniques on the moving paper web are often seen in the mill environment. The monitoring of appearance properties is possible using either reflected or transmitted visible radiation [23,27]. Figure 23 describes the layout of a spectrophotometric system of a general nature applicable to both transmittance and reflectance measurements. A digital proces-

Optical and Appearance Properties   29

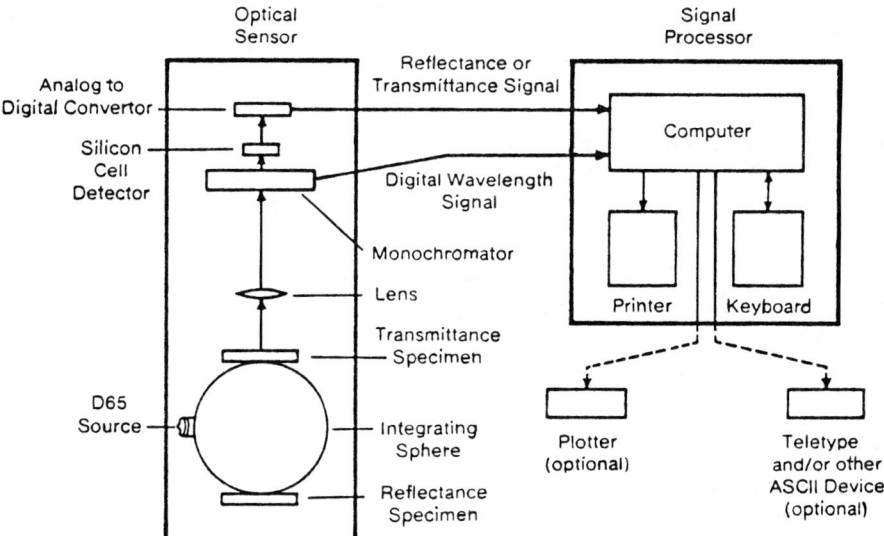

Fig. 23 On-line processor control of transmittance and reflectance. (From Ref. 23.)

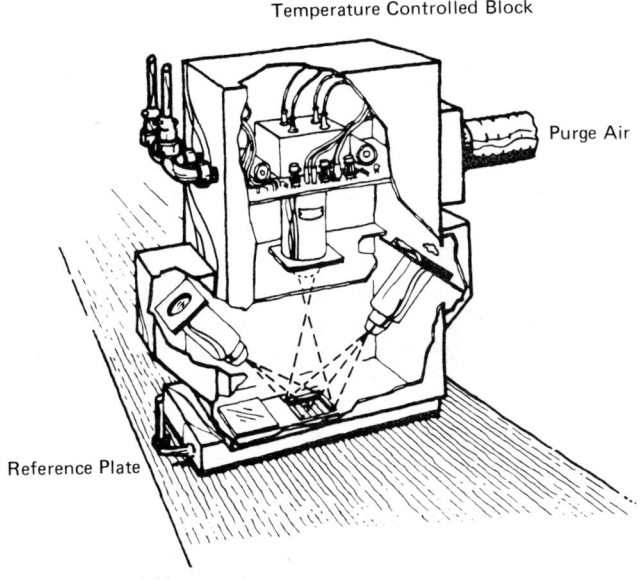

Fig. 24 On-line reflectance sensor. (From Ref. 23.)

sor controls the instrument and provides either fast readout for operator control (open-loop system) or interface for automatic adjustment necessary for off-specification corrections (closed-loop system) [168]. Colorimetric systems require the correct placement of the monitoring sensor in order to obtain accurately the shade of dry, "conditioned" paper [43]. A drawback is that this creates relatively long delay times for closed-loop systems [120].

Present sensors available for the control and testing laboratory can in many instances be modified for on-line use (Fig. 24). Here it is paramount that the design can withstand the mill environment with minimal maintenance efforts. Solid state technology similar to that applied in the newer types of laboratory instruments has greatly improved the accuracy and reliability of on-line measurement of color characteristics [23,155,165,168]. Readings obtained with the more advanced instrument types have been shown to compare favorably with laboratory instrument values [118].

C. Calibration

Reflectometers All commercially available reflectometers provide relative reflectance values, that is, the measurement is obtained relative to the reflectance of a standard sample. The nature of the reference standard must be given if the measurement is to be meaningful.

ISO specifies the following reference standards (ISO 2469):

ISO Reference Standard of Level 1 (IR 1) IR 1 is the perfect reflecting diffuser according to CIE 45-20-195. This ideal uniform diffuser is defined to have a reflectance equal to 1. Magnesium oxide has traditionally been the primary standard against which reflectance values have been compared in reflectometry [78]. Unfortunately, samples of magnesium oxide are not sufficiently stable for extended use, nor are they easy to prepare in a reproducible manner; ISO has therefore adopted an imaginary primary standard [19,145].

ISO Reference Standard of Level 2 (IR 2) This is a standard whose reflectance factor has been determined by a standardizing laboratory in relation to the IR 1. These standards are used by the authorized laboratories for the calibration of their reference instrument. Barium sulfate is a suitable IR 2 standard (ISO 2469). Figure 25 shows the absolute reflectance of barium sulfate using methods and instrumentation specifically designed for absolute rather than comparative light measurements [46].

ISO Reference Standard of Level 3 (IR 3) This standard is one whose reflectance factor has been determined by an authorized laboratory in relation to the IR 2. These standards are used by the working laboratories for the calibration of their instruments. They are often paper sheets that have been characterized using the reference instrument at the issuing laboratory. They require adequate storage to maintain their optical characteristics during any extended period of usage. Dark storage at low temperature and humidity minimizes the effect of aging [151].

A list of standardizing and authorized laboratories at the time of issue of ISO 2469–2471 (1976) is given in Table 4. Reflectance standards are also available from instrument makers and laboratories engaged in optics and paper research [102,159].

Table 4 Standard Issuing ISO Laboratories

Standardizing Laboratories

National Research Council
Ottawa, Canada

Physikalisch-Technische Bundesanstalt
Braunschweig, Federal Republic of Germany

Authorized Laboratories

Bundesanstalt für Materialprüfung
Berlin, Federal Republic of Germany

Centre Technique de l'Industrie des Papiers,
 Cartons et Celluloses
Grenoble, France

Ente Nazionale per la Cellulosa e per la Carta
Rome, Italy

Finnish Pulp and Paper Research Institute
Helsinki, Finland

U. S. Department of Commerce
National Bureau of Standards
Washington, D.C.

Pulp and Paper Research Institute of Canada
Pointe Claire, Quebec, Canada

Research Association for the Paper and Board
 Printing and Packaging Industries
Leatherhead, England

Swedish Forest Products Research Laboratory
Stockholm, Sweden

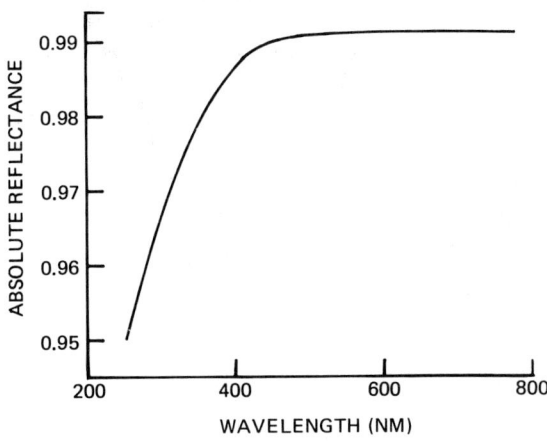

Fig. 25 Absolute reflectance of Eastman white reflectance standard (barium sulfate). (From Ref. 46.)

Glossmeters, Goniophotometers, and Microdensitometers In contrast to diffuse reflectance measurements, gloss values like those shown in Fig. 14 are often obtained relative to perfect mirror reflections or black glass standards. For example, the theoretical specular gloss standard of TAPPI T 480 is assigned 384.4 gloss units using a mirror of 100 gloss units with a glass standard of refractive index equal to 1.540.

The scanning type instruments (goniophotometers, microdensitometers, etc.) are generally required to measure variation in intensity with angle or sample position. Various standards are employed to assure instrument stability comparable to that of a spectrophotometer, for which the measurement procedure is time consuming. Hunter [70] has described standardization of appearance-measuring instruments in detail.

IV. INTERPRETATION OF OPTICAL MEASUREMENTS

A. Kubelka-Munk Theory

Principle The industry relies on the *Kubelka-Munk theory* for more detailed analyses of reflectance measurements [58,83,138]. The theory allows the interconversion of light reflectance and transmittance data. Furthermore, it permits the derivation of parameters that are independent of sheet caliper and that can be added up for the purpose of evaluating the optical effect of discrete papermaking components. Details of Kubelka-Munk calculations are given in nearly any textbook dealing with color and reflectance [78]. Its application to paper optics in general has been described by Stenius [139], Van den Akker [157,162], and Casey [21].

Homogeneous Sheets The Kubelka-Munk scattering theory is a phenomenological approach to the description of light interactions in a paper sheet. Each increment in basis weight dW absorbs, transmits, and back-scatters parts of the incident light (Fig. 26). The change in light transmittance due

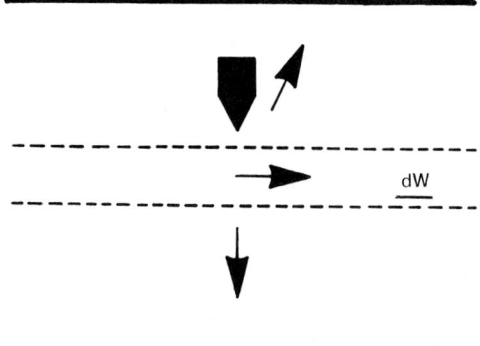

Fig. 26 Kubelka-Munk scattering element.

to scattering is characterized by the *scattering coefficient s* or *scattering power sdW*. The change due to absorption is characterized by the *absorption coefficient k* or *absorption power kdW*.

For homogeneous sheets (all elements dW are identical through the cross-directional sheet structure), differential equations valid for the element dW can be integrated to yield the following relationships for the reflectance R_0 and the transmittance T of the sheet:

$$R_0 = \frac{\sinh(bsW)}{a \sinh(bsW) + b \cosh(bsW)} \tag{14}$$

$$T = \frac{b}{a \sinh(bsW) + b \cosh(bsW)} \tag{15}$$

where

$$a = \frac{1/R_\infty + R_\infty}{2} \tag{16}$$

$$b = \frac{1/R_\infty - R_\infty}{2} \tag{17}$$

$R_\infty = R_0$ in cases where W is so large that T = 0 and sinh bsW and cosh bsW are hyperbolic functions of bsW [81].

The sheet scattering power sW can be calculated from measurements of R_0 and R_∞ using Eq. (14), which can be transformed from its hyperbolic dependency to the more convenient form

$$sW = \frac{1}{2b} \ln \frac{1 - R_0/R_\infty}{1 - R_0/R_\infty} \tag{18}$$

The sheet absorption power kW is obtained from measurement of R_∞ and calculated values of sW, since

$$\frac{k}{s} = \frac{kW}{sW} = \frac{(1 - R_\infty)^2}{2R_\infty} \tag{19}$$

Further evaluation of sheet scattering and absorption powers in terms of the values of sW and kW for the individual components of multicomponent scattering structures is feasible if the following additivities may be assumed:

$$sW = s_a W_a + s_b W_b + s_c W_c + \cdots \tag{20}$$

$$kW = k_a W_a + k_b W_b + k_c W_c + \cdots \tag{21}$$

where s_a, k_a, and W_a are coefficients and weight of component a; where s_b, k_b, and W_b are those of component b; and so on [1,57,63]. The additive equations are valid for homogeneous mixtures (pulp and coating mixtures, pigment-loaded sheet structures, etc.). Further optical interaction must be accounted for by including extra interaction terms in the equation system [162]. Gross heterogeneity in the cross direction of the sheet structure necessitates the calculation procedures for heterogeneous sheets or a summation procedure like that for layer calculations, both shown below.

In principle, the light interactions for the Kubelka-Munk element shown in Fig. 26 are similar to those shown for the whole sheet in Fig. 2. Kubelka-Munk calculations are quantifications of the reflectance, transmittance, and absorptance processes in the sheet elements. The integration procedure produces quantitative relationships that are valid for changes in light fluxes through the total sheet cross section. Note that the Kubelka-Munk coefficients do not account for scattering particle size, refractive index, or absorption coefficient in their derivation. The specific coefficients s and k are merely expressions of scattering and absorption powers per basis weight unit. Attempts to correlate the scattering coefficient s with fiber size or specific sheet surface area have been described by Rennel [122]. By use of the layer calculations described on p. 39, such correlations for pulp sheets have been shown to be related to the fiber refractive index [129].

The hyperbolic nature of the reflectance and transmittance equations Eqs. (14) and (15), was not apparent when the theory was originally applied to paper but was derived later by Kubelka [81]. Meanwhile, extensive graphic aids had been developed to provide the speed and convenience desired by users who were not familiar with or interested in the mathematical detail (TAPPI T 425 and TIS 017-2) [21]. Today's pocket calculators provide both sufficient programmability and the more specialized mathematical functions that are necessary to calculate any of the numerous relationships presented in Kubelka's second paper [75a,105,115a,116,143]. A summary of the equations used for paper optics calculations has been presented by Robinson [124].

Heterogeneous Sheets Paper coatings seldom have the same scattering and absorption coefficients as does the underlying base stock. Therefore, optical calculations for coated paper sheets are more complex than those for uncoated sheets.

The calculations of the reflectance and transmittance of multilayer structures has also been considered by Kubelka [82]. His approach has been applied to coated papers by Clark and Ramsay [24] and to coated board by Ramsay [121]. The principle of light interactions in two-layer structures is shown in Fig. 27. The incoming light beam interacts consecutively with both upper layer (coating) and lower layer (base stock). The reflectance $(R_0)_{1,2}$ and transmittance $T_{1,2}$ of the composite structure are given through the following equations:

$$(R_0)_{1,2} = R_{01} + T_1^2 R_{02}(1 + R_{01}R_{02} + R_{01}^2 R_{02}^2 + R_{01}^3 R_{02}^3 + \cdots$$

$$= R_{01} + \frac{T_1^2 R_{02}}{1 - R_{01}R_{02}} \tag{22}$$

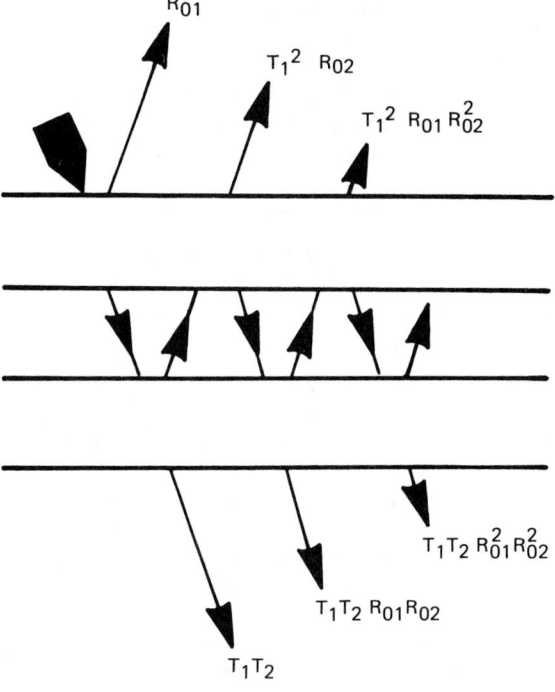

Fig. 27 Light interactions in a two-layer structure.

$$T_{1,2} = T_1 T_2 (1 + R_{01} R_{02} + R_{01}^2 R_{02}^2 + R_{01}^3 R_{02}^3 + \cdots)$$

$$= \frac{T_1 T_2}{1 - R_{01} R_{02}} \tag{23}$$

where R_{01}, T_1 and R_{02}, T_2 characterize reflectance-transmittance of coating and base stock, respectively.

In practice, it is not necessary to calculate transmittance to obtain $(R_0)_{1,2}$ since it can be shown that

$$(R_0)_{1,2} = \frac{(R_{01} - R_{\infty 1})/R_{\infty 1} - R_{\infty 1}(R_{02} - 1/R_{\infty 1}) e^{2s_1 W_1 b_1}}{(R_{02} - R_{\infty 1}) - (R_{02} - 1/R_{\infty 1}) e^{2s_1 W_1 b_1}} \tag{24}$$

where
  b_1 is defined according to Eq. (17) using reflectance factor $R_{\infty 1}$ of coating
  s_1 = scattering coefficient of the coating,
  W_1 = weight of applied coating.

This equation [24,124] is similar to that used when homogeneous sheets are characterized on the basis of reflectance measurements over "white" and "black." Robinson has described methods suitable for characterizing both one-side and two-side coated stock [124].

Error Sources The Kubelka-Munk approach is relatively simple both in principle and in application compared to other scattering theories, such as the Mie theory described on pp. 39–40. Nevertheless, its application to paper optics has not always been successful. There are two main reasons why the mathematics may fail in describing appearance properties adequately:

> The sheet structure does not interact with light "as intended" due to gross heterogeneity in the Kubelka-Munk defined layers.
> Reflectance values cannot be measured with sufficient accuracy for calculation purposes.

Migration of pigment particles is a well-known phenomenon in both uncoated paper sheets and in paper coatings. Sheet reflectance may change, since the scattering power of added pigment is different from that anticipated on the basis of its scattering coefficient and the amount added. Similarly, pigment particles will produce optical effects that depend on whether they are added into or onto the sheet structure. Filler pigments are often effective light scatterers because of their good dispersion in the bulk sheet structure. Coating pigment particles can aggregate into less effective, fewer and larger aggregates—especially when pigment and binder concentrations are changed [57]. Nevertheless, theoretically, a high brightness pigment (one with high scattering coefficient and low absorption coefficient) will produce maximum-reflectance when applied in coating layers rather than added to the base stock, assuming that particle aggregation does not alter its scattering ability (Fig. 28) [13,15].

In view of the differences in reflectance shown in Fig. 28 it becomes understandable that the application of Kubelka-Munk calculations to optical

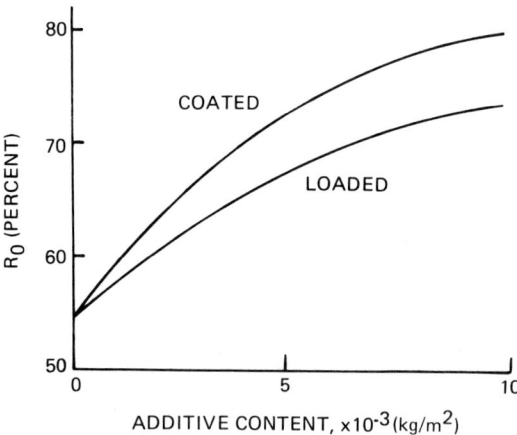

Fig. 28 Reflectance of coated sheet compared with loaded sheet for high brightness additive. (From Ref. 15.)

sheet characterization should always take into account physical sheet structure. Gross heterogeneity in the distribution of scattering particles requires modified calculation procedures as demonstrated theoretically by Van den Akker [161]. Paper reflectance measurements [37,112] and diffuse reflectance spectroscopy of powders consisting of particles similar in size to paper fibers [80] indicate that Kubelka-Munk calculations are not suitable for samples with high light absorbing characteristics. The increase in k values over the wavelength range of light absorption produces a decrease in s values, as shown in Fig. 29 for data obtained by Nordman et al. [112]. Consequently, reflectance at high absorption levels cannot be accurately predicted on the basis of measurements of s at low absorption levels.

Dissatisfaction with reflectance values predicted versus those measured experimentally and the calculation work involved in obtaining predicted values—especially for coated sheets—have in some instances led to calculation shortcuts. For example, Luey [93,136] has postulated an opacity factor concept that may lead to better reflectance predictions for coated boxboard. Bauer [5] has proposed that the light scattering coefficients of paper coatings can in some instances be determined by the method valid for filled papers. Use of this proposal would make the two curves shown in Fig. 28 coincide [15]. Robinson and Linke [125] have attempted to account for pigment dispersion characteristics when predicting coating opacities. At high binder levels, the pigment "film" on the paper surface becomes continuous and it is then necessary to account for Fresnel reflected light distributions as defined by the refractive index change at the coating surface [Eqs. (12) and (13)] [128].

The application of the Kubelka-Munk theory requires that at least two reflectance measurements be obtained for the purpose of determining both scattering and absorption coefficients. It is necessary to minimize the error by which both reflectance values are measured to obtain the necessary accu-

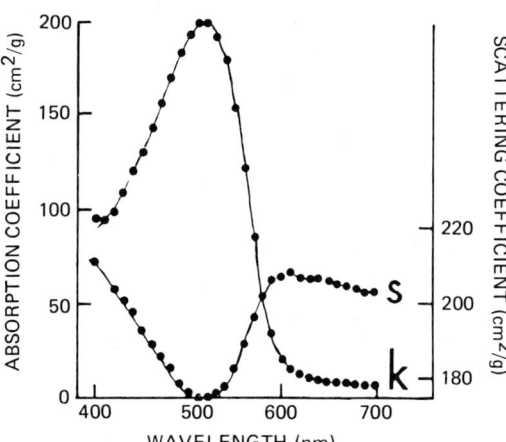

Fig. 29 Variations in scattering and absorption coefficients for sheets made from red-dyed pulp. (From Ref. 112.)

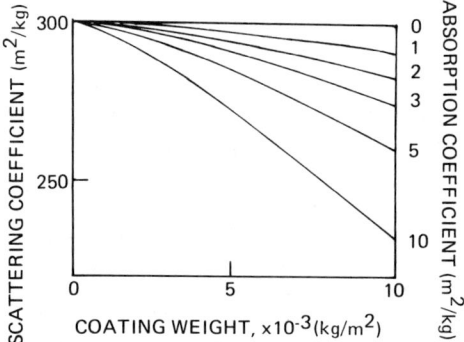

Fig. 30 Decrease in scattering coefficient using black glass techniques for increased levels of coating weight and absorption coefficient. (From Ref. 13.)

racy in calculated coefficient values. Similarly, for coated sheets, the substrate reflectance R_{02} must be sufficiently different from the reflectance factor $R_{\infty 1}$ to assure reasonable accuracy in calculated coating scattering coefficient values [13,137]. The effect of substrate reflectance can be ignored when a translucent substrate is used for successive coating applications [137].

High Brightness Sheets For uncoated sheets of high reflectance factor (R_∞), the necessity of obtaining two sufficiently different reflectance measurements for calculations of s can be avoided by using the following relationship [78]:

$$sW = \frac{R_0}{1 - R_0} \qquad (25)$$

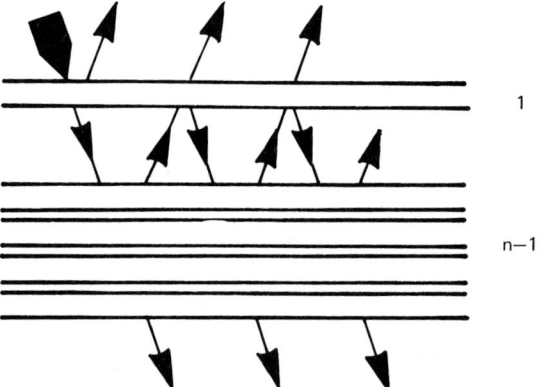

Fig. 31 Light interactions in n-layer structure. (From Ref. 129.)

This much faster procedure is also useful for the analysis of high brightness coating pigments applied on light absorbing substrates (black glass technique) [150]. Here, the absorption coefficient is not known and will, in addition to the coating weight, determine how much the scattering coefficient is lowered relative to that obtained using exact analysis (Fig. 30) [13]. Neglecting the absorption coefficient also implies that the scattering power sW is, in fact, the ratio between sheet reflectance and transmittance ($T = 1 - R_0$; $sW = R_0/T$). Lathrop [86] has pointed out that this relationship may be useful for lightweight sheets of low absorption. Unfortunately, the transmitted light is difficult to measure accurately as discussed on pp. 20–21.

B. Particulate Scattering Approaches

Layer Calculations The lack of physical reality inherent in the Kubelka-Munk approach has led to reflectance analysis procedures that are more closely related to the manner in which light interacts with granular or fibrous structures [26,80]. Layerlike models for light interactions with granular powders have been proposed by Johnson [76] and Melamed [103]. For paper sheets, the approach of considering consecutive light interactions between layers (Stokes calculations) [146] has been demonstrated for two layers (see Fig. 27). The concept can be extended to any number of layers (Fig. 31). The layers may be different or identical; they may consist, for example, of separate paper sheets [132]. The principle is to interpret the optical properties of the stack in terms of single layer properties, that is,

$$R_0 = R_n = R_1 + \frac{R_{n-1} T_1^2}{1 - R_1 R_{n-1}} \tag{26}$$

where n = 2,3, ... [129]. This equation is similar to Eq. (22).

Ultimately, the pile of layers can be considered to represent the stack of bonded fibers that constitute the sheet itself when viewed in cross section [129,130]. As such, cell wall reflectance, transmittance, and degree of bonding can in principle be related to sheet reflectance and transmittance, as shown by Scallan and Borch [130].

Mie Theory The scattering from single spherical or cylindrical particles of submicron size has been rigorously calculated by Mie [79,104,156]. As shown in Fig. 32, the scattered intensity is strongest in the forward direc-

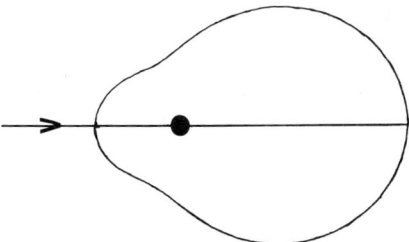

Fig. 32 Mie scattering from submicron sphere. (From Ref. 127.)

tion, around which it is centrosymmetrically distributed for spheres [127]. Scattering from paper and textile fibers is asymmetric [94]. This asymmetry has created efforts to develop light transmission methods as a means of deriving fiber orientation in paper sheets, as described in Chap. 24.

It has generally been assumed that Mie scattering calculations cannot be applied to paper reflectance [57]. The reason for this assumption is that the particles are so numerous and closely situated that details in scattered intensity distributions related to single particle properties are lost. While it is certainly true that the scattering particles interact with each other (multiple scattering), recent developments in both paint and paper optics indicate that single particle properties are reflected to some degree in overall scattered intensity [14,126,127]. The approach relies on the analysis for pigmented paints by Ross [126]. Ross found that the following formula holds for the scattering coefficient s_d of a pigment of low volume concentration in a paint layer:

$$s_d = C(1 - \cos \langle \theta \rangle_{av}) \frac{Q_{sca}}{D} \qquad (27)$$

where

D - particle diameter
Q_{sca} = Mie scattering efficiency
$1 - \cos \langle \theta \rangle_{av}$ = a factor necessary to account for the asymmetry of the scattering envelope (Fig. 32)
C = a proportionality constant dependent upon pigment volume concentration and instrumentation constants

Both Q_{sca} and $\cos \langle \theta \rangle_{av}$ are functions of particle refractive index and $x = \pi D / \lambda$ where λ is the wavelength of the light of radiation. The scattering coefficient is given in reciprocal units of paint thickness d. It can be related to the reflectance R_0 or the scattering coefficient on a weight basis s_w through Eq. (25) [14,126]. Both Q_{sca} and $\cos \langle \theta \rangle_{av}$ have been defined through tables or computer programs for spherical particles of selected size-wavelength ratios at a number of refractive indices [14,79]. Consequently, it is possible to obtain the relative variations in s with either size or wavelength for pigments added to paper coatings, as demonstrated by Borch and Lepoutre [14].

Calculations for spherically shaped plastic pigments have shown that the scattering coefficient is strongly dependent on both the size and the refractive index in the submicron size range (Fig. 33) [25]. The particle size for maximum scattering is larger than half the radiation wavelength, in contrast to what is usually found experimentally for pigments of higher refractive index [97,162]. Calculations for more complex void-particle systems have shown unexpected increases in hiding power for microvoid paints [133]. The potentiality exists to demonstrate and achieve similar effects for paper coatings.

C. Interpretation of Gloss

Specular Contra Diffuse Reflectance The interpretation of gloss and the influence of gloss on other appearance properties are among the most severe

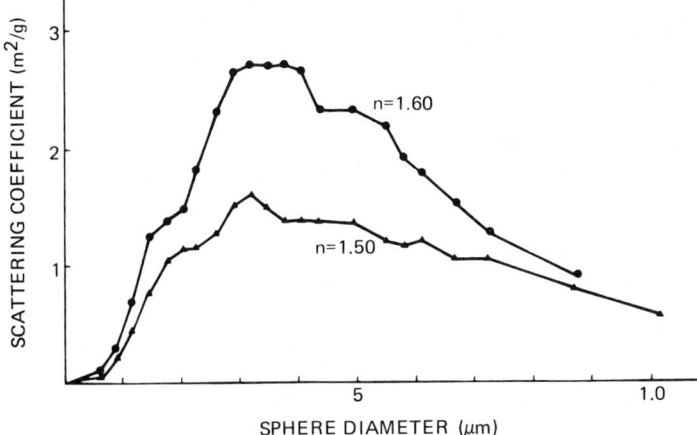

Fig. 33 Calculated variation of the scattering coefficient with spherical pigment size and refractive index. (From Ref. 14.)

challenges that face the papermaker within the area of appearance properties. The types of gloss described in Table 3 are based upon visual perception rather than corresponding physically measurable properties [64]. The variations in glossmeter design and the number of more sophisticated goniophotometers that have been proposed for gloss characterization [53,55,70,107], reflect the uncertainty with which gloss characterization is connected.

Any sheet structure, including uncoated plain paper, will surface reflect light, and there is no accurate method by which surface-reflected light can be completely distinguished from light that has been internally reflected. Nevertheless, considerable attempts have been made to quantify paper reflectance in terms of specular and diffuse light components (elementary mirror concept) [4,53]. Barkas [4] originally proposed that surfaces showing low gloss can be modeled as containing small elementary facets which may be set at any angle to the mean surface. They reflect the light either diffusely or specularly according to Fresnel's theories [for example, Eqs. (12) and (13)]. At low gloss levels, adequate agreement with reflectance measurements is attained. Angular increases in peak reflectance position can be explained as being caused by random tilts in Fresnel reflecting surface parts that produce excess light at higher angles [47,78]. However, further analysis at high gloss levels produces irrational conclusions with regard to diffuse and specular light fractions as shown by Kurita et al. [85] who employed goniophotometry and Tanaka [148,149] who used polarized light analysis at three reflectance angles. Therefore, this approach has had limited use, as discussed in detail by Harrison [53].

Surface Roughness A different approach is to relate specular and diffuse light components to surface roughness [42,75]. Gate et al. [42] have demonstrated that surface roughness in the submicron range correlates with TAPPI gloss. Light is surface-reflected in a manner similar to the scattering from submicron-size particles (described on pp. 39-40) for which Freshnel reflection does not occur because of the relationship between size and wavelength of irradiation. For example, the theories applicable to scattering of IR radiation and micro- and radio-waves from rough surfaces predict that reflectance in the specular direction ρ is the sum of specular and diffuse components such that [166]:

$$\rho = \rho_0 \exp\left[-\left(\frac{K_1 \sigma \cos\theta}{\lambda}\right)^2\right] + \rho_0 K_2 m \Psi \left(\frac{\sigma}{\lambda}\right)^4 \quad (28)$$

where
ρ_0 = reflectance of a smooth surface under identical conditions,
σ = root mean square (r.m.s.) roughness
m includes other roughness factors
θ = angle of incidence
λ = wavelength of radiation
Ψ includes instrument factors
K_1 and K_2 are constants

Consequently, the σ/λ ratio must be sufficiently small to create measurable specular reflectance [first term in Eq. (28)]. The equation describes a perfectly conducting surface that does not transmit radiation. Visible light is both transmitted and reflected by paper. Nevertheless, the formula indicates why glossy papers are usually those that are coated or supercalendered to a roughness level appreciably below that of the fiber structure itself.

Effect on Color Since routine color measurements are obtained using instrumentation of fixed optical design, paper color measurements seldom reflect the appreciable variation in surface color with viewing angle for both glossy and matte sheets. Diffusely scattered radiation from the internal sheet structure reflects the object color of the sheet. Specularly reflected light from the sheet surface reflects the color of the light source [47,78,109]. The color shift with viewing angle has been described by Nishiwaki [111] and Gunji, Nihira, and Tsuboi [47,109]. For both colored glossy and colored dull papers, the tristimulus values increase for increasing incidence-receiving angle [47]. In addition, this study illustrates a number of paper reflectance properties described earlier in this chapter: light reversal does not produce identical color values; angle of maximum reflectance is not necessarily that of incidence; and paper smoothness has a profound effect on paper reflectance.

Present color characterization is based upon separate, well-defined measurements using concise instrumentation design, as shown above. In practical paper applications, illumination and viewing geometries are seldom well controlled; there is a need for more sophisticated instrumentation design that is capable of simulating the visual combinations of specular and diffuse reflectance in a more universal manner.

D. Formation Theory

Paper Structure Corte and coworkers [28,30] have applied statistical geometry to the description of paper structure. The sheet is conceived as a network of fibers each of which is characterized by its center of gravity and average fiber length. For a sheet composed of randomly deposited fibers, the center of gravity distribution is such that the fractional area p(n) that is covered by n fibers can be calculated through:

$$p(n) = \frac{\langle n \rangle_{av}^n}{n!} \exp(-\langle n \rangle_{av}) \tag{29}$$

where $\langle n \rangle_{av}$ is the average number of fibers at any point within the gross area of the sheet structure. Formation testing generally measures the variation in mass density for regions of submillimeter size such that the theoretical variations on fiber size levels defined by this equation must be transformed to the larger area. This technique affords a means of comparing experimentally measured formation variations with the theoretically defined degree of randomness [30,115].

The effect of sheet formation on sheet reflectance is one where the number of layers vary over the irradiated part of the surface. Consequently,

$$R_0 = R_{\langle n \rangle_{av}}$$

$$= \exp(-\langle n \rangle_{av}) \left(\frac{\langle n \rangle_{av}^1}{1!} R_1 + \frac{\langle n \rangle_{av}^2}{2!} R_2 + \frac{\langle n \rangle_{av}^3}{3!} R_3 + \cdots \right) \tag{30}$$

where $R_1, R_2, R_3, \ldots$ are the reflectances of sheets composed of 1,2,3... layers [16]. The expression permits the calculation of the correct reflectance variation irrespective of variations in s and k at high absorption levels. This is shown in Fig. 34 where sheets similar to those described in Fig. 29 were

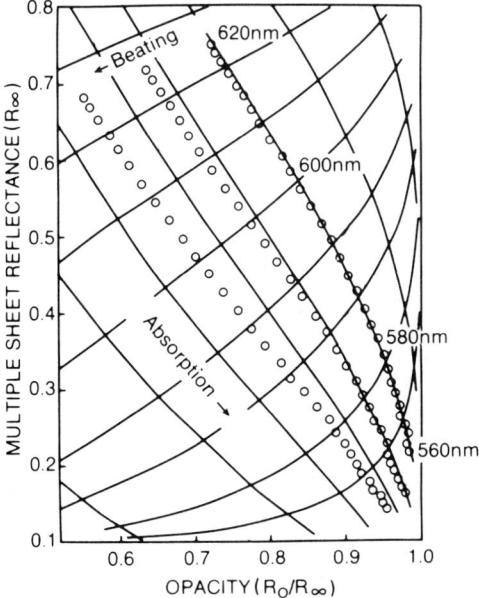

Fig. 34 Variations in opacity and reflectance factor for sheets made from red-dyed pulp. Increased absorption produces reflectance change as predicted by layer calculations modified for sheet formation (shifts along constant layer thickness curves). (From Ref. 16.)

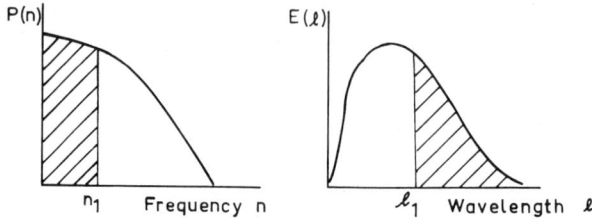

Fig. 35 Frequency and wavelength spectra concepts applied to paper formation (Norman and Wahren [113], reproduced with permission). (From Ref. 49.)

measured for reflectance and the results compared to the theoretically defined variations in opacity and reflectance factor [16]. Similarly, the more detailed reflectance analysis allows the quantitative consideration of cell wall reflectance and transmittance of different pulps as demonstrated by Hasuike et al. [54].

Wavelength Spectra Norman and Wahren have developed the concept of wavelength spectra to formation testing [113]. Here, the sheet is considered to consist of flocs of different sizes and weights. The spectral density represents the intensity of various floc sizes (Fig. 35). For a given floc in the sheet, this means that its wavelength is equal to twice its geometrical size, and measured wavelength spectra can be compared to calculated distributions for given scanning size [49].

An extensive account of the concept and interpretation of wavelength spectra and their applicability to formation measurements has been presented in connection with a general review of formation testing [115].

REFERENCES

1. Adrian, A. P. (1944). New procedure for evaluating the opacifying properties of pigments. *Tech. Assn. Papers* 27:465–471.
2. Alinče, B., and Lepoutre, P. (1980). Light-scattering of coatings formed from polystyrene pigment particles. *J. Coll. Interface Sci.* 76(1):182–187.
3. Anders, G., and Ganz, E. (1979). Metameric white samples for testing the relative UV content of light sources and of natural daylight. *J. Appl. Opt.* 18(7):1067–1072.
4. Barkas, W. W. (1939). Analysis of ligth scattered from a surface of low gloss into its specular and diffuse components. *Proc. Phys. Soc. (London)* 51:274–292.
5. Bauer, R. W. (1973). A simple method for calculation of light scattering coefficients for coated papers and paper coatings. *Tappi* 56(11):113–116.
6. Billmeyer, F. W., Jr. (1962). Caution required in absolute color measurement with colorimeters. *Official Digest* 34(455):1333–1342.
7. Billmeyer, F. W., Jr. (1968). The significance of recent CIE recommendations for color measurement. *Color Eng.* (Jan.-Feb.) 34–38.

8. Billmeyer, F. W., Jr. (1969). Current American practice in color measurement. *Appl. Opt.* 8(4):737–750.
9. Billmeyer, F. W., Jr. (1969). Comparative performance of color-measuring instruments. *Appl. Opt.* 8(4):775–783.
10. Billmeyer, F. W., Jr., Campbell, E. D., and Marcus, R. T. (1974). Comparative performance of color-measuring instruments: Second report. *Appl. Opt.* 13(6):1510–1518.
11. Billmeyer, F. W., Jr., and Marcus, R. T. (1969). Effect of illuminating and viewing geometry on the color coordinates of samples with various surface textures. *Appl. Opt.* 8(4):763–768.
12. Billmeyer, F. W., Jr., and Smith, R. (1967). Optimized equations for MacAdam color differences. *Color Eng.* (Nov.-Dec.):28-29.
13. Borch, J. (1978). Optical properties of paper coatings. Paper presented at the 1978 Summer Conference in Polymer Science and Technology, New Paltz, New York, June 26–28.
14. Borch, J., and Lepoutre, P. (1978). Light reflectance of spherical pigments in paper coatings. A comparison with theory. *Tappi* 61(2):45–48.
15. Borch, J., and Scallan, A. M. (1975). A comparison between the optical properties of coated and loaded paper sheets. *Tappi* 58(2):135-135.
16. Borch, J., and Scallan, A. M. (1976). An interpretation of paper reflectance based upon morphology. The effect of mass distribution. *Tappi* 59(10):102–105.
17. Bryntse, G., and Norman, B. (1976). A method to measure variations of surface and diffuse reflection from printed and unprinted paper samples. *Tappi* 59(4):102–106.
18. Budde, W., and Chapman, S. M. (1968). The calibration of standards for "absolute brightness" measurements with the Elrepho. *Pulp Paper Mag. Can.* 69(7):153–156.
19. Budde, W., and Chapman, S. (1975). Measurement of brightness and opacity according to ISO-standards. *CPPA Trans. Tech. Sect.* 1(2):61–64.
20. Burkhard, G., Wrist, P. E., and Mounce, G. R. (1960). A formation tester which graphically records paper structure. *Pulp Paper Mag. Can.* 61(6):T319–T333.
21. Casey, J. P. (1961). *Paper Testing and Converting. Pulp and Paper Chemistry and Chemical Technology*, 2nd ed., vol. 3, Interscience, New York, pp. 1362–1434.
22. Chickering, K. D. (1967). Optimization of the MacAdam-modified 1965 Friele color-difference formula. *J. Opt. Soc. Am.* 57(4):537-541.
23. Christie, J. S. (1977). On-machine measurement of the chromatic aspects of appearance. *Tappi* 60(2):119–121.
24. Clark, H. B., and Ramsay, H. L. (1965). Predicting optical properties of coated papers. *Tappi* 48(11):609–612.
25. Clydesdale, F. M., and Podlesny, C. H., Jr. (1968). A computer program for the interconversion of color data. *Color Eng.* (May-June):55-56.

26. Companion, A. L. (1965). Theory and applications of diffuse reflectance spectroscopy. In *Developments in Applied Spectroscopy*, vol. 4 (E. N. Davis, ed.), Plenum, New York, pp. 221–234.
27. Cook, A. J. (1973). On-line appearance measurement. A status report. *Tappi 56*(2):55–58.
28. Corte, H. (1970). On the distribution of the mass density in paper. II. *Das Papier 24*(5):261–271 (in German).
29. Corte, H. (1976). Perception of the optical properties of paper. In *The Fundamental Properties of Paper Related to Its Uses* (F. Bolam, ed.), British Paper and Board Industry Federation, London, pp. 626–658.
30. Corte, H., and Dodson, C. T. J. (1969). On the distribution of the mass density in paper. *Das Papier 23*(7):381–393 (in German).
31. Dearth, L. R. (1963). Numerical evaluation of color employing suitable instrumentation. *Tappi 46*(10):146A–151A.
32. Dearth, L. R., Shillcox, W. M., and Van den Akker, J. A. (1960). Instrumentation studies, No. 85. A study of photoelectric instruments for the measurement of color: reflectance and transmittance. 12. The Elrepho (photoelectric reflectance photometer). *Tappi 43*(2):230A–239A.
33. Dearth, L. R., Shillcox, W. M., and Van den Akker, J. A. (1963). Instrumentation studies, No. 87. A study of photoelectric instruments for the measurement of color: reflectance and transmittance. 14. The standard brightness tester as a four-filter colorimeter. *Tappi 46*(1):179A–188A.
34. Dearth, L. R., Shillcox, W. M., and Van den Akker, J. A. (1967). A study of photoelectric instruments for the measurements of color, reflectance, and transmittance. 16. Automatic color-brightness tester. *Tappi 50*(2):51A–58A.
35. Emerton, H. W., Page, D. H., and Hale, W. H. (1962). Structure of papers as seen in their surfaces. In *The Formation and Structure of Paper* (F. Bolam, ed.), British Paper and Board Industry Federation, London, pp. 53–99.
36. Foote, W. J. (1939). An investigation of the optical scattering and absorption coefficients of dyed handsheets and the application of the I.C.I. system of color specification to these handsheets. *Paper Trade J. 108*(10):TS125–TS132.
37. Foote, W. J. (1939). An investigation of the fundamental scattering and absorption coefficients of dyed handsheets. *Paper Trade J. 109*(25):TS333–TS340.
38. Frei, R. W., and MacNeil, J. D. (1973). *Diffuse Reflectance Spectroscopy in Environmental Problem-Solving*, CRC Press, Cleveland, pp. 31–79.
39. Ganz, E. (1976). Whiteness: Photometric specification and colorimetric evaluation. *Appl. Opt. 15*(9):2039–2058.
40. Ganz, E. (1977). Assessment of the ultraviolet range of artificial light sources for the best fit to standard illuminant D_{65}. *Appl. Opt. 16*(4):806.
40a. Ganz, E. (1979). Whiteness formulas: A selection. *J. Appl. Opt. 18*(7):1073–1078.

40b. Ganz, E. (1979). Whiteness perception: individual differences and common trends. *J. Appl. Opt. 18*(17):2963–2970.
41. Gartaganis, P. A., Heintze, H. U., and Gordon, R. W. (1976). From printograph to videoscanner. *Tappi 59*(12):113–117.
42. Gate, L., Windle, W., and Hine, H. (1973). The relationship between gloss and surface microtexture of coatings. *Tappi 56*(3):61–65.
43. Gill, J. P. (1969). Continuous on-machine color control of paper. *Tappi 52*(2):232–236.
44. Grum, F., and Costa, L. (1977). Color evaluation by fluorescence measurement without the need for multiple illumination sources. *Tappi 60*(8):119–121.
45. Grum, F., and Patek, J. M. (1965). Evaluation of whiteness using relative spectral radiance measurements. *Tappi 48*(6):357–362.
46. Grum, F., and Wightman, T. E. (1977). Absolute reflectance of Eastman white reflectance standard. *Appl. Opt. 16*(11):2775-2776.
47. Gunji, T., Nihira, K., and Tsuboi, T. (1973). Goniophotometric color measurement for papers. *J. Soc. Fiber Sci. Technol. (Japan) (Sen-i Gakkaishi) 28*(11):462–469 (in Japanese).
48. Haddad, S. F. (1967). Measurement of optical brightener in paper. *Tappi 50*(12):90A-91A.
49. Haglund, L., Norman, B., and Wahren, D. (1974). Mass distribution in random sheets: Theoretical evaluation and comparison with real sheets. *Svensk Papperstidn. 77*(10):362–370.
50. Hall, R. J. (1976). Instrumental quantification of subjective formation. In *The Fundamental Properties of Paper Related to Its Uses* (F. Bolam, ed.), British Paper and Board Industry Federation, London, pp. 662–677.
51. Hammond, H. K. and Hsia, J. J. (1975). Evaluation of instrument tolerances for 75° gloss. *Tappi 58*(11):143-144.
52. Harrison, V. G. W. (1940). Measurement of Opacity of Paper: A Theoretical Survey. *Proc. Tech. Sect. British Paper and Broad Makers Assn. 21*:67–173.
53. Harrison, V. G. W. (1962). Optical properties of paper. In *The Formation and Structure of Paper* (F. Bolam, ed.), British Paper and Board Industry Federation, London, pp. 467–485.
54. Hasuike, M., Shingai, M., Murakami, K., and Imamura, R. (1979). Evaluation of the optical properties of thin- and thick-walled fiber sheets. *J. Japan Wood Res. Soc. (Mokuzai Gakkaishi) 25*(2):132–138 (in Japanese).
55. Hasunuma, H., and Nara, J. (1955). On the specular gloss of paper. *J. Appl. Phys. (Japan) 24*(2):74–79 (in Japanese).
56. Hellawell, J. M. (1973). Analysis of the small-scale distribution of mass density in paper by beta-radiography. *Paper Technol. 14*(1):24–32.
57. Hemstock, G. A. (1962). The effect of clays upon the optical properties of paper. *Tappi 45*(2):158A-159A.
58. Hillend, W. J. (1966). Opacity problems in printing papers: Kubelka-Munk theory gives good, quick answers. *Tappi 49*(7):41A–47A.

59. Hoban, R. F. (1974). Fluorescence measurements by the two-mode method. *Tappi* 57(7):97–100.
60. Hopkins, L. F. (1969). The standardization of color measurement of paper. *Printing Technol.* 13(3):114–122.
61. Huey, S. J. (1972). Scientific approach to visual color comparison. *J. Paint Technol.* 44(573):83–89.
62. Huey, S. J. (1972). Standard practices for visual examination of small color differences. *J. Color Appearance* 1(4):24–26.
63. Hughes, D. A. (1962). Methods of obtaining the optical properties of papers containing titanium dioxide and mixtures of titanium dioxide and other fillers. *Tappi* 45(2):159A–163A.
64. Hunter, R. S. (1952). Gloss evaluation of materials. *ASTM Bull.* no. 186, pp. 48–55.
65. Hunter, R. S. (1958). Photoelectric color difference meter. *J. Opt. Soc. Am.* 48(12):985–995.
66. Hunter, R. S. (1958). Standardization of test for specular gloss of paper at 75°. *Tappi* 41(8):385–396.
67. Hunter, R. S. (1962). Measurements of the appearance of paper. *Tappi* 45(9):203A–209A.
68. Hunter, R. S. (1965). Meeting the paper industry's requirements for color instrumentation. *Tappi* 48(2):63A–65A.
69. Hunter, R. S. (1968). High resolution goniophotometer and its use to measure appearance properties and light-scattering phenomena. In *Modern Aspects of Reflectance Spectroscopy* (W. Wendlandt, ed.), Plenum, New York, pp. 226–241.
70. Hunter, R. S. (1975). *The Measurement of Appearance*, John Wiley and Sons, New York.
71. Hunter, R. S. (1977). Proposed revision of standard T 524. *Tappi* 60(10):45–46.
72. Ingersoll, L. R. (1921). The glarimeter. *J. Opt. Soc. Am.* 5(3):213–217.
73. Institute of Paper Chemistry (1939). Instrumentation studies, 30. A study of instruments for the measurement of opacity of paper. I: A general discussion of opacity and methods for the measurement of opacity. *Paper Trade J.* 109(4):TS31–TS36.
74. International Commission on Illumination (1970). *Colorimetry*, CIE Publication no. 15 (E-1.3.1), Bureau Central de la CIE, Paris.
75. Ito, Y., Murakami, I., and Saito, N. (1966). Studies of print qualities. 1: Influence of paper and ink properties on the print gloss. *Res. Bull. Printing. Bur. (Japan)* 1: no. 1, pp. 1–20 (in Japanese).
75a. Johnson, J. W. D. (1979). Newsprint properties: Opacity, percent saturation, dominant wavelength, and scattering coefficient. *Tappi* 62(12):115-116.
76. Johnson, P. D. (1952). Absolute optical absorption from diffuse reflectance. *J. Opt. Soc. Am.* 42(12):978–981.
77. Jones, C. L., and Budinger, A. B. (1965). Equipment and techniques for measurement and characterization of paper reflectance variation. *Tappi* 48(1):54–59.

78. Judd, D. B., and Wyszecki, G. (1975). *Color in Business, Science, and Industry,* 3rd. ed., John Wiley and Sons, New York.
79. Kerker, M. (1969). *The Scattering of Light and Other Electromagnetic Radiation,* Academic, New York.
80. Kortüm, G. (1969). *Reflectance Spectroscopy,* Springer-Verlag, Berlin.
81. Kubelka, P. (1948). New contributions to the optics of intensely light-scattering materials, 1. *J. Opt. Soc. Am. 38*(5):448–457, 1067.
82. Kubelka, P. (1954). New contributions to the optics of intensely light-scattering materials. 2. Non-homogeneous layers. *J. Opt. Soc. Am. 44*(4):330–335.
83. Kubelka, P., and Munk, F. (1931). A contribution to the optics of colorant layers. *Z. Tech. Physik 12*(11a):593–601 (in German).
84. Kuehni, R. G. (1976). Color-tolerance data and the tentative CIE 1976 L*a*b* formula. *J. Opt. Soc. Am. 66*(5):497–500.
85. Kurita, T., Yano, H., Nara, J., and Hasunuma, H. (1955). Barkas' analysis for the light scattered by surface of paper. *J. Appl. Phys. (Japan) 24*(8):318–325 (in Japanese).
86. Lathrop, A. L. (1964). A Kubelka-Munk calculation chart for light-scattering materials. *Tappi 47*(12):789–791.
87. Leekley, R. M., Denzer, C. W., and Tyler, R. F. (1970). Measurement of surface reflection from papers and prints. *Tappi 53*(4):615–621.
88. Leekley, R. M., and Tyler, R. F. (1975). The measurement of optical unevenness. *Tappi 58*(3):124–127.
89. Leekley, R. M., Tyler, R. F., and Hultman, J. D. (1978). Effect of paper on color quality of prints. *Tappi 61*(10):108–111.
90. Liebert, E. (1973). Testing of paper, paperboard, and board: Measurement of gloss. *Das Papier 27*(2):56–60 (in German).
91. Lodzinski, F. P. (1973). Experience with a transmittance-type on-line opacimeter for monitoring and controlling opacity. *Tappi 56*(2):78–82.
92. Loof, H. (1967). Definition, measurement and valuation of the white color of paper. *Das Papier 21*(6):297–309 (in German).
93. Lucy, A. T. (1971). Predicting the brightness of clay-coated boxboard using the opacity factor. *Tappi 54*(2):252–254.
94. Lynch, L. J., and Thomas, N. (1971). Optical diffraction profiles of single fibers. *Textile Res. J. 41*:568–572.
95. Maas, O. (1936). Opacity measurements. *Pulp Paper Mag. Can. 37*(11):689–695.
96. MacAdam, D. L. (1955). Color measurement and whiteness. *Tappi 38*(2):78–87.
96a. Mangin, P. J., and Lyne, M. E. (1979). Color analysis of newsprint in the dominant wavelength range 570–585 nm. *Tappi 62*(5):94–96.
96b. Mangin, P. J., and Lyne, M. B. (1979). Color analysis of newsprint: Hunter L a b and L* a* b* (CIELAB). *Tappi 62*(8):129-130.
97. Marchetti, F. R., and Willets, W. R. (1960). The optical efficiency of titanium pigments in paper coatings. *Tappi 43*(3):273–276.

98. Marcus, R. T. (1978). Long-term repeatability of color-measuring instrumentation: Storing numerical standards. *Color Res. Applic.* 3(1):29–33.
99. Marcus, R. T., and Billmeyer, F. W., Jr. (1974). Statistical study of color-measurement instrumentation. *Appl. Opt.* 13(6):1519–1530.
100. Marsh, R. S. (1965). Instrumental color specifications as they pertain to the paper manufacturer. *Tappi* 48(2):59A.
101. McConnell, D. J. (1973). Optical tests for paper and board. *Paper Technol.* 14(3-4):226–228.
102. McLean, J. D. (1964). Reproducibility of standardization of an Elrepho reflectance meter. *Pulp Paper Mag. Can.* 65(10):T434–T436.
103. Melamed, N. T. (1963). Optical properties of powders. 1. Optical absorption coefficients and the absolute value of the diffuse reflectance. 2. Properties of luminescent powders. *J. Appl. Phys.* 34(3):560–570.
104. Mie, G. (1908). Contributions to the optics of diffuse media. *Ann. Physik* 25:377–445 (in German).
105. Mosher, C. (1978). Optical properties of paper calculated from single-sheet reflectance data R_0, R, and R_g. *Tappi* 61 (10):123-124.
106. Munsell, A. H. (1905). *A Color Notation*, Ellis, Boston; 11th ed., Munsell Color Co., Baltimore, 1961.
107. Nara, J. (1957). Gloss meter by "distinctness-of-image." *J. Appl. Phys. (Japan)* 26(9):452–454 (in Japanese).
108. Nickerson, D. (1977). History of the OSA committee on uniform color scales. *Optics News* (Winter):8–17.
109. Nihira, K., Gunji, T., and Tsuboi, T. (1976). Study of goniophotometric curve by color measurement for papers. *J. Soc. Fiber Sci. Technol. (Japan) (Sen-i Gakkaishi)* 32(1):T23–T31 (in Japanese).
110. Nishiwaki, J. (1957). Gloss of machine glazed papers. *J. Phys. Soc. (Japan)* 12(1):53–57.
111. Nishiwaki, J. (1959). Effect of color on visual gloss, 1 and 2. *J. Appl. Phys. (Japan)* 28(5):267–276.
112. Nordman, L., Aaltonen, P., and Makkonen, T. (1966). Relationships between mechanical and optical properties of paper affected by web consolidation. In *Consolidation of the Paper Web* (F. Bolam, ed.), The British Paper and Board Industry Federation, London, pp. 909–927.
113. Norman, B., and Wahren, D. (1972). A comprehensive method for the description of mass distribution in sheets and flocculation and turbulence in suspensions. *Svensk Papperstidn.* 75(20):807–818.
114. Norman, B., and Wahren, D. (1974). The measurement of mass distribution in paper sheets using a beta radiographic method. *Svensk Papperstidn.* 77(11):397–406.
115. Norman, B., and Wahren, D. (1976). Mass distribution and sheet properties of paper. In *The Fundamental Properties of Paper Related to Its Uses* (F. Bolam, ed.), British Paper and Board Industry Federation, London, pp. 7–70.

115a. Olf, H. G. (1980). Printing opacity as a function of basis weight. *Tappi* 63(9):149-150.
116. Ostromecki, R. R. (1978). Optical properties of paper calculated from multisheet TAPPI brightness and opacity. *Tappi* 61(10):123, 125.
117. Peltz, G., Heiszler, L., and Hilke, E. (1973). Optical properties of optical-character-recognition paper for electronic data processing. *Papier* 27(6):217−224 (in German).
118. Popson, J. (1974). On-line measurement of paper color, brightness, and opacity. *Paper Trade J.* 158(31):24−27.
119. Poulter, S. R. C. (1968). Measurement of print unevenness. *Tappi* 51(8):87A−91A.
120. Presgrave, J. E. (1975). On-line control of fluorescent white papers. *Paper Technol.* 16(1):34−38, 42-43.
120a. Proposal for study of color spaces and color difference equations (1974). *J. Opt. Soc. Am.* 64(6):896-897.
121. Ramsay, H. L. (1966). Simplified calculation for predicting optical properties of coated board. *Tappi* 49(12):116A−118A.
122. Rennel, J. (1969). Opacity in relation to strength properties of pulps, 3. Light-scattering coefficient of sheets of model fibers. *Tappi* 52(10):1943−1947.
123. Rich, D. C. and Billmeyer, F. W., Jr. (1979). Practical aspects of current color-measurement. Instrumentation for coatings technology. *J. Coatings Technol.* 51(3):650−656.
124. Robinson, J. V. (1975). A summary of reflectance equations for applications of the Kubelka-Munk theory to optical properties of paper. *Tappi* 58(10):152-153.
125. Robinson, J. V., and Linke, E. G. (1963). Theory of the opacity of films of coating pigment and adhesive: A method for calculating the opacity of coatings. *Tappi* 46(6):384−390.
126. Ross, W. D. (1971). Theoretical computation of light scattering power: Comparison between TiO_2 and air bubbles. *J. Paint. Technol.* 43(563):50−66.
127. Ross, W. D. (1974). Theoretical light-scattering power of TiO_2 and microvoids. *Ind. Eng. Chem., Prod. Res. Devt.* 13(1):45−49.
128. Ruckdeschel, F. R. (1979). Light scattering and spectral properties of white cast-coated enamel paper. *Tappi* 62(1):61−64.
129. Scallan, A. M., and Borch, J. (1972). An interpretation of paper reflectance based upon morphology. 1. Initial considerations. *Tappi* 55(4):583−588.
130. Scallan, A. M., and Borch, J. (1976). Fundamental parameters affecting the opacity and brightness of uncoated paper. In *The Fundamental Properties of Paper Related to Its Uses* (F. Bolam, ed.), British Paper and Board Industry Federation, London, pp. 152−163.
131. Schlegel, M., and Sterl, W. (1975). Opacity measurement: With green or blue filter? *Zellstoff Papier* 24(10):303−306 (in German).
132. Schmidt, G. (1960). The paper stack theory. *Das Papier* 14(10):445−452 (in German).

133. Seiner, J. A. (1977). Microvoid coatings: Material and energy savers? *J. Oil Col. Chem. Assn.* 60:335–347.
134. Sinclair, R. S., and Wright, W. D. (1969). Color measurement in Europe. *Appl. Opt.* 8(4):751–756.
135. Springer, G. (1971). A light transmission type on-line opacity meter. *Tappi* 54(3):411-412.
136. Starr, R. E., and Young, R. H. (1975). An improvement in the determination of R_∞ and scattering coefficients for paper, pigments, and coatings. *Tappi* 58(5):75–78.
137. Starr, R. E., and Young, R. H. (1978). Paper coating formulations. A study of limitations involved in the determination and use of the Kubelka-Munk constants. *Tappi* 61(6):78–80.
138. Steele, F. A. (1935). The optical characteristics of paper. 1. The mathematical relationships between basis weight, reflectance, contrast ratio, and other optical properties. *Paper Trade J.* 100(12):37–42.
139. Stenius, Å. S. (1951). The application of the Kubelka-Munk theory to the diffuse reflection of light from paper. 1. A critical study. *Svensk Papperstidn.* 54(19):663–709.
140. Stenius, Å. S. (1965). SCAN-test brightness measuring system. *Tappi* 48(12):45A–52A.
141. Stenius, Å. S. (1972). Brightness, a deprecated term. *Svensk Papperstidn.* 75(12):473-474 (in Swedish).
142. Stenius, Å. S. (1972). Fluorescence in paper. *J. Color Appearance* 1(6):8–10.
143. Stenius, Å. S. (1979). Multi- or single-sheet light scattering and absorption coefficients. *Tappi* 62(1):89–91.
143a. Stenius, Å. S. (1979). Tristimulus values and chromaticity coordinates from spectral reflectance values. *Tappi* 62(11):123.
144. Stenius, Å. S., Kyrklund, B., and Loras, V. (1966). Color: Its graphical representation, its measurement and evaluation. *Svensk Papperstidn.* 69(5):150–158.
145. Stenius, Å. S., Rydberg, J., and Söderhjelm, L. (1975). ISO brightness: The new brightness value. *Svensk Papperstidn.* 78(11):403–408.
146. Stokes, G. G. (1862). On the intensity of the light reflected from or transmitted through a pile of plates. *Proc. Roy. Soc. (London)* 11:454–556.
147. Sundstrom, F. O., and Stearns, E. I. (1950). Practical art of color matching on paper. *Paper Mill News* (July 1):12–15.
148. Tanaka, S. (1956). Measurement of reflection characteristics of paper using polarized light. *J. Appl. Phys. (Japan)* 25(5):207–213 (in Japanese).
149. Tanaka, S. (1958). Measurement of reflection characteristics of paper using polarized light, 2. *J. Appl. Phys. (Japan)* 27(10):600–604.
149a. TAPPI RC 332 (1966). Specification and ranking of whites and near-whites. *Tappi* 49(9):167A-168A.
150. Trader, C. D. (1971). Laboratory studies relating coating structure and coating performance. *Tappi* 54(10):1709–1713.

151. Trosset, S. W., Jr. (1966). A method of storage for maintaining brightness and color of white standard samples. *Tappi* 49(4):61A–65A.
152. Troster, M. J., Williams, W. F., Dearth, L. R., and Wink, W. A. (1967). Evaluation of linerboard finish by means of low-angle gloss. *Tappi* 50(2):83A–86A.
153. Tsuchida, K., Handa, Y., Kadoya, T., and Murata, M. (1967). Photoelectric measurement of wire mark and sheet formation. 2. The quantitative representation of sheet formation. *Japan Tappi* 21(1):43–50 (in Japanese).
154. Vaeck, S. V. (1966). New method of whiteness determination: Correlation with independent visual gradings. *Ann. Sci. Textiles Belges* 1:95–117.
155. Van Brimer, R. H. and Howard, R. C. (1967). A noncontact contrast-ratio opacimeter for on-machine measurements. *Tappi* 50(2):65A–70A.
156. Van de Hulst, H. C. (1957). *Light Scattering by Small Particles,* John Wiley and Sons, Inc., New York.
157. Van den Akker, J. A. (1963). Theory of the optical properties of pulp. In *the Bleaching of Pulp* (W. H. Rapson, ed.), TAPPI monograph no. 27, pp. 17–39.
158. Van den Akker, J. A. (1965). Developments in spectrophotometry and papermaking. *Tappi* 48(2):57A.
159. Van den Akker, J. A. (1965). Standard brightness, color, and spectrophotometry with emphasis on recent information. *Tappi* 48(12):57A–62A.
160. Van den Akker, J. A. (1967). The meaning and measurement of opacity. *Tappi* 50(5):41A–43A.
161. Van den Akker, J. A. (1968). Theory of some of the discrepancies observed in application of the Kubelka-Munk equations to particulate systems. In *Modern Aspects of Reflectance Spectroscopy* (W. Wendlandt, ed.), Plenum, New York, pp. 27–46.
162. Van den Akker, J. A. (1977). Optical aspects of coating pigments. In *Physical Chemistry of Pigments in Paper Coatings* (C. L. Garey, ed.), TAPPI Press, Atlanta.
163. Van den Akker, J. A., and Sears, G. R. (1964). Gloss of coated paper and paperboard. *Tappi* 47(11):179A–182A.
164. Wahren, D., and Bryntse, G. (1976). An improved model of the reflectance properties of uneven solid prints. In *The Fundamental Properties of Paper Related to Its Uses* (F. Bolam, ed.), British Paper and Board Industry Federation, London, pp. 616–621.
165. Ward, J. W. (1969). Production experience with a continuous color monitor. *Tappi* 52(2):239–244.
166. Warren, C. A., and Peel, J. D. (1973). The scattering of infrared radiation as a method of measuring paper roughness, 1. *Paper Technol.* 14(4):91–97.
167. Wendlandt, W. W. (1968). Reflectometers, colorimeters, and reflectance attachments, 1 and 2. *J. Chem. Ed.* 45(11):A861–A876; 45(12):A947–A958.

168. Wickstrom, W. A., and Horner, M. (1970). Closed-loop color control for printing paper. *Tappi* 53(5):784–791.
169. Windle, W., and Gate, L. F. (1968). Brightness Measurement. *Tappi* 51(12):545–551.
170. Wink, W. A., Delevanti, C. H., Jr., and Van den Akker, J. A. (1953). Instrumentation studies, No. 77. Study on Gloss. 1. A. goniophotometric study of high-gloss papers. *Tappi* 36(12):163A–172A.
171. Wurzburg, F. L., Jr. (1963). Survey of instruments for color specification. *Tappi* 46(7):155A–159A.
172. Wyszecki, G. (1966). The measurement of brightness and color. *Metrologia* 2(3):111–125.
173. Wyszecki, G. (1978). Colorimetry. In *Handbook of Optics* (W. G. Driscoll and W. Vaughan, eds.), McGraw-Hill, New York, section 9.
173a. Young, R. H., and Drexel, R. J. (1980). Another approach to CIELAB coordinates. *Tappi* 63(7):121-122.
174. Yule, J. A. C., Howe, D. J., and Altman, J. H. (1967). The effect of the spread-function of paper on halftone reproduction. *Tappi* 50(7):337–344.

INTERACTIONS WITH LIQUIDS AND GASES

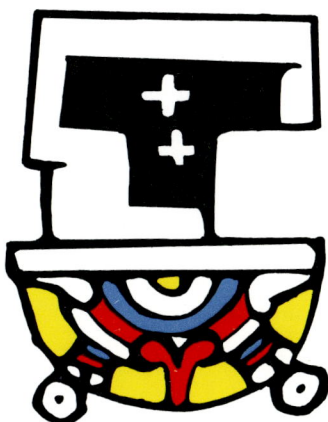

Emblem of Teotlilan

Many of the inks used by early Mesoamericans for writing and painting were derived specially for priestly usage. Teotlilan, the "Place Where Sacred Ink Is Made," was responsible for producing such materials. The sacredness of the location is symbolized by the sun, revered by the Aztecs, surmounted by a temple. Its name derives from Teyotl (sacred), tlilli (ink), and an (one of the variant forms of designating a place where something is found or is done).

17
POROSITY AND GAS PERMEABILITY

KOJI MURAKAMI
RIKIZO IMAMURA*

Faculty of Agriculture
Kyoto University
Kyoto, Japan

I.	Internal Porous Structure	57
	A. Description and Definition	57
	B. Measurement of Porosity	59
	C. Measurement of Specific Surface	61
	D. Measurement of Pore Size Distribution	64
II.	Gas Permeability	73
	A. Fundamental Description	73
	B. Measurement of Gas Permeability	82
	C. Measurement of Water Vapor Permeability	88
	Symbols	92
	References	94

I. INTERNAL POROUS STRUCTURE

A. Description and Definition

Pore space forms a considerable amount of the volume contained within a piece of paper or paperboard. The pores range in size from interfiber gaps to molecular interstices. Most of the pore spaces connect with each other to form complicated three-dimensional channels through the sheet.

---

*Currently Professor Emeritus of Kyoto University.

Since paper is anisotropic because of preferential fiber orientation and is deformable in its bulk dimensions under an external load or upon the sorption of moisture, the structure of the pore system is also anisotropic and deformable.

The pore spaces in paper can be classified according to their origin:

- The pore spaces that originally occurred in papermaking fibers or converting materials such as fillers, coating materials and so on
- ● The pore spaces made by the papermaking or converting process

Interfiber gaps, interstices around fiber-to-fiber bonds or fiber-to-filler contacts, and pores in coatings are examples of process-related pore spaces.

However, the pore spaces in porous materials can also be classified by considering whether they are interconnected with each other [37,89]. That is, *interconnected* pores are accessible from the outside through one or both ends. If accessible from only one end, the pore is referred to as a *dead-end* pore. *Noninterconnected* pores are not accessible at all. The interconnected part of a pore system is called the *effective pore space* of a porous medium.

It is possible to quantify some aspects of the porous structure of paper, employing space-related terms such as porosity, specific surface, pore size or pore size distribution, and others. Using these terms, many attempts have been made to elucidate the porous properties of paper and paperboard [11,28,75,101].

Porosity is a generic term that comprises several properties closely connected with the geometry of porous structure. Sometimes the term is used for a specified property such as pore volume or air permeability. Since it seems desirable to avoid this confusion, porosity in this chapter means pore volume ratio. Porosity of paper may be defined as the ratio of pore volume to total volume of a specimen and is expressed either as a fraction or in percent.

Specific internal surface area of paper may be defined as the internal surface area per unit mass of a specimen because basis weight is a more acceptable fundamental quantity for paper than bulk volume. It should be noted that the ratio of the internal surface area to the bulk volume is customarily used in other porous materials.

Pore size or *pore size distribution* of paper would be a very desirable quantity for characterizing porous structure if it could be ascertained with precision. Unfortunately, there is no acceptable experimental method for obtaining a geometrically defined quantity describing the size of an irregularly shaped pore. In spite of this uncertainty, the term *pore radius* or *pore diameter* as an expression of pore size is widely used in paper science with tacit understanding of the difference from the actual features of pores. Pore size or pore size distribution expressed in terms of pore radius can be defined only by both the experimental technique and the analytical method used to determine this quantity. Therefore, one must pay careful attention when comparing data resulting from two different test methods.

Another attempt to describe porous structure is an approach from statistical geometry developed by Corte and Kallmes [30] and Corte and Lloyd [31]. Although this theory—which assumes for paper a multiplanar model consisting of a pile of ultrathin two-dimensional sheets—intends to describe the effective pore size distribution and the maximum pore size in terms of fiber and sheet dimensions, it requires further development to interpret the porous properties of paper sufficiently.

Much information on the porous structure of paper can be obtained from light and electron micrographs. During the past decade, remarkable developments have been achieved in the theory and equipment for quantitative analysis of complicated images [34,103]. Recently, Climpson and Taylor [25] applied this technique to pore size distribution in clay particles and structures.

B. Measurement of Porosity

General Methods Porosity, as defined on p. 58, is the ratio of pore volume to total volume of paper, that is,

$$\text{Porosity} = \frac{\text{pore volume of paper}}{\text{total volume of paper}}$$

$$= 1 - \frac{\text{bulk density of paper}}{\text{solid density of paper}} \tag{1}$$

Therefore, the methods for determination involve either volume or density measurement. The values obtained, however, vary slightly depending upon the method and the conditions used because of the compressible nature of paper and also because of surface effects.

Caliper Method Bulk density of paper can be conventionally obtained by calculating from the basis weight and the thickness measured by a standard paper caliper as described in Chap. 26. If the solid density of a specimen is known or calculated from the density and the fractional ratio of each material making up the specimen, the porosity may be calculated.

Mercury Method Another method to measure the bulk density of paper involves using mercury as a nonwetting liquid for paper substances. Several methods, such as the mercury pycnometry [91] and the mercury buoyancy technique [59,69,109], are available for this purpose. The values of bulk density obtained from these methods may be slightly higher than that obtained from the caliper method because the caliper method yields a greater thickness (and therefore volume). Readers may consult the chapter on dimensional properties measurement (Chap. 26) for greater detail concerning the interpretation of thickness.

Helium Displacement Method The effective pore volume of paper can be measured directly by the helium displacement method or the mercury intrusion method as pore volume per unit mass. These methods will be discussed later in detail (pp. 60-61 and 66-70). If the bulk density of a specimen is known, this value can be used to convert to porosity on a volume basis, as implied by Eq. (1).

Displacement Methods The density of the solid fraction of a porous medium can be measured accurately by the displacement method. The basic principle of this method is direct measurement of the volume of gas or liquid that is contained in the pore spaces.

In the case of paper and paperboard, it is necessary to take care in the selection of the displacement fluid. The following requirements need to be met:

The size of the fluid molecule must be sufficiently small compared with that of the pore spaces.
The fluid must be inert to the sample, that is, chemical and physical effects must be absent.
The liquid surface tension must be small enough.

While the most desirable fluid is helium gas, it has been reported that values comparable to that in helium gas can be obtained with the use of nonpolar liquids, such as heptane, benzene, toluene, and so on [50].

Liquid Displacement Method   When the measurement is made, the pore spaces of the specimen must be filled completely with the displacement liquid. For this purpose, it is advisable to use a pycnometer with a device for degassing and filling under vacuum [110,112].

The solid density of a specimen ρ_S can be calculated by

$$\rho_S = \frac{m_S \rho_L}{m_L - (m_{LS} - m_S)} \qquad (2)$$

where

m_S = ovendry mass of specimen
m_L = mass of displacement liquid required to fill pycnometer
m_{LS} = mass of specimen + liquid required to fill pycnometer
ρ_L = density of liquid at temperature used

Gas Displacement Method   The method involves measuring the change either in pressure or in volume to maintain a given pressure at a constant temperature, when a known amount of gas is introduced into a sample chamber that contains a known mass of the specimen. The solid volume of the specimen can be calculated from the ideal gas law equation. For example, if an evacuated sample chamber is connected to a gas reservoir containing a known amount of gas, the solid volume of a specimen V_S can be calculated by

$$V_S = V_1 + V_2 - \frac{p_1 V_1}{p_{12}} \qquad (3)$$

where

V_1 = volume in gas reservoir
p_1 = pressure in gas reservoir
V_2 = volume of sample chamber
p_{12} = pressure when gas reservoir is connected to sample chamber

Helium is considered the most desirable displacement gas because of the absence of molecular packing effects during measurement and because of its small molecular size. A variety of apparatus using helium gas have been designed [33,90,92,112]. A relatively large amount of specimen material is required to obtain accurate results. However, an apparatus designed by Schumb and Rittner [90] allows accurate measurement on small specimens, for example, 0.5 cm³ of specimen having about 50% porosity with an accuracy of ± 0.1%. An apparatus for measurement of adsorption phenomena can also be used for this purpose. With this procedure, it is essential to completely remove any moisture and air that might be adsorbed on the specimen and to allow enough time for the helium molecules to penetrate throughout the micropores of the specimen.

C. Measurement of Specific Surface

Methods

Gas Adsorption Method If the area occupied by a molecule adsorbed on a solid surface σ_m is known, the specific surface area can be calculated from the following equation by measuring the amount of gas v_m required to cover the entire internal surface of a specimen as a monomolecular layer. That is,

$$S = \frac{v_m N_A \sigma_m}{m_S} \tag{4}$$

where

S = specific surface area
N_A = Avogadro number
m_S = mass of specimen

Methods to determine the amount of gas adsorbed can be classified into two groups. The first is a volumetric method that accurately measures the pressure and volume of the gas before and after adsorption at a constant temperature (in practice this is achieved either at a given constant volume or at a given constant pressure). The second is a method to directly measure the mass of gas adsorbed using a special balance such as an electric balance. A wide variety of suitable apparatus and modifications have been devised; readers are referred to the books of Young and Crowell [121] and Gregg and Sing [44] for further details. In practice, these methods require rather elaborate high vacuum apparatus in which very precise pressure-volume measurements must be made. As improvements, Nelsen and Eggertsen [68] and, later, Stone and Nickerson [95] developed dynamic nitrogen adsorption methods using conventional thermal conductivity cells. These methods will be described later in this chapter (pp. 62-64).

The amount of gas adsorbed as a monolayer can be calculated from the dependence of the amount of adsorbed gas upon the pressure at a constant temperature, that is, the adsorption isotherm. Although there are many theoretical controversies over the calculation methods employed for this purpose, the BET method [15] is generally used in practice in preference to others, such as the B-point method [14], the relative method by Harkins-Jura [47], and the Fu-Bartell method [10,40].

Today one widely accepted method to measure the internal surface area of paper is a combination of the determination of the adsorption isotherm at liquid nitrogen temperature with calculation by the BET equation. Haselton [48] studied several gases as adsorbates and considered the analytical methods for determining the surface area. He concluded that the aforementioned combination could be recommended for surface area studies on pulp and paper.

Solution Adsorption Method The internal surface area can also be measured, using the same principle as the gas adsorption method, by adsorption of a solute from its solution in a nonswelling solvent. Stamm [94] applied this method to evaluate the internal surface area of wood and filter paper using stearic acid solution in benzene.

Optical Method It has been well recognized that the specific scattering coefficient calculated by the Kubelka-Munk equation from reflectance measurements of pulp sheets is proportional to the BET surface area [49]. This method has been widely used in relative comparisons of the internal surface area of various pulp sheets. Readers may consult the chapter on optical and appearance properties (Chap. 16) for greater detail concerning the interpretation of reflectance measurements.

Other Methods Attempts have been made to determine the hydrodynamic surface area of pulp mats from fluid permeability measurements by application of the Kozeny-Carman equation or its modifications [9,12,87]. This method has some problems that need to be overcome in order to be applicable to paper and paperboard.

For wetted pulp fibers or pulp sheets, the silvering-catalytic method of Clark [13,24] is available for determining surface area.

Dynamic Nitrogen Adsorption Method Nelsen and Eggertsen [68] proposed a dynamic nitrogen adsorption method to determine the surface area of a powder. It was shown by Stone and Nickerson [95] and later Rennel [77] that this method is satisfactorily applicable for the internal surface area determination of paper.

Principle The method involves measuring the change in the composition of a helium-nitrogen gas mixture passed through a sample tube by means of a thermal conductivity cell such as that used in gas chromatography. Thus, when a sample tube is cooled in liquid nitrogen, a part of the nitrogen gas is adsorbed from a helium-nitrogen gas mixture onto the surface of the specimen and a peak is recorded on a recorder chart. Using a calibration curve, the area under the peak can be converted into the amount of nitrogen adsorbed. The validity of the adsorption measurement can be checked by the desorption measurement that is achieved by recording another peak when the sample tube is warmed to room temperature.

The amount of nitrogen required for a monomolecular layer can be estimated by the use of the BET method from nitrogen adsorption experiments at at least three different partial pressures of nitrogen (refer to calculation, p. 64).

Apparatus A schematic diagram of the apparatus developed by Stone and Nickerson [95] is shown in Fig. 1. It is advisable to provide three separate supply cylinders of premixed helium and nitrogen, each containing the required partial pressure of nitrogen [85]. For the BET plot, it is necessary to maintain the ratio of the partial pressure of nitrogen to the total pressure in the range of 0.05 to 0.35. It is convenient in practice to use mixtures of about 5, 15, and 25% nitrogen in helium [95].

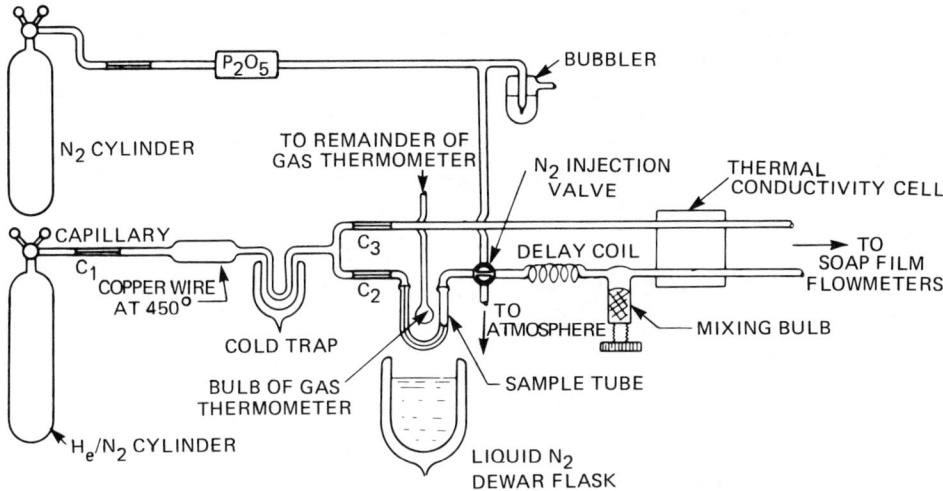

Fig. 1 Dynamic nitrogen adsorption apparatus for surface area measurements. (From Ref. 95, Technical Section, CPPA.)

The flow rate of the gas mixture through the apparatus is maintained at a constant rate, about 20 cm³/min, by means of capillary restrictions before and after the purification train, and by adjusting the pressure regulator on the gas cylinder. The actual flow rate does not matter particularly, but it must remain constant during measurement since the chart area is inversely proportional to the flow rate.

The sample tube is a simple glass U-tube having a sample chamber of 5 to 6 mm in inside diameter and 19 cm in length on the downstream side. Depending on the bulkiness of a sample, its 0.2 to 0.5 g can be packed in without difficulty.

The purpose of the delay coil is to prevent a sudden change in flow rate, such as may occur upon warming the sample tube and releasing the adsorbed nitrogen or upon injecting a known amount of nitrogen for the calibration. The mixing bulb is used to vary the peak height and width on the chart.

A conductivity cell is connected to the recorder through a Wheatstone bridge in order to detect any change in the electroconductivity of the gas stream that passes through the cell.

Procedure

1. Packing the specimen. Strips of the sample material 10 cm long and about 4 mm wide are advisable, in order to leave passage for an unrestricted flow of gas through the tube; a glass rod should be used to fill the space of the sample chamber above the specimen.
2. Drying the specimen. The sample tube is connected to the apparatus as shown in Fig. 1, and gas flows through the sample tube heated at 378 K (105°C) for 30 min.
3. Adsorption. Gas flows through the apparatus until the recorder shows a stationary state. The sample tube is then cooled very slowly in

liquid nitrogen, and the adsorption peak is recorded. Then a known amount of nitrogen is introduced to the apparatus from the nitrogen injection valve for calibration.
4. Desorption. The sample tube is warmed as rapidly as possible to room temperature, and the desorption peak is recorded. A calibration peak is produced by injecting nitrogen gas.
5. Replication. Procedures 2 to 4 are repeated at two other nitrogen partial pressures, and finally the specimen used is weighed exactly.

Note that (1) it is advisable to make a periodic analysis of the nitrogen content in helium-nitrogen mixtures; (2) in the most accurate work, the temperature of the liquid nitrogen used as a coolant should be measured in each measurement.

Calculation One of the most useful forms of the Brunauer, Emmett, and Teller equation [15] is

$$\frac{x}{v(1-x)} = \frac{(c-1)x}{cv_m} + \frac{1}{cv_m} \tag{5}$$

where

$$x = \frac{\text{partial pressure of } N_2 \text{ in gas mixture}}{\text{saturation pressure of } N_2 \text{ at liquid nitrogen temperature}}$$

v = amount of nitrogen adsorbed
c = a constant

A plot of $x/v(1-x)$ against x gives a straight line, where $(c-1)/cv_m$ is the slope of this line and $1/cv_m$ is the intercept at $x = 0$. From these values, the constant c can be eliminated and the amount of nitrogen required to form a monolayer, v_m, is calculated. The specific surface area of a specimen can be calculated from Eq. (4), in which the value of σ_m for a nitrogen molecule at 77.4 K (liquid nitrogen temperature) is equal to 16.2×10^{-20} m^2.

It has been reported that materials of low specific surface area, less than 1 m^2/g, can be measured with a precision of ± 1% at the 95% confidence level [95].

D. Measurement of Pore Size Distribution

General Methods Although the pore size of paper, as well as its distribution, is generally expressed in terms of pore radius or pore diameter, one should keep in mind that the pore radius specifies only an experimentally determined value and not a geometrically defined value, as discussed in Sec. I.A.

X-Ray Small Angle Scattering Method This method, based on the measurement of the X-ray scattering pattern corresponding to the electron density distribution in the solid part of a sample, is useful for detecting not only interconnected pores but also noninterconnected pores. It may be more suitable for detection of micropores less than about 0.1 μm in diameter than for accurate determination of pore size distribution. Herman et al. [51] applied this method in a study on micropores in cellulosic fibers.

Method Based on Gas or Vapor Sorption Below the critical temperature of a gas, gas molecules are adsorbed on a solid surface to form one or more layers, the number of layers increasing in relation to gas pressure. If micropores are present, capillary condensation can also occur simultaneously. In theory, these relations can be used to determine the pore size distribution. Very accurate results would be expected, but controversy exists concerning the methods of analysis employed on the experimental data in connection with the adsorption theory [16,37]. Since the measurable range is limited to effective pores of less than several nm decades in radius, this method does not detect the bulk pore spaces of paper and paperboard. Nitrogen gas [72,73,96,100] and benzene vapor [66,72] have been used as adsorbates for pulp fibers and papers.

Methods Based on Capillarity Acceptable methods to provide quantitative information on the bulk pore spaces of paper are as follows:

- Mercury intrusion method [79]. This method involves injection of mercury as a nonwetting fluid into an evacuated porous medium. The volume of injected mercury and the pressure applied are measured simultaneously.
- • Gas drive methods, such as Corte's dioxane method [28]. These methods involve injection of a nonwetting phase such as nitrogen gas into a porous medium saturated with a wetting liquid. The flow rate and the applied pressure are recorded.

According to basic capillary theory, the pressure difference across a curved interface between two fluid phases, Δp, is

$$\Delta p = \gamma \left(\frac{1}{r_1} + \frac{1}{r_2} \right) \tag{6}$$

where

γ = interfacial surface tension
r_1 and r_2 = principal radii of curvature of an interfacial meniscus

If the meniscus is confined in a circular capillary, Eq. (6) becomes, for an imperfect wetting fluid,

$$\Delta p = \frac{2(\gamma \cos \theta)}{r} \tag{7}$$

where

r = capillary radius
θ = contact angle between a fluid and a solid body

If the pore space of a sample is considered to be a group of circular capillaries having various sizes, the pore size distribution can be measured by varying the pressure difference applied over a sufficiently wide range.

There is another method, called the capillary suction method [45], based on this principle. White and Marceau [111] applied this method to determine the pore size distribution of wetted blotter sheets. This method is applicable to dry sheets if a nonswelling liquid is used as a saturant, although one cannot expect it to yield accurate results. Finally, the capillary rise method [101], which is based on an analogous principle, will be described in Chap. 19.

Other Methods Banacki and Bowers [3] advanced a unique method to determine pore size within filter papers. The method involves passing a nonswelling liquid, in which a large number of spherical beads having different sizes are suspended, through a specimen. The beads passed through the specimen are analyzed for size and number. The pore radius obtained by this method implies that of a circle inscribed in the narrowest part of the path through which the nonswelling liquid has passed.

Mercury Intrusion Method Since the mercury intrusion method was first proposed by Washburn [108] in 1921, the technique has been widely applied on incompressible, inorganic porous materials [36,79] and also on compressible, organic materials such as cigarette filters [27], fibers, and papers [66]. There is an extensive literature on this technique, and it has now become one of the most popular methods of determining the pore size distribution. The method is applicable to a wide range of pore sizes, including the whole spectrum of pore sizes found in paper.

Principle and Limitations The method involves filling an evacuated sample chamber with mercury and applying pressure to the meniscus. Mercury penetrates into pores having capillary pressures equivalent to the applied pressure. If each change in the meniscus level of the mercury is measured versus stepwise rising or descending pressures, a cumulative volume-capillary pressure curve with penetration and retraction branches can be obtained. Assuming a porous system consists of a number of cylindrical capillaries, the capillary pressure is inversely proportional to the capillary radius, as shown in Eq. (7). Using this relation, the penetration branch of the cumulative volume-capillary pressure curve can be converted to the cumulative pore size distribution curve.

Since mercury penetrates through the most accessible pore path, regardless of its direction and tortuosity, this method is effective only for the interconnected pores. In addition, information about anisotropy of a porous structure cannot be obtained, except for special cases such as the internal pore structure of synthetic fibers [74]. Furthermore, the method will not detect the dimension of those pores that are accessible only through necks narrower than the pore itself; those pores are assigned the diameter of the narrower neck.

In applying this method to paper and paperboard, one item of concern is whether or not compression of a specimen and collapse of the voids takes place during measurement under high pressure. Although there is no experimental solution for this problem, data suggesting the occurrence of significant compression have not been reported except at very high pressures—a region in which wood fibers are destroyed. The mechanism of compression in the mercury intrusion method would be quite different from that in compression testing of paper. For example, mercury penetrates into pore spaces having equivalent radii of above 3.8 μm at the applied pressure of about 196 kPa (2 kg $\cdot$ cm^{-2}) and those above 1.5 μm at about 490 kPa (5 kg $\cdot$ cm^{-2}). The intruded mercury will not act to collapse the specimen. Rather, it would tend to support the porous structure. Levlin and Nordman [63] recommended using an applied pressure of below about 2.5 MPa (25 kp · cm^{-2}, above 0.3 µm in equivalent radius), because the compressive effects of high pressure on the fiber network could then be neglected.

At pressures above approximately 7.3 MPa (74 kg · cm^{-2}), McKnight et al. [66] reported that some penetration of mercury into pores less than 100 nm radius may be apparent because of a significant collapse of voids in the fiber and compression of the fiber itself. In contrast to this, Stone et al. [97], who used a maximum pressure of about 83 MPa (840 kg · cm^{-2} or 12,000 psi), did not report any data suggesting presence of the compression effects. Chiodi and Silvy [23] obtained data suggesting that a large fraction of the voids in the coating layer cannot be compressed even at some 20 MPa (about 200 kg · cm^{-2}) of external force in the Z direction.

Finally, the contact angle that mercury makes with a solid surface plays an important role in determining the capillary pressure, as seen in Eq. (7), but the true value of the contact angle is a rather elusive quantity. In practice, assuming that the contact angle is constant and independent of the applied pressure, one particular value between 130° and 142° is generally used to convert into the equivalent pore radius from the capillary pressure. For example, Ritter and Drake [79] and others [66, 74] assumed the value of 140° for the contact angle of mercury as an average for most organic materials [1]. On the other hand, Rootare [81] recommended using the value of 130° because of the good agreement between porosimeter and BET surface areas for inorganic powders having low-energy surfaces [82,83]. This value is used also in paper materials [42]. Since the value of pore radius may be uncertain, it would seem advisable to express the result by simultaneous use of capillary pressure and equivalent pore radius.

Apparatus Various types of commercial instruments are now available for the measurement. A typical instrument consists of a chamber for the sample, with an accessory to measure the volume change in mercury, a mercury reservoir, a pressure generator, a vacuum pump, and a series of pressure gages. The measured volume change in the mercury caused by its penetration into pore spaces is often determined electrically because the volume is so small.

Procedure

1. Placement of sample. A known mass of dry specimen material is placed carefully into the sample chamber. It is essential to avoid the formation of pore spaces that might result from overlapping of the specimen surfaces [70].
2. Evacuation. The sample chamber is evacuated completely to remove gases and vapors adsorbed in the specimen.
3. Intrusion of mercury. The space within the sample chamber is filled with mercury, and then the applied pressure and the volume change in the mercury are simultaneously measured during stepwise pressuring or depressuring.

Note that in very high pressure regions, the value of the volume intruded into voids should be corrected for the compressibility of mercury itself.

Calculation for Differential Pore Size Distribution If dV is the pore volume in the range of pore radii between r and r + dr, we may write

$$dV = D(r)\, dr \qquad (8)$$

where $D(r)$ is a volume distribution function of pore radii. If the mercury contact angle and its surface tension are assumed to be constant, it follows from Eqs. (7) and (8) that

$$D(r) = \frac{p^2}{2(\gamma \cos \theta)} \frac{dV}{dp} \quad \text{or} \quad D(r) = \frac{p}{2(\gamma \cos \theta)} \frac{dV}{d \log p} \qquad (9)$$

where the right-hand terms of both equations above consist of known or measurable quantities. That is, if the cumulative volume of mercury intruded is plotted as a function of the applied pressure, the slope of this curve is equal to the value of dV/dp. Thus, a derivative function $D(r)$ is easily obtained by graphic differentiation, and a plot of the value of $D(r)$ as a function of r gives the differential pore size distribution curve. Recently, Garey et al. [42] recommended using a computer method involving the quadratic smoothing of the raw data and differentiation to avoid the erratic results that often arise from experimental data scatter. A computer program for this purpose has been developed at the Institute of Paper Chemistry, Appleton, Wisconsin. Another computer program for the pore volume and pore area distribution calculations from mercury porosimeter data was also reported by Rootare and Spencer [84].

Expression of Result The result obtained is usually expressed as a volume distribution curve of either pore radius or capillary pressure in a cumulative or differential form. If the specific volume displaced by the specimen at a given applied pressure can be calculated, it is possible to plot the specific displacement volume as a function of capillary pressure or pore radius. This expression, which is called *Stone-Scallan's plot*, may be very useful, for example, in investigations on microvoids such as occur in wood and pulp fiber walls [97].

Corte [28] has shown that the pore size distribution of paper can be fitted approximately by a logarithmic normal distribution and, further, that a comparison of various papers showed linear relationships between three parameters determining the distribution—the extrapolated smallest value, the mean value, and the variation. Assuming a logarithmic normal distribution for base sheets and coatings, Masaki et al. [64] were able to separate the distribution curve of a coated paper into curves for each individual layer, that is, base sheet and coatings.

Other Information Concerning Porous Structure As mentioned earlier, the penetration curve in mercury porosimetry will not provide any information concerning the dimensions of bottleneck or ink-bottle pores that are accessible only through necks narrower than the pore itself. A comparison of the pore volume distribution curve from mercury porosimetry with results from the photomicrographic method was made by Dullien and Dhawan [38] for a sandstone and by Climpson and Taylor [25] for coating clays. They found, as expected, that the curve from mercury porosimetry has a much lower mean value and is narrower in width than that from the photomicrographic method. This result may suggest that information on the size of

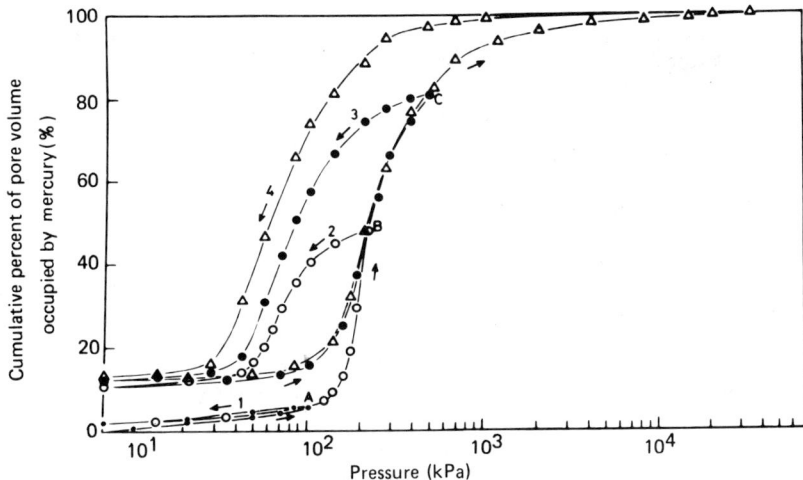

Fig. 2 Cumulative percent-pressure curves of four cycle experiments for a pulp sheet by mercury porosimetry. (From Ref. 118.)

bottleneck or ink-bottle pores is needed in order to better understand the porous structure of paper. The additional information may be obtained from the retraction branch of a capillary pressure curve.

An example of the penetration and retraction branches of a capillary pressure curve is shown in Fig. 2 [118]. The penetration branch rises steeply in the range of relatively low pressure where mercury is intruded into relatively large, interconnected pore channels of the paper sheet. The retraction branch shifts toward the lower pressure side compared with the penetration branch. Alternate penetration and retraction of mercury will produce a hysteresis loop. The pore system expressed by a closed hysteresis loop is the system of effective interconnected pores for mercury porosimetry and may be reasonably assumed to be the pore paths accessible to the flow of gases [119] and liquids [43]. The remaining pores, in which mercury is retained after retraction, may be assigned to those pores created by inhomogeneities of the porous structure, such as pit chambers and lumens in pulp fibers [70] or severely disordered regions in the fiber network.

Although possible mechanisms of the hysteresis phenomena are not well formulated at this time, many investigations [6,60,65,78,98] have been made on the subject. These results logically lead to the following conclusion [67]: the apparent capillary pressure on the retraction curve may be affected by three factors. These are (1) the size of ink-bottle pores, (2) the difference in the advancing and receding contact angles of mercury, and (3) the geometry of the micropores of a sample. If a series of samples have a similar surface nature, the effect of the second factor will become constant for each sample. Bell [6] suggested the possibility that the third effect will be eliminated by the choice of experimental conditions, but this is still under study. Using various wood fiber sheets and latex-impregnated sheets, Yamauchi et al. [118,120] obtained data suggesting that the retraction curve may provide information regarding the size of ink-bottle pores in paper sheets.

Gas Drive Method The gas drive method, which was first demonstrated by Corte [28] as the dioxane method, is one of a few methods to determine the pore size distribution of paper. This method has been applied for a variety of papers, such as wood fiber sheets [9,87], glassine and greaseproof papers [28], and coatings [42].

Principle The determination is conducted by covering a test specimen with a wetting liquid and subsequently applying gas pressure below the specimen to eject the liquid from the pore structure. Any path that passes through the sheet will be opened for gas flow when the applied pressure exceeds the capillary pressure developed by the most constricted part of the path. Assuming the paths in the specimen consist of a group of cylindrical capillaries having equivalent capillary pressures and the same length (and if the flow rate of gas through opened paths can be described by the Hagen-Poiseuille equation for a cylindrical capillary), the number pore size distribution can be obtained from the flow rate-capillary pressure data.

For analysis of the data, Corte [29] has proposed two models—the *saturation* and the *liquid head* models. Which type of flow rate-capillary pressure data can be obtained depends on the experimental conditions [29] and on the porous nature of the sample [42], as on pp. 72-73.

Apparatus The apparatus consists of a gas supply reservoir with a pressure regulator, a sample chamber, two manometers to measure the pressure drop across the specimen and the back pressure on the chamber, and a flowmeter to measure the flow rate of the permeating gas. The sample chamber for this purpose requires special design to avoid (1) deformation of the specimen during measurement and (2) entrapment of the wetting liquid. Various types of gas drive chambers have been reported. One, for example, has a chamber that supports the upper surface of the specimen with a 20-mesh stainless steel screen [8]; another holds the specimen between an upper grid plate and a lower plate consisting of a screen of 2 mm spacing [29].

The desirable properties of a wetting liquid are as follows:

- It should be nonswelling for paper substances.
- It should have a relatively low surface tension (because a small pore may be penetrated at relatively low pressure).
- It should have a low vapor pressure.

A number of liquids may be used in the gas drive test, such as dioxane [28], hexyl alcohol [9,87], and carbon tetrachloride [29] for papers, and a hydrocarbon, Magie 500 oil [42], for pigment coatings. Dry air [29] or dry nitrogen gas may be used as the driving gas.

Procedure

1. Introduction of specimen, gas and wetting liquid. A specimen of known thickness is clamped in the gas drive chamber and conditioned

by passing a gas stream through it slowly. Then a prescribed amount of wetting liquid is introduced into the upper part of the chamber and brought to equilibrium with the specimen.
2. Determination of flow parameters. The gas pressure beneath the specimen is increased slowly until the first bubbles of gas appear in the liquid. At this time the following quantities are recorded: the volume rate of flow, the pressure drop across the chamber, the back pressure on the chamber, and the liquid temperature. By stepwise increase of the applied pressure, the flow rate-pressure drop data are obtained.
3. Mass determination. Finally, the ovendry mass of the specimen is measured.

Theory and Calculation Corte [29] has proposed the two limiting models to describe the behavior of gas passed through each opened pore and has established the method to analyze flow rate-pressure drop data according to each model.

- Saturation model: once a capillary has become opened to gas flow, it conducts gas at the attained pressure without further effect on the liquid.
- • Liquid head model: the gas issues from the end of a capillary in the form of discrete bubbles; a back pressure accompanying the growing bubbles acts in the direction opposite the flow.

The flow rate of gas driven through each opened pore can be described by the Hagen-Poiseuille equation for a cylindrical capillary,

$$q_i = \frac{\pi r_i^4 p}{8\eta L} \quad \text{for saturation model} \tag{10}$$

$$q_i = \frac{\pi r_i^4}{8\eta L} \left(p - \frac{2\gamma \cos \theta}{r_i} \right) \quad \text{for liquid head model} \tag{11}$$

where

q_i = volume rate of gas passing through i-th capillary
p = pressure drop across capillary
η = viscosity of the gas
r_i = radius of the i-th capillary
L = length of i-th capillary

An assumption is made that the pore paths through which gas flows are equivalent to an equal length of cylindrical capillaries having the same capillary pressures. If $n(r)\,dr$ is the number of capillaries in the range of radii between r and $r + dr$, it is expressed, for the saturation model by

$$n(r)\,dr = -\frac{\eta L}{2\pi(\gamma \cos \theta)^4} p^3 \left(\frac{dQ}{dp} - \frac{Q}{p} \right) dp \tag{12}$$

and, for the liquid head model,

$$n(r) \, dr = - \frac{\eta L}{2\pi(\gamma \cos \theta)^4} p^4 \frac{d^2Q}{dp^2} \, dp \qquad (13)$$

where Q is the volume rate of flow and the negative sign in the right-hand terms of the above equations results from the inverse relation of pore radius to increasing pressure. The right-hand terms of both equations consist of known or measurable quantities. That is, if the volume rate of flow is plotted as a function of the applied pressure (Fig. 3), dQ/dp is the tangential slope of this curve and d^2Q/dp^2 the increment of tangential slope. Thus, the values of $n(r) \, dr$ are obtained by graphical differentiation, and a plot of the values of $n(r) \, dr$ as a function of r yields a number pore size distribution curve of the differential type. A normalized number distribution can be derived from $n(r) \, dr$ by dividing by $\Sigma \, n(r) \, dr$; if this is done, the factor determined by the liquid used, $\eta L/[2\pi(\gamma \cos \theta)^4]$ in Eqs. (12) and (13), is eliminated. Garey et al. [42] recommended using a cumulative number distribution—a plot of dQ/dp versus p or r—for the liquid head form of gas drive because the data are not sufficiently precise to permit calculation of consistent values for the second derivative.

The two models lead to flow rate-pressure drop curves of entirely different shape. Figure 3 shows experimental results for a coating base paper of 62 g · m^{-2} in basis weight reported by Corte [29]. Curve 1 was obtained

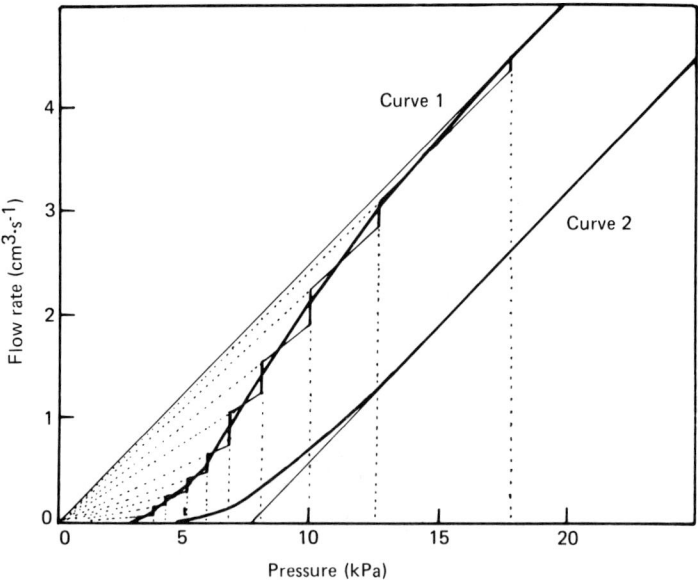

Fig. 3 Volume rate-pressure curves of coating base paper by gas drive experiments. [From Ref. 29, reprinted from the publication *Das Papier*, Eduard Roether Verlag, Berliner Allee 56, D-6100 Darmstadt, *19*(7):346–351 (1965).]

by a determination in which only the essential amount required to saturate the specimen by the wetting liquid was used; it shows that the saturation form of gas drive occurred. As shown in Eq. (10), when no new pores are being opened at a given pressure range, the flow rate increases in proportion to the applied pressure. Curve 2 was obtained by forming a liquid column over the specimen and shows that the liquid head form of gas drive occurred. As shown in Eq. (11), the back pressure of a bubble growing from the end of the i-th capillary is the same as the pressure required to empty this capillary, p_i. Consequently, the flow starts when the applied pressure p exceeds p_i. When no new pores are being opened at a given pressure range, the flow rate is proportional to $p - p_i$.

Garey et al. [42] applied this method to a variety of porous materials, and they found that the nature of the porous material rather than the experimental procedure determines the curve shape. That is, if r_0 is the radius of a pore path at the end of the downstream side and r_c is the radius of the narrowest channel in the pore path, a curve of saturation form should be obtained when $r_0 > r_c$ and one of liquid head form when $r_0 \cong r_c$. As an example, they reported that the flow curve obtained with a coated paper is of the liquid head form even when a very small amount of liquid is used.

It has been pointed out that the number pore size distribution estimated by this method contains several uncertainties introduced by oversimplified assumptions and models [42]. In spite of this, it is also true that this method can provide valuable information on the pore paths passing through paper.

II. GAS PERMEABILITY

A. Fundamental Description

The gas permeability of paper and other sheet materials is the ability to allow the flow of gas or vapor through the structure of a sheet under a pressure gradient. It is expressed by a permeability coefficient or by various equivalent quantities in arbitrary units. For air, the terms *air permeability* and *air resistance* are used for such equivalent quantities. These are important properties from the fundamental structural standpoint and also as practical descriptive measurements for such papers as filter papers, packaging papers, and so on.

The gas flow through a pore path of porous paper is treated as a laminar or viscous flow, while gas permeation through nonporous papers, such as barrier-coated papers and film-laminated papers, is treated as a diffusion-type permeation. The diffusion-type permeation is also observed in water vapor permeation of paper. But the behavior of water vapor through paper is quite complicated, as described later (p. 81), because of the great affinity for water and water vapor shown by the fiber polysaccharides.

Darcy's Law A quantitative description of flow through porous media begins with an experimental rule found by Darcy [32]. This rule, in which the rate of flow passing through a powder bed is proportional to the pressure gradient of the flow, has been extended to a variety of porous media. There are various equations to express this rule and its modifications, known collectively as Darcy's law [89].

We consider the flow to be of a viscous streamline or laminar type. If a fluid flows across a cross-sectional area A at a volume rate Q, Darcy's law is described by

$$\frac{Q}{A} = -K\frac{\partial p}{\partial x} = K\frac{\Delta p}{L} \tag{14}$$

if the flow is assumed to be parallel to the X axis and Δp is the total pressure drop across a sample of length L. The proportionality constant K defined by Eq. (14) depends on both the properties of the porous medium and the nature of the fluid; it is called the *volume permeability coefficient* [18] or permeability constant [89], and its dimensions are $L^3 tm^{-1}$.

Another permeability coefficient, one that depends only on the structure of the porous medium, can be defined by introducing a relation that the rate of flow is inversely proportional to the viscosity of the fluid. That is, for incompressible fluids (a liquid, for example),

$$\frac{Q}{A} = -K_v \frac{\partial p}{\eta \partial x} = K_v \frac{\Delta p}{\eta L} \tag{15}$$

where η is the viscosity of the fluid. The proportionality constant K_v is called the permeability coefficient for viscous flow or specific permeability [89], and its dimensions are L^2. Although mixed units, "darcy" or "cm · s^{-1} for water," are widely employed in petroleum technology and groundwater hydrology, consistent SI units, m^2, are conventionally used for a wide variety of porous sheet materials.

In the case of gases, the volume rate of flow changes continuously as the pressure varies. However, the mass rate of flow—the product of the volume rate and the density at a given point—must remain constant, and further, for isothermal flow the product of the volume rate and the pressure also remains constant. Equation (15) becomes

$$p_1 Q_1 = pQ = -K_v \frac{p}{\eta} \frac{\partial p}{\partial x} A \quad \text{or} \quad \frac{Q_1}{A} = \frac{K_v}{\eta} \frac{p_{av}}{p_1} \frac{\Delta p}{L} \tag{16}$$

where

Q_1 = apparent volume flow rate at either surface of a specimen at which the pressure is p_1
p_{av} = mean pressure of system

If Δp is very small compared with p_1, the ratio p_{av}/p_1 is near unity. In such cases, the compressibility of the gas can be ignored. For example, if the volume change in compression can be held to within less than 1%, in the case of air at STP (273.15 K = 0°C, 1.01325×10^5 Pa = 760 mm Hg), Eq. (15) can be applied instead of Eq. (16) for pressure drops in the range of less than 2 kPa (approximately 20 cm H$_2$O).

Relation Between Permeability and Porous Structure The behavior of incompressible viscous flow passing through a cylindrical capillary is described by the Hagen-Poiseuille equation. Thus, the volume rate q through a capillary is

$$q = \frac{\pi r^4 \Delta p}{8\eta L} \tag{17}$$

where

r = capillary radius
L = length of capillary

The velocity of the flow is

$$\frac{q}{\pi r^2} = \frac{r^2 \Delta p}{8\eta L} \tag{18}$$

If we assume that a porous medium consists of a group of cylindrical capillaries having uniform radii, where all of both ends of the capillaries are open to the direction of flow, Eq. (18) becomes

$$\frac{Q}{A} = \frac{\varepsilon}{\beta} \frac{r^2 \Delta p}{8\eta L} \tag{19}$$

where

ε = porosity of medium
β = tortuosity ratio = $\dfrac{\text{effective length of capillary}}{\text{thickness of the specimen}}$

Combining Eq. (19) with Eqs. (14) and (15) yields

$$K_v = K\eta = \frac{\varepsilon}{\beta} \frac{r^2}{8} \tag{20}$$

This equation shows that the coefficient K_v is proportional to both porosity and the second power of capillary radius and that it is inversely proportional to the tortuosity ratio. Unfortunately, there is no method available to determine the value of tortuosity ratio directly in the case of paper. Carson [21] reported a method to determine the mean pore radius of paper by measuring the permeability in a slip flow region.

The above relation can be extended to media having pores of noncircular cross-sectional shape by introducing the hydraulic radius concept. The hydraulic radius is defined as the ratio of the perimeter to the cross-sectional area of a pore path through which fluid passes. The hydraulic radius for a circular cross section is equal to r/2, in which the factor 2 is called the *shape factor*. The shape factor k_0 varies with the shape of the cross section. If it is assumed that a porous medium consists of a group of capillaries having complex cross-sectional shapes and that all of the capillaries lie parallel to the direction of flow, the mean hydraulic radius of this medium $m_{h \cdot av}$ is expressed by

$$m_{h \cdot av} = \frac{\text{Volume occupied by fluid} \div \text{total volume}}{\text{Surface area exposed to flow}} = \frac{\varepsilon}{S_0} \tag{21}$$

where S_0 is the internal surface area per unit volume.

The actual mean velocity of fluid passing through capillaries is equal to $(Q/A)(\beta/\varepsilon)$. Considering this relation, Eq. (18) may be rewritten

$$\frac{Q}{A} = \frac{\varepsilon^3 \Delta p}{k_0 \beta^2 S_0^2 \eta L} \tag{22}$$

in which the product $k_0 \beta^2$ is called the *Kozeny factor* and is expressed by k. For an unconsolidated bed of packed spheres, it was found experimentally that $k \simeq 5.0$ [18].

The internal surface area per unit volume S_0 may be expressed on a mass basis,

$$S_0 = S_h \frac{1 - \varepsilon}{v_h} \tag{23}$$

where

S_h = hydrodynamic surface area per unit mass
v_h = hydrodynamic specific volume for a solid body

Thus Eq. (22) may be written

$$\frac{Q}{A} = \frac{\varepsilon^3 v_h^2 \Delta p}{k S_h^2 (1 - \varepsilon)^2 \eta L} \tag{24}$$

Equation (24) may be a useful form of the Kozeny-Carman equation. When Eq. (24) is combined with Eqs. (14) and (15), one obtains

$$K_v = K\eta = \frac{\varepsilon^3 v_h^2}{k S_h^2 (1 - \varepsilon)^2} \tag{25}$$

It should be noted that the Kozeny-Carman equation has validity only under certain conditions:

- The range of pore shapes is such that k_0 is reasonably constant.
- The tortuosity ratio is not very susceptible to variations in pore geometry.
- The pore size is uniform.

Although it has been confirmed experimentally that the Kozeny-Carman equation is valid for unconsolidated media with a random pore structure, there remain a number of problems—particularly concerning the shape factor and the tortuosity factor—to be solved for application of this theory to complicated porous media such as pulp fiber mats and papers. Many attempts to apply this theory for such porous media have been made by Brown [12] (refer to p. 78), Bliesner [9], Labrecque [61], and Peterson [71]. On the other hand, Scheidegger [88] and later Higgins and De Yong [52] pointed out that such an application of this theory is dubious because of unreasonably large values of the tortuosity ratio.

Limitations of Darcy's Law Darcy's law is valid only for viscous streamline or laminar flow. Various effects that cause deviation from Darcy's law have been reported. Two effects that may occur at high flow rate and under reduced pressure require special caution when gas permeability experiments are performed. These are (a) turbulence, and (b) molecular effects.

Turbulence If the rate of flow through a pore path becomes large, an inertial force acts on the flow in addition to the viscous force and the manner of flow turns from laminar to turbulent. The Reynolds number* at which turbulence starts is called the critical Reynolds number. The value of the critical Reynolds number varies, dependent on the shape and the structure of pore paths. A value of around 2000 for a straight, circular tube and values ranging from 0.1 to 75 for porous rocks and ceramics [89] have been reported.

Molecular Effects If the pressure of a gas flowing in a capillary is reduced until its mean free path is an appreciable fraction of the capillary radius, the gas molecules slip at the capillary walls to cause deviation from Darcy's law. If the pressure is further decreased until the mean free path is much greater than the capillary radius, viscosity plays no part in the flow, and such flow is called the *free molecular flow* or *Knudsen flow*.

The mean free path of a gas† is inversely proportional to the molecular density. For example, the mean free path of air is about 60 nm under a pressure of 101 kPa (1 atm) at 273 K (0°C) and about 4.5 μm under 1.3 kPa (1 cm Hg) at 273 K. Since the pore spaces between fibers in a paper sheet may range in size from 0.1 μm to several μm [21], special attention should be paid in evaluating results of gas permeability experiments at below atmospheric pressure. Bublitz et al. [17] found that the values for permeability coefficient determined from air permeation were quite large in comparison with values obtained from oil permeation, and they suggested that the difference was due to the presence of molecular slippage in air permeation tests, even at high pressures.

Correction for Slip Flow Since a pressure gradient is also a concentration gradient, a mass permeability coefficient K_m, instead of a volume permeability coefficient K, can be defined for a gas in terms of a gas density or a concentration gradient, that is,

$$\frac{w}{A} = -K_m \frac{\partial \rho}{\partial x} \quad \text{or} \quad \frac{n}{A} = -K_m \frac{\partial c}{\partial x} \tag{26}$$

where

w = mass rate of gas flow across the cross-sectional area A
n = number of molecules permeated in unit time across A

---

*Reynolds number is a dimensionless quantity, and is defined by the equation, $R_e = r u_{av} \rho / \eta$, where r is the radius of the tube, u_{av} is the mean velocity of flow, and ρ is the density of the fluid.
†Mean free path of a gas molecule can be described by the equation, $\lambda = (2\sqrt{2} \pi N d^2)^{-1}$, where d is the diameter of the gas molecule, N is the gas density, and λ is the mean free path.

If the coefficient K_m is independent of c or ρ in a small enough interval, Eq. (26) becomes

$$K_m = \frac{nL}{\Delta cA} = \frac{Q_1 p_1 L}{\Delta p A} \qquad (27)$$

The dimensions of K_m are $L^2 t^{-1}$, that is, the same as for the diffusion coefficient. For purely viscous flow, the coefficient K_m is equal to Kp_{av}; that is, it does vary with p_{av} or c_{av}, even if K does not.

In the range in which the mean free path of a gas is large in comparison with the pore radius, a plot of K_m against p_{av} gives a linear relation,

$$K_m = \frac{Q_1 p_1 L}{pA} = ap_{av} + b \qquad (28)$$

where a and b are constants dependent on the porous media and the nature of the gas; they can be evaluated experimentally. The first term in the right side of Eq. (28) corresponds to the viscous term, which disappears in Knudsen flow; the second term corresponds to molecular slippage. The constant a is equal to K_v/η. The constant b should be inversely proportional to the external friction coefficient between gas molecules and pore walls and hence is proportional to $(T/M)^{1/2}$ in which M is the molecular weight of the gas. Many attempts to explain the significance of the constants a and b have been made theoretically [18,89]. Brown [12] derived a solution for the constants a and b from the Kozeny-Carman equation [Eq. (24)] and proposed a method to determine the internal surface area of pulp fiber mats.

Fick's Law Consider a system in which two mixed gas phases having different compositions are in contact with each other at a constant pressure and temperature. There will be a transfer of gas molecules from the high-concentration side to the low-concentration side of the interface plane. If gas molecules transfer a distance x along the X axis perpendicular to the plane, the number of gas molecules that pass through the plane is proportional to a partial concentration gradient,

$$\frac{n}{A} = -D \frac{\partial c}{\partial x} \qquad (29)$$

where

n = number of gas molecules permeated in unit time across an area A of the plane
c = concentration of gas

The proportionality constant D is called the *diffusion coefficient*, and its dimensions are $L^2 t^{-1}$. The negative sign of the right side of the equation indicates that the molecules transfer to the low-concentration side.

If n_1 is the number of gas molecules flowing through the plane at position x, and n_2 is the number flowing through the plane at position x + dx, the concentration gradient across the thickness dx is

$$\frac{\partial c}{\partial t} dx = \frac{n_1 - n_2}{A} = -D \frac{\partial c}{\partial x} + D \frac{\partial}{\partial x}\left(c + \frac{\partial c}{\partial x} dx\right) \qquad (30)$$

therefore,

$$\frac{\partial c}{\partial t} = D \frac{\partial^2 c}{\partial x^2} \qquad (30a)$$

if D is a variable dependent on c, it can be expressed by

$$\frac{\partial c}{\partial t} = \frac{\partial}{\partial x}\left[D(c) \frac{\partial c}{\partial x}\right] \qquad (30b)$$

Equations (29) and (30) are called, respectively, the first and the second laws of Fick.

Diffusion-Type Permeation

Diffusion-Type Permeation Through the Porous Structure of Paper Here we deal with a system in which there is the same total gas pressure on both sides of the sheet, but in which there are also partial pressure differences for particular components of the gas. Gas molecules are able to diffuse freely through the pore paths within the paper. If the diffusion coefficient is independent of the concentration, $\partial c/\partial t$ in Eq. (30a) becomes zero under a steady state. The number of gas molecules n passing in unit time through a specimen having area A and thickness L can be described from Eq. (29) by

$$\frac{n}{A} = -D \frac{\partial c}{\partial x} = D \frac{\Delta c_A}{L} \qquad (31)$$

If p_A is the partial pressure of a gas, the concentration c_A is equal to p_A/RT. Therefore,

$$\frac{n}{A} = D \frac{\Delta p_A/RT}{L} \qquad (32)$$

where R is the gas constant. The diffusion coefficient D includes both the effects of porous structure and the nature of the gas for diffusional resistances.

Diffusion-Type Permeation Through Nonporous Paper For simple explanation, we consider a polymer film that has no cracks or pinholes. When there is a difference in total or partial pressure between the two sides, gas molecules will transfer across the film. The permeation of a gas or vapor through a polymer film typically includes the following phenomena: (1) the gas or vapor dissolves in the film at one surface, (2) the dissolved molecules diffuse through the film under a concentration gradient, and (3) the diffused molecules evaporate from the surface on the low-concentration side.

In a manner similar to Eq. (31), if we assume that the diffusion coefficient is independent of concentration, we find for the steady state condition that

$$\frac{n}{A} = -D\frac{\partial c}{\partial x} = D\frac{\Delta c}{L} = D\frac{\Delta p/RT}{L} \qquad (33)$$

If Henry's law can be expected to hold at both surfaces of the film, the concentration of gas or vapor at the film surface will be proportional to its partial pressure. Therefore, since c is equal to n/Q, Eq. (33) becomes

$$\frac{Q}{A} = DS_{sol}\frac{\Delta p}{L} \qquad (34)$$

The proportionality constant S_{sol} is called the *solubility coefficient*, and its dimensions are Lt^2m^{-1}.

If Eqs. (34) and (14) are combined, the relation

$$K = DS_{sol} \qquad (35)$$

can be obtained. Coefficients D and S are both temperature-dependent and obey the Arrhenius relationship [80];

$$D = D_0 \exp(-E_d/RT) \qquad (36a)$$

$$S_{sol} = S_{sol0} \exp(-\Delta H/RT) \qquad (36b)$$

where

E_d = activation energy for the diffusion process
ΔH = heat of solution
D_0 and S_{sol0} = diffusion and solubility constants irrespective of temperature, respectively

Now, we consider a laminate consisting of n layers of film or other sheet materials. The number of gas molecules that pass through each layer is uniform under a steady state condition [7]. If Δp_i is the pressure difference across the i-th layer having a permeability coefficient K_i and a thickness L_i, this relation can be expressed on a volume basis by

$$\frac{Q}{A} = K_1\frac{\Delta p_1}{L_1} = K_2\frac{\Delta p_2}{L_2} = \cdots = K_n\frac{\Delta p_n}{L_n} \qquad (37)$$

The permeability coefficient for the laminate overall is defined by

$$\frac{Q}{A} = K\frac{\Delta p}{L} \qquad (38)$$

where

$$\Delta p = \sum_{i=1}^{i=n} p_i$$

$$L = \sum_{i=1}^{i=n} L_i$$

From Eqs. (37) and (38), one obtains

$$\frac{1}{K} = \sum_{i=1}^{i=n} \frac{L_i}{L} \frac{1}{K_i} \qquad (39)$$

Water Vapor Permeation Through Paper When a partial pressure gradient of water vapor is maintained across a paper sheet, water molecules pass through the paper. Vollmer [106] determined the permeability of dense paper for air and water vapor and found that the permeability for water vapor is much larger than that expected on the basis of air permeability measurements. The amount of water vapor transferred to the zero relative humidity side of a sheet increases exponentially with increasing relative humidity at the other side [86,93].

Part of the water vapor transport through paper is the result of gaseous diffusion through relatively large pore paths and the Knudsen flow through micropore paths. The remaining water vapor transport is of the diffusional transport type, caused by the interaction between water vapor and cellulosic fibers.

If gas molecules are adsorbed and retained on a solid surface to produce a concentration gradient on the surface, the molecules can diffuse along this concentration gradient. This type of diffusion is called surface diffusion, but it is probably absent in a water vapor-cellulosic fiber system.

Water molecules adsorbed onto the surface of a cellulosic fiber are often subsequently adsorbed or dissolved into the matrix of polysaccharides in the fiber. They thus penetrate the fiber wall progressively. In a manner similar to that described for the case of gas permeation through a polymer film, the water vapor dissolves in the polysaccharide matrix at or near the cellulosic fiber surface on one side of the sheet, diffuses through the matrix under a concentration gradient, and evaporates from the other surface at the lower concentration. This process is called *solid-solution diffusion*.

If the relative humidity becomes very high, some of the microvoids in a sheet of paper will be filled by capillary condensation water. It is suggested that the transport of such condensed water results from a gradient of the capillary pressure [2].

It is true that permeation of water vapor through paper or paperboard occurs by the additive effects of the above-mentioned mechanisms, but there is an additional effect that results from the swellability of the fibers by water. Although a number of theoretical investigations have been made [2,86,93], a complete explanation of water vapor permeation through paper is not well formulated at this time because of the complexities involved in the structure of paper and in the interaction between water and the structural components of paper.

B. Measurement of Gas Permeability

Method for Gas Permeability Measurement The permeability of a sample of paper or paperboard to the flow of an inert gas can be measured by an apparatus that provides the pressure levels needed to achieve flow rates that would be expected in actual use. For relatively porous papers, a flow-pressure drop apparatus is generally used. On the other hand, it is advisable to use a system based on low or negative pressures for nonporous, low permeability papers such as barrier-coated papers and film-laminated papers.

Measurement of Gas Permeation Through Porous Papers

Principle and Calculation The principle of the measurement is quite simple. It can be achieved by simultaneous measurements of the pressure drop and the flow rate when a gas flows through a specimen at a constant temperature. The permeability, measured as a permeability coefficient or other arbitrary unit, can be calculated in one of the three following ways: (1) from Eqs. (14) and (15) if the compressibility of the gas can be ignored, that is, $p_1/p_{av} \cong 1$, (2) from Eq. (16) if compressibility cannot be ignored, and (3) from Eq. (28) if slip flow occurs.

Apparatus The apparatus consists of three parts—the gas supply system, the permeability cell, and the flow rate measurement system. In the construction of the apparatus, care should be taken to avoid any harmful pressure drop (across a narrow path such as in a stopcock or connection) or leakage of gas.

Gas Supply System. The gas supply system consists of a high-pressure gas source and pressure regulator. Also, if suction is required, a vacuum pump will be connected to the return line of the flow rate measurement system. In the case of air, it is convenient to use the pressure of the atmosphere as a constant-pressure gas source [19]. If needed, columns for drying and purification of the gas can be inserted either before or after the pressure regulator.

Permeability Cell. It is advisable to use a cell with single [17,19] or double [8,87] guard rings to avoid specimen edge effects, especially gas leakage. The cell should be equipped with manometers to measure the pressure drop across the specimen and the absolute pressure in the cell. It should also have a thermometer to record the temperature of the permeating gas.

Flow Rate Measurement System. A capillary flowmeter [17,19] or a rotameter is used to measure the rate of flow. The capillary flowmeter is based on the principle that the rate of flow is proportional to the pressure drop across a standard (calibrated) capillary. A rotameter consists of a graduated glass tube containing an unattached float that balances between the buoyancy caused by the flow and the opposing gravitational force. Which type of flow rate measurement system should be used depends on the flow rate region to be measured. The use of a capillary flowmeter is often preferable for the measurement of relatively low flow rates.

Procedure

1. The specimen is brought to equilibrium at the desired experimental temperature and humidity conditions.
2. The specimen is clamped onto the permeability cell, and then a gas

stream is introduced into the apparatus under the prescribed pressure condition.
3. The following quantities are measured: the flow rate and the pressure drop across the specimen, the absolute pressure in the cell, and the temperature of the flowing gas.

Measurement of Diffusion-Type Permeation

Principle If the volume on the low-pressure side of the specimen in the apparatus is maintained constant during the measurement, the amount of gas that passes through the specimen can be measured as a pressure change. There are two types of methods:

- The low-vacuum methods such as the ASTM D 1434 method, in which the volume in the low-pressure side in the apparatus is maintained as small as possible
- • The high-vacuum methods, in which very small pressure changes in a high vacuum are measured

The latter may be preferable for high accuracy. In what follows, one of the high-vacuum methods, such as used by Rogers et al. [80], will be described.

Apparatus A permeability apparatus consists of a gas reservoir, gas purification traps, a permeability cell, a solvent trap, and a McLeod gage; the apparatus is connected to a high-vacuum system and a gas supply. Since several days are required for a measurement, airtightness of the apparatus is particularly important. The permeability cell consists of two halves; these are assembled with a guard ring, which allows a mercury seal to form between the outer wall of the upper half and the inner wall of the lower half. The pressure at the high-pressure side of the cell is measured by a manometer. If a condensable vapor such as water vapor is used, it is advisable to use a special McLeod gage in which the compression capillary is jacketed with a glass tube for warming [35].

Procedure
1. The specimen is clamped in the cell, and then evacuation of the apparatus is repeated several times.
2. A prescribed pressure is applied on the high-pressure side of the cell.
3. At this moment, the measurement is started. The following quantities should be recorded as a function of time: the pressure at the high-pressure side of the cell by the manometer, the pressure at the low-pressure side by the McLeod gage, and the cell temperature. The total volume on the low-pressure side of the cell should also have been measured previously.

Calculation From Eqs. (34) and (35), the volume permeability coefficient K is expressed by

$$K = \frac{QL}{(p_1 - p_2)A} \quad (40)$$

where

p_1 = pressure at high-pressure side
p_2 = pressure at low-pressure side

In Eq. (40), Q stands for the volume of gas at STP (273.15 K = 0 °C, 1.01325 × 10^5 Pa = 760 mm Hg) transmitted in unit time in a steady state. From a plot of p_2 as a function of the time, $\Delta p_2 / \Delta t$ in the steady state can be obtained. If V_2 is the total volume on the low-pressure side, K can be calculated by

$$K \doteq \frac{\Delta p_2}{\Delta t} \frac{273 V_2 L}{1.013 \times 10^5 T p_1 A} \qquad (41)$$

where p_2 is very small in comparison with p_1. The units of K are, of course, $m^4 \cdot N^{-1} \cdot s^{-1} = m^3 \cdot s \cdot kg^{-1}$, but other units are often used for barrier materials, namely, cubic centimeters (STP) per second, per square centimeter of cross section, per millimeter of thickness, per centimeter of mercury pressure drop across the specimen [1 cm^3(STP) · mm · s^{-1} · cm Hg^{-1} · cm^{-2} $\doteq$ 7.5 × 10^{-9} · m^4 · N^{-1} · s^{-1} = 7.5 × 10^3 · mm^4 · N^{-1} · s^{-1}].

If the diffusion coefficient is independent of the concentration, the diffusion coefficient can be calculated by the following equation [4],

$$D = \frac{L^2}{6\tau} \qquad (42)$$

where τ is the time lag. The time lag is equal to the intercept on the time axis when the linear part of the above mentioned pressure-time plot is extrapolated to $p_2 = 0$.

Methods for Routine Control In a region of laminar flow, the behavior of air passing through the pore paths of paper may be described by Darcy's law. As expected from Eq. (14), the volume of air that permeates is proportional to the pressure difference across the sheet, the effective area of the sheet, and the time; the volume is inversely proportional to the sheet thickness. Among these variables, the thickness is a specific property of the paper, and the pressure difference and the effective area are quantities determined by the instrument used. Therefore, if one of the remaining variables is measured as the volume of permeating air in a given period of time (rate of flow) or the permeation time for a given volume of air, a relative value that specifies the air permeability of a sample can be obtained.

A number of instruments have been designed for the quality control of paper and paperboard [20], and some of them are listed in Table 1. The majority of them are also devised to be able to measure the surface smoothness of paper (by changing their testing heads). The pressure conditions are specified for each instrument. In the majority of these instruments, atmospheric pressure is used as the constant pressure source on one of the sides of the specimen, and then positive or negative pressure is applied on the other side. The results from these instruments are expressed by the respective arbitrary units specified for each instrument or method. These values are often interchangeable theoretically [107] and experimentally [22], and some examples are listed in Table 2.

Method of Permeation Time Measurement This method involves measuring the time in seconds required for a constant volume of air to pass across a specimen of constant area under a specified pressure gradient.

Table 1 List of Air Permeability Testers for Routine Control

Instrument	Pressure difference (kPa)	Units	Standard method and reference
Method by permeation time			
Gurley densometer	1.22 (124 mmH$_2$O)	s/100 cm^3	TAPPI T-460, CPPA-TS D14, SCAN-P19, JIS P8117, etc.
Gurley-Hill S-P-S tester	1.22 (124 mmH$_2$O)	s/100 cm^3	[57]
Bekk tester	64.3 to 37.6 (482 to 282 mmHg)	s/100 cm^3	[5]
Williams tester	—	s/25 cm^3	[56,113,114]
Volumetric method			
Schopper densometer	0.981 (100 mmH$_2$O)	cm^3/60 s	DIN DVM 3413, [107]
Emiel Greiner porosimeter	5.35 to (546 mmH$_2$O to)	cm^3/15 s	JIS C2111
Method by flowmeter			
Bendtsen porosity tester	1.47 (150 mmH$_2$O)	cm^3/60 s	TAPPI RC-303
Sheffield porosimeter	10.3 or 68.9 (1.5 or 10 psi)	Sheffield units (by calibrated rotameter)	TAPPI Useful Method 524
Emanueli porosity tester	Compressed air (depends on the sample)	Emanueli units (pressure difference in capillary flowmeter)	[39]
Oken type denso-asperometer (back pressure type air micrometer)	Less than 4.90 (500 mmH$_2$O) (depends on the sample)	mmH$_2$O (pressure difference in capillary flowmeter)	Japan TAPPI No. 5, [117]
Smooster (vacuum air micrometer)	Suction by vacuum pump (depends on the sample)	mmHg (pressure difference in capillary flowmeter)	Japan TAPPI No. 5, [53,54]

Table 2 List of Conversion Equations Between Air Permeability Testers

Instrument	Conversion equation	Symbol	Reference
Gurley densometer vs. Schopper densometer	$a_G = \dfrac{24096.38}{b_S}$	a_G = Gurley reading in s/100 cm^3 b_S = Schopper reading in cm^3/300 s	[107]
Gurley densometer vs. Emiel Greiner porosimeter	$b_e = \dfrac{83046.6}{0.803\, a_G + 513.34}$ and $b_{EG} = \dfrac{(979 + 0.675\, b_e)}{1033} b_e$	b_e = Uncorrected Emier Greiner reading in cm^3/300 s b_{EG} = Corrected Emiel Greiner value in cm^3/300 s	[107]
Gurley-Hill S-P-S tester vs. Sheffield porosimeter	$\log a_{GH} = 4.3393 - 1.16 \log b_{Sh}$ $\log a_{GH} = 3.70975 - 1.134 \log b_{Sh}$	a_{GH} = Gurley-Hill reading in s/100 cm^3 b_{Sh} = Sheffield reading in the gage	[22] TAPPI useful method 524
Gurley densometer vs. Emanueli porosity tester	$b_E = \dfrac{43.478\, a_G}{L}$ or $a_G = 523.023\, \dfrac{P_1}{P_2}$	b_E = Emanueli reading in the gage L = Thickness of the specimen in cm P_1 = Pressure drop across a standard capillary P_2 = Pressure drop across the specimen	[107]
Gurley densometer vs. Oken type denso-asperometer	$\log a_G = 0.965 \log b_0 + 0.0818$	b_0 = Oken reading in s/100 cm^3	[107]

One of the most widely used instruments is the Gurley densometer, which has been adopted for use in the standard methods in many countries. The result obtained by this tester is often called the air resistance or Gurley seconds. This tester can be applied in cases where the papers have permeation times ranging from 2 to 1800 s for 100 cm^3 of air. For more impervious papers, an instrument in which the rate of flow is measured directly is more desirable; the Gurley densometer measurements tend to be time consuming. It is another limitation of this tester that it cannot be used for papers having rough surfaces such as creped and corrugated papers.

The Gurley-Hill S-P-S tester [57] is quite similar to the Gurley densometer; the two instruments differ only in the manner of clamping the specimen. In the Gurley-Hill S-P-S tester, the specimen is held with a constant clamping pressure. Results from the two testers usually correlate almost perfectly, but for papers having poor smoothness one observes some differences caused by air leakage at the clamps.

The Williams tester [56,113,114] is also a constant pressure type tester. This tester was designed originally to evaluate the surface smoothness of paper by measuring the air leakage from the gap between adjacent surfaces formed when a specimen is folded in two. The value measured must be corrected for the volume of air that passes through the internal pore paths of the specimen. The measurement for the correction is accomplished by inserting a soft rubber plate with a circular hole between the two surfaces of the specimen. The value obtained by this procedure provides a measure of the permeability in the direction perpendicular to the paper thickness.

The Bekk tester [5] is a differential pressure gradient type tester. A relatively high pressure gradient is used in this tester.

Volumetric Method In this method, the volume of air passed in a constant interval of time in seconds across a constant area of the specimen under a specified pressure gradient is measured. The instruments for this purpose may be classified into two types, specifically,

- Constant pressure gradient type tester such as the Schopper air permeability meter,
- Differential pressure gradient type tester such as the Emiel Greiner porosity meter. This tester may be used to measure the relatively dense papers such as condenser papers.

Method by Flowmeter The flow rate of air passing across a constant specimen area under a specified pressure gradient is determined. Instruments designed for this method provide rapid measurements and can be used for a wide variety of papers having various air permeabilities.

The Bendtsen porosity tester and the Sheffield porosimeter [46] are quite similar. In both instruments, a group of rotameters (two to four), covering an appropriate range of flow rates, is inserted on the line from a constant-pressure source of compressed air to a sample holder. The rotameters are selected according to the air permeability of the sample.

Another type of instrument is made by applying the principle to a capillary flowmeter. In this type of tester, a standard capillary is inserted on the air line on one side of the cell, and then the pressure drop across the calibrated standard capillary is measured. Two types of instruments are available. One type uses compressed air as a pressure source; examples are the Emanueli porosity tester [39] and the Oken-type denso-asperometer [117]. The other type, of which the Smooster [53,54] is an example, uses a vacuum pump as a suction source. In both types of testers, the pressure difference across a specimen depends on the permeability of the specimen, and, therefore, will become larger in the case of denser paper.

In SCAN-P26 a method is given to evaluate the air permeability of dense paper by measuring the pressure drop and the rate of flow across a constant area of specimen under conditions of constant compressed air pressure. The pressure drop across the specimen is measured by a mercury manometer, and the rate of low is measured by one of three rotameters in series or by a capillary flowmeter. The air permeability in this method, P, is defined by

$$P = \frac{u_{av}}{A \Delta p} \tag{43}$$

where

u_{av} = mean flow rate
A = test area
Δp = pressure difference across specimen

The air permeability can be calculated by

$$P = 1500 \frac{a}{b} \tag{44}$$

where a is the flow rate in $cm^3 \cdot s^{-1}$, b is the difference between the mercury levels of the manometer in mm, and the units of P are $mm^3 \cdot N^{-1} \cdot s^{-1}$.*

For highly porous sheet materials (such as porous papers, fabrics, and pulp handsheets), another air permeability is defined in TAPPI T 251. The method involves measuring the rate of flow through the specimen under a specified pressure drop, that is, 124.4 Pa (12.7 mm H_2O, 0.50 in. H_2O). The air permeability is usually expressed as cubic centimeters per second per square centimeter, or cubic feet per minute per square foot, of area at 12.7 mm (0.50 in.) water pressure.

C. Measurement of Water Vapor Permeability

The mechanism of water vapor permeation through paper is much more complicated in comparison with that of air, as mentioned previously. The reason for the complication is the interaction between paper substances and water vapor. The water vapor permeability varies not only with the partial pressure difference across a specimen but also with the concentration of water vapor, (i.e., the relative humidity) and with the temperature. In addition to these factors, some papers and heavy paperboards require very long periods until a steady state has been established; for example, a barrier paper coated on both sides with a hydrophobic material usually requires several weeks or longer [55].

---

*In Eq. (43), u_{av} = a $cm^3 \cdot s^{-1}$ = a × 10^{-3} $m^3 \cdot s^{-1}$, A = 50 cm^2 = 50 × 10^{-4} m^2, and Δp = b/(7.50 × 10^{-3}) $N \cdot m^{-2}$.

Water vapor permeability is a very important property in terms of practical uses for such materials as packaging papers and building papers. Accordingly, more than one test condition will normally be specified; tests are not only performed under standard conditions but might also include, for example, the temperature extremes that might be present under actual use conditions. In TAPPI standards the following three conditions are designated for measurement of the water vapor transmission rate:

- At 293 K (23°C, 73°F) with an atmosphere of 50% RH on one side and a desiccant on the other (TAPPI T-448)
- At 310.8 K (37.8°C, 100°F) with an atmosphere of 90% RH on one side and a desiccant on the other (TAPPI T-464)
- At 255.2 K (− 17.8°C, 0°F) with an atmosphere in equilibrium with cracked ice on one side and a desiccant on the other (TAPPI-482)

One also finds that water vapor transmission rate tests have special significance in certain cases, such as creased papers and papers that have been subjected to twisting and compression in the Gelbo flex test [105].

For homogeneous sheet materials, water vapor permeability is defined by Eq. (40); results are customarily expressed in units of $g \cdot cm \cdot cm^{-2} \cdot s^{-1} \cdot cm\ Hg^{-1}$ or $g \cdot cm \cdot m^{-2} \cdot 24h^{-1} \cdot mm\ Hg^{-1}$. On the other hand, paper is a nonhomogeneous material and its water vapor permeability varies markedly, depending on the test conditions. Therefore, the water vapor permeability of paper and paperboard is often expressed as the mass rate of water vapor transfer through a unit area of a specimen under the test conditions. This value (under a steady state) is called the *water vapor transmission* or the water vapor transmission rate and is expressed in $g \cdot m^{-2} \cdot d^{-1}$.

Methods to measure the water vapor permeability may be classified into the following three types:

- Pressure gradient method
- Gravimetric method
- Hygrometric method

The first method has already been described on pp. 83-84. In this case water vapor is used instead of an inert gas.

Gravimetric Method

Principle This method involves the permeation of water vapor across a constant area of a specimen under a specified humidity gradient, as well as the measurement of the amount of water vapor permeated under a steady state as the weight change in either a moisture absorbent or the moisture supply. Although many devices and modifications for practical measurement have been proposed [62,122,123], a simple method called the cup test, as described in TAPPI T-448, SCAN-P22 and so on, will be covered here. This method requires several days for an experiment, but it enables simultaneous measurement for a number of samples. For papers of very low water vapor permeability, it may not be a very desirable method.

Apparatus It is necessary for this purpose to provide a light, shallow, nonpermeable test dish, such as a Vapometer cup, and a cabinet or room maintained at prescribed temperature and relative humidity. The humidity gradient is determined by a combination of water vapor pressure in the cabinet and water vapor pressure of a material that either supplies moisture or absorbs it, placed in the test dish. Hence, the humidity gradient can be varied to a wide extent by putting a desiccant, a saturated salts solution, or distilled water into the test dish.

Procedure

1. The specimen is placed on the top of the test dish containing the moisture-absorbing or -supplying material and is sealed by wax at the edge around the cup, leaving a test area with a diameter equal to that of the top of the cup.
2. The dish, assembled with contents and specimen, is placed in a cabinet maintained at prescribed conditions; after a steady state is achieved, the weight change of the assembled dish is recorded as a function of time.

Calculation The water vapor transmission rate X can be calculated by

$$X = 240 \frac{a}{b} \tag{45}$$

where

a = increase in mass per unit time, mg · h^{-1}
b = exposed area, cm^2

Note that the hygroscopicity of the test dish itself should be checked.

Methods by Measurement of Water Vapor Concentration The following methods are considered here:

• Closed cell method
•• Sweep gas method
••• Comparison method

A detailed comparison of these three types has been given by Van den Akker [104].

Each of the first two methods involves measurement of the change in the concentration of water vapor at one side of a cell and the calculation of the average water vapor transfer rate from the concentration change as a function of the permeation time. These methods are often preferred for papers having low water vapor permeability, such as water vapor barrier materials, and they provide rapid measurements. Although the theoretical relationships to convert the values from these dynamic methods into those from the standard cup method under a given steady state have not been established yet, calibration curves can be made by testing a series of specimens by both dynamic and gravimetric procedures. The third method itemized above is a steady-state process and requires the establishment of equilibrium. A value comparable to that from the standard cup method is obtained. In the case of composite materials, the measurement by each of the three methods must be made with the relatively impermeable surface of the specimen facing the low-humidity side of the cell [41,116].

Closed Cell Method The tester consists of two half-cells that are bolted together with a specimen between them. The down side of the cell is maintained at a prescribed high humidity by means of distilled water or a saturated salts solution. The specimen is conditioned by passing a stream of dry air through the upper side of the cell for a prescribed period; then the cell is closed. The humidity change in the upper side of the cell is measured by means of a hygrometer using an electrical resistance element [76,104] or an infrared instrument [55]. The average water vapor transfer rate X' can be calculated by

$$X' = 2.4 \frac{\Delta rh \rho_{sat} a}{b} \qquad (46)$$

where

Δrh = relative humidity change per unit time at low-humidity side of the cell, % · h^{-1}
ρ_{sat} = saturated water vapor density at temperature used
a = volume of low-humidity side, cm^3
b = exposed area, cm^2

Since the permeated water vapor accumulates in the low-humidity side of the cell in this method, one might not expect to obtain an accurate result. However, in the case of a sample having relatively low water vapor permeability, the moisture content of the atmosphere in the low-humidity side of the cell would be maintained at less than 0.5% RH during the measurement period [55].

Sweep Gas Method The specimen divides the test cell into two parts, the part on one side of the specimen being at low humidity and the other at high humidity. The method involves passing a dry gas stream at a slow rate through the low-humidity side of the cell. The relative humidity difference between the effluent gas and the influent gas is measured by means of an electrical hygrometer [104,115] or an electrolytic cell [26,58,102]. The mean water vapor transfer rate can be calculated by

$$X' = a\Delta rh \rho_{sat} b \qquad (47)$$

where

a = a constant determined by apparatus used
Δrh = relative humidity difference between effluent and influent gases
ρ_{sat} = saturated water vapor density at temperature used
b = total volume of gas passed through low-humidity side in unit time

Another measurement can be achieved by passing two gases having equal pressure simultaneously through both sides of the cell at the same flow rate. If two streams of nitrogen gas containing different concentration of water vapor are used, the water vapor concentration of the effluent gases can be measured by a calibrated thermal conductivity cell. An accurate measurement of water vapor permeation can be made by recording the following quantities: the flow rate and pressure of the influent gases, the water vapor concentration of the influent and the effluent gases, and the temperature used [2]. This method is also generally applicable for gases and vapors other than water vapor.

Note: In experimental work, it is particularly important that there be (1) prevention of adsorption or condensation of water vapor on the cell walls,

(2) application of exactly the same pressure on both sides of the cell, and (3) assurance that the gas composition in the cell is homogeneous.

Comparison Method As in the previous case, the test cell is divided by the specimen into two halves. The down side of the cell is maintained at a constant high humidity. The upper part of the cell is further divided into two chambers by a partition with a standardized humidity resistor (such as glass capillary tubing, flared and covered at each end with thin permeable sheeting [104]) that possesses a known resistance to water vapor transfer. The chamber on one side of the specimen is equipped with an electrical hygrometer; the chamber on the other side is the dry side, and it contains a desiccant. At equilibrium, the rate of water vapor transfer through the specimen is equal to that of water vapor passing through a standardized resistor. If the rate of water vapor transfer through the specimen is directly proportional to the relative humidity differential across the specimen, the water vapor transfer rate can be calculated by

$$X' = X'_s \frac{\Delta rh_2}{\Delta rh_1} \tag{48}$$

where

X'_s = water vapor transfer rate of standardized resistor
Δrh_1 = relative humidity difference across specimen
Δrh_2 = relative humidity difference across standardized resistor

Since certain materials such as paper and paperboard have a water vapor permeability that depends on relative humidity itself at one side of the material, the standardized resistor should be selected from a set of resistors covering an appropriate range to give a prescribed relative humidity in the middle chamber between the specimen and the resistor. A plastic film having a known humidity-water vapor permeability relation is also used as a resistor [99].

SYMBOLS

All quantities are in consistent SI units. Symbols will be specified by subscript, and they are defined in the immediate context of their use.

Symbol	Definition	Unit
A	Area	m^2
a,b,c	A constant or a quantity	—
c	Concentration	$mol \cdot m^{-3}$
D	Diffusion coefficient	$m^2 \cdot s^{-1}$
d	Diameter of molecule	m
E_d	Activation energy	$kg \cdot m^2 \cdot s^{-2} = J$

Symbol	Definition	Unit
ΔH	Heat of solution	J
K	Volume permeability coefficient	$m^3 \cdot s \cdot kg^{-1}$
K_m	Mass permeability coefficient	$m^2 \cdot s^{-1}$
K_v	Permeability coefficient for viscous flow	m^2
k	Kozeny factor	1
k_0	Shape factor	—
L	Length	m
M	Molecular weight (relative molecular mass of a substance)	1
m	Mass	kg
m_h	Mean hydraulic radius	m
N	Number molecules in unit volume	m^{-3}
N_A	Avogadro number, 6.022169×10^{23}	mol^{-1}
n	Mol flow rate	$mol \cdot s^{-1}$
P	Air permeability defined in SCAN-P26	$m^3 \cdot N^{-1} \cdot s^{-1} = m^2 \cdot s \cdot kg^{-1}$
p	Pressure	$kg \cdot m^{-1} \cdot s^{-2} = N \cdot m^{-2} = Pa$
Q	Volume flow rate	$m^3 \cdot s^{-1}$
R	Gas constant, 8.3134	$J \cdot K^{-1} \cdot mol^{-1}$
R_e	Reynolds number	1
r	Radius	m
rh	Relative humidity	1
S	Specific surface area	$m^2 \cdot kg^{-1}$
S_h	Specific hydrodynamic surface area	$m^2 \cdot kg^{-1}$
S_0	Surface area per unit volume	m^{-1}
S_{sol}	Solubility coefficient	$m \cdot s^2 \cdot kg^{-1}$
T	Temperature	K
t	Time	s
u_{av}	Mean velocity of flow	$m \cdot s^{-1}$
V	Volume	m^3
v	Amount of gas adsorbed	mol
v_h	Specific hydrodynamic volume of solid	$m^3 \cdot kg^{-1}$

Symbol	Definition	Unit
w	Mass rate of flow	$kg \cdot s^{-1}$
X	Water vapor transmission rate	$g \cdot m^{-2} \cdot d^{-1}$
x	Distance	m
x	A ratio	1
β	Tortuosity ratio	1
γ	Surface tension	$kg \cdot s^{-2} = N \cdot m^{-1}$
ε	Porosity	1
λ	Mean free path	m
η	Viscosity of gas	$kg \cdot m^{-1} \cdot s^{-1}$
ρ	Mass density	$kg \cdot m^{-3}$
σ	Area occupied by a molecule	m^2
τ	Time lag	s
θ	Angle of contact	°

REFERENCES

1. Adam, N. K. (1941). *Physics and Chemistry of Surfaces*. Oxford Univ. Press, p. 185.
2. Ahlen, A. T. (1970). Diffusion of sorbed water vapor through paper and cellulose film. *Tappi* 53(7):1320–1326.
3. Banacki, W. J., Jr., and Bowers, R. L. (1962). Pore size testing of filter papers. *Tappi* 45(10):805–807.
4. Barrer, R. M. (1939). Permeation, diffusion and solution of gases in organic polymers. *Trans. Faraday Soc.* 35:628–643.
5. Bekk, J. (1932). Apparatus for measuring smoothness of paper surfaces. *Paper Trade J.* 94(26):41-42.
6. Bell, W. K. (1972). Mercury penetration and retraction hysteresis in closely packed spheres: A non-independent domain system. Ph.D. thesis, Univ. Technology, Delft.
7. Bhargava, R., Rogers, C. E., Stannett, V., and Szwarc, M. (1957). Studies in the gas and vapor permeability of plastic films and coated papers, 4. Effect of a paper substrate. *Tappi* 40(7):564–567.
8. Bliesner, W. C. (1963). A study of the porous structure of fibrous sheets using permeability techniques. Ph.D. thesis, Institute of Paper Chemistry, Appleton, Wis.
9. Bliesner, W. C. (1964). A study of the porous structure of fibrous sheets using permeability techniques. *Tappi* 47(7):392–400.
10. Boyd, G. E., and Livingston, H. K. (1942). Adsorption and the energy changes at crystalline solid surfaces. *J. Amer. Chem. Soc.* 64(10):2383–2388.

11. Brecht, W. (1962). Effect of structure on major aspects of paper behaviour with fluids. In *The Formation and Structure of Paper,* vol. 1 (F. Bolam, ed.), British Paper and Board Makers Association, pp. 427–460.
12. Brown, J. C. (1950). Determination of the exposed specific surface of pulp fibers from air permeability measurements. *Tappi* 33(3):130–137.
13. Browning, B. L. (1950). The specific surface of pulps by the silvering method. *Tappi* 33(8):410–412.
14. Brunauer, S., and Emmett, P. H. (1935). The use of van der Waals adsorption isotherms in determining the surface area of iron synthetic ammonia catalysts. *J. Amer. Chem. Soc.* 57(9):1754-1755.
15. Brunauer, S., Emmett, P. H., and Teller, E. (1938). Adsorption of gases in multimolecular layers. *J. Amer. Chem. Soc.* 60(2):309–319.
16. Brunauer, S., Mikhall, R. S., and Bodor, E. E. (1967). Pore structure analysis without a pore shape model. *J. Coll. Interface Sci.* 24:451–463.
17. Bublitz, W. J., Dappen, J. W., and Van den Akker, J. A. (1948). An investigation of the relationship between the permeabilities of paper to oil and air. *Tech. Assn. Papers* 31:305–315.
18. Carman, P. C. (1956). *Flow of Gases Through Porous Media.* Butterworths, London.
19. Carson, F. T. (1934). Effect of experimental conditions on the measurement of air permeability of paper. *Paper Trade J.* 99(11): 107–116; *J. Res. Nat. Bur. Stand.* 12:587–608.
20. Carson, F. T. (1934). A sensitive instrument for measuring the air permeability of paper and other sheet materials. *Paper Trade J.* 99(16):44–52 (TAPPI Sect. 196–204).
21. Carson, F. T. (1940). Some observations on determining the size of pores in paper. *J. Res. Nat. Bur. Stand.* 24:435–442.
22. Chesley, K. G., Jones, E. D., and Truax, R. L. (1959). Correlation of the Sheffield porosimeter and the Gurley densometer. *Tappi* 42(4):299-300.
23. Chiodi, R., and Silvy, J. (1971). Porosity of the paper, variation of the porosity and of the mean dimension of the pores in function of compressibility of the sheet. *Revue A.T.I.P.* 25(1):81–96 (in French).
24. Clark, J. D'A. (1942). A new method for measuring specific surface of fibers. *Paper Trade J.* 115(1):32–39.
25. Climpson, N. A., and Taylor, J. H. (1976). Pore size distributions and optical scattering coefficients of clay structures. *Tappi* 59(7):89–92.
26. Cole, L. G., Czuha, M., and Mosley, R. W. (1959). Continuous coulometric determination of parts per million of moisture in organic liquids. *Anal. Chem.* 31(12):2048–2050.
27. Corte, H. (1955). Contribution to porosity analysis as exemplified by a study of cigaret filters. *Das Papier* 9(7):290–295 (in German).
28. Corte, H. (1958). The porous structure of paper: Its measurement, its importance and its modification by beating. In *Fundamentals of*

Papermaking Fibers (F. Bolam, ed.), British Paper and Board Makers Association, Kenley, England, pp. 301–331.

29. Corte, H. (1965). Determination of the pore size distribution of paper. *Das Papier* 19(7):346–351 (in German).
30. Corte, H., and Kallmes, O. J. (1962). The interpretation of paper properties in terms of structure. In *The Formation and Structure of Paper*, vol. 1 (F. Bolam, ed.), British Paper and Board Makers Association, London, pp. 351–368.
31. Corte, H. K., and Lloyd, H. (1966). Fluid flow through paper and sheet structure. In *Consolidation of the Paper Web*, vol. 2 (F. Bolam, ed.), British Paper and Board Makers Association, London, pp. 981–1009.
32. Darcy, H. (1856). Les fontaines publiques de la ville de Dijon. (The public fountains of the city of Dijon). Dalmont, Paris (in French). Original not seen.
33. Davidson, G. F. (1927). The specific volume of cottom cellulose. *J. Textile Inst.* 18:175T–186T.
34. Dehoff, R. T., and Rhines, F. N., eds. (1968). *Quantitative Microscopy*. McGraw-Hill, New York.
35. Doty, P. M., Aiken, W. H., and Mark, H. (1944). Water vapor permeability of organic films. *Ind. Eng. Chem., Anal. Ed.* 16(11):686–690.
36. Drake, L. C., and Ritter, H. L. (1945). Macropore-size distributions in some typical porous substances. *Ind. Eng. Chem., Anal. Ed.* 17(12):787–791.
37. Dullien, F. A. L., and Batra, V. K. (1970). Determination of the structure of porous media. *Ind. Eng. Chem.* 62(10):25–56.
38. Dullien, F. A. L., and Dhawan, G. K. (1974). Characterization of pore structure by a combination of quantitative photomicrography and mercury porosimetry. *J. Coll. Interface Sci.* 47(2):337–349.
39. Emanueli, L. (1927). The Emanueli porosity tester, an instrument for measuring the porosity of paper. *Paper Trade J.* 85(10):48-50 (TAPPI Sect. 98–100).
40. Fu, Y., and Bartell, F. E. (1951). Surface area of porous adsorbents. *J. Phys. Coll. Chem.* 55(5):662–675.
41. Galbraith, A. D., and Kitchen, E. A. (1962). Permeability theory vs. practice in packaging liquid products. *Tappi* 45(2):173–176.
42. Garey, C. L., Leekley, R. M., Hultman, J. D., and Nagel, S. C. (1973). Determination of pore size distribution of pigment coatings. *Tappi* 56(11):134–138.
43. Gate, L. F., and Windle, W. (1976). Absorption of oil into porous coatings. In *The Fundamental Properties of Paper Related to Its Uses*, vol. 2 (F. Bolam, ed.), British Paper and Board Industry Federation, London, pp. 438–451.
44. Gregg, S. J., and Sing, K. S. W. (1967). *Adsorption, Surface Area and Porosity*. Academic, London.
45. Haines, W. B. (1925). Studies in the physical properties of soils, 3. Observations on the electrical conductivity of soils. *J. Agri. Sci.* 15:536–543.
46. Hardacker, K. W., Bobb, F. C., and Wink, W. A. (1958). Instru-

mentation studies, no. 82. The Sheffield porosimeter. *Tappi* 41(8):231A–238A.
47. Harkins, W. D., and Jura, G. (1944). Surfaces of solids, 13. A vapor adsorption method for the determination of the area of a solid without the assumption of a molecular area, and the area occupied by nitrogen and other molecules on the surface of a solid. *J. Amer. Chem. Soc.* 66(8):1366–1373.
48. Haselton, W. R. (1954). Gas adsorption by wood, pulp, and papers, 1. The low-temperature adsorption of nitrogen, butane, and carbon dioxide by sprucewood and its components. *Tappi* 37(9):404–412.
49. Haselton, W. R. (1955). Gas adsorption by wood, pulp, and papers, 2. The application of gas adsorption techniques to the study of the area and structure of pulps and the unbonded and bonded area of paper. *Tappi* 38(12):716–723.
50. Hermans, P. H. (1949). *Physics and Chemistry of Cellulose Fibers*. Elsevier, New York.
51. Hermans, P. H., Heikens, D., and Weidinger, A. (1959). A quantitative investigation on the X-ray small angle scattering of cellulose fibers, 2. The scattering power of various cellulose fibers. *J. Polymer Sci.* 35(128):145–165.
52. Higgins, H. G., and De Yong, J. (1966). Visco-elasticity and consolidation of the fiber network during free water drainage. In *Consolidation of the Paper Web*, vol. 1 (F. Bolam, ed.), British Paper and Board Makers Association, London, pp. 242–267.
53. Horii, T. (1958). Trial manufacture of the smoothness tester applied vacuum air micrometer. *Japan TAPPI* 12(86):318–323, 334 (in Japanese).
54. Horii, T., and Chino, T. (1960). Trial manufacture of the smoothness tester applied vacuum air micrometer, 2. *Japan TAPPI* 14(110):312–321, 327 (in Japanese).
55. Husband, R. M., and Petter, P. J. (1966). An infrared instrument for the rapid measurement of water vapor permeation through barrier webs. *Tappi* 49(12):565–572.
56. Institute of Paper Chemistry (1938). Instrumentation studies, no. 25. The Williams smoothness tester. *Paper Trade J.* 106(4):38–42 (TAPPI Sect. 34–38).
57. Institute of Paper Chemistry (1940). Instrumentation studies, no. 36. The Gurley-Hill S-P-S tester. *Paper Trade J.* 110(23):303–309.
58. Keidel, F. A. (1959). Determination of water by direct amperometric measurement. *Anal. Chem.* 31(12):2043–2048.
59. Kimura, M., Oda, M., Iwasaki, Y., and Kadoya, T. (1979). Study on determination of paper thickness by mercury buoyancy method. *J. Japan Wood. Res. Soc. (Mokuzai Gakkaishi)* 25(2):139–144 (in Japanese).
60. Kruyer, S. (1958). The penetration of mercury and capillary condensation in packed spheres. *Trans. Faraday Soc.* 54:1758–1767.
61. Labrecque, R. P. (1968). The effects of fiber cross-sectional shape on the resistance to the flow of fluids through fiber mats. *Tappi* 51(1):8–15.

62. Lelie, H. J. (1964). Modified Patra method for WVT. *Modern Packaging* 37(7):145ff. (Original not seen.)
63. Levlin, J-E., and Nordman, L. (1967). On the penetration of ink into paper. In *Paper in the Printing Process*, Advances in Printing Science and Technology (W. H. Banks, ed.), vol. 4, Pergamon, Oxford, pp. 33–55.
64. Masaki, E., Yanagawa, A., and Takahashi, T. (1970). Porous structure of coated paper. *Res. Bull. Printing Bur. (Japan)* no. 2, pp. 35–45 (in Japanese).
65. Mayer, R. P., and Stowe, R. A. (1966). Mercury porosimetry: filling of toroidal void volume following breakthrough between packed spheres. *J. Phys. Chem.* 70(12):3867–3873.
66. McKnight, T. S., Marchessault, R. H., and Mason, S. G. (1958). The distribution of pore sizes in wood-pulp fibers and paper. *Pulp Paper Mag. Can.* 59(2):81–88.
67. Murakami, K. (1979). Porous structure of paper and its evaluation. *Japan TAPPI* 33(4):243–249 (in Japanese).
68. Nelsen, F. M., and Eggertsen, F. T. (1958). Determination of surface area, adsorption measurements by a continuous flow method. *Anal. Chem.* 30(8):1387–1390.
69. Oda, M., Kadoya, T., Usuda, M., and Kimura, M. (1979). Stiffness of paper, 1. Subjective judgments and Clark's stiffness measurements. *Japan TAPPI* 33(3):214–219 (in Japanese).
70. Onogi, Y., Yamauchi, T., Murakami, K., and Imamura, R. (1974). The porous structure of paper: An approach from mercury porosimetry. *Japan TAPPI* 28(3):99–107 (in Japanese).
71. Peterson, R. M. (1970). Two-dimensional flow of incompressible fluids through deformable porous media. *Tappi* 53(1):71–77.
72. Pierce, C. (1953). Computation of pore sizes from physical adsorption data. *J. Phys. Chem.* 57(2):149–152.
73. Pierce, C. (1959). Effects of interparticle condensation on heats of adsorption and isotherms of powder samples. *J. Phys. Chem.* 63(7):1076–1079.
74. Quynn, R. G. (1963). Internal volume in fibers. *Textile Res. J.* 33(1):21–34.
75. Rance, H. F. (1958). The porous structure of paper. In *The Structure and Properties of Porous Materials* (D. H. Everett and F. S. Stone, eds.), Butterworths, London, pp. 302–321.
76. Ranger, H. O., and Gluckman, M. J. (1964). A faster WVT method. *Modern Packaging* 37(11):153–156, 202.
77. Rennel, J. (1969). Opacity in relation to strength properties of pulps, 1. Method for producing unbonded fibers and determining their light-scattering coefficient and surface area. *Svensk Papperstidn.* 72(1):1–8.
78. Reverberi, A., Ferraiolo, G., and Peloso, A. (1966). Experimental determination of the distribution function of cylindrical macropores and "ink bottles" in porous systems. *Ann. Chim.* 56:1552–1561 (in Italian, ref. *Chem. Abstr.* 66:98804a). (Original not seen.)
79. Ritter, H. L., and Drake, L. C. (1945). Pore-size distribution in porous materials, pressure porosimeter and determination of complete

macropore-size distributions. *Ind. Eng. Chem., Anal. Ed.* 17(12): 783—786.
80. Rogers, C. E., Meyer, J. A., Stannett, V., and Szwarc, M. (1956). Studies in the gas vapor permeability of plastic films and coated papers, 1. Determination of the permeability constant. *Tappi* 39(11):737—741.
81. Rootare, H. M. (1968). A short literature review of mercury porosimetry as a method of measuring pore-size distributions in materials, and a discussion of possible sources of errors in this method. *Aminco Lab. News* 24(3):4A—4H. (Original not seen.)
82. Rootare, H. M., and Nyce, A. C. (1971). The use of porosimetry in the measurement of pore size distribution in porous materials. *Intern. J. Powder Met.* 7:3—11.
83. Rootare, H. M., and Prenzlow, C. F. (1967). Surface area from mercury porosimeter measurements. *J. Phys. Chem.* 71(8):2733—2736.
84. Rootare, H. M., and Spencer, J. (1971). A computer program for pore volume and pore area distribution calculations from mercury porosimeter data on particulate or porous materials. *Powder Technol.* 6:17—23.
85. Roth, J. F., and Ellwood, R. J. (1959). Determination of surface area using a gas chromatograph. *Anal. Chem.* 31(10):1738-1739.
86. Rounsley, R. R. (1964). Vapor transport through paper. *Tappi* 47(2):95—98.
87. Sanborn, I. B. (1962). A study of irreversible, stress-induced changes in the macrostructure of paper. *Tappi* 45(6):465—474.
88. Scheidegger, A. E. (1962). On the validity of the Kozeny equation. In *The Formation and Structure of Paper*, vol. 2 (F. Bolam, ed.), British Paper and Board Makers Association, London, pp. 829—831.
89. Scheidegger, A. E. (1974). *The Physics of Flow Through Porous Media*, 3rd. ed., Univ. Toronto Press.
90. Schumb, W. C., and Rittner, E. S. (1943). A helium densitometer for use with powdered materials. *J. Amer. Chem. Soc.* 65(9):1692—1695.
91. Setterholm, V. C. (1974). A new concept in paper thickness measurement. *Tappi* 57(3):164.
92. Stamm, A. J. (1929). Density of wood substance, adsorption by wood, and permeability of wood. *J. Phys. Chem.* 33(3):398—414.
93. Stamm, A. J. (1956). Diffusion of water into uncoated cellophane, 2. From steady state diffusion measurements. *J. Phys. Chem.* 60(1):83—86.
94. Stamm, A. J., and Millett, M. A. (1941). The internal surface of cellulosic materials. *J. Phys. Chem.* 45(1):43—54.
95. Stone, J. E., and Nickerson, L. F. (1963). A dynamic nitrogen adsorption method for surface area measurements of paper. *Pulp Paper Mag. Can.* 64(3):T155—T161.
96. Stone, J. E., and Scallan, A. M. (1965). A study of cell wall structure by nitrogen adsorption. *Pulp Paper Mag. Can.* 66(8):T407—T414.

97. Stone, J. E., Scallan, A. M., and Aberson, M. A. (1966). The wall density of native cellulose fibers. *Pulp Paper Mag. Can.* 67(5):T263−T268.
98. Svata, M. (1971). Determination of pore size and shape distribution from porosimetric hysteresis curves. *Powder Technol.* 5:345−349.
99. Takahashi, F., and Naito, Y. (1965). An electric hygrometer method of test for water vapor transmission rate of coated paper. *Japan TAPPI* 19(5):245−250 (in Japanese).
100. Thode, E. F., Swanson, J. W., and Becher, J. J. (1958). Nitrogen adsorption on solvent-exchanged wood cellulose fibers, indications of "total" surface area and pore size distribution. *J. Phys. Chem.* 62:1036−1039.
101. Tollenaar, D. (1967). Capillarity and wetting in paper structures: Properties of porous systems. In *Surface and Coatings Related to Paper and Wood* (R. H. Marchessault and C. Skaar, eds.), Syracuse Univ. Press, pp. 195−219.
102. Toren, P. E. (1965). Application of electrolytic moisture meter to measurement of water vapor transmission through plastic films. *Anal. Chem.* 37(7):922-923.
103. Underwood, E. E. (1970). *Quantitative Stereology.* Addison-Wesley, Boston.
104. Van den Akker, J. A. (1948). Application of the electric hygrometer to the determintion of water vapor permeability at low temperature. *Paper Trade J.* 126(1):32−37.
105. Van Ness, R. T. (1975). Special extrusion coating tests. *Tappi* 58(4):115−118.
106. Vollmer, W. (1954). Transport of gas and vapor in paper. *Chem. Ing. Tech.* 26(2):90−94 (in German).
107. Warashina, Y., Takeshita, H., and Take, Y. (1962). Studies on the measurement method of porosity of paper. *Japan TAPPI* 16(137):589−600 (in Japanese).
108. Washburn, E. W. (1921). Note on a method of determining the distribution of pore sizes in a porous material. *Proc. Nat. Acad. Sci.* 7:115-116.
109. Wasser, R. B. (1974). A mercury buoyancy technique for determining apparent density and thickness of paper. *Tappi* 57(3):166.
110. Weatherwax, R. C., and Tarkow, H. (1968). Cell wall density of dry wood. *Forest Products J.* 18(2):83−85.
111. White, R. E., and Marceau, W. E. (1962). The capillary behavior of paper. *Tappi* 45(4):279−284.
112. Wilfong, J. G. (1966). Specific gravity of wood substance. *Forest Products J.* 16(1):55−61.
113. Williams, F. M. (1935). The relation of smoothness and plane porosity to printing qualities of paper. *Paper Trade J.* 100(23):37−39 (TAPPI Sect. 291−293).
114. Williams, F. M. (1937). Studies upon smoothness, porosity and printability of paper with new methods for determination. *Paper Trade J.* 105(8):43−47 (TAPPI Sect. 123−127).
115. Wink, W. A., and Dearth, L. R. (1949). Measurement of water-vapor permeability at low temperature: The hygrometric sweep gas method. *Tappi* 32(3):232−238.

116. Winrich, K. M. (1971). Dynamic WVTR testing, an experience report on the St. Regis-Honeywell WVTR tester. *Tappi* 54(8):1302–1304.
117. Yamamoto, K., Kaida, K., and Iwasaki, M. (1966). On Bekk's smoothness of paper. *Japan TAPPI* 20(2):81–88 (in Japanese).
118. Yamauchi, T., Murakami, K., and Imamaura, R. (1975). The porous structure of paper, analytical research on the behavior of mercury penetration and retraction. *Japan TAPPI* 29(9):492–497 (in Japanese).
119. Yamauchi, T., Murakami, K., and Imamura, R. (1976). The air permeability of paper related to the porous structure. *Japan TAPPI* 30(5):273–280 (in Japanese).
120. Yamauchi, T., Murakami, K., and Imamura, R. (1978). Studies on latex impregnated paper, 1. The porous structure and rubber distribution of latex impregnated handsheets. *Japan TAPPI* 32(9):534–540 (in Japanese).
121. Young, D. M., and Crowell, A. D. (1962). *Physical Adsorption of Gases.* Butterworths, London.
122. Ziegler, R. D. (1957). A procedure for measuring the water vapor permeability of insulation board. *Tappi* 40(11):881–884.
123. Ziegler, R. D. (1960). The effect of temperature and moisture content on the moisture permeability of building boards. *Tappi* 43(11):913–918.

18
WETTING AND THE PENETRATION OF AQUEOUS LIQUIDS

M. BRUCE LYNE

Pulp and Paper Research Institute of Canada
Pointe Claire, Quebec, Canada

I.	Introduction	103
II.	Theory of Wetting and Aqueous Absorption	104
III.	Test Methods	107
	A. Wetting	108
	B. Rate of Penetration of Aqueous Liquids	113
	C. Total Uptake of Aqueous Liquids	117
IV.	Concluding Remarks	118
	References	119

I. INTRODUCTION

Wetting is truly a surface phenomenon; therefore, it is unsatisfactory to view paper as simply being composed of cellulose, hemicelluloses, and lignin, since the chemical composition of surface layers down to monomolecular thicknesses determines the wetting characteristics of paper. In particular, low surface energy resin and fatty acids present in all species of wood used for papermaking tend to spread over the surface of paper, rendering it more hydrophobic.

Mechanically prepared wood pulp, such as that used in newsprint, also tends to have a very heterogeneous surface chemistry. In mechanical pulping, fibers are liberated from wood by physical degradation of middle lamella between the fibers. In order to increase opacity and promote interfiber bonding in paper, the surface area of the fibers is further "developed" by mechanically peeling the outer layers of the fiber wall. Lignin is more concentrated in the outer layers of the fiber. Thus paper made from mechanical pulp can vary in local lignin concentration due to variation in the degree to which the outer fiber layers have been stripped away and according to the location of the resulting debris, or fines. Since lignin is more hydrophobic than either cellulose [15,16] or the hemicelluloses, the surface chemistry varies locally with lignin concentration. The same can be said for the local concentration of resin-bearing ray cell fines in the paper sheet.

Surface morphology also plays a role in wetting. Drops of nonwetting liquids tend to exhibit higher contact angles on rough surfaces and tend to extend more readily along grooves or fibers than across them [20]. Therefore, even paper made from chemically prepared wood pulps having relatively homogeneous fiber surface chemistries will exhibit local nonuniformities and anisotropy in wetting behavior due to variation and orientation in fiber and network morphology.

The penetration of aqueous liquids into paper is further complicated by absorption into fiber walls and consequent increases in fiber wall thickness. It is though that swelling occurs as aqueous liquids break and replace interchain hydrogen bonds in cellulose. Swelling appears to be proportional to the amount of liquid absorbed [8,12,28] and swelling of the fiber walls generally tends to close voids in the fiber surfaces while enlarging interfiber voids in the fiber network. Thus the rate of capillary imbibition is generally altered by the absorption of aqueous liquids [17].

Finally, it should be mentioned that wetting, absorption, and capillary imbibition can be altered by the addition of hydrophobic agents, or size, during the manufacture of paper (typically rosin internally, and starch, casein, polyvinyl alcohol, or wax emulsions externally). When added internally or added to the surface, they act chiefly to prevent absorption into the fiber walls and consequent swelling. Hard sizing can also be used to cause complete hydrophobicity or repellence.

Techniques for measuring wetting and the penetration of aqueous liquids into paper have traditionally not separated the surface chemistry, morphological, and swelling aspects—nor have they attempted to replicate the highly dynamic conditions that pertain in most industrial processes where wetting and penetration are important. Therefore, this chapter does deal with the more traditional techniques, such as the Cobb test, contact angle measurement, and drop penetration tests, but it concentrates on more modern techniques. Readers particularly interested in penetration of tissue and toweling by water will find additional coverage in Hollmark's chapter on absorbency, Chap. 20.

II. THEORY OF WETTING AND AQUEOUS ABSORPTION

Liquids will wet and spread over the surface of a solid if the total surface free energy will thereby be reduced:

$$S = \gamma_S - \gamma_{LS} - \gamma_L \qquad (1)$$

That is, spreading will occur when S is positive and the surface free energy of the solid γ_S exceeds that of the liquid γ_L plus its interfacial tension between the solid and liquid γ_{LS}. Spreading can also be described as occurring when the work of adhesion between the liquid and solid $W_{A(SL)}$ exceeds the work of cohesion in the liquid $W_{C(L)}$:

$$W_{A(SL)} = \gamma_S + \gamma_L - \gamma_{LS} \tag{2}$$

$$W_{C(L)} = 2\gamma_L \tag{3}$$

$$S = W_{A(SL)} - W_{C(L)} = \gamma_S - \gamma_{LS} - \gamma_L \tag{4}$$

Since spreading occurs only when S is positive, any process that reduces the surface free energy of the solid γ_S sufficiently will prevent wetting. The sizing and the spreading of the low surface free energy resin and fatty acids over the surface of paper (or self-sizing) essentially cause hydrophobicity by this process.

More specifically, the surface free energy of a liquid (or a solid) can be considered in terms of a London dispersion component α_L and a polar component β_L:

$$\gamma_L = \alpha_L^2 + \beta_L^2 \tag{5}$$

The work of adhesion between the liquid and solid may then be expressed as

$$W_{A(SL)} = 2(\alpha_L \alpha_S + \beta_L \beta_S) \tag{6}$$

where α_S and β_S are the dispersion and polar components of the surface free energy of the solid, respectively. Referring back to Eq. (4), it can be seen that increasing the polar character of the solid or reducing the polar character of the liquid will tend to enhance wetting and spreading.

When cellulosic or lignin-containing films absorb moisture, their surfaces progressively approach the more polar and higher surface free energy of a surface water layer [4,15,16]. Conversely, when a surfactant is added to water it becomes less polar and exhibits a lower surface energy. Both tend to facilitate wetting and increase spreading.

However, the effect on absorption and capillary imbibition of increasing the moisture content of cellulose and adding surfactants to water is not as simple. Consider a liquid in a capillary which wets the surface of capillary walls. The liquid spreads parallel to the surface of the capillary walls, but the surface of the bulk liquid lies perpendicular to the capillary walls. Therefore, a curvature must exist to the surface of the liquid. This curvature, or meniscus, is a reflection of the capillary pressure exerted by the liquid spreading on the pore walls. The Laplace equation describes this capillary pressure P in terms of the contact angle θ between the liquid and the capillary walls (Fig. 1), the surface free energy of the liquid γ_L, and the radius of the capillary r:

$$P = \frac{2\gamma_L \cos \theta}{r} \tag{7}$$

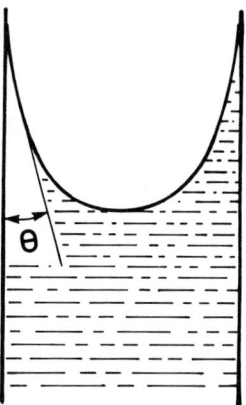

Fig. 1 Capillary showing contact angle θ with rising liquid.

This capillary pressure results in a net flow of liquid along the capillary. In the case of a horizontal capillary, once the flow has been initiated the rate of penetration dh/dt can be described by the Lucas-Washburn equation:

$$\frac{dh}{dt} = \frac{r^2 P}{8\eta h} = \frac{r\gamma_L \cos\theta}{4\eta h} \tag{8}$$

where

h = length of filled portion of capillary
η = viscosity of the liquid.

If a surfactant is added to an aqueous liquid, both γ_L and θ will decrease. The penetration rate dh/dt will be altered according to whether $\gamma_L \cos\theta$ is increased or decreased. For example, if the addition of a surfactant reduces the contact angle of water to zero and reduces the surface free energy from 72 to 30 dynes/cm the penetration rate will increase only if 30 cos 0° > 72 cos θ_ω —or the contact angle for water θ_ω > 65°.

It can also be appreciated from the Lucas-Washburn equation that swelling of fibers as the result of the absorption of moisture by paper will cause the penetration rate to increase because the interfiber capillary radii will increase with the attendant expansion of the fiber network.

However, moisture also acts as a plasticizer for cellulose, making paper more compressible. Therefore, under compression (such as in the nip of a printing press) paper may actually exhibit a slower penetration rate when moisturized—and thus, reduction of the interfiber capillary radii—because of increased compression.

Finally, aqueous liquids penetrate dense and sized papers more by diffusion than by capillary imbibition [12,15]. Fick's second law describes the change in concentration of a diffusing liquid in time at any point x along the direction of diffusion as a function of a diffusion coefficient D:

$$\frac{\partial c}{\partial t} = D \frac{\partial^2 c}{\partial x^2} \qquad (9)$$

Hoyland [12] proposed that the following solution of Fick's law should describe the penetration of aqueous liquids into paper by diffusion:

$$\frac{\Delta Z}{\Delta Z_{max}} = \frac{m_t - m_0}{m_\infty - m_0} = \frac{2}{L}\left(\frac{Dt}{\pi}\right)^{1/2} \qquad (10)$$

where

$\frac{\Delta Z}{\Delta Z_{max}}$ = change in thickness of paper as aqueous liquid diffuses into fiber walls, relative to the maximum thickness change at fiber saturation

m_0, m_t, m_∞ = moistures absorbed at times 0, t, and fiber saturation, respectively

t = time elapsed from introduction of liquid

L = initial thickness of paper

Thus, the diffusion coefficient D can be calculated by measuring the thickness change as a function of time after introducing the aqueous liquid to the paper surface. Also, the change in the diffusion coefficient with depth of diffusion into the paper web can be studied with Hoyland's apparatus [12,13].

III. TEST METHODS

From the foregoing, it is apparent that wetting and absorption of aqueous liquids depend on the initial moisture content of paper. Temperature also affects chemical reactivity, viscosity, contact angle [4], and paper rheology. Therefore it is important that all tests be made on conditioned paper samples in an atmosphere at 50% relative humidity and 23°C (ISO standard atmosphere).

Further, minor impurities in water can have considerable influence on absorption. It should not be expected that tests made with tap water can be replicated, since dissolved inorganics and pH generally vary considerably with time and location. Where process conditions require tests with tap water the amount of dissolved solids and the pH should be recorded, or if possible, controlled. Bristow [6] has shown that for kraft liner, absorption rate is strongly influenced by the pH of the aqueous liquid and by the pH of the paper. More rigorous testing may also require deionized or distilled water; in some cases, the water must be deaerated to insure reproducibility.

Finally, test apparatus should be made of nonreactive materials such as stainless steel, glass, or gold-plated brass. Every effort should be made to keep all test surfaces free from contamination—particularly by low surface free energy liquids such as oils and silicones.

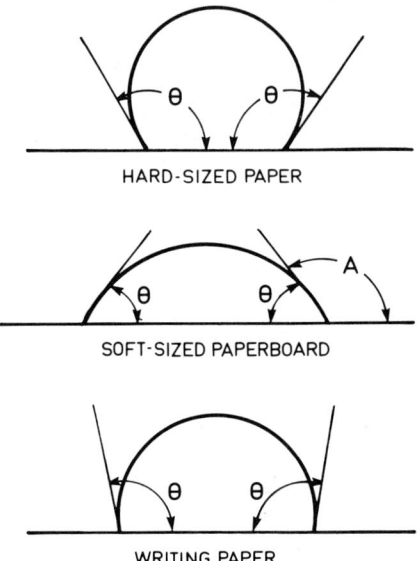

Fig. 2 Contact angle between sessile drop and flat surface. (From Ref. 22. Copyright © 1970. TAPPI. Reprinted with permission.)

A. Wetting

Contact Angle Measurement Contact angle measurements between sessile drops and the surface of paper have long been used as an indication of wettability. As shown in Fig. 2, the contact angle θ for a drop of aqueous liquid increases as the paper surface becomes more hydrophobic and, conversely, decreases as the surface becomes more wettable. Methods for measuring static contact angles are described in TAPPI standard T 458 os-70, CPPA F.3, and SCAN-P18. Essentially, these methods involve depositing a drop of liquid from a small hypodermic needle (1 ml) onto the paper surface and measuring the static contact angle after 5 s with the aid of a microscope or projection apparatus. This presupposes that the contact angle does not change rapidly as wetting and absorption proceed and that surface roughness does not make estimation of the contact angle impossible.

Caution should also be exercised in the interpretation of the contact angles of sessile drops. As mentioned earlier, surface topography of the paper plays a role in the static contact angle—rougher surfaces cause higher contact angles for nonwetting liquids [20]. Further, dynamic processes involve *dynamic* contact angles, which can be much larger than for the static case.

More sophisticated dynamic contact angle methods are used in other industries. One example is the measurement of rise-canceling velocity in the textile industry: a filament to be tested is drawn through a liquid bath, and the velocity at which the contact angle becomes 90° is recorded (Fig. 3). Wetting force measurements are also performed on textile filaments and single fibers using a capillograph (Fig. 4) [2,19]. Advancing and receding contact

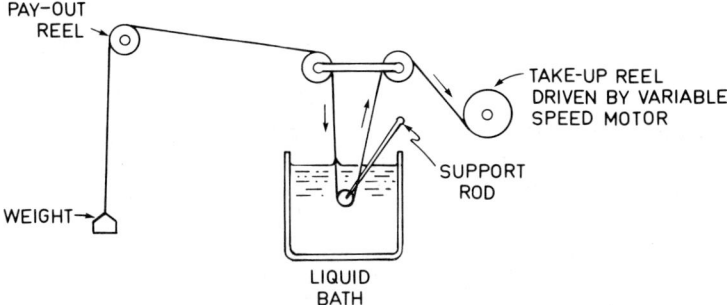

Fig. 3 Apparatus for measurement of rise-canceling velocity in textile filaments. (From Ref. 18.)

angles (θ_a and θ_r) can be calculated from wetting force measurements with the liquid advancing F(a), receding, F(r), and leaving the tip of the fiber [18]. The last gives the maximum force F(m) at near zero contact angle (neglecting any buoyancy forces):

$$F(m) = \gamma_L P \qquad (11)$$

where P is the fiber perimeter

$$F(a) = F(m) \cos \theta_a \qquad (12)$$

$$F(r) = F(m) \cos \theta_r \qquad (13)$$

Aberson [1] proposed a method for calculating dynamic contact angles between water and paper in which two measurements are performed: the times required for water and water with a small amount of surfactant (e.g., 0.3% Teepol solution) to travel a preset distance h in the plane of the paper web. If the dynamic contact angle for water with surfactant is assumed to be zero,

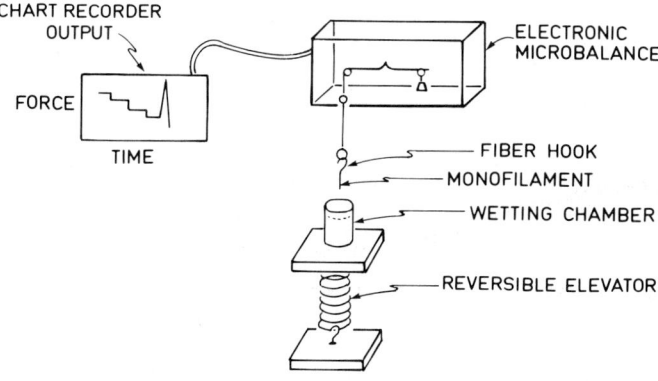

Fig. 4 Schematic of capillograph.

the Lucas-Washburn equation reduces to

$$h = \sqrt{\frac{r\gamma_L t}{2\eta}} \qquad (14)$$

It is a simple matter to calculate the effective capillary radius r, if the surface energy γ_L of the water-surfactant solution (e.g., 32.8 dynes/cm for the 0.3% Teepol in water) and the viscosity of the liquid (essentially that of water) are known.

Having calculated the effective capillary radius r, it is then possible to calculate the contact angle θ for the case of pure water.

$$\cos\theta = \frac{2h^2 \eta}{r\gamma_L t} \qquad (15)$$

While the method obviates the difficulty of observing dynamic contact angles, it does depend on the assumption that the advancing contact angle of the surfactant-water solution is close to 0°.

Dynamic Wetting Instruments Useful information about wetting and absorption can be gained without reference to contact angles or effective capillary radii. Hawkes and Bedford [11], and later Bristow [5] designed instruments to study dynamic penetration and wetting of paper. Figure 5 shows Bristow's instrument. A paper strip is mounted on a rotating wheel and is drawn past a miniature headbox which is deadweighted so that it rests on the paper strip with a pressure of 0.1 MPa. The headbox is filled with a known amount of liquid which has time available for absorption determined by the peripheral speed V and the width of the slice opening (L = 1 mm). Since this speed may be varied, the relationship between the amount of liquid transferred and the time available for absorption may be established as in Fig. 6.

The amount of liquid that transfers to the paper in time t may be described by

$$K_r + K_a t^{1/2}$$

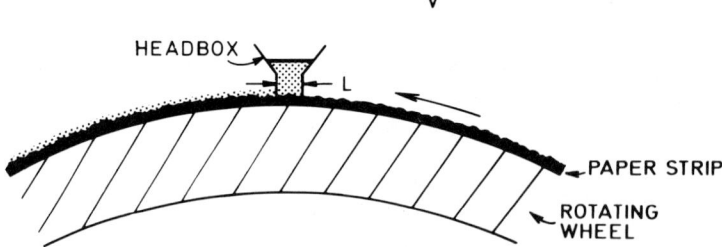

Fig. 5 Bristow's instrument for dynamic wetting and absorption studies. (From Ref. 17.)

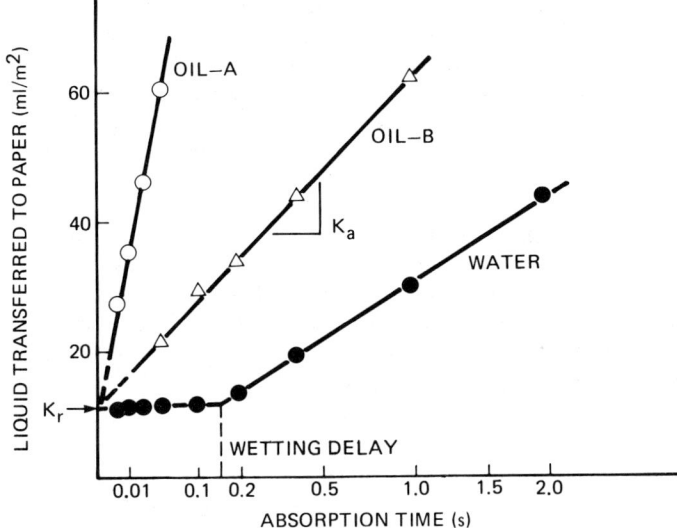

Fig. 6 Amount of liquid transferred to paper on a square-root time axis. (From Ref. 5.)

where

K_r = roughness index, or amount of liquid that theoretically would fill the surface of the paper at time zero
K_a = absorption coefficient or slope of the curve

It is apparent from the units for liquid transferred that K_r is effectively the mean depth to which the fluid may penetrate before capillary forces are initiated (i.e., a mean topographic depth from the plane of the headbox opening).

It is also apparent in Fig. 6 that the curves for water show a measurable wetting time while those for oil do not. This wetting delay for polar liquids has been observed by several researchers [5,12,17,29], but the mechanism has never been clearly elucidated. Bristow [5] concluded that the wetting time was a two-stage process involving the time for water to run into the surface topography and the chemical wetting time of the fiber surfaces.

Wetting times for water may vary from about 10 ms for chemically prepared fiber networks of cellulose, for example, filter paper [12], to about 50 ms for newsprint made from fresh mechanical pulp [17], to several seconds for sized papers (hydrophobic agent added to the network) [5,7]. In the case of self-sized papers made from mechanical pulp, wetting times of about 1 s have been observed [17]. Extraction with acetone is necessary to remove physically absorbed resin and fatty acids [25,26]. There is evidence to suggest that chemisorption may occur when paper is treated with very small amounts of stearic acid (0.003%) or natural resin and fatty acids and heated to above 100°C [1]. Sodium methoxide extraction has been found to eliminate hydrophobicity in these cases, while acetone extraction was found ineffective [26].

Repellency Tests Water repellency is a desirable property for many grades of paper and board. A simple test of the effectiveness of surface sizings or coatings in creating water repellency is described in TAPPI T 513 su-69. As shown in Fig. 7, the test paper is held at an angle of 25° to the horizontal and a 1 mm inside diameter capillary tube is used to deliver 8.0 ± 0.5 ml per minute of water (in droplets) to the surface of the test specimen. The time required for a continuous stream of water to flow down the specimen surface is taken as a measure of water repellency.

Repellency can also be a negative property. In particular, it is necessary that water (or fountain solution) in nonimage areas of lithographic printing be absorbed rapidly. In multicolor printing, ink is often printed in areas that received water in a previous printing unit. The oil-based ink will not wet the surface of the paper unless this water film has been absorbed. The IGT [10a] and Prüfbau [18a] laboratory print testers have wetting ac-

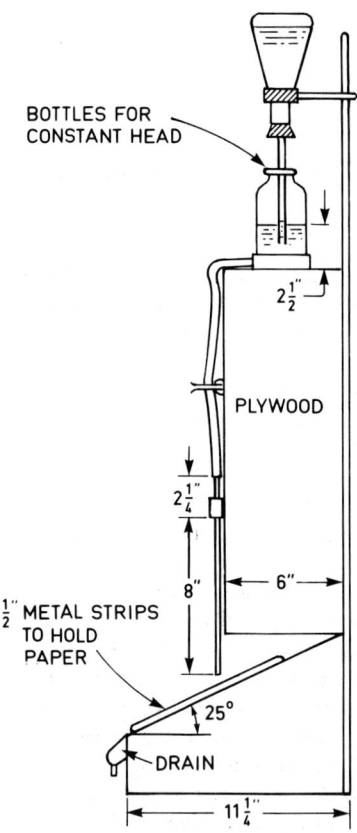

Fig. 7 Water drop repellency test. (From Ref. 23. Copyright © 1969. TAPPI. Reprinted with permission.)

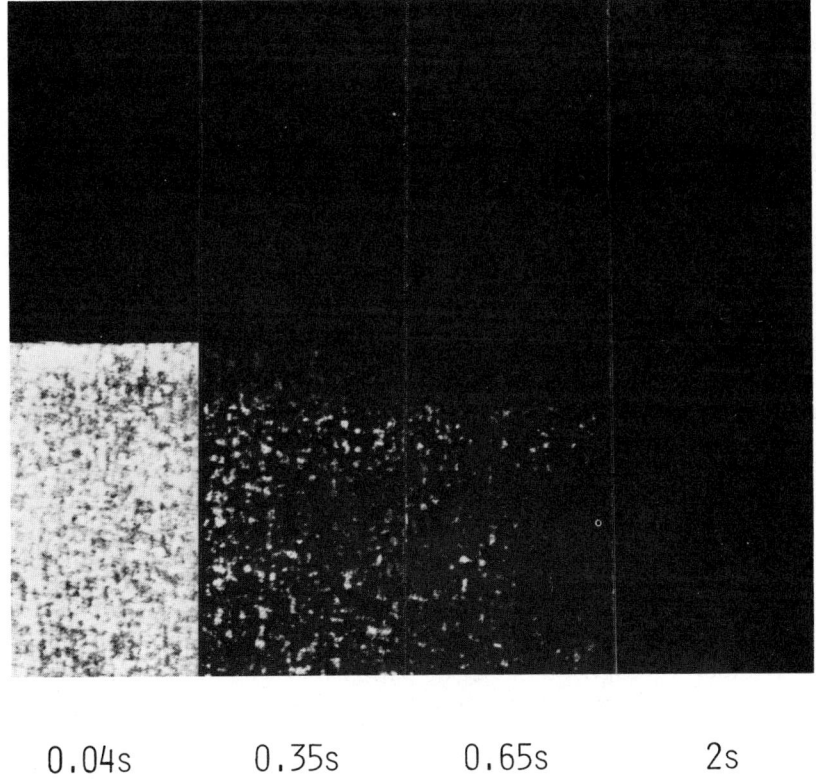

0.04s   0.35s   0.65s   2s

Fig. 8 Ink repellence as a function of time between the application of water and printing ink.

cessories that allow a metered film of water to be applied to the surface of a test paper strip prior to printing with oil-based ink. As shown in Fig. 8, the interval between the application of the water and the printing can be varied in order to assess the time required for absorption of the water film. However, the water film thickness is hard to adjust to that in offset printing.

B. Rate of Penetration of Aqueous Liquids

As described on pp. 110-111 above, the Bristow type of instrument can be used to directly measure the rate of penetration (absorption coefficient) for oil or aqueous liquids [5]. However, most methods of measuring penetration rates for aqueous liquids rely on detection of changes in paper properties as a function of liquid penetration. These include measuring changes in electrical conductivity [3,12,13,21], ultrasonic wave propagation [9,27], infrared absorption [14], and surface polarity. The last is most easily measured by the IGT [10a] and Prüfbau [18a] tests mentioned above, where the time required for absorption of a water film is detected by overprinting with an oil-based ink at varied time intervals.

The simplest penetration test is doubtless the measurement of the time required for a drop of test liquid to be absorbed after it is allowed to fall onto the test paper. As specified in TAPPI T 492 pm-6 the drop volume is controlled by using a burette and the endpoint is judged by the sudden loss in gloss of the drop upon absorption. In a variation of this test, the burette can be brought into contact with the paper and opened. The rate of penetration can then be calculated by noting the level of the liquid in the burette at various time intervals. However, the column of liquid in the burette causes a hydraulic head that decreases with the falling liquid level. This problem can be circumvented by introducing the liquid to the underside of the paper sample from an orifice level with the surface of a large reservoir of the liquid. In this case the absorption of the liquid must be monitored gravimetrically.

The simplest method of measuring the rate at which an aqueous liquid penetrates through paper is to float the paper sample on the liquid and measure the time required for an indicator dye to be wetted on the upper surface of the sample. As described in CPPA F.1, a 45/5/1 mixture of powdered sugar, soluble starch, and methyl-violet dye is sprinkled on the back side of the paper sample. The edges of the paper sample are sealed in molten wax and a watch glass is placed over the powder mixture to prevent vapor and penetration in the plane of the paper web from interfering in the test. Alternatively, a flotation device with a desiccator can be used.

However, the powder containing the dye can be triggered by the penetration of water vapor through the sheet of paper ahead of the penetration of the liquid. Also, liquid penetration may not be uniform, making subjective judgment of the penetration time difficult. These shortcomings can be overcome by using a fluorescent dye that is relatively insensitive to water vapor or by using a dyed water solution and measuring the mean drop in reflectance from the reverse side of the test sheet. The Hercules size test employs this optical measuring technique. The dye solution may also contain formic acid to speed the penetration of hard-sized papers. However, this practice should be avoided, since the acid changes the chemistry of the sorption process.

Below is a description of more sophisticated methods that are not subject to industrial standards.

Conductivity Change Method Tap water contains sufficient ionic material to be electrically conductive. Therefore, conductivity between electrodes placed on either side of a paper sheet can be used as an indicator of penetration of water from one side of a sheet of paper to the other [3,12,13,21]. Low voltage DC current (e.g., 9 volts) can be used in conjunction with a standard conductance meter. In order to avoid polarizing the liquid a low frequency (e.g., 1000 Hz) AC current can also be used. The upper electrode can be used to introduce the aqueous liquid and electrodes can be positioned at various locations under the paper sheet to test penetration times not only through the thickness of the sheet but in any set direction in the plane of the paper sheet (Fig. 9).

One shortcoming of the conductance method for detection of penetration through the thickness of paper is that pinholes in paper may cause conduction paths to be created ahead of bulk penetration.

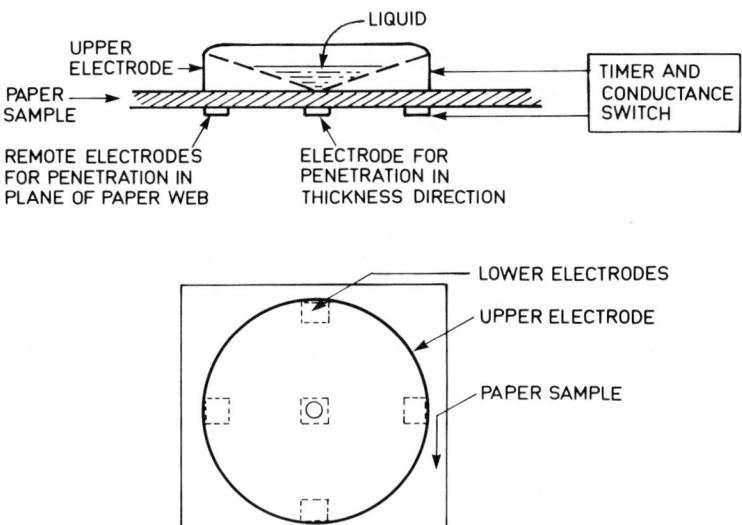

Fig. 9 Possible configuration for electrodes for measurement of penetration in the thickness direction and in the plane of a paper sheet.

Multiple Internal Reflection Spectroscopy In certain frequency bands infrared light is strongly absorbed by water. In fact, the concentration of water in aqueous liquids such as adhesive solutions can be assessed by the degree to which they absorb infrared light. Multiple internal reflection is a means of measuring this absorption over a broad sample area and to a depth not exceeding that of a film of adhesive [14].

As shown in Fig. 10, if light is directed across an interface between two media (e.g., glass and an adhesive film) above the critical angle for total internal reflection, the light will be reflected. However, there will be a certain depth of penetration d into the second medium described by

$$d = \frac{\lambda}{2\pi\sqrt{\sin^2\theta - n_{21}^2}} \qquad (16)$$

where

λ = wavelength in infrared (3400 cm^{-1}, or about 3 µm)
θ = angle of incidence
n_{21} = refractive index from medium 1 to medium 2

Using a configuration as in Fig. 11 it is possible to cause multiple internal reflections, and hence, multiple penetrations of the light into the second medium. If the second medium is an infrared absorbing liquid, the light will be attenuated each time it passes through the liquid. In practice, the internal reflection plate is made of germanium or sapphire with edges beveled to allow incidence and detection at the appropriate angles.

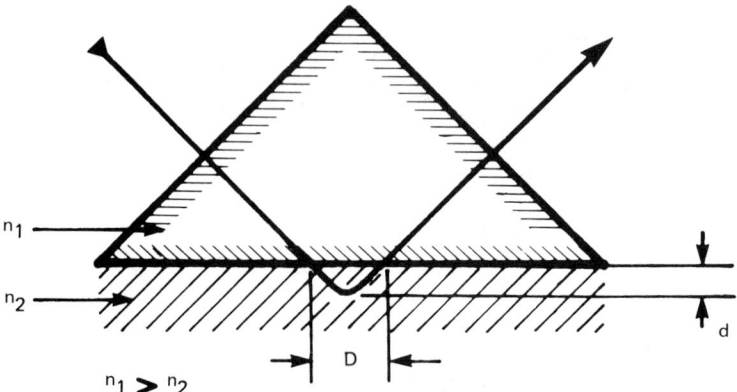

Fig. 10 Total internal reflection in a glass prism. (From Ref. 14.)

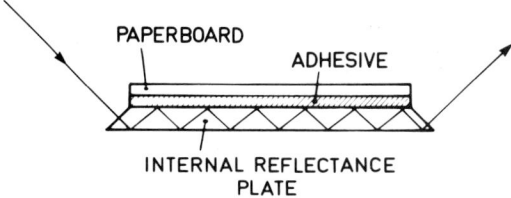

Fig. 11 Multiple internal reflections in a plate of germanium or sapphire. (From Ref. 14.)

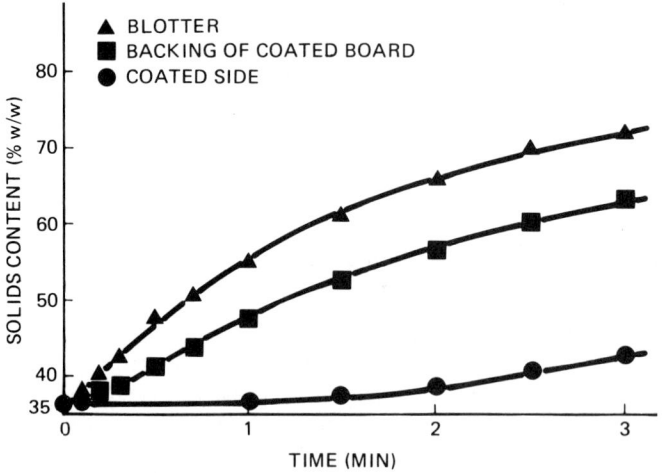

Fig. 12 Solids content in adhesive film as a function of time after application to various types of paper. (From Ref. 14.)

The instrument compares the ratio of the light intensities exiting the internal reflection plate when it is in contact with the liquid film and when the surface of the plate is dry (no attenuation). Calibration of this ratio of light intensities to moisture content is made using liquid films of known moisture content. The instrument can then be used to monitor the penetration of aqueous liquids into paper and paperboard (Fig. 12).

One shortcoming of this technique is that during the penetration of the aqueous portion of the adhesive into the paperboard, a concentration gradient is established across the adhesive layer. Therefore, there is a delay in detecting the penetration which depends on the rate of diffusion of water across the adhesive film.

Ultrasonic Propagation Technique In this technique [9,27], transducers are attached to two ends of the test strip of paper. The rate of propagation of ultrasonic wave trains between the transducers is measured. As water is absorbed into the fibers in the paper strip, the speed of propagation decreases. Thus, the rate of propagation can be calibrated against moisture content in the paper strip. Alternatively, if the aqueous liquid is introduced to one side of the paper strip, the mean rate of penetration of the liquid across the paper can be monitored.

C. Total Uptake of Aqueous Liquids

Immersion Test An immersion test consists of submerging a sample of paper or paperboard under a fixed height of aqueous liquid (e.g., 7.6 cm in TAPPI T 491 su-63), holding it down with a 12-mesh wire screen for a set time interval (e.g., 10 min), and then blotting both sides of the sample with a blotter having a standardized rate of absorption. The weight increase of the sample as the result of immersion is taken to be a measure of the amount of liquid absorbed.

Since the blotting action of the standardized blotter is crucial to the final weight increase of the sample, it must be assumed that the blotter acts to remove all free liquid—leaving only the absorbed liquid. The absorptiveness of the blotter is checked by cutting machine and cross-machine strips 15 mm × 200 mm and immersing one end of the strips to the depth of 10 mm in deionized water. The height of rise after 10 min should be between 50 and 100 mm in both directions. However, this method (described in SCAN-P13:64) does not guarantee that the water uptake in a blotting action (in the thickness direction rather than in the plane of the blotter) will be constant, nor does it guarantee that all free water will be removed from the sample. According to Eq. (7), capillaries in the paper sample smaller than those in the blotter would be expected to retain their free water by virtue of higher capillary pressures.

Van den Akker has recently proposed a water-holding capacity test that obviates the need for blotting paper. A 3 × 3 in. sample is cut and weighed to the nearest milligram. The sample is immersed in water for 60 s then removed and placed on a gridlike water extraction device. It exerts suction on the sample equivalent to a 5 mm column of water for 15 s, after which the sample is reweighed and the water uptake calculated.

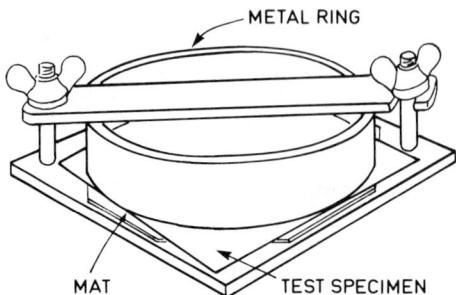

Fig. 13 Schematic of apparatus for the Cobb test. (From Ref. 24. Copyright © 1977. TAPPI. Reprinted with permission.)

Finally, air bubbles trapped on the under surface of the sample may also restrict access to the aqueous fluid. While a light brushing with the liquid prior to immersion of the sample may remove most of these bubbles, their presence does make the test harder to standardize.

One Side Immersion: Cobb Test As shown in Fig. 13, the Cobb test [10] amounts to sealing one side of the paper sample to a rubber mat via a cylindrical ring clamped over the upper surface of the sample. The cylindrical ring has an inside area of 100 cm^2 (11.28 cm inside diameter) and a height of 2.5 cm. The aqueous liquid is poured into the ring to a depth of 1 cm (100 cm^3), and 10 s before the end of a specified time interval (120 s in TAPPI T 441 os-77) the liquid is poured off. The sample is removed and blotted with a standardized blotting paper (see p. 117). Exactly at the end of the preset time interval, a second blotter is put on top of the first and a 10 kg roller is passed over the three sheets to extract the free water. The weight increase in the paper sample is taken to be the amount of liquid absorbed in the preset time interval.

Slightly different test geometries and procedures are followed in CPPA F.2 and SCAN-P12:64. The criticisms mentioned in relation to the immersion test apply to the Cobb test as well, and Bristow [7] has further criticized the Cobb test.

IV. CONCLUDING REMARKS

Modern industrial processes, such as lithographic printing, printing with water-based inks, gluing operations in paper and paperboard converting, and water-based coating of paper, require wetting and penetration by aqueous liquids at rates measured in tens of milliseconds. Contact angle measurements on sessile drops, the Cobb test, and similar industry standard tests operate on time scales that are orders of magnitude too long to be useful in these applications.

Newer tests have been reviewed which have the necessary time resolution. For example, the PPRIC version of the Bristow apparatus is designed to measure absorption and wetting delays down to five milliseconds. The conductivity change method and the ultrasonic propagation technique can measure penetration times to a similar resolution These techniques should be considered by the paper industry, since it must cope with ever increasing process speeds and product applications requiring rapid wetting and absorption by aqueous liquids.

The methods that have been reviewed have been selected because they are of potential use in industrial laboratories. There are, of course, more sophisticated approaches to analyzing wetting and absorption. For example, electron spectroscopy for chemical analysis (ESCA) can be used to probe the chemical composition of the immediate surface (< 5 nm) of papermaking fibers. Fiber capillography can give a detailed physical characterization of the surface of single fibers. Inverse gas chromatography (GC) is also used to measure certain thermodynamic properties of cellulose, cellulose derivatives, and paper. Among these properties are the heats of wetting and absorption.

However, techniques such as ESCA, capillography, and inverse GC are used primarily for fundamental research into the surface chemistry of materials. They are usually applied under strictly controlled conditions, and experiments must be carried out with care. In contrast, the methods reviewed in this report are meant to be applicable to the problems encountered in the high-speed processes used in commercial production. As the process problems become clearly defined, the more fundamental techniques of surface chemistry may be used to find the origin of the problems and to suggest approaches for their solution.

REFERENCES

1. Aberson, G. M. (1969). The water absorbencey of pads of dry unbonded fibers. TAPPI Special Technical Association publication no. 8 (D. H. Page, ed.), pp. 282–305.
2. Bendure, R. L. (1973). Dynamic adhesion tension measurement. *J. Coll. Interface Sci.* 42:137–144.
3. Bohmer, E., and Lute, J. (1966). Adhesive migration and water retention with reference to blade coating. *Svensk Papperstidn.* 69(18):610–618.
4. Borgin, K. (1959). The properties and nature of the surface of cellulose. 1. Cellulose in contact with water: Experimental results and their interpretation. *Norsk Skogind.* 13(11):429–442.
5. Bristow, J. A. (1967). Liquid absorption into paper during short time intervals. *Svensk Papperstidn.* 70(19):623–629.
6. Bristow, J. A. (1968). The absorption of alkaline solutions by paper. *Paper and Timber (Finland)* 11:639–646.
7. Bristow, J. A. (1968). The absorption of water by sized papers. *Svensk Papperstidn.* 71(2):33–39.
8. Bristow, J. A. (1971). The swelling of paper during short time intervals. *Svensk Papperstidn.* 74(20):645–652.
9. Chatterjee, P. K. (1971). The sonic velocity response during the absorption of water in paper. *Svensk Papperstidn.* 74(17):503–508.
10. Cobb, R. M., and Lowe, D. V. (1934). A sizing test and sizing theory. *Tech. Assn. Papers* 17:213–216.

10a. Determination of the wet pick and wet repellance by means of the IGT damping unit (1969). IGT Information Leaflet no. W32, Amsterdam.

11. Hawkes, C. V., and Bedford, T. (1963). *The Absorption Characteristics of Paper*, I. PATRA Lab Report no. 51.

12. Hoyland, R. W. (1978). Swelling during the penetration of aqueous liquids into paper. In *Fiber-Water Interactions in Paper-Making* (F. Bolam, ed.), British Paper and Board Industry Federation, London, pp. 557–577.

13. Hoyland, R. W., Howarth, P., and Field, R. (1976). Fundamental parameters relating to performance of paper as a base of aqueous coating. In *The Fundamental Properties of Paper Related to Its Uses* (F. Bolam, ed.), British Paper and Board Industry Federation, London, pp. 464–510.

14. Huynh, H. K., Lancaster, P. E., Lepoutre, P., and Robertson, A. A. (1978). The setting of aqueous adhesives on paper. *Tappi* 61(12):63–65.

15. Lee, S. B., and Luner, P. (1972). The wetting and interfacial properties of lignin. *Tappi* 55(1):116–121.

16. Luner, P., and Sandell, M. (1969). The wetting of cellulose and wood hemicelluloses. *J. Polymer Sci.*, C 28:115–142.

17. Lyne, M. B. (1978). The effect of moisture and moisture gradients on the calendering of paper. In *Fiber-Water Interactions in Paper-Making*, (F. Bolam, ed.), British Paper and Board Industry Federation, London, pp. 641–665.

18. Miller, B., and Young, R. A. (1975). Methodology for studying the wettability of filaments. *Textile Res. J.* 45(5):359–365.

18a. Niesser, G. (1972). The interaction of printing ink and paper, 2. *Druck* 10:687 (in German).

19. Okagawa, A., and Mason, S. G. (1978). Capillarography: A new surface probe. In *Fiber-Water Interactions in Paper-Making*, (F. Bolam, ed.), British Paper and Board Industry Federation, London, pp. 581–586.

20. Oliver, J. F., and Mason, S. G. (1976). Scanning electron microscope studies of spreading of liquids on paper. In *The Fundamental Properties of Paper Related to Its Uses* (F. Bolam, ed.), British Paper and Board Industry Federation, London, p. 209.

21. Stinchfield, J. C., Cliff, R. A., and Thomas, J. J. (1958). The water retention test in evaluating coating color. *Tappi* 41(2):77–79.

22. TAPPI T 458, Surface Wettability of Paper (Angle of Contact Method), 1970.

23. TAPPI T 513, Water Repellency of Paper and Boards, 1969.

24. TAPPI T 441, Water Absorptiveness of Sized (Non-Bibulous) Paper and Paperboard (Cobb Test), 1977.

25. Takeyama, S., and Gray, D. G. (1979). Surface analysis of some sulphite pulps by ESCA. *Trans. Tech. Sect. Can. Pulp Paper Assn.* 6(3):61–64.

26. Takeyama, S., and Gray, D. G. (1982). An ESCA study of the

chemisorption of stearic acid on cellulose. *Cellulose Chem. and Technol.* *16*(2):133–142.
27. Taylor, D. L., and Dill, D. R. (1967). Water retention of coating colors: A study of the sonic velocity method and the effect of color composition on water retention. *Tappi* *50*(11):536–541.
28. Tremaine, P. R., Mohlin, U., and Gray, D. G. (1977). The adsorption of n-decane on the surface of water-swollen cellulose fibers. *J. Coll. Interface Sci.* *60*(3):548–554.
29. Windle, W., Beazley, K. M., and Climpson, M. (1970). Liquid migration from coating colors. *Tappi* *53*(12):2232–2242.

19
THE PENETRATION OF NONAQUEOUS LIQUIDS

TAKASHI KADOYA
MAKOTO USUDA

The University of Tokyo
Tokyo, Japan

I.	Introduction	124
II.	Forced Penetration of Liquids into Paper	124
	A. Dynamic Compressibility of Paper and Void Fraction Under Mechanical and Hydraulic Load	124
	B. Penetration of Liquids into Paper Under Pressure	126
	C. Penetration of Inks into Paper During Printing Compression	127
	D. Phase Separation in Inks and Coating During Penetration	128
	E. Penetration of Adhesives into Paper	129
III.	Laboratory Methods and Equipment for Measurement of the Penetration of Nonaqueous Liquids into Paper	130
	A. Receptivity of Paper to Printing Ink	130
	B. Reflection Method of Measuring Depth of Ink Penetration into Printing Paper	130
	C. Optical Transmittance of Oil-Penetrated Paper	131
	D. Ink Penetration Immediately After Printing	132
	E. Organic Liquid Penetration Measurement	133
IV.	Measurement of Surface Pore Size Distribution	135
	A. Distribution of Surface Pore Depth by Stylus Method	135
	B. Rate of Lateral Absorption	136
	C. Magnetic Scanning Technique	138
	D. Dynamic Measuring Instrument for Void Fraction	139
	References	140

I. INTRODUCTION

Paper is a multilayer structure of fibers with numerous micro-scale pores. Almost all the pores in paper consist of tortuous capillarylike tubes that are continuously interconnected. Therefore, paper is permeable and can absorb liquids and gases. The most obvious utilizations of the tortuous capillary structure of paper are filter paper, blotting paper, and so on. Furthermore, paper-converting operations involve penetration into the voids of paper by coatings, adhesives, and inks. In these processes, liquids such as wax, paraffins, silicones, and oil-based ink are applied to paper in non-aqueous form, whereas latex, silicon emulsions, and water-based inks and most adhesives are basically aqueous liquids.

The fundamental theory of the wetting and penetration phenomena of paper holds for both aqueous and nonaqueous liquids. However, in the aqueous case, fiber swelling and wetting cause deviations from the classic theory of the penetration of porous media. This chapter includes a description of the penetration of oil-based inks into paper as an example of the typical penetration phenomenon for nonaqueous liquids. The wetting and penetration of aqueous liquids are discussed in Chap. 18.

II. FORCED PENETRATION OF LIQUIDS INTO PAPER

In the printing and converting processes of paper, when liquids are transferred to the sheet surface, the liquids are forced to penetrate into the paper under the pressure exerted by the nip rolls. Therefore, first it is important to examine the relationship between the dynamic compressibility and surface profile of paper as well as the depth of penetration of liquids under compressive conditions.

A. Dynamic Compressibility of Paper and Void Fraction Under Mechanical and Hydraulic Load

The compression of paper in a calendering nip has been expressed by Colley and Peel [5] and checked by Robertson and Haglund [21] and Baumgarten and Göttsching [1].

$$\frac{\Delta t}{t_0} = \frac{t_0 - t}{t_0} = A(1 + \tanh \mu) \tag{1}$$

where

t_0 = Initial thickness
t = thickness under pressure
A = empirical constant

and μ is expressed in the following form:

$$\mu = \alpha \log p_{max} + \beta \log T + \gamma M + \delta \theta + \varepsilon \tag{2}$$

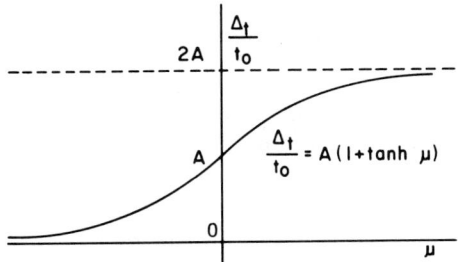

Fig. 1 Master creep relationship for the compression of paper. (Modified from Ref. 16a.)

where α, β, γ, δ, ε are positive constants that vary according to the type of paper.

Equation (2) shows that the variable μ is a function of the maximum applied pressure p_{max}, dwell time T, sheet moisture M, and sheet temperature θ. Figure 1 shows the relationship between μ and $(t_0 - t)/t_0$. The permanent compression $(t_0 - t)/t_0$ is represented by a curve that is asymptotic to zero (no compression because μ is too small) and 2A (no compression upon increasing μ because the web has reached its limiting density). In theory, the same form of compression occurs in a printing press nip.

Mardon [17] has developed two dynamic compressibility testers for the purpose of relating the void fraction of paper to dynamic compression load. The compressibility and compliance of paper depend upon the fraction of sheet volume represented by the void space in the paper and upon the stiffness of the network of fibers. The compliance of paper is defined as the reciprocal of elastic modulus in compression. From a knowledge of the proportions and densities of the paper constituents and of the caliper of the paper, the approximate void space can be calculated for any sheet. The compliance is estimated by the electrical signal from the dynamic

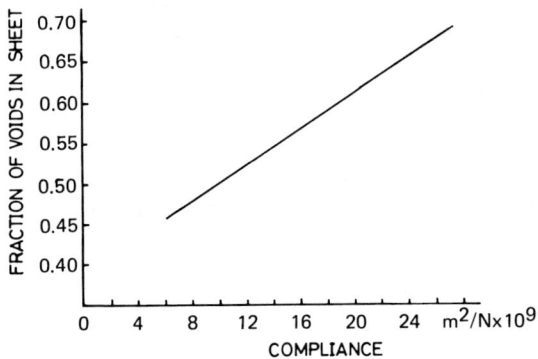

Fig. 2 Compliance as a function of sheet void space. (Modified from Ref. 17.)

compressibility tester as the average slope of the initial compression part of the curve.

The relationship between the fraction of voids present and compliance is shown in Fig. 2 as a common line through the points obtained from testing paper from successive nips of 10 different calender stacks on grades ranging from paper containing 70% groundwood through 70% hardwood kraft to 70% softwood kraft fiber. The slope of the common line shows that as the fraction of void decreases, the fiber network becomes progressively less compliant. From this general tendency and with a knowledge of the void fraction of a given sheet, the compliance can be predicted.

B. Penetration of Liquids into Paper Under Pressure

In order to calculate the penetration depth of liquids into paper under pressure, Poiseuille's law is applied to the penetration phenomenon, as in the case of the derivation of the Lucas-Washburn equation. If an external pressure P is applied, the penetration of a liquid into the porous structure is expressed as follows:

$$\frac{dh}{dt} = \frac{r^2 \Delta P}{8 \eta h} \tag{3}$$

Solving Eq. (3) yields

$$h = \left(\frac{r^2 t \Delta P}{4 \eta} \right)^{1/2} \tag{4}$$

Equation (4) is the Lucas-Washburn equation,

where

    r = radius of capillary
    h = penetration depth
    η = viscosity of the liquid
    t = penetration time

For cases where t and h are small in value, Davidson [7] has modified Eq. (3) for increments dh and dt and obtained the following equation:

$$(\Delta h)^2 = \frac{r^2 \Delta P \Delta t}{8 \eta} \tag{5}$$

Since the Lucas-Washburn equation is not suitable for penetration of liquids into paper under pressure, another equation is derived by combining Darcy's law with Kozeny-Carman's equation,

$$h^2 = \frac{2 r^2 t P}{\eta \kappa} \tag{6}$$

where κ (constant) is approximately equal to 6 for paper.

Equation (6) is similar to Eq. (4). It means that the penetration depth of a liquid into paper under external pressure is proportional to the radius of the capillary and to the square root of penetration time and pressure applied; the penetration depth is inversely proportional to the square root of the viscosity. (A discussion of effective radii of capillaries may also be found in Chap. 17.)

C. Penetration of Inks into Paper During Printing Compression

The penetration of inks into uncoated paper was studied by Hsu [9] using a platen printing device, by which ink was pressed against paper with pressures up to 10^7 Pa and dwell times down to 0.01 s. Hsu found that the penetration changed with viscosity and dwell time, as would be expected for laminar flow. The effect of pressure was similarly explained, except that a correction for compression of the paper was required. If Darcy's law is applied to penetration of ink into paper, then for a rolling nip it is found that

$$B^2 = \frac{2\Pi\varepsilon}{\eta} \int_{-\infty}^{\infty} p \, dt \qquad (7)$$

where

B = penetration
p = local pressure in ink film over any point on paper surface at any instant t during its passage through nip
Π = intrinsic permeability
ε = porosity of paper surface
η = viscosity of ink

The line load L in the nip is given by

$$L = \int_{-\infty}^{\infty} p \, d\ell \qquad (8)$$

where ℓ is the distance measured through the nip.
Since we may write $(dt/d\ell) \, d\ell$ for dt, then

$$B^2 = \frac{2\Pi\varepsilon L}{\eta u} \qquad (9)$$

where u is the printing speed.

It is possible to estimate the intrinsic permeability Π in Eq. (9) from the Bendtsen air permeability [19]. Karttunen [11] presented ink transfer parameters for four inks printed at 1 m/s and 4 m/s onto newsprint, but he did not quote the intrinsic permeability Π. It is interesting to find—by working back from the apparent density of the paper (0.753 g/cm^3), the basis weight (60.2 g/m^2), and the values assumed in the above analysis—that the Bendtsen permeability is roughly 42 mℓ/min, which is low but not unreasonable [20].

D. Phase Separation in Inks and Coating During Penetration

In letterpress printing, when the volume of ink on the printing plate is sufficient, the simplest form of ink transfer equation may be written

$$Y = b + f(x - b) \tag{10}$$

where

Y = ink transferred to paper (g/m^2)
f = parameter describing ink film split
b = parameter describing ink immobilization
x = initial amount of ink on printing plate (g/m^2)

The above equation implies that when a print is made using thickness x of ink on the printing plate or coating roll, a constant quantity b is immobilized in or on the paper surface, and a fraction f of the remaining ink also splits so as to remain on the paper, giving total transfer Y. This holds only for films of ink that are thick enough to completely separate the plate or roll from the paper throughout impression. Under such conditions the entire impression pressure is transmitted to the paper by the ink film. Such a thick ink film not only makes the immobilization independent of the ink film thickness, but also ensures that the impression pressure is applied uniformly to both the troughs and the crests of the paper surface. The fractional ink transfer Y/x is described as follows:

$$\frac{Y}{x} = \frac{fx}{x} + \frac{b(1-f)}{x} \tag{11}$$

When x is large, Eq. (11) tends to f.

After the transfer of ink or coating color onto the paper, the vehicle penetrates into the paper due to the surface tension between liquid and substrate. The penetration continues until the remaining liquid is immobilized in the substrate. In the case of roll coating, paper is compressed by the nip pressure of the roll, resulting in decreased pore sizes and volume. After the paper passes through the roll nip, the pore sizes again increase, owing to the elastic recovery of the paper, and the liquid that has penetrated remains on the pore surfaces as thin films. This is evidenced by the fact that pigment can penetrate up to 20 to 30% of the thickness of paper. After printing or coating, most of the pigment remains in the penetrated layers and the vehicle continues to spread into the paper structure by capillary force [13]. The penetration depth T of coating liquids into porous structures is theoretically expressed as follows [18]:

$$T^2 = \left(\frac{V}{\pi \bar{r}^2}\right)^2 = \frac{2\bar{r}\gamma \cos\theta + P_c \bar{r}^2}{4\eta} t \tag{12}$$

where

> V = volume of coating liquid that has penetrated at time t
> $\bar{r}$ = mean effective pore radius of structure
> γ = surface tension of coating liquid
> θ = contact angle between coating liquid and structure
> P_c = pressure on coating liquid
> η = viscosity of liquid
> t = penetration time

If $P_c = 0$, the penetration depth without pressure can be calculated from Eq. (12). Note, however, that Eq. (12) does not take into account the influence of adhesive and gravitational forces.

The rate of penetration of a liquid depends on wettability unless the pressure applied to the liquid is sufficiently large. In the latter circumstance, Eq. (12) becomes independent of wettability.

$$V = \frac{1}{8} \frac{\pi r^4}{\ell} \frac{\Delta P}{\eta} \tag{13}$$

where

> ΔP = pressure difference
> η = Newtonian viscosity
> ℓ = length of capillary
> r = radius of capillary

E. Penetration of Adhesives into Paper

Many properties of an adhesive are important from a practical standpoint. Some of these are solids content, wetting properties, type of solvent, alkalinity, rheological behavior, and setting rate. The penetrating property of the adhesive is very important. The properties that promote penetration are long wet life, small angle of contact, low viscosity, and low surface tension. A simple test for measuring the penetration of an adhesive is based upon the extent of penetration through a stack of filter paper under pressure [4].

A measured quantity of adhesive is placed on a stack of four or five sheets of Whatman no. 4 filter paper and the stack is then pressed at 686 kPa (7 kg/cm^2) for 15 min. Wetting agents increase the speed of wetting but do not necessarily increase the penetration. Clays are sometimes used in adhesives to increase solids content, but at the same time they decrease tack and thus must not be used in excess. Clays tend to increase viscosity, but specially modified clays, which have been treated to make the particles more hydrophilic, can decrease viscosity.

III. LABORATORY METHODS AND EQUIPMENT FOR MEASUREMENT OF THE PENETRATION OF NONAQUEOUS LIQUIDS INTO PAPER

A. Receptivity of Paper to Printing Ink

A procedure described in TAPPI T 462 su-72 (Fig. 3) provides a measure of the absorbancy rate of paper for printing inks having oil vehicles. However, the method is normally used only for very permeable papers such as newsprint, book paper, and mimeo bond. This test was historically one of the first to show correlations between printing and penetration behavior for uncoated papers.

B. Reflection Method of Measuring Depth of Ink Penetration into Printing Paper

In the case of paper printed with black ink on one side, the ink penetration depth can be calculated by continuously measuring reverse-side reflectance [9]. If the reflectance from the printed paper surface R_g is zero and the thickness of the unpenetrated fraction of the paper is χ, the reflectance R of the unprinted paper surface is

$$R = \frac{1}{a + b \coth bS\chi} \qquad (14)$$

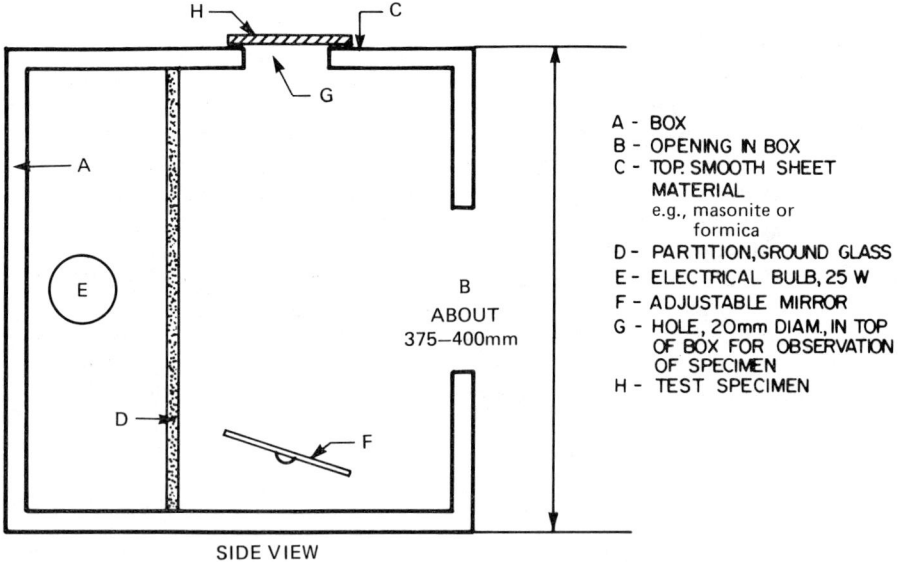

Fig. 3 Apparatus for printing ink penetration test of paper. (From Ref. 21a. Copyright © 1972, TAPPI. Reprinted with permission.)

where

$$a = 1/2 \left(R_\infty + \frac{1}{R_\infty}\right)$$

$$b = -1/2 \left(R_\infty - \frac{1}{R_\infty}\right)$$

$$\coth u = \frac{(e^u + 1/e^u)}{(e^u - 1/e^u)}$$

S = light scattering coefficient of paper

When R_0 is the reflectance of paper backed with a standard black plate of $R_g = 0$, S is obtained from Eq. (15).

$$S = \frac{1}{bc} \coth^{-1} \frac{(1 - aR_0)}{bR_0} \qquad (15)$$

Then χ is calculated from Eq. (14) for a given R.

$$c - \chi = h$$

where

    c = thickness of original paper
    h = ink penetration depth into paper

C. Optical Transmittance of Oil-Penetrated Paper

It is difficult to measure the depth of penetration of liquid from the surface of paper because the thickness of paper is only about 100 μm. However, Larocque [12] has calculated the penetration depth of an oil film of uniform thickness D_0 on the surface of the sheet, assuming that it starts to penetrate into the paper at time 0 and penetrates to depth h at time t (Fig. 4).

An oil layer of thickness D remains on the surface, that is, the original thickness D_0 is separated into D and h after t seconds, where h is the thickness of the oil-penetrated layer of the paper. Then the relationship between the intensity of incident and transmitted light (I_0 and I, respectively) can be described by the Lambert-Beer law:

$$I = I_0 \exp[-\mu_1(c - h) - \mu_2 h - \mu_3 D] \qquad (16)$$

where

    c = caliper of paper
    μ_1 = light absorption coefficient for paper
    μ_2 = light absorption coefficient for oil-penetrated paper
    μ_3 = light absorption coefficient for oil

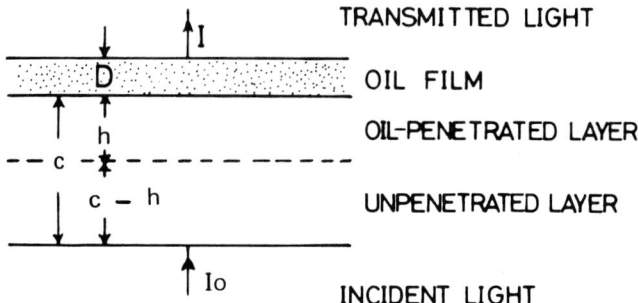

Fig. 4 Optical transmittance of oil-penetrated paper.

On the other hand, the depth of oil penetration h can be shown by the Lucas-Washburn equation,

$$h = Kt^{1/2} \tag{17}$$

Let W be the void fraction; then we obtain the oil film thickness remaining on the surface, D.

$$D = D_0 - WKt^{1/2} \tag{18}$$

From Eqs. (16) and (17)

$$K \doteq c \frac{d(\ln I)/d(t^{1/2})}{\ln I_{p'} - \ln I_p} \tag{19}$$

where

I_p = transmitted light intensity of paper without oil absorption
$I_{p'}$ = transmitted light intensity of paper with oil absorption

$d(\ln I)/d(t^{1/2})$ can be calculated from the slope of the plotted curve of $\ln I$ versus $t^{1/2}$.

D. Ink Penetration Immediately After Printing

The penetration into paper of pigment particles in ink is often hindered by the filtration mechanism of the paper structure. The rate of oil is not always the same as that of ink. Especially when the oil is more quickly absorbed into the paper from the ink, the concentration of pigment becomes so high that the pigment remains behind the oil front. Therefore, many workers stress that it is important to measure the penetration rate of ink immediately after printing.

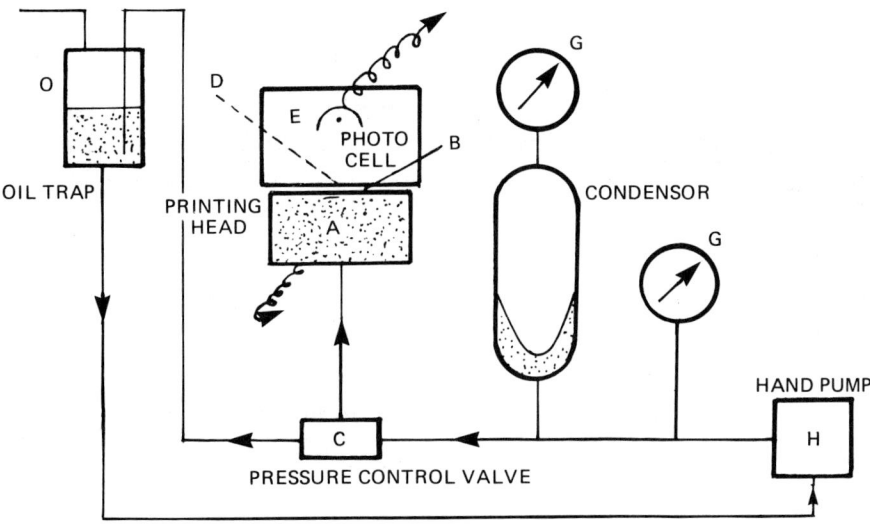

Fig. 5 Apparatus for measurement of ink penetration. (Modified from Ref. 6.)

Coupe and Hsu [6] developed an instrument that measures the penetration of varnish and ink into paper immediately after printing. Figure 5 shows the apparatus for the measurement of ink penetration. A is a printing head covered with a rubber blanket charged with ink. The paper strip B is attached to a glass plate and pressed by compressed oil. The paper is printed very quickly under the action of the oil pressure. D is incident light that strikes the unprinted paper surface at 45°. The reflected light is detected by a photocell E. The signal from the photocell and a printing pressure sensor can be measured simultaneously by a dual channel oscilloscope. Maximum printing pressure and minimum pressing time of the apparatus are 8.33 MPa (85 kg/cm^2) and 0.02 s, respectively.

E. Organic Liquid Penetration Measurement

The penetration volumeter [22] has been developed to measure penetration in a layered structure (Fig. 6). The apparatus consists of a calibrated capillary F in which a drop of colored alcohol G is free to move. The paper sample B is clamped on top of a container A by a wirelike holder C. The container is half filled with an oil of petroleum base of known viscosity and surface tension. As soon as the container is turned upside down (180°) by means of the handle D, the oil E spreads over the surface of paper sample and the penetration process begins. Simultaneously, a stopwatch is started. The air expelled by the penetrating oil drives the alcohol drop deeper into the capillary F, and its position G is measured at different time intervals. In

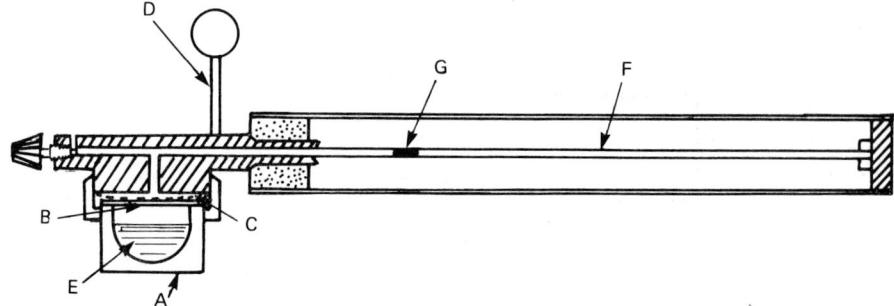

Fig. 6 Schematic of penetration volumeter. (Modified from Ref. 22.)

this way the volume of oil that has penetrated can be plotted against the square root of time. The relationship between oil penetration volume V and time t can be described as follows:

$$V = K \left(\frac{\gamma t}{\eta} \right)^{1/2} \qquad (20)$$

where

K = a constant for number and size of capillaries in paper
γ = surface tension of oil
η = viscosity of oil

The oil-penetration characteristics of the paper are described by the K value. If the penetrating oil meets a layer of different porous structure in paper, the curve of V versus $t^{1/2}$ will show a kink.

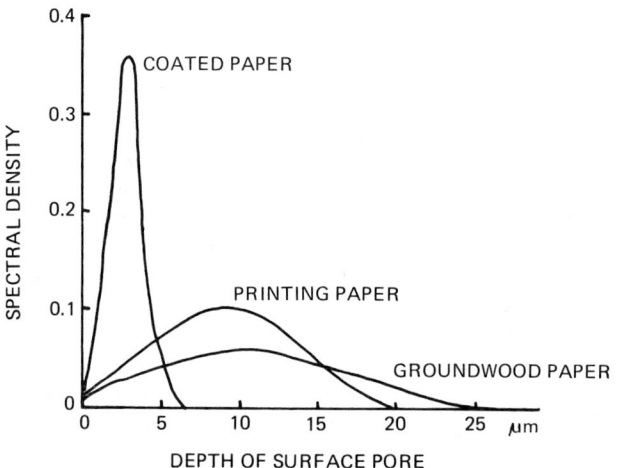

Fig. 7 Distribution of surface pore depths.

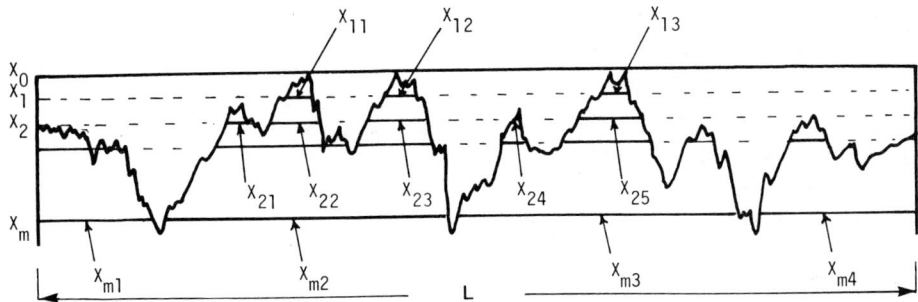

Fig. 8 Schematic to calculate distribution of pore depths.

IV. MEASUREMENT OF SURFACE PORE SIZE DISTRIBUTION

A. Distribution of Surface Pore Depth by Stylus Method

The distribution curve of surface pore depth (Fig. 7) can be calculated from the surface roughness of paper measured by the stylus method [10]. When the pore depth distribution is expected to be normal, the surface roughness diagram is divided by some parallel and equidistant lines $(x_1, x_2, x_3, \ldots, x_m)$, as shown in Fig. 8. The base line x_0 is positioned to correspond to the highest peak of the curve. The term x_{ij} describes the width of the j-th peak cut with line x_i, and Σx_{ij} divided by length L gives y_i, the relative cumulative frequency of surface pore depths.

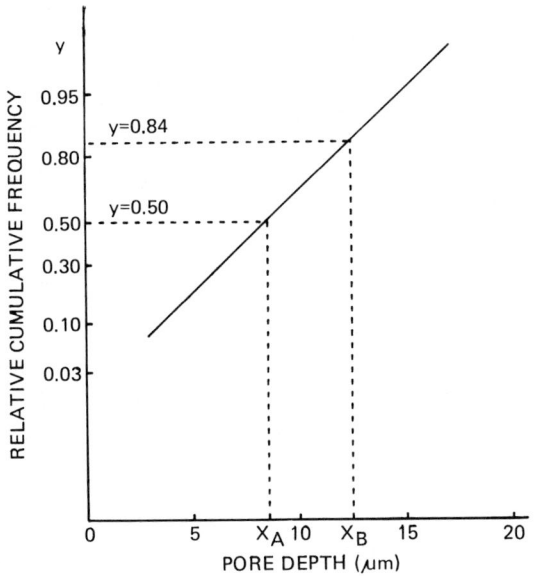

Fig. 9 Surface pore depth distribution of fine paper.

$$y_i = \frac{1}{L} \sum_{j=1}^{n} x_{ij} = \frac{(x_{i1} + x_{i2} + x_{i3} + \cdots + x_{in})}{L} \qquad (21)$$

When the data are replotted using a probability scale, they are in linear form as in Fig. 9. From this diagram, we can read both x_a and x_b corresponding to y = 0.50 and 0.84, respectively. If one takes $\mu = x_a$ and $\sigma = x_b - x_a$, a distribution function $N(\mu, \sigma^2)$ can be obtained where μ is the average of the distribution and σ^2 is the variance.

When the stylus pressure is moderate, the thickness of paper is measured rather than the surface roughness. Furthermore, as paper has resiliency, the surface pore depth distribution of paper under printing pressure differs from that of paper free from printing pressure.

A method that uses an optical contact device such as the Chapman tester [3] is useful for detecting the presence of surface pores. The method, however, is apt to place too much emphasis on small pores because it does not take into account the depth of the pores.

B. Rate of Lateral Absorption

Figure 10 shows an apparatus developed by Ernst and Tollenaar [8], in which the liquid ascent can be observed. A sample strip B is clamped to the holder A that is equipped with two glass scales C and C'. The stopwatch D is set up so as to start at the instant the sample is dipped into the liquid by

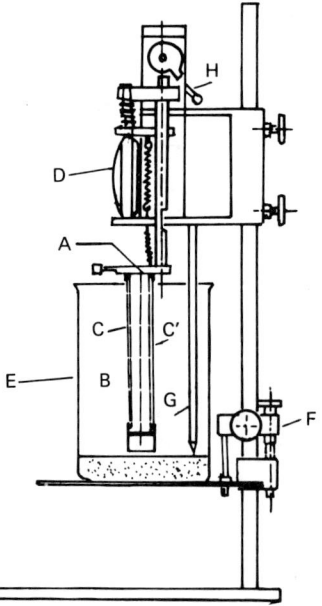

Fig. 10 Apparatus for the measurement of height of liquid rise in paper. (Modified from Ref. 22.)

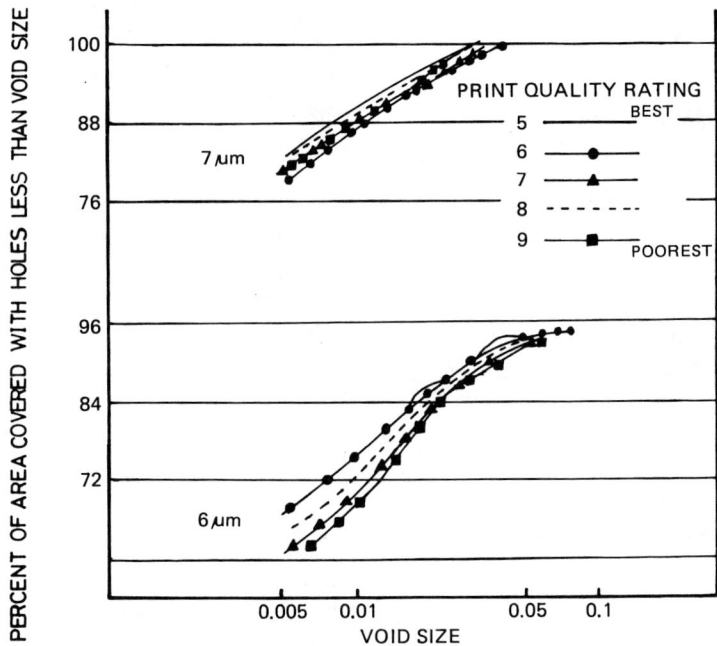

Fig. 11 Void distribution of newsprint with print quality ratings 9 to 5 on wire side. [From Ref. 15]

lowering the lever H. The height h of liquid rise at time t is read out from the glass scales.

As it has been observed that the relationship between h and t obeys the Lucas-Washburn equation (22), a plot of h versus $t^{1/2}$ is linear.

$$h = \left[\frac{\bar{r}\gamma(\cos\theta)t}{2\eta}\right]^{1/2} \quad (22)$$

where

$\bar{r}$ = mean pore radius
γ = surface tension
θ = contact angle
η = viscosity of liquid

The slope of this curve is defined as the rate of absorption. When oil is used, $\cos\theta$ has a value of nearly 1. The mean pore radius $\bar{r}$ of the capillary system can in that case be obtained from Eq. (23).

$$\bar{r} = \frac{2\eta h^2}{\gamma} \quad (23)$$

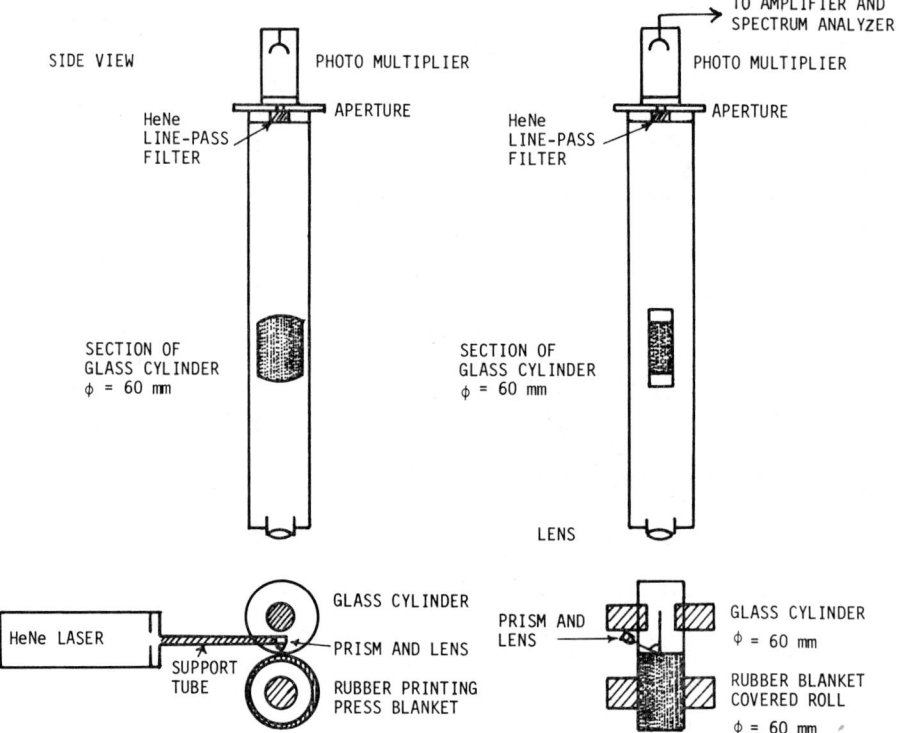

Fig. 12 Optical system of instrument for surface void size distribution measurement. (From Ref. 16.)

C. Magnetic Scanning Technique

Lyne and Copeland [14,15] have developed a new apparatus for measuring the voids in the surface of a sheet under compression. The magnetic scanning technique can be summarized as follows.

The surface voids in paper can be filled with magnetic ink by means of a drawdown blade or blades under local pressure corresponding to those in the printing nip. The ink used must flow readily under shear but become immobilized once the shear force is removed. After the ink drawdown, the specimens of paper in this apparatus pass under a write head of a special tape recorder at low velocity. A low frequency pulse is recorded in the ink. The specimen is then passed under a read head at higher velocity, and the pulses recorded in the ink are displayed on an oscilloscope or are fed to an electronic analyser which prints out the desired parameters.

The height of pulses above the zero level is proportional to the depth of magnetic ink in the surface void. However, as manual analysis of the oscilloscope traces is too time consuming to make the method practical, an electronic digital counter is used.

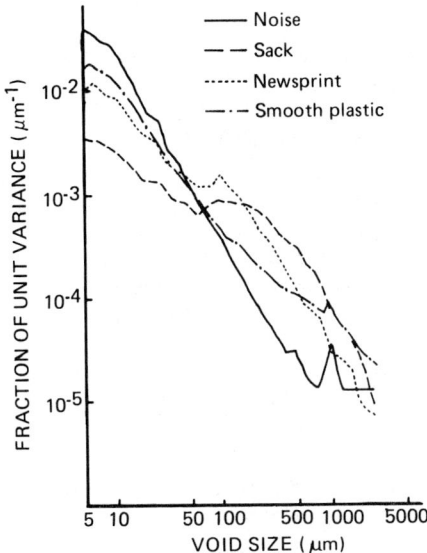

Fig. 13 Spectral diagram of void sizes for sack paper, newsprint, and smooth plastic. The spike at 1000 µm in the white noise curve is a 50 Hz disturbance in the white noise generator. (From Ref. 16.)

With a reproducible scanning technique established many applications of the method have appeared. For example, it has been applied to the wire side of newsprint samples which were printed and rated by subjective print quality evaluation as 9, 8, 7, 6, and 5 (the highest number being the poorest print). Figure 11 shows the relation of void distribution for 6 and 7 µm below the surface.

D. Dynamic Measuring Instrument for Void Fraction

Lyne [16] and Blokhuis [2] have developed dynamic measuring instruments for the optical contact of paper under printing compression. Lyne's instrument (Fig. 12) also permits measurement of the void size (width) distribution. It continuously scans paper during its passage between a glass roll and steel roll covered with a printing press blanket material. Laser illumination of the contact region in the nip is arranged such that light is reflected to an optical sensor from those areas of the paper surface that are in contact with the glass cylinder. The signal from a scanning spot 5 µm in diameter is fed to a spectrum analyzer and computer for reduction to a display of the variance distribution of the lengths of voids lying along the scan line.

The portion of the paper surface that contacts the glass roll, or dynamic contact fraction, can also be obtained simultaneously if the signal is first normalized to one volt between contact and noncontact. The time average amplitude of the signal is then equal to the dynamic contact fraction. Typical results for sack paper, newsprint, and smooth plastic are illustrated in Fig. 13.

REFERENCES

1. Baumgarten, H. L., and Göttsching, L. (1975). The influence of calendering parameters on the thickness changes of some printing papers. Preprint. Symposium on the Calendering and Supercalendering of Paper at University of Manchester Institute of Science and Technology, organized by J. D. Peel, R. J. Kerekes, and H. L. Baumgarten, Manchester, England, Sept. 1975.
2. Blokhuis, G., and Kalff, P. J. (1976). Dynamic smoothness measurements of papers and print unevenness. *Tappi* 59(8):107–110.
3. Chapman, S. (1954). The Chapman printing smoothness tester. 1. Basic development and recent modifications. *Pulp Paper Mag. Can.* 55(4):88–93.
4. Claxton, A. W. (1957). The functional behavior of clays in adhesive. *Tappi* 40(6):180A–186A.
5. Colley, J., and Peel, J. D. (1972). Calendering processes and the compressibility of paper. *Paper Technol.* 13(10):350–357.
6. Coupe, R. R., and Hsu, B. (1960). Penetration of varnishes and inks into paper under pressure. *J. Oil Color Chem. Assn.* 43:720–736.
7. Davidson, J. W. (1971). The effect of ink pressure on the sorption of ink into newsprint. In *Recent Developments in Graphic Arts Research*, Advances in Printing Science and Technology, (W. H. Banks, ed.), vol. 6, Pergamon, Oxford, pp. 185–190.
8. Ernst, P. A. H., and Tollenaar, D. (1965). Studies of the absorption of liquid into uncoated paper. *Das Papier* 19(9):515–517 (in German).
9. Hsu, G. (1962). Some observations on the ink-paper relationship during printing. In *Problems in High Speed Printing*, Advances in Printing Science and Technology, (W. H. Banks, ed.), vol. 2, Pergamon, Oxford, pp. 1–10.
10. Ito, Y., Murakami, I., and Saito, N. (1966). Studies of print quality. 1. Influence of paper and ink properties on the print gloss. *Res. Bull. Printing Bur. (Japan)* no. 1, pp. 1–20.
11. Karttunen, O., and Oittinen, P. (1972). Models for wet-on-wet ink transfer. *Graphic Arts in Finland* 1:1, 9–22.
12. Larocque, G. L. (1938). General laws governing the oil resistance of paper in printability. *Pulp Paper Mag. Can.* 39:C106–C119.
13. Levlin, J-E., and Nordman, L. (1967). On the penetration of ink into paper. In *Paper in the Printing Process*, Advances in Printing Science and Technology (W. H. Banks, ed.), vol. 4, Pergamon, Oxford, pp. 33–55.
14. Lyne, L. M., and Copeland, D. E. (1968). Print quality evaluation by magnetic scanning. *Tappi* 51(8):363–370.
15. Lyne, L. M., and Copeland, D. E. (1971). Improvements in the magnetic scanning technique for printability. *Pulp Paper Mag. Can.* 72(10):67–72.
16. Lyne, M. B. (1976). Measurement of the distribution of surface void sizes in paper. *Tappi* 59(7):102–105.

16a. Lyne, M. B. (1978). The effect of moisture and moisture gradients on the calendering of paper. In *Fibre-Water Interactions in Paper-Making*, (F. Bolam, ed.), British Paper and Board Industry Federation, London, pp. 641–665.
17. Mardon, J. W. (1966). Dynamic consolidation of paper during calendering. In *Consolidation of the Paper Web* (F. Bolam, ed.), British Paper and Board Makers Association, London, pp. 576–626.
18. Olsson, I., and Phil, L. (1954). Penetration of ink into paper. *Int. Bull. Printing Allied Trades* 67(Jan.):19–24.
19. Parker, J. R. (1966). Effects of dry pressing on printing properties of uncoated paper webs. In *Consolidation of the Paper Web* (F. Bolam, ed.), British Paper and Board Makers Association, London, pp. 959–976.
20. Parker, J. R. (1976). Fundamental paper properties in relation to printability. In *The Fundamental Properties of Paper Related to Its Uses* (F. Bolam, ed.), British Paper and Board Industry Federation, London, pp. 517–543.
21. Robertson, G., and Haglund, L. (1974). Local thickness reduction in a calender nip. *Svensk Papperstidn.* 77(14):521–531.
21a. TAPPI T 462 su-72, Castor-Oil Penetration Test for Paper, 1972.
22. Tollenaar, D. (1967). Capillarity and wetting in paper structures: Properties of porous systems. In *Surface and Coatings Related to Paper and Wood* (R. H. Marchessault and C. Skaar, eds.), Syracuse Univ. Press, pp. 195–219.

20
ABSORBENCY OF TISSUE AND TOWELING

HOLGER HOLLMARK

Swedish Forest Products Research Laboratory
Stockholm, Sweden

I.	Introduction	143
II.	Theory	144
	A. Conditions for Wetting	145
	B. Wetting Kinetics	148
III.	Methods of Evaluation	152
	A. Absorption Rate	152
	B. Absorption Capacity	161
IV.	Factors Affecting Absorption Rate	162
	A. Pulp Manufacture and Stock Preparation	162
	B. Sheet Forming and Creping	164
	C. Aging	164
	References	165

I. INTRODUCTION

The quality requirements for tissue products have to be considered in relation to the particular type of product that is being evaluated. The importance of various properties may vary from product to product. Apart from the fact that the converting operation creates some relatively well defined quality requirements based on runnability criteria, the consumer's point of view is the most important criterion.

Disposable paper products are generally considered to be substitutes for durable items. Accordingly, sanitary tissue paper is thought of as a substitute for textile products. Facial tissue, kitchen towels, napkins, and paper handkerchiefs are examples. Therefore, irrespective of their convenience in handling, sanitary tissue products have to compete with the inherent advantages of textile materials. One of the features of a textile material is that it retains its strength upon wetting. Once wet, a textile cloth may exhibit a very rapid uptake of water. A disposable paper product has a limited wet strength and has to be used initially dry. Therefore, the rate of absorption is often limited because the wetting time is extended. However, when the two products are compared on a dry basis, the disposable paper generally shows a faster rate of wetting.

Apparently, the absorbency properties of a sanitary tissue are to a large extent governed by the surface chemistry of its fibers. Thus the manufacturer of absorbent tissue must match the following elements in order to achieve optimum absorbency of the product:

- Pulp furnish
- - Chemistry of the paper machine white water
- - - Additives

In the ensuing pages, absorption rate and absorption capacity are discussed as two separate concepts. Absorption rate is considered to be the more important of the two properties. For example, the effectiveness of products used for wiping (e.g., kitchen towels) should mainly be related to the rate of absorption. On the other hand, a certain minimum absorption capacity is considered necessary to maintain desired wipe-dry properties of a cloth.

Most of the considerations in this chapter refer to the absorption of water. Absorption of other liquids (e.g., organic compounds) might be of practical interest in some cases. The principles of surface chemistry will generally predict a different behavior for the absorption of low surface tension liquids compared with that of water.

II. THEORY

A brief summary of the general theory of wetting and absorption will be given in this chapter. For general insight into the phenomena associated with the penetration of liquids into paper, readers are referred to Chaps. 18 and 19. Emphasis here will be placed on the areas that are particularly relevant to the absorption characteristics of tissue and similar products; we will also discuss some theoretical problems that are incompletely solved and may have some bearing on tissue absorbency.

A porous body is characterized by a nonhomogeneous distribution of solid and fluid material. The porous body is said to contain pores; the porosity is the ratio of pore volume to the overall volume of the body. Various types and categories of pore systems have been treated in this book by Murakami and Imamura (Chap. 17) and elsewhere in the literature [31,46]. In some types of porous materials, the pores are interconnected in such a way that transport of fluid material in one or more directions through the entire body is possible. Many practical applications exist of liquid or gaseous flow through porous bodies. Since the absorption of liquid into paper is the subject of this chapter, we shall restrict ourselves to the case where a gas is displaced by a liquid in paper, particularly absorbent paper.

When flow through porous media is discussed, it is often practical to consider forced flow and spontaneous flow separately. In the first case, an external pressure differential is applied which will cause the flow of the liquid. In the latter case, which is referred to as capillary flow, no external pressure differential is applied. The driving force of spontaneous flow comes from the interactions between the molecules of the various phases present. The conditions for spontaneous flow are given by

- The surface energies of the phases present
- • The geometrical structure of the pore system
- • • Gravity

The first two factors will be further treated below. In practice, capillary flow often occurs against gravity. The effect of gravity will be less pronounced when the mean cross-sectional areas of the flow channels are small. This is so because the driving force, which is proportional to the area of interface between the solid and fluid materials, is then larger in relation to the weight of raised liquid volume. In the absence of gravity, spontaneous flow will occur irrespective of the dimensions of the available flow channels [48].

A. Conditions for Wetting

In a three-phase system, where two fluid phases are in contact with a solid surface, a contact angle referring to one of the fluid phases can be defined, provided that the solid surface is smooth and without discontinuities (Fig. 1). The contact angle is considered to be a measure of the interaction between the molecules of the phases present. The contact angle will be small for a fluid phase whose molecules have a high affinity for the solid surface compared both with internal molecular cohesion and with the affinity of the second fluid phase to the solid surface [12]. It has long been anticipated that the surface tensions of the free energies of the interfaces can be related to each other and to the contact angle by means of the Young equation:

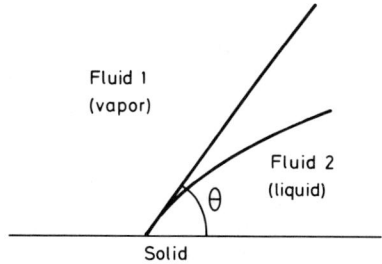

Fig. 1 Equilibrium configuration around a three-phase boundary line. (Shape of liquid/vapor interface influenced by gravity.)

$$\gamma_{SV} = \gamma_{SL} + \gamma_{LV} \cos \theta \qquad (1)$$

where

γ_{SV} = surface tension of solid/vapor interface
γ_{SL} = surface tension of solid/liquid interface
γ_{LV} = surface tension of liquid/vapor interface
θ = contact angle

Various attempts have been made to relate the surface energies to forces between molecules. The interfacial tension γ_{12} is thought to be composed of the cohesive tensions in the two phases and the dispersion forces between different molecules in the following manner [12]:

$$\gamma_{12} = \gamma_1 + \gamma_2 - 2\sqrt{\gamma_1^d \gamma_2^d} \qquad (2)$$

An adsorbed gas will affect the surface energy of the solid phase. Therefore a *spreading pressure* π_e is defined with reference to the Gibbs adsorption isotherm:

$$\pi_e = \gamma_S - \gamma_{SV} = RT \int_0^{P_0} d \ln p \qquad (3)$$

where

p = vapor pressure of the adsorbate
γ_S = surface tension of the solid in vacuum

Another useful quantity is the spreading coefficient S. Consider the liquid phase in Fig. 1. Suppose that the liquid has a tendency to spread on the solid phase. Work is done against the interfacial tensions γ_{LV} and γ_{SL}, but along with the tension γ_S (in the absence of a gas phase)

$$S = \gamma_S - \gamma_{SL} - \gamma_{LV} \qquad (4)$$

Equations (2), (3), and (4) give

$$S = \pi_e + \gamma_{LV}(\cos \theta - 1) \qquad (5)$$

The spreading pressure π_e is very small for water on a hydrophobic surface. At the same time, $\cos \theta$ in this case is close to zero or negative so that the spreading coefficient S will be negative and no spreading will occur.

Wetting of Capillary Walls The capillary is the simplest model of a porous body. A meniscus within it can usually be assumed to have a spherical shape. If the capillary has parallel walls, the shape of the meniscus will remain constant during the wetting process. When spreading occurs (i.e., the contact angle $\theta = 0°$), the meniscus can be approximated by a half-sphere (see Fig. 1 of Chap. 18). Its radius of curvature will then equal the capillary radius, a fact that will simplify the calculations.

It was stated above that the driving force for the spreading of a liquid on a solid surface originates from the attraction between molecules of the two phases. The forces of attraction will govern the contact angle and generally also the curvature of the interface. Based on the original treatment of Gibbs, it may be derived thermodynamically [32] that for every interface where the sum of the main curvatures is not zero, that is,

$$k_1 + k_2 \neq 0 \tag{6}$$

there exists a pressure difference ΔP, which according to the Laplace equation is

$$\Delta P = \gamma(k_1 + k_2) \tag{7}$$

where γ is the interfacial tension. However, as soon as the system to be described deviates from the most simple models, the solution of the Laplace equation will present serious difficulties. For the simple capillary model, Eq. (7) reduces to

$$\Delta P = \frac{2\gamma}{R} \tag{8}$$

where R is the radius of the capillary.

At equilibrium, the force acting on the meniscus is balanced by the weight of the liquid column. This gives the well-known expression of the height of the liquid column h,

$$h = \frac{2\gamma}{R\rho g} \tag{9}$$

where

ρ = density of the liquid
g = constant of gravity

Alternatively, it is possible, without presuming any particular shape of the meniscus, to arrive at Eq. (9) simply by assuming that a zero contact angle is equivalent to the existence of an adsorbed film of liquid adjacent to the bulk liquid. The surface tension of the liquid tends to reduce the liquid surface area, including the adsorbed film on the capillary wall. This reduction will cause the liquid to rise in the capillary. An elevation dh means that the total liquid/air interface is reduced by $2\pi R\, dh$ and that the free surface energy G is reduced by

$$dG_s = \gamma 2\pi R\, dh \tag{10}$$

The same amount of energy is required to raise the liquid column dh against gravity

$$dG_p = \pi R^2 \rho g h\, dh = dG_s \tag{11}$$

The combination of Eqs. (10) and (11) yields Eq. (9).

Wetting of a Plane Surface The spreading coefficient S can be used to investigate whether a liquid will spread on a plane surface or not. If it does, the spreading will, in the ideal case, be unlimited, so that eventually the entire solid surface will be covered by a liquid film; the thickness of the film will depend on the amount of liquid available.

In many cases, the contact angle will assume a constant, finite value. There are theories for the equilibrium configuration of the liquid in such cases. The degree of wetting can then be evaluated for individual droplets on a plane surface and also for dry patches in an unlimited film of liquid [18].

The effect of surface roughness on wetting has been studied by several researchers [11,23]. Recently, Mason and coworkers [17] have undertaken a rigorous treatment—theoretical as well as experimental—of the influence of surface roughness on spreading behavior. They were able to demonstrate that small irregularities and sharp-edged discontinuities in the surface have a much larger effect on the shape of the contact line than was previously realized.

Wetting of Cylindrical Rods Calculation procedures for the shape of the meniscus around one or more cylindrical rods have been suggested [18,35,36]. A film of liquid covering a solid cylindrical rod is not stable if its thickness exceeds a certain critical value in relation to the diameter of the rod. The liquid then tends to contract itself into a number of droplets with unduloid shape [33,38].

In those cases where the amount of liquid is very large in relation to the diameter of the rod, the influence of the contact angle on the configuration of the liquid is small [43]. A certain amount of liquid has to be present between the droplets—apparently a very thin film. The thickness of this film is comparable to an adsorbed layer. The phenomenon may be of some importance for the adsorption of liquid into porous systems made of, for instance, very thin fibers. A quantitative penetration may then be prevented, even though the contact angle is zero and spreading may occur. The phenomenon is probably partly responsible for the fact that some plants and animals that are covered by hair or feathers may show extreme water repellency.

B. Wetting Kinetics

Hydrodynamic Approach The flow of liquid in a capillary tube is generally described by the Poiseuille equation:

$$\frac{dh}{dt} = \frac{R^2 \Delta P}{8 \eta h} \tag{12}$$

where

$\frac{dh}{dt}$ = linear flow rate

R = capillary radius

ΔP = pressure drop across meniscus

h = height of rise
η = viscosity of the liquid

If this equation is combined with Eq. (8), we obtain at zero contact angle

$$\frac{dh}{dt} = \frac{R\gamma}{4\eta h}$$

or

$$h = \sqrt{\frac{R\gamma t}{2\eta}} \qquad (13)$$

This equation can be made more complicated through the introduction of a finite contact angle (the Lucas-Washburn equation—see Eq. (4) of Chap. 19) and by accounting for gravity forces. If the effect of gravity is introduced and the contact angles is still zero we get

$$\frac{dh}{dt} = \frac{1}{8\eta}\left(\frac{2R\gamma}{h} - R^2 g\rho\right) \qquad (14)$$

From this equation, it is evident that for any height of rise h, there exists an optimal capillary radius for which the rate of liquid rise is maximum [1].

Although the combination of Eqs. (8) and (12) yields results that reasonably match the available experimental data, the theoretical justification for these formulations is dubious, which has been pointed out by Schultze [47]. The derivation of the Poiseuille equation [Eq. (12)] assumes an externally applied pressure gradient, whereas the origin of the capillary forces is at the solid/liquid interface.

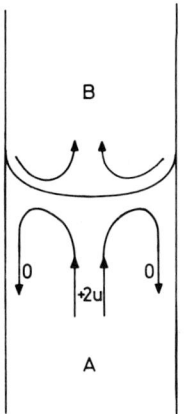

Fig. 2 Movement of a capillary meniscus. The "fountain" effect is shown.

Boundary Layer Flow Real liquids are characterized by a finite viscosity, that is, resistance to shear deformation. In all treatments of liquid flow, the velocity profile perpendicular to the streamlines is such that the velocity close to a solid wall is zero (no-slip).

At interfaces separating two fluids, the conditions are probably different. Apparently, interfacial tangential flow is common. It has been advocated [41] that such flow will not occur in a capillary where the two fluids have the same viscosity and density, and reference has been made to the so-called fountain effect. The fountain effect is demonstrated in Fig. 2 where two immiscible liquids A and B are introduced in a capillary. Suppose that the meniscus moves upward with the velocity +u. It seems to a viewer whose own motion relative to the meniscus is zero that liquid in the center of the capillary moves upward with velocity +u but that liquid near the capillary wall moves downward with the velocity −u. In the area near the interface separating the two phases, there will be a radial flow toward the capillary wall in phase A and away from the wall in phase B.

If the no-slip principle is adhered to, the stream pattern may be confusing. It has been shown [14,19] that even if the no-slip principle is applied only at the solid wall, anomalous flow will result in the region close to the boundary line between the three phases. It is possible that some of this inconsistency is reflected in reality by the occurrence of unpredictably high shear rates in this region. It is therefore also possible that the viscosity of real liquids may be the limiting factor for the rate of capillary flow to a greater extent than is usually expected.

Levine et al. [28] have developed a modified theory of capillary rise, in which they assume that the no-slip principle is not adhered to in the vicinity of the contact line. Instead, the slip velocity is taken to be proportional to the shear stress exerted on the solid wall. A slightly different and more basic approach has been used by Ruckenstein and Dunn [42], who define a chemical potential gradient (on a molecular scale) through which a slip velocity can be calculated. The slip velocity is found to be dependent on the contact angle.

Interfacial flow can be induced by surface tension gradients. This phenomenon is usually called the Marangoni effect. Surface tension gradients may occur in multicomponent systems, for instance through selective evaporation or adsorption. In single-component systems, it may be a result of a temperature gradient.

In a system where a surface tension gradient is present, flow will occur from a region of low surface tension to a region of high surface tension. The flow rate is highest at the interface. However, the interfacial flow is accompanied by a bulk flow penetrating to a certain depth in the liquid. The result of the Marangoni effect is often that a considerable quantity of liquid is transported on a solid surface, even against gravity [30].

Surface tension gradients may occur in many practical cases. Consider, for instance, a cellulosic fiber on the surface of which has been deposited an excess of some surface active substance. When the fiber end is dipped into water, a certain portion of the surface active agent is transferred to the water. The concentration of the surface active agent in the water/air interface is then, momentarily greatest in the vicinity of the fiber. Therefore, a surface tension gradient will occur, directed away from the fiber and generating a flow in the same direction. This flow will—at least temporarily—counteract the spreading of water on the surface of the fiber (which, incidentally, might have been the purpose for the application of the surface active agent.)

Lately, an increasing amount of attention has been directed toward the presence of surface tension gradients and their effect on the spreading of liquids on solid surfaces. This is particularly the case as far as liquids of low surface tension are concerned [4,15]. The observation has been made that two forms of spreading occur, primary and secondary. Primary spreading is represented by a very thin film ($\sim$ 1 µm) of liquid present along the boundry line of an amount of bulk liquid. Secondary spreading consists of considerably larger quantities of liquid that follow the primary spreading. A surface tension gradient is supposed to exist in the transition zone between the primary and secondary spreading areas. Derjaguin and Churaev [8] have developed a theory for the transition region between an adsorbed film and the liquid meniscus. Recently, an analytical treatment of the meniscus shape close to the transition region has been presented [40].

The cause of primary spreading and the cause of the surface tension difference in the two areas (primary and secondary) are a matter of some dispute. Some defend the view that primary spreading occurs through transport in the gas phase [15], while others also advocate the presence of surface diffusion [3]. It has been pointed out that the thickness in the primary spreading area is sufficiently small for some interaction between molecules in the liquid/air interface and the solid wall to take place. This could induce a higher surface tension in the primary spreading area as compared with the secondary spreading area and the surface of the bulk liquid. Some doubt has been expressed, however, that molecular interaction may persist over such long distances, at least for some studied organic liquids [4]. On the other hand, it is well known that long-reaching molecular interaction occurs in water [9].

Liquid Absorption in Composed Systems The most common model of a porous body is that consisting of a number of parallel capillaries. The rate phenomena are subsequently treated according to the Lucas-Washburn equation (p. 149). For the purpose of obtaining a better agreement between theory and experimental results, various models incorporating a pore radius distribution have been tried. In doing this, two approaches, involving 2 different assumptions, can be followed:

- Capillaries of various radii are running parallel [20].
- The radius varies for each capillary along its length [10,49].

Peak and McLean [37] made the observation that a model incorporating a pore radius distribution $\phi(r)$ gave improved agreement between theory and experiments. They assumed a simple pore radius distribution

$$\phi(r) = C \text{ for } R_1 < r < R_2$$

$$\phi(r) = 0 \text{ for } r < R_1 \text{ and } r > R_2$$

with which they were able to calculate the penetration rate of various liquids in paper with great accuracy.

Other models of porous systems in which the material is thought to be composed of spheres or cylindrical rods do not lend themselves very easily to quantitative treatment. On the other hand, a sphere or cylindrical rod model generally represents the real system far better than the capillary model.

Some attempts have been made to explain certain observations by applying a spherical or cylindrical rod model. In an experiment involving penetration into a system of spherical particles wherein the contact angle between the liquid and solid material approaches zero, Brakel and Heertjes [5] observed that there is no sharp borderline between the filled and unfilled portions of the system. Instead, there is a zone of decreasing degree of saturation. When the rising liquid reaches a point of contact between spherical particles, a ring of liquid will be formed around the contact point. If the spreading tendency of the liquid is high, it may be transported on the surface of the solid material between contact points more rapidly than the liquid will fill up the void volume.

In the expression for the spreading coefficient of the liquid [Eq. (4)], the liquid surface tension has a negative sign. This means that low surface tension liquids spread very easily according to the procedure described above. The effect is also promoted by a high surface energy of the solid material. As the spreading coefficient is diminished, surface spreading becomes relatively less a factor than the filling of the void volume between the particles.

III. METHODS OF EVALUATION

A. Absorption Rate

When water is absorbed in paper, the time dependency of the process generally has the appearance shown in Fig. 3. The rate of absorption is highest in the beginning and will subsequently decline. The reason for this will be discussed later. The rate process can be followed by noting the volume rate or by measuring the linear rate of displacement of the liquid/air interface. There seem to be three main causes for the declining rate of absorption:

> Liquid friction. The total flow resistance increases as more liquid is absorbed.
> Gravity. If the penetration of liquid proceeds upward, the height of the liquid will counteract the process so that the rate is decreased.
> Narrow section. If the liquid is supplied to a limited area of the paper and is allowed to spread in all directions, the velocity of the liquid front will decrease rapidly.

When one attempts to measure the rate of water absorption in paper, it is important that the distinction between volume flow rate and linear flow be made clear. It is also advisable that the various causes of rate decrease be identified.

In some cases, the main interest behind the measurement is to achieve a maximum total absorption, that is, the amount of water absorbed after "infinite" time. This quantity is termed absorption capacity. For other applications, one may want to know the rate of absorption. To cite an example, one may need to know the maximum amount of water that can be absorbed during the shortest possible time, or one may need to know the reverse—when a minimum absorption is desired after a certain time of exposure. In

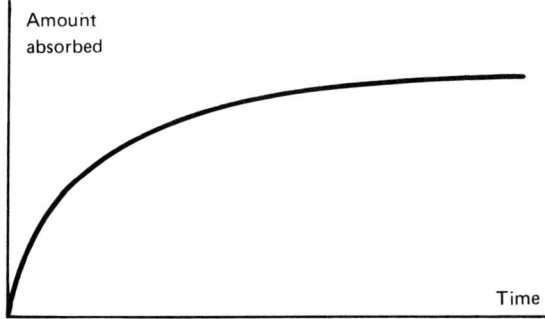

Fig. 3 General shape of absorption curve.

this chapter our discussion centers on the absorption properties of tissue paper and similar products as distinct from other grades of paper. Tissue paper is characterized by a high bulk and a very open, porous structure. This structure is created by various processing techniques, for example, (1) creping, (2) maintaining a low basis weight (which gives a low apparent density as well as distinct wire marks), (3) light beating, and (4) putting two or more plies together. The rate of absorbency is generally the most relevant absorption property to be considered in relation to the use of the material, particularly for wiping. A high bulk will favor a rapid water uptake by presenting larger flow channels and minimizing the flow resistance.

Practically all the water is transported in the interstices between fibers. The driving force for water uptake is governed by the surface free energy of the fibers and the available surface area. The surface area will eventually decrease with increasing bulk of the material. For that reason, a very high bulk will counteract a rapid absorption—the rate of absorption will pass a maximum as the bulk is increased. Conventional tissue products generally have a bulk lower than the optimum value, whereas air-laid products can be made with a bulk that even exceeds the optimum.

There are great difficulties in performing experiments relating rate of absorption with bulk because a bulky sheet will collapse upon wetting. Sheet collapse is in fact a major problem encountered with all of the measurement methods listed below. Not only will it jeopardize the reliability of the bulk value as a process parameter, it will also (in combination with fiber swelling) affect the flow resistance of the material and in some cases quite considerably inhibit the continuous supply of liquid.

There exist a number of methods for measuring the rate of water absorption in tissue paper. In the following discussion, an attempt is made to group these into specific categories. Very few of the prevailing methods have been officially standardized, and laboratories concerned with the examination of tissue paper have to a considerable extent developed their own methods. Therefore, very few standardized methods will be described and discussed here. On the other hand, the principles underlying the various approaches will be elucidated and their potentials with regard to both their theoretical backgrounds and their practical fitness will be considered.

Existing methods may be grouped into the following categories:

- Orifice methods. Water is supplied through a capillary or a small orifice and is allowed to spread in all directions of the sample. Water can be supplied either from the top side or the bottom side of a paper sheet.
- Porous plate methods. Water is supplied through a porous plate that is parallel to and in contact with the sample. The porous plate is located beneath the sample, with the water level adjusted to the upper surface of the porous plate or slightly below the upper surface.
- Floating time methods. Several varieties of this method exist. The sample is formed into a certain configuration and placed to float in a water basin. The time from the introduction of the sample into the basin until it sinks or gets completely wet is recorded. The sample may be contained in a wire basket, for instance.
- Capillary rise. The paper is cut into a strip, one end of which is dipped into a reservoir of water. The strip is generally arranged vertically but may also be aligned horizontally in order to avoid the effect of gravity. Also, the capillary rise in a pile of sheets has been considered.
- Inclined plane methods. The paper is placed on an inclined plane and a fixed amount of water is applied to the paper. The drop then travels downward and the distance (length of the track) required for the water to be completely absorbed is noted.
- Wipe-dry methods. Wiping of a wet surface is simulated in the test methods. Under standardized conditions, the amount of liquid left on the surface after wiping is determined.

In the following subsections, a few methods representative of the six groups outlined above will be described and discussed.

Orifice Methods In this type of method, the sheet or sheets of paper are generally oriented horizontally. Liquid is applied to a relatively small area in the center of the specimen, either from above or from beneath. The liquid is then allowed to spread radially outward (in the X- and Y-directions) from the zone of application. The principle of the method can be regarded as a special case of a three-dimensional process in which the liquid is spread in X-, Y- and Z-directions. Winch [56] calls this case an increasing total mass flow rate process. Since the total mass flow rate is increased by increasing the area of interface across which mass transfer occurs, the total mass flow rate tends to increase during the wetting process. On the other hand, the flow rate close to the zone of application will soon become very high, and the viscous drag in this area will be the rate-controlling factor.

Choksi [7] has found, using a method similar to the one by S. G. Reid (described below), that the increasing area of interface and the increasing flow resistance as the wetted region becomes larger cancel each other out so that the volume flow rate will be constant over a considerable period of time, at least when the absorbed liquid is water.

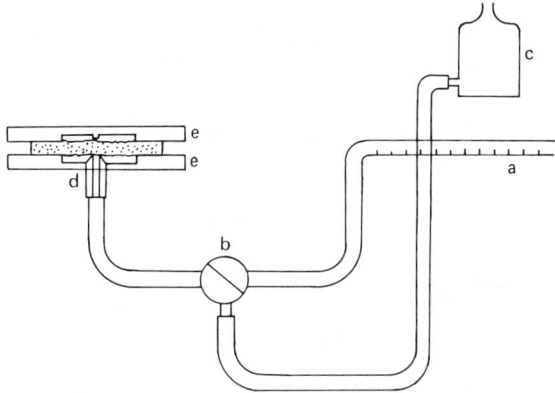

Fig. 4 Apparatus for method of S. G. Reid. (From Ref. 39.)

Many testing methods of this type use the same number of plies as intended for the material in service. When more than one ply is used in these methods, lamellar flow channels are created between the plies, which considerably reduces the viscous flow resistance. According to Reid [39], capillary or orifice methods tend to yield results different for single-ply products than for multiple-ply products by at least a factor of 10. This discrepancy is far too great to reflect the actual difference between single- and multiple-ply products in practical situations.

The flow rate in orifice methods can be recorded in various ways. The two main, and substantially different, ways are (1) recording the absorption time for a certain quantity of liquid and (2) measuring the linear rate of expansion of the wetted area.

Method of S. G. Reid The apparatus for Reid's [39] method is schematically illustrated in Fig. 4 and consists of the following parts:

A 0.1 ml pipette graduated in 0.01 ml divisions (a)
A 3-way stopcock (b)
An aspirator bottle containing water (c)
0.5 mm capillary glass tubing (d)
Two acrylic plates positioned as in the figure (e)

The acrylic plates are recessed so that the wet tissue does not touch the surface during a test. The plates are designed to give contact of tissue and capillary tip throughout the test without compression of the paper. The paper to be tested is placed on the lower plate and covered with the top plate. The time taken for a fixed volume of water (usually 0.01 ml) to flow from the pipette is recorded.

The apparatus is said to work well with high and low basis weight creped tissue and toweling. Heavily embossed products do not function well because of difficulty in achieving and maintaining contact of the paper and the capillary tip.

As is usually the case with capillary or orifice type methods, the rate of absorption is extremely sensitive to the type of backing of the specimen. Whereas this method is designed without any rigid backing, that is, the sample is held freely in air, the choice still exists whether to measure on one or more plies. In a typical experiment, the absorption of 0.01 ml by one ply of a particular tissue took 26 s. Two plies of the same tissue, tested together, gave a rate of 0.01 ml in 2 s.

Drop absorption time (TAPPI T 432) This method [53] determines the rate at which an absorbent paper, such as tissue or toweling, absorbs a specified quantity of water. The specimen is placed on a support consisting of a square metal plate (about 10 × 10 cm) with a 4 cm diameter hole in its center. Water is delivered to the center of the specimen by means of a buret, pipet, or syringe, depending on the quantity of water desired.

With a stopwatch the time of absorption is measured from the start of the flow of water until its complete absorption as indicated by the disappearance of the specular light reflection. For papers classified as tissue a water volume of 0.01 ml is recommended, whereas the volume 0.1 ml is recommended for toweling.

The absorption process in this method is similar to that in S. G. Reid's method, previously described. The liquid is applied in the center of the apparatus and is allowed to spread in the X- and Y-directions of the sheet or sheets.

WAT (Water Absorbency Tester) The design of this instrument has been developed at the Swedish Forest Products Research Laboratory (STFI). It uses the principle of electrical detection of liquid penetration. Two limitations are implicit in its use: (1) only liquids with a certain electric conductivity can be detected (for instance, distilled water); (2) the paper sample must be dry prior to the measurement.

The instrument measures the time from the moment the liquid starts penetrating the sample until the conductivity across the paper sheet exceeds a certain value, that is, the paper is permeated with liquid. By means of a particular arrangement of the electrodes (Fig. 5A) the penetration times in the X, Y, and Z directions can be recorded individually.

A 50 to 60 Hz AC voltage of 10 V is applied across the measuring cell and a 1 kΩ resistor (Fig. 5B). A comparator picks up the voltage across the resistor and sends impulses to the start/stop unit. The photodetector receives an impulse of light from the black/white fluorocarbon strip covering the orifice of the upper electrode (container) and gives a start impulse to the start/stop unit. The start/stop unit also receives information from a 1 MHz frequency crystal. The penetration times in the X, Y, and Z directions are displayed individually.

Demand Wettability The apparatus for the demand wettability method of B. M. Lichstein [29] is depicted in Fig. 6. It uses a sample plate (a) with a circular orifice of about 3 mm diameter. The orifice is connected via a tube with a buret (b) which has an air bleed (c) at its lower end. The buret is adjusted so that the air bleed and the orifice are at the same height.

When there is no sample on the plate and the main valve (d) is open, the liquid is at equilibrium, balanced by the atmospheric pressure at both ends. During measurement, capillary forces in the sample pull liquid out of the buret and a corresponding amount of air bleeds in. When the sample ceases to demand liquid, the liquid ceases to flow. Liquid can be applied to either side of the sample, and the plate may be tilted to enable testing at various angles.

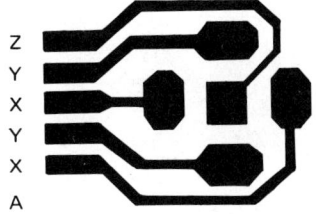
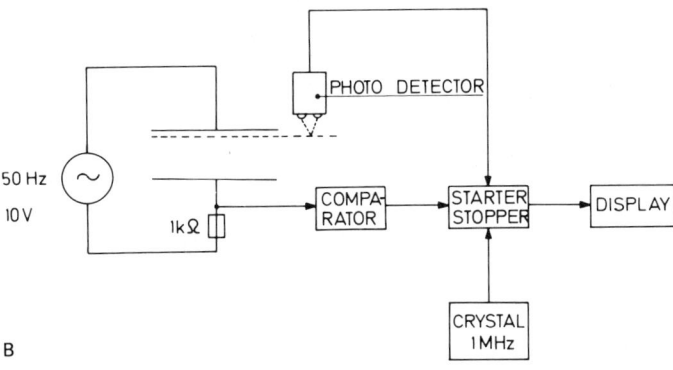

Fig. 5 (A) Arrangement of base plate electrodes in WAT. Center electrode (Z) located beneath point of water application. (B) Schematic of operation of WAT.

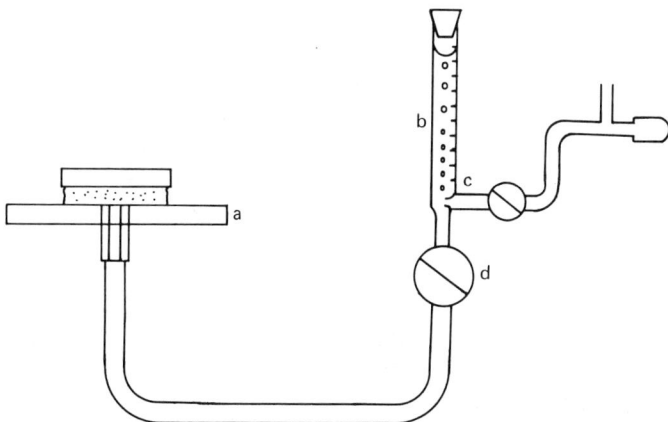

Fig. 6 Apparatus for demand wettability method. (From Ref. 29.)

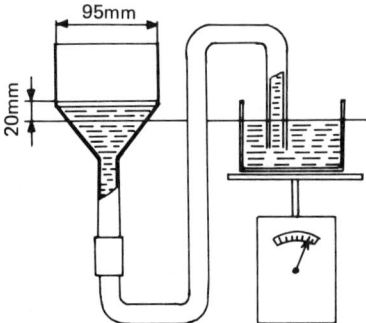

Fig. 7 Schematic of apparatus for standard method SIS 25 12 28. (From Ref. 52.)

Porous Plate Methods There are several methods, most of them developed within the textile area, that measure rate of absorption in terms of mass flow rate. In these methods the liquid is supplied through a porous plate that is in contact with the sample. The porous plate acts at the same time as a support for the sample.

The porous plate principle is supposed to possess the advantage of a very simple flow pattern. A liquid front progresses through the sample in the direction perpendicular to the surface of the porous plate. The sample can be one or more layers of a sheetlike material such as tissue or textile or a pad of absorbent fluff. However, it is somewhat doubtful that the flow velocity can be kept constant across the area of the sample. The variation in flow pattern with sample structure seems to be a disturbing factor.

Method of Swedish Textile Research Institute This method [52] uses a standard-type glass filter as a liquid supply means and as a support. The liquid flow rate is determined gravimetrically (Fig. 7).

Method of E. C. Jackson and E. R. Roper This method [21] uses a glass filter similar to the preceding method. However, the liquid is supplied through a horizontal buret connected with the porous plate and placed at the same height or below the upper surface of the plate. The amount of liquid in the system is adjusted by sucking out excessive quantities so that the water level is close to the upper surface of the porous plate and the zero mark of the buret. The distance between the buret and the porous plate is supposed to determine the degree of wetness of the plate and thus the rate of absorption. It is assumed that this distance must be constant for any set of comparisons between samples.

Floating Time Methods Floating time methods have found wide application for certain grades of tissue, nonwoven fabrics, and textiles. These are mostly used for quality control purposes where the span in physical appearance of the sample is fairly narrow, since methods of this type are sensitive to the mode of spatial arrangement of the sample prior to measurement.

In an interesting study of surface wetting rates using a continuous filament or yarn, Kimmel and Steiger [24] found that a floating time method may be a suitable method at low wetting rates, although at high wetting rates (above 0.5 cm/s) the method becomes insensitive to wettability variations.

ANSI/ASTM [2] The sample is inserted, in form of a loosely packed roll, in a cylindrical wire basket that is open at one end. Its dimensions are 50 mm in diameter and 80 mm high, and its weight is nominally 3 g. The basket is dropped on its side from a height of 25 mm into water at room temperature. The time required for complete wetting is measured by means of a stop watch.

Capillary Rise Methods Capillary rise or wicking methods have only limited importance for use in connection with tissue. The methods cannot be applied on multiple-ply tissues, and data for a single ply are not relevant if the material is to be used in the form of a two-ply or multiple-ply product. However, wicking methods have found some applications in connection with absorbent products of moderate to high basis weight.

Klemm's method, now Scandinavian method SCAN-P13-64 [44] for the measurement of capillary rise in paper is a standardized method largely used with unsized paper having a high absorbency. The height of rise is generally determined visually. Similar techniques are in use with materials other than paper, for example, textiles. In one variation of the method, an electrical determination of the height of rise has been conceived [25]. In this case, the wetted portion of the sample constitutes one electrode in a capacitor. The area of the sample electrode then changes in proportion to the rate of wetting. It is claimed that this technique is superior to other electrical approaches wherein the sample constitutes the dielectric, the properties of which are dependent on the degree of wetting.

It is generally considered that capillary rise methods, in applications where they are justified, provide a simple and reliable way of characterizing the absorbency. Precautions, however, should be taken; the apparatus should be placed in a cabinet to minimize variations due to evaporation.

There is no simple relationship between the height of capillary rise and the amount of water absorbed. Experimental studies have revealed that in most cases there is a saturation gradient in the specimen along the height of rise. Fujita [13] has referred to experiments by Takahashi on water distribution in chromatography paper (Fig. 8A) and applied a hydrodynamic model that makes a distinction between free and fixed liquid. Based on a different approach, van Brakel and Heertjes [5] presented a similar liquid distribution (Fig. 8B), obtained with packed polystyrene spheres. They found that the width of the gradient was dependent on the contact angle between the liquid and the solid.

Absorbency of Cellulose Wadding This method [55] is most conveniently characterized as a capillary rise method, although the sample is not in the form of a paper strip but a pile of circular coupons. It differs also from ordinary capillary rise methods in that the volume of water, rather than the height of rise, is measured.

Piles containing 100 sheets of 30 mm diameter are punched out of a pile of tissue (or wadding) sheets. Since the upper sheets tend to have too large a diameter caused by the punching operation, the upper 25 sheets are discarded.

A pile is inserted in a graduated stand that is subsequently placed in a water reservoir so that the water surface touches the lower end of the pile. Three water containers are used in series; it is recommended that the

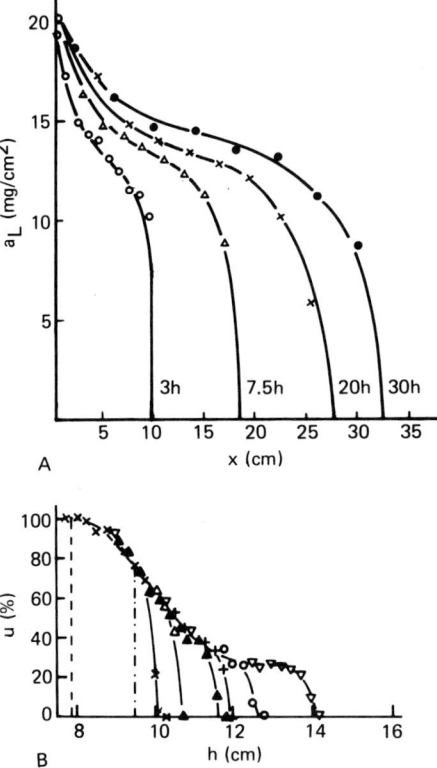

Fig. 8 A saturation gradient is established during capillary rise of water in paper. (A) (From Ref. 5, reprinted by permission from *Nature*, 254, pp. 585-586. Copyright © 1975 Macmillan Journals Limited.) (B) (From Ref. 13, reprinted with permission from H. Fujita, *J. Phys. Chem.* 56(5):625-629. Copyright © 1952 American Chemical Society.)

stand with the pile be held in the three containers for 5, 25, and 150 s, respectively. The decrease in water content in each of the three containers is measured.

Although this method is somewhat tedious for the operator, it is one of very few methods that is sensitive to the rate of transverse flow through a pile of tissue sheets. Such flow conditions exist under practical conditions.

Inclined Plane Methods In a typical method of this type, the characterization of the absorbency of a tissue sample is based upon the following principles.

- The absorbency of a tissue sample is characterized by the length of the track required for the absorption of a certain amount of liquid during conditions stated by the description of the method.

- • The absorption time is the time required for the drop to travel the full length of the track and to be completely absorbed by the paper.

The paper is adhered to an inclined plane, generally at an angle of 60° relative to the horizontal. A buret is mounted close to the upper end of the plane. A liquid volume of 1 ml is expelled during a time interval of 2.4 s. Measurement is made of either or both of the two quantities, track length and absorption time.

The advantages of this method are claimed to be the following:

- A mean value is obtained that is representative of the area covered by the drop along its track

- ● The method is not sensitive to the number of plies (one or two) as the orifice or capillary methods. Because no narrow section for liquid flow is created close to the point of application, the flow rate for single-ply products will not level off.

Wipe-Dry Methods Only one wipe-dry method is known to the author, although several may be in use by the major producers of tissue. The wipe-dry principle is the one closest to requirements based on the practical use of the material. Although the wipe-dry method described below seems very useful in this respect, it is worth noting that it does not take into account the effects of sheet flexibility and softness. In actual practice a soft and flexible wiping tissue generally wipes drier than a stiff and rigid one because of its ability to conform more effectively to the surface to be wiped.

Wipe-Dry Method This method is an internal method of Scott Paper Co., developed by E. Kasinkas. In a design similar to a phonograph turntable, a flat, 40 cm diameter phenolic disc serves as the surface to be wiped. The tissue sample is attached to a sleigh mounted on a moving arm. During the test, the sleigh moves from the center of the disc outward. Immediately before the test, 0.5 cm^3 of water is deposited in the center of the disc. The sleigh with the sample is lowered and its motion outward is initiated. The pressure of the sleigh against the surface and the effective area of the sample are both adjusted to real values in practical handling of materials used for wiping.

The wipe-dry effectiveness of a sample is judged from the wetness of the disc after the sleigh has traversed. The disc is marked with equidistant circles and the residual wetness is judged from the rate of disappearance of the liquid film. It is important that the instrument be placed in a room with constant temperature and humidity so that the rate of evaporation of the liquid film is constant.

It is apparent that the approach represented by the wipe-dry method attempts to fill an important gap between theory and pactice. For several products, the wipe-dry properties are the most important characteristics in practical use. In many cases, the wipe-dry effectiveness of a product cannot be derived from other tests.

B. Absorption Capacity

Some of the methods for measuring rate of absorption will also provide an appropriate measure of the absorption capacity. In theory, the absorption capacity equals the amount of liquid absorbed after infinite time. Basically, the amount of liquid absorbed in a sample can be determined either gravimetrically or volumetrically. In the former case, either the sample can be weighed before and after the absorption process, or the liquid can be supplied from a tared reservoir. In the latter case, the liquid may be supplied by means of a graduated buret. All of these methods are or have been in use.

Of the methods described in Sec. III.A, porous plate methods and floating time methods are used in particular, not only for detection of absorption rate but for absorption capacity as well. Of the other types of described methods, the demand wettability [29] and absorbency of cellulosic wadding [55] tests have been reported to be suitable for absorption capacity measurement.

A short description of the floating time technique for measuring absorption capacity is given below.

ANSI/ASTM Floating Time Method [2] The same equipment as described on p. 159 is used. After the time required for complete wetting has been determined, the basket with the sample is allowed to remain submerged in the water for 10 s. Subsequently it is drained for 10 s and then weighed. The absorption capacity is reported as the ratio of the water held by the specimen to the weight of the dry sample.

IV. FACTORS AFFECTING ABSORPTION RATE

The factors controlling the rate of absorption in tissue and similar products can roughly be divided into two groups: (1) topochemical factors and (2) structural factors. We shall now follow briefly the sequence of manufacture of a typical tissue product to see where and to what extent the potential absorbency of the product can be altered, either by topochemical or structural changes.

A. Pulp Manufacture and Stock Preparation

Typical pulps used for tissue manufacture may be the following:

- Bleached softwood, kraft, or sulfite
- Bleached hardwood
- Mechanical pulp
- Chemi-mechanical pulp
- Reclaimed fibers

The pulp fibers may contribute to the absorbency of the final product through their chemical composition and the distribution of the chemical components. Also, the dimensions and mechanical properties of the fibers may be of some importance.

Different pulps behave differently at various pH values [54]. Tests on laboratory sheets have indicated that the neutral pH range gives the poorest absorption properties for the most common pulps used for tissue manufacture in Europe. All of the pulps tested except unbleached spruce sulfite showed rapid absorption at pH $\sim$ 9.

Certain components in wood or cellulosic fibers are more or less hydrophobic. Extractives (most, but not all) are generally very hydrophobic [50, 51]. Carbohydrates on the other hand, and also lignin, are considered hydrophilic in the chemically modified (extracted) state [27], even though their

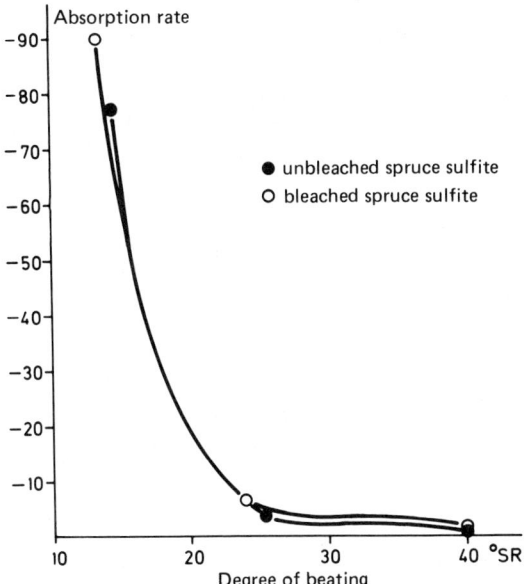

Fig. 9 Influence of beating on absorption rate for sulfite pulp. The absorption rate is expressed as the rate of change in contact angle (degrees/sec). (From Ref. 54.)

physical state and function in unaltered wood or fibers is more hydrophobic in nature.

There are only a few systematic investigations that have been reported as to the influence of fiber geometry and the mechanical properties of the fibers on absorbency [1]. However, some conclusions can be drawn intuitively. It is realized that for a high rate of absorption, the porosity should be neither too high nor too low. If it becomes too high, there will be an increased effect of gravity and a reduced driving force due to insufficient fiber/liquid interface. On the other hand, if the porosity becomes too low, the flow resistance will reduce the rate of absorption. For these reasons, beating must be done with some care because increased beating may reduce the porosity rapidly (Fig. 9) [26,54].

In the practical range of beating pulp for tissue, the effect of beating on the absorbency of the creped sheet is virtually nonexistent. Table 1 shows the relationship of refiner power to absorption time compiled at a mill trial.

It seems beneficial, in terms of absorbency, that the fibers possess a certain wet stiffness in order for the structure not to collapse upon wetting. A collapse of the structure seems to have a dual effect in that it decreases the suction power simultaneously as it causes a tighter packing of the material and increases the friction.

A superior structure is generally considered to be one that is made up of more than one constituent. Various components can then be chosen with reference to their particular advantages. For instance, a stiff backbone of mechanical (most likely thermomechanical) fibers to prevent collapse of the

Table 1 Refiner Power and Related Absorption Times Taken from a Tissue Machine in Commercial Operation

	Power consumption (kW)	Absorption time (s)
Refiner type 1	250	3.5
	300	2.8
	350	3.1
Refiner type 2	200	2.8
	250	2.7
	300	3.3

structure may be mixed with some more hydrophilic chemical fibers that will exert the main suction power [22].

B. Sheet Forming and Creping

Very little information is available on the influence of the papermaking variables on the absorbency of creped tissue or toweling. Variables of this kind assumed to have some effect on the absorbency are (1) fiber orientation (2) crepe structure, and (3) degree of crepe. A pronounced degree of fiber orientation (see Chap. 24) is correlated with anisotropic flow of water, the most rapid flow occurring in the direction parallel with the aligned fibers [1]. However, this effect is usually too small to be detected in practical cases, apparently because the effect of crepe structure is more dominant. The crepe effect can be compared by some theoretical models that treat nonplanar surfaces as being composed of arrays of ridges and valleys [34]. It has been demonstrated that the valleys act like capillaries, promoting a rapid transport of liquid. The net effect is that the crepe enhances penetration in the cross-machine direction. This phenomenon is very easily demonstrated for virtually all creped tissue and toweling materials simply by applying a drop of water and noting the shape of the wetted area. In the case of multiple-ply sheets, the effect is more pronounced, since in that case actual channels are created along the crepe ridges, considerably facilitating the transport of liquid in the cross-machine direction.

The dimensions of the crepe ridges can be altered by means of variations in the creping technique and by changing the degree of crepe, that is, the degree of pull-out after the creping operation. The maximum bulk of the structure is attained at a degree of crepe around 40%, which also corresponds to the maximum rate of water absorption. This is a higher than usual degree of crepe, which means that—in most practical cases—absorption rate will increase with increasing degree of crepe.

C. Aging

As was stated on page 153, a high wettability and rate of absorption is favored by a high surface free energy of the fibers. High energy surfaces have a short lifetime, because the ease with which substances adhere to a a surface increases with its free energy. Therefore, the wettability of a cellulosic material will always decline with time. The rate with which the wettability will decline is extremely difficult to predict; it will vary with the type of environment and conditions of storage. This fact is very often the reason why measurements of absorbency of paper tend to show considerable scatter. The surface energy as well as wettability can be partially restored through activation by means of a corona discharge or a microwave plasma [6].

If a fiber rich in extractives is stored, the resinous components, which have low surface tensions, will spread over the surface of the fibers. Also, a vapor phase diffusion of some low-boiling fatty acids has been detected [51]. These processes, which are termed self-sizing, are particularly noticeable if aluminum ions are present, because extremely hydrophobic aluminum-resin soaps are formed. The presence of aluminum ions—and to a minor degree also calcium and magnesium ions—together with the fatty acids and resin acids naturally occurring in mechanical pulp, is harmful for absorbency and creates conditions of sizing. In a mill trial where toweling containing a large amount of mechanical pulp was produced, the water absorption time was reduced from 500 to 2 s when sodium hydroxide was exchanged for alum as a creping aid [16].

The rate of water absorption into a sheet of tissue paper or similar product changes drastically upon heat treatment of the material, which can be thought of as an accelerated aging. Suitable combinations of time and temperature for an accelerated aging treatment have been suggested. For tissue products, the duration of 1 h at 90°C has been found to correspond to 3 months of storage at room temperature [54]. A standard method (SCAN-C 32) for fluff pulp [45] prescribes a heat treatment for 3 h at 105°C.

REFERENCES

1. Aberson, G. M. (1970). The water absorbency of pads of dry, unbonded fibers. In *The Physics and Chemistry of Wood Pulp Fibers*, TAPPI Special Technical Association publication no. 8, New York, pp. 282–305.
2. ASTM D 1117-71. Standard Methods of Testing Nonwoven Fabrics (1977).
3. Bangham, D. H., and Saweris, Z. (1938). The behaviour of liquid drops and adsorbed films at cleavage surfaces of mica. *Trans. Faraday Soc.* 34:554–570.
4. Bascom, W. D., Cottington, R. L., and Singleterry, C. R. (1964). Dynamic surface phenomena in the spontaneous spreading of oils on solids. In *Contact Angles, Wettability and Adhesion* (R. F. Gould, ed.), Advances in Chemistry, vol. 43, American Chemical Society, Washington, D.C., 1964, pp. 355–379.
5. van Brakel, J., and Heertjes, P. M. (1975). Capillary rise in porous media. *Nature* 254:585–586.
6. Chan Tang, T. W., and Bosisio, R. G. (1980). Enhanced wettability of cellulose strips treated in a microwave plasma. *Tappi* 63(3): 111–113.

7. Choksi, P. V., Spaeth, E. E., and Shiff, J. A. (1977). Adsorption characteristics of woven and nonwoven laparotomy sponges as measured by a novel device. INDA Technical Symposium on Nonwoven Technology, Washington, D.C., 1977, 5:29–41.
8. Derjaguin, B. V., and Churaev, N. V. (1976). Polymolecular adsorption and capillary condensation in narrow slit pores. *J. Coll. Interface Sci.* 54(2):157–175.
9. Derjaguin, B. V., and Zorin, Z. M. (1957). Optical study of the absorption and surface condensation of vapours in the vicinity of saturation on a smooth surface. Proc. 2nd Intern. Congress on Surface Activity, London, 1957, 11:145–152.
10. Dullien, F. A. L., El-Sayed, M. S., and Batra, V. K. (1977). Rate of capillary rise in porous media with nonuniform pores. *J. Coll. Interface Sci.* 60(3):497–506.
11. Eick, J. D., Good, R. J., and Neumann, A. W. (1975). Thermodynamics of contact angles. 2. Rough solid surfaces. *J. Coll. Interface Sci.* 53(2):235–248.
12. Fowkes, F. M. (1967). Surface chemistry. In *Treatise on Adhesion and Adhesives*, vol. 1. (R. L. Patrick, ed.), Edward Arnold Ltd/Marcel Dekker, Inc., London/NewYork, pp. 325–449.
13. Fujita, H. (1952). On the distribution of liquid ascending in a filter paper. *J. Phys. Chem.* 56(5):625–629.
14. Hansen, R. J., and Toong, T. Y. (1971). Dynamic contact angle and its relationship to forces of hydrodynamic origin. *J. Coll. Interface Sci.* 37(1):196–207.
15. Hardy, W. B. (1936). *Collected Works*, Cambridge Univ. Press, pp. 711–717.
16. Hollmark, H. (1977). The use of thermochemical pulp in tissue and toweling, *STFI-meddelande ser. D, no. 28* (in Swedish).
17. Huh, C., and Mason, S. G. (1977). Effects of surface roughness on wetting (theoretical). *J. Coll. Interface Sci.* 60(1):11–38.
18. Huh, C., and Scriven, L. E. (1969). Shapes and axisymmetric fluid interfaces of unbounded extent. *J. Coll. Interface Sci.* 30(3):323–337.
19. Huh, C., and Scriven, L. E. (1971). Hydrodynamic model of steady movement of a solid/liquid/fluid contact line. *J. Coll. Interface Sci.* 35(1):85–101.
20. Institute of Paper Chemistry (1940). Instrumentation studies, No. 34. Penetration of papers by liquids and solutions. 1. General discussion of the phenomena. *Paper Trade J.* 110: (Tappi Section), 42–48.
21. Jackson, E. C., and Roper, E. R. (1949). A water absorbency apparatus. *Amer. Dyestuff Reptr.* 38(10):397–401.
22. Jackson, M. (1974). Fluff absorbency testing and related wood pulp properties. TAPPI Testing/Paper Synthetics Conference, Boston, pp. 237–241.
23. Johnson, R. E., Jr., and Dettre, R. H. (1964). Contact angle hysteresis. 1. Study of an idealized rough surface. In *Contact Angles, Wettability and Adhesion* (R. F. Gould, ed.), Advances in Chemistry, vol. 43, American Chemical Society, Washington, D.C., pp. 112–144.

24. Kimmel, J. M., and Steiger, F. H. (1970). Effect of temperature on the surface wetting rate of cellulose fibers. *Ind. Eng. Chem. Prod. Res. Devt.* 9(2):259–264.
25. Klauer, H. (1968). Wicking in textiles. *Chemiefasern* 12:928–933 (in German).
26. Kuniak, L. (1962). Studies on the capillary properties of pulp sheets. *Svensk Papperstidn.* 65(19):760–766 (in German).
27. Lee, S. B., and Luner, P. (1972). The wetting and interfacial properties of lignin. *Tappi* 55(1):116–121.
28. Levine, S., Lowndes, J., Watson, E. J., and Neale, G. (1980). A theory of capillary rise of a liquid in a vertical cylindrical tube and in a parallel-plate channel. *J. Coll. Interface Sci.* 73(1):136–151.
29. Lichstein, B. M. (1974). Demand wettability, a new method for measuring absorbency characteristics of fabrics. INDA Technical Symposium 1974, pp. 129–142.
30. Ludviksson, V., and Lightfoot, E. N. (1968). Deformation of advancing menisci. *AIChE J.* 14(4):674–677.
31. Manegold, E. (1955). Kapillarsysteme (Capillary Systems). vols. 1 and 2. Strassenbau, Chemie und Technik Verlagsgesellschaft, Heidelberg, pp. 12–28, 205–481 (in German).
32. Melrose, J. C. (1968). Thermodynamic aspects of capillary. *Ind. Eng. Chem.* 60(3):53–70.
33. Minor, F. W., Schwartz, A. M., Wulkow, E. A., and Buckles, L. C. (1959). The migration of liquids in textile assemblies. 3. The behaviour of liquids on single textile fibers. *Textile Res. J.* 29(12):940–949.
34. Oliver, J. F., and Mason, S. G. (1977). Microspreading studies on rough surfaces by scanning electron microscopy. *J. Coll. Interface Sci.* 60(3):480–487.
35. Orr, F. M., Jr., Brown, R. A., and Scriven, L. E. (1977). Three-dimensional menisci: Numerical simulation by finite elements. *J. Coll. Interface Sci.* 60(1):137–147.
36. Orr, F. M., Jr., Scriven, L. E., and Rivas, A. P. (1975). Menisci in arrays of cylinders: Numerical simulation by finite elements. *J. Coll. Interface Sci.* 52(3):602–610.
37. Peek, R. L., Jr., and McLean, D. A. (1934). Capillary penetration of fibrous materials. *Ind. Eng. Chem.* 6(2):85–90.
38. Princen, H. M. (1970). Capillary phenomena in assemblies of parallel cylinders. 3. Liquid columns between horizontal parallel cylinders. *J. Coll. Interface Sci.* 31(2):171–184.
39. Reid, S. G. (1967). A method for measuring the rate of absorption of water by creped tissue paper. *Pulp Paper Mag. Can.* 68 (Convention Issue):T-115–117.
40. Renk, F., Wayner, P. C., Jr., and Homsy, G. M. (1978). On the transition between a wetting film and a capillary meniscus. *J. Coll. Interface Sci.* 67(3):408–414.
41. Rose, W. (1961). Fluid-fluid interfaces in steady motion. *Nature* 191:242-243.
42. Ruckenstein, E., and Dunn, C. S. (1977). Slip velocity during wetting of solids. *J. Coll. Interface Sci.* 59(1):135–138.

43. Ryong-Joon Roe (1975). Wetting of fine wires and fibers by a liquid film. *J. Coll. Interface Sci.* 50(1):70–79.
44. Scandinavian Pulp, Paper and Board Testing Committee (1964). SCAN-P13:64. Capillary rise of water in paper and paperboard by the Klemm method (in Swedish).
45. Scandinavian Pulp, Paper and Board Testing Committee (1978). SCAN-C32:78. Fluff. Specific volume and absorption properties (in Swedish).
46. Scheidegger, A. E. (1957). The physics of flow through porous media. Univ. of Toronto Press, pp. 5–8, 114–124.
47. Schultze, K. (1955). Wettability as an effect of real capillarity. *Kolloid-Zeitschrift,* 141(1):11–20 (in German).
48. Siegel, R. (1961). Transient capillary rise in reduced and zero-gravity fields. *J. Appl. Mech.* 28(6):165–170.
49. Skawinski, R., and Lasowska, A. (1974). Flow of water solutions in a quartz capillary tube with periodically variable diameter. *Bulletin de l'Academie Polonaise des Sciences 22(6):307-311.*
50. Soteland, N., and Lovås, V. (1976). Self-sizing of mechanical pulp. *Svensk Papperstidn.* 79(7):203-202 (in German).
51. Swanson, J. W., and Cordingly, S. (1959). Surface Chemical Studies on Pitch. 2. The mechanism of the loss of absorbency and development of self-sizing in papers made from wood pulps. *Tappi* 42(10):812–819.
52. Swedish Standard SIS 25 12 28 (1971). Textiles. Determination of water absorbency (in Swedish).
53. TAPPI Standard T 432 ts-64 (1964). Water Absorbency of Bibulous Papers.
54. Vahtila, M. (1973). Absorption properties of sanitary crepe papers: The aging phenomenon and factors that influence it. SCAN-Forsk report no. 36 (in Swedish).
55. Verein der Zellstoff- und Papier-Chemiker und -Ingenieure. (German Society of Pulp and Paper Chemists and Engineers) (1949). Subcommittee on Pulp Analysis report 27. Determination of the absorption properties of cellulosic wadding (in German).
56. Winch, A. R. (1959). Theoretical analysis of rate of liquid absorption by capillary absorbing media. *Textile Res. J.* 29(3):193–199.

ELECTRICAL AND THERMAL INTRACTIONS

Emblem of Cuauhxomulco

Although it was not used as a papermaking fiber, the Piñon pine (*Pinus edulis* Engelm.) was an important source of edible seeds used as tribute to the Aztec rulers. The tribute town Cuauhxomulco derived its name from its location in a corner (whence the bent branch) of a Piñon pine forest. The tree was called cuauhuitl. Xomulco designated a sheltered corner in a forest or grove. Pine foliage was drawn in a highly stylized manner.

21
ELECTRICAL PROPERTIES: I. THEORY

GARY A. BAUM
The Institute of Paper Chemistry
Appleton, Wisconsin

I.	Introduction	171
II.	Definitions	172
III.	Fundamental Considerations	175
	A. Direct Current Conductivity	176
	B. Dielectric Constant	184
	C. Dielectric Anisotropy	190
	Symbols	195
	References	196

I. INTRODUCTION

The electrical properties of paper and board are important from a number of points of view. As a dielectric in capacitors or as an insulator for wires, coil windings, or cables, paper provides good dielectric properties, together with desirable mechanical properties and a relatively low cost. In combination with other materials, for example, saturating oils or resins, it is used in transformers, electronic circuit boards, and numerous other applications.

The electrical properties are also important in many reprographic processes. For example, in direct electrophotographic and dielectric printing methods, the paper must meet certain minimum electrical conduction requirements. Usually it must be rendered conductive by the application of special materials prior to being coated with the photoconductive or dielectric film.

The importance of the electrical characteristics is obvious in the applications mentioned above. Perhaps not so obvious is the need to understand these properties of paper (or fibers) with respect to other processes. For example, electrostatically assisted rotogravure, radio frequency or microwave driers, and radio frequency or microwave moisture gages all involve the interaction of electromagnetic fields with paper and water systems. A fundamental understanding of this interaction is essential to the proper use of these devices. The same understanding is needed with respect to fibers in certain air-lay processes or solid-solid separation processes.

Although much research has been carried out over the years with respect to the electrical properties of paper and board, there is still an incomplete understanding of such basic phenomena as the nature of the electrical conduction process itself, the role played by water and the other polar groups in the dielectric response of cellulose, time and temperature effects on the conduction process, and the effects of additives or sheet process variables.

A greater awareness is needed of the fact that the electric properties of paper are anisotropic. Just as paper must be considered an orthotropic medium in terms of its elastic properties, it should also be considered an orthotropic medium in terms of its dielectric properties. While this may not be very important in terms of the more conventional electrical uses of paper as an insulator, it has much significance in terms of electromagnetic driers, moisture gages, and other devices.

II. DEFINITIONS

The electrical properties of interest with respect to paper and board include volume and surface conductivities (or resistivities), dielectric constants and dielectric loss, permittivities, and dielectric breakdown strength. When paper is used as an insulator, the requirements would be for low conductivity, low dielectric loss, and high dielectric breakdown strength. In other applications, such as certain reprographic processes, the paper must have relatively high volume or surface conductivities.

The *volume conductivity* σ is the ratio of the current density J to the applied electric field strength E or

$$J = \sigma E \tag{1}$$

Both E (newtons/coulomb) and J (amps/m^2) are vector quantities. In a homogeneous isotropic material, the conductivity (ohm-cm)$^{-1}$ will be a scalar. In a homogeneous anisotropic material, where the properties vary in a different manner along different directions at a point, the conductivity σ forms a symmetric second rank tensor. In terms of mechanical properties, paper and board behave as (three-dimensional) orthotropic materials in which there are three mutually perpendicular symmetry planes. The elastic properties are different in the three directions. A similar behavior is expected for the electrical properties. Three independent conductivities are anticipated, one along each of the three principal directions in the paper (the machine direction, cross-machine direction, and thickness or Z direction). Although there do not appear to be any measurements reported in the literature for conductivities of paper, Lin [36,37] did study this anisotropy in wood.

In most uses of paper where its electrical properties are important, the direction of concern is the thickness direction, and values reported for the volume conductivity of paper are measured in this direction. In many applications, the in-plane conductivities may be of little importance. As discussed later, a similar anisotropy is expected (and observed) for the dielectric constant.

The *volume resistivity* ρ in a given direction in a material is the reciprocal of the volume conductivity in the same direction. Both ρ and σ are material properties, independent of the specimen and electrode geometries. The *volume resistance* R is defined between two electrodes contacting the specimen as the ratio of the applied voltage to the volume current. It is a function of both the resistivity and the geometry. The volume resistivity and R are related by $R = \rho L/A$ (ohms), where L is the distance between electrodes and A is the cross-sectional area of the specimen between electrodes.

It is often useful to speak in terms of a *surface resistivity*, especially in those situations where the surface of the material has been altered in some manner. The surface resistivity ρ_s is taken as the (surface) resistance measured between the opposite sides of a square on the surface. It is independent of the size of the square and is expressed in ohms. The *surface conductivity* σ_s is the reciprocal of ρ_s. The surface conductivity is frequently used to characterize reprographic papers where the surface has been rendered conductive by the application of special resins.

The *insulation resistance* of a material is defined as the ratio of the applied voltage to the resultant current, and it involves both the volume and the surface resistances.

Actual measurements of any of the above parameters may be made using either direct current (DC) or alternating current (AC) equipment. To avoid confusion, the quantity involved is usually prefaced with the letters AC or DC, for example, AC conductivity or DC conductivity.

In electromagnetic theory, the proportionality constant between the electric displacement vector D (coulombs/m^2) and the electric field strength E in a material, is called the *permittivity* ε (farads/m) or

$$D = \varepsilon \cdot E \tag{2}$$

Just as in the case of the conductivity, ε is a scalar if the material is homogeneous and isotropic. If the material is anisotropic, ε is a symmetric second rank tensor. For paper, assuming orthotropic symmetry, this permittivity tensor has three independent nonzero components. Each component relates the electric displacement to field strength in one of the three principal directions.

The permittivity itself is usually represented as a complex number. The significance of this is perhaps most easily understood by considering a parallel plate capacitor. If a capacitor with a vacuum dielectric is connected to a sinusoidal voltage source, $V = V_0 \exp(iwt)$, it will store a charge (coulombs), $Q = C_0 V$. In the above expression V_0 (volts) is the voltage at time t equal to zero, w is the angular frequency ($= 2\pi f$), and $i = (-1)^{1/2}$. The quantity C_0 (farads) is the geometric (or vacuum) capacitance of the capacitor, and neglecting any fringing fields, is equal to $\varepsilon_0 A/d$, where ε_0 is the permittivity of free space, A the surface area of one capacitor plate, and d the distance between plates. The charging current I_c (amperes) is given by $I_c = iwC_0V$. If the vacuum capacitor is now filled with some material having a permittivity ε, two things happen. The capacitance increases to $C = \varepsilon'A/d$, where ε' is the real part of the permittivity of the substance, and at the same time energy losses occur because the material is not a perfect insulator. The substance-filled capacitor must be characterized by both a capacitance parameter and a loss parameter. Thus, in addition to the charging current, we now must include a loss current I_L, so the total current is $I_T = I_c + I_L$. The loss current might arise because of ion migration or because dipoles, induced in the material as a result of the impressed voltage, are hindered in their attempt to follow or align themselves with the applied field. Such loss mechanisms may be represented as either a parallel or a series combination of a resistance and a capacitance.

In the parallel circuit of Fig. 1A, R represents the loss component and C the pure capacitive component. The loss current is $I_L = V/R$, so that the total current is $I_T = iwCV + V/R$. A vector diagram for this circuit is shown in Fig. 1B. In the figure, ϕ is the phase angle and δ is the dielectric loss angle. The *dissipation factor* D is defined as the cotangent of the phase angle or

$$D = \cot \phi = \tan \delta = \frac{1}{wRC} \tag{3}$$

This quantity is also referred to as the loss tangent.

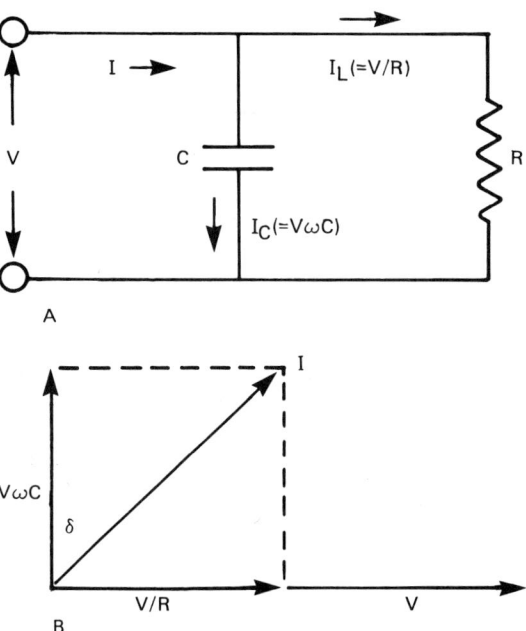

Fig. 1 (A) Schematic circuit depicting charging and loss currents. (B) Vector representation of Fig. 1A.

It is customary to describe the charging current and the loss current by the introduction of a complex permittivity,

$$\varepsilon = \varepsilon' - i\varepsilon'' \qquad (4)$$

or a complex *dielectric constant* (dimensionless),

$$k = \frac{\varepsilon}{\varepsilon_0} = k' - ik'' \qquad (5)$$

where ε'' and k'' represent the loss parts of the permittivity and dielectric constant, respectively. In terms of the complex dielectric constant, the total current may now be written $I_T - iwkC_0V$, and the *loss tangent* becomes

$$\tan \delta = \frac{\varepsilon''}{\varepsilon'} = \frac{k''}{k'} \qquad (6)$$

In the case of paper, the dielectric parameters reported are usually the real and imaginary parts of the dielectric constant k' and k'', although the latter is usually called the loss factor. It is also common, however, to see reference to the dissipation factor D or loss tangent $\tan \delta$. The quantity $R(= 1/wC \tan \delta)$ is referred to as the AC *resistance*.

In addition to the above parameters, there is another that is important in describing the electrical properties of paper. This is the *dielectric strength*. The dielectric strength is a measure of the ability of the dielectric to withstand high voltages and maintain its high resistance to current flow. It is usually defined as the maximum electric field strength that may be applied without causing an irreversible failure of the material. If the dielectric strength is exceeded, the material may break down, meaning that the material has lost its high resistance to current flow. That is, it becomes conductive.

III. FUNDAMENTAL CONSIDERATIONS

In a literature review on electrical papers in 1972, Raman and Walker [54] concluded that even though a considerable amount of research had been carried out, there was a distinct lack of detailed understanding of many aspects concerning the electrical characteristics of papers. Although there has been some progress made in recent years, the general conclusion still appears to be valid today.

With respect to its electrical properties, paper has some interesting and somewhat unusual characteristics. This section is intended to acquaint the reader with those characteristics and attempts to provide an overview of our current understanding of electrical phenomena in paper and board.

A word of caution is in order. Much of the work reported in the literature refers to "cellulose." This term may encompass a number of substances and cellulose-based materials, with varying amounts of lignin and hemicelluloses; therefore, it is sometimes difficult to directly compare experimental results reported in the literature. In this summary we attempt to include as complete a description of the specimen as possible.

A. Direct Current Conductivity

Introduction If a DC potential is suddenly applied to paper, the initial current is a function of time, eventually decaying to some steady state (time-independent) value. Such behavior is not unique to paper and can be attributed to electrolytic polarization and high contact resistances, among other things. This time dependence is discussed later. It does give rise to differences between measured AC and DC conductivities in paper. Friedrich and Chiu [19], studying surface conductivities, observed that at long times (several minutes) the DC conductivity approached the AC value (from the high side). The following discussion refers to the steady state DC conductivity.

The very early literature dealing with the electrical properties of paper and related materials has been reviewed elsewhere [5,28,49,54] and will not be presented in detail here, since our objective is to provide an overview of current understanding concerning these properties. A series of papers published in 1947 and 1948 by O'Sullivan [49–53] considered the conduction of electricity in cellulosic materials from a fundamental viewpoint. He demonstrated that conduction in cotton or regenerated cellulose was dependent on an association of moisture and naturally occurring electrolytes in the material. If the electrolytes were removed, the conductivity at a given relative humidity decreased. For a fixed salt content, the conductivity was an exponential function of the conditioning relative humidity. In addition, at salt contents greater than about 1%, the conductivity was largely governed by the moisture content, being insensitive to the amount or type of salt present. At low salt concentrations (< 1%), the conductivity was affected by the amount present (at a fixed relative humidity). O'Sullivan showed that ions could migrate between electrodes and concluded that the conduction process in cellulose was ionic, rather than electronic. He strengthened that argument by further demonstrating that the moisture dependence observed for conduction in the cellulosic sheet was similar to the dependence found for the mobility of ions as a function of moisture content. Similar observations were made by Hearle [23–27] in textile materials.

Phenomenologically, the conductivity of any material may be expressed in terms of the density of charge carriers n, the electronic charge on the carrier e, and the mobility of the carrier in the solid μ, as

$$\sigma = ne\mu \tag{7}$$

The mobility is defined as the average drift velocity of the carrier per unit electric field strength. If more than one type of ion is present,

$$\sigma = \sum_i n_i e_i \mu_i \tag{8}$$

where the subscript i refers to a particular ion species.

In paper, the alkali metal ions are often considered to be the predominant current carriers, although anions may also play an important role [56]. The metal ions are assumed to be bound at sites along the molecule but capable of being dissociated. In this case, n_i would be expected to be proportional to the salt content, according to a Boltzmann expression:

$$n_i = N_i \exp\left(\frac{-U_i}{2k_B T}\right) \qquad (9)$$

where

N_i = concentration of ionizable sites of species i
U_i = energy required to dissociate the ion
k_B and T = Boltzmann's constant and absolute temperature, respectively

The effect of changing moisture on n_i (or μ_i) is not clear at this point, but O'Sullivan's observation that the conductivity and mobility (measured separately) varied in the same manner with increasing moisture implies that the mobility is the dominant factor controlling the conductivity in the case of high salt contents. At low salt contents, the concentration of charge carriers is limited, and this factor must outweigh the strong effect of moisture on the mobility. This picture is probably too oversimplified, however, in that U_i in Eq. (9) would be expected to be a function of the moisture content because the energy required for dissociation will depend upon the local conditions at the ionizable site, and in particular on the dielectric constant at the site. The dielectric constant will change as moisture is added, because the dielectric constant of water is much greater than that of dry paper or other cellulosic materials. The increasing local dielectric constant at the ionizable site, as moisture is added, causes a decrease in the energy required for dissociation, with a subsequent increase in n_i (assuming that all sites are not already ionized). While the same type of argument may apply to the observed increase of mobility with increasing moisture, there does not appear to be a straightforward method of separating the contributions due to mobility and carrier concentration.

The effect of changing moisture content on the conductivity is very dramatic. Figure 2, for example, shows that the conductivity of a natural cellulosic (cotton) varies over 12 orders of magnitude as the moisture content is varied from bone dry to saturation. Murphy [42] studied this phenomenon in some detail. He found that the conductivity could be expressed in terms of the moisture content M (expressed in percent of the dry weight) as

$$\sigma = \sigma_{sm} \left(\frac{M}{M_s}\right)^m \qquad (10)$$

where

M_s = moisture content at saturation
σ_{sm} = conductivity at the saturation water content

For his cotton samples, Murphy found m to be 9.3. To explain this empirical result, he proposed that there were metal ion generating sites that could be ionized according to Eq. (9). These sites were assumed to be periodically distributed throughout the material and connected by a continuous chain of water adsorption sites. Murphy argued that a contribution to the conductivity occurred when there was a complete chain of water molecules adsorbed between the ion generating sites. The probability that only one water adsorption site is occupied is M/M_s. The probability that m of these would be

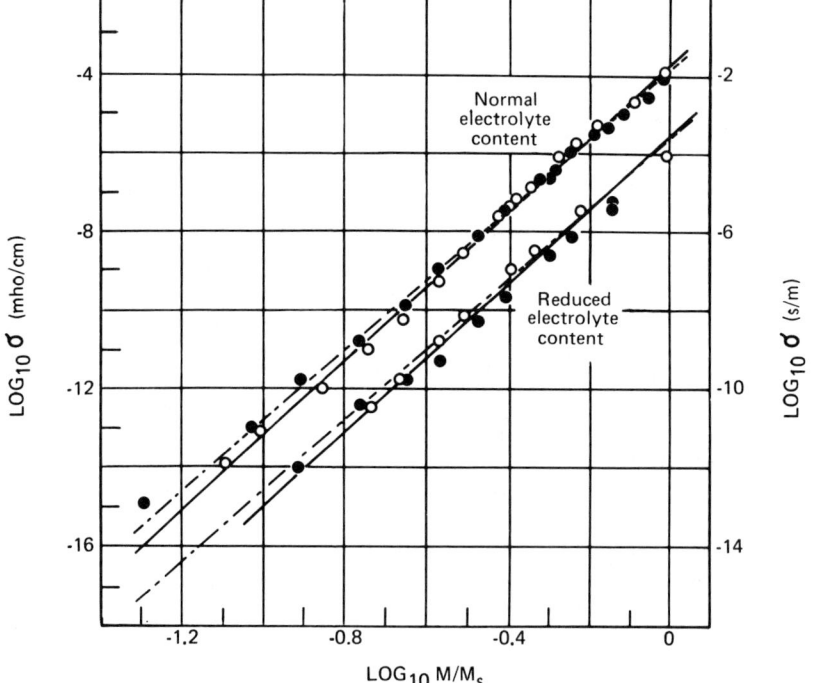

Fig. 2 Conductivity of a natural cellulosic (cotton) as a function of relative water content, M/M_s. Closed circles refer to an increasing sequence of relative humidities, open ones to a decreasing sequence. Solid lines are the empirical curves corresponding to the power 9.3 in Eq. (10). Broken lines correspond to the power 9 given by Murphy's model when m is restricted to integer values. (From Ref. 42.)

occupied simultaneously is $(M/M_s)^m$. Hence, in cotton, where experimentally m = 9.3, Murphy proposed there were nine water adsorption sites between each neighboring pair of ion generating sites.

Barker and Thomas [2], studying alkali-halide-doped cellulose acetate, suggested some modifications to Murphy's model which better explain their results and which provide a different viewpoint as to the role of the water. They assumed that the sample could be divided into small cells, each containing a site that can generate an ion if the appropriate number of water molecules enter the cell. Thus, in this case, m is the number of water molecules necessary for ion generation. If the distribution of water molecules among the cells is individually and collectively random, Poisson's distribution applies, leading to the expression

$$\sigma = A\sigma_{sm} \left(\frac{M}{M_s}\right)^m \exp\left[m_s\left(1 - \frac{M}{M_s}\right)\right] \quad (11)$$

where

> A = constant
> m_s = number of water molecules in the cell at saturation

In this model, m is perhaps related to the hydration numbers of the ions.

Although the mechanism by which moisture influences the conductivity may not be certain, the notion that discrete water molecules can affect the electrical properties is consistent with the phenomenological argument as presented earlier.

Ionic Conductivity and Other Models A useful semiquantitative model of the conduction process in paper is similar to one described by Seitz [59] for ionic conduction in alkali halides and which has also been applied to polymers [1,33]. This model assumes ions are generated according to Eq. (9), and that, additionally, the ions experience potential barriers due to the other constituents of the material. [The ions can recombine with a site of opposite sign, of course. Equation (9) describes the net concentration of carriers at equilibrium.] Some of the ions will have sufficient energy to overcome these barriers and thus migrate through the material. In the absence of an electric field, there will be as many ions moving to the right as to the left. Ion migration is assumed to be equally probable in any direction. When an electric field is present, however, the patter of migration of the ions is altered such that more (positive) charge carriers move parallel to the field than antiparallel. The net movement of ions constitutes a current that is detectable in the external circuit. The drift velocity of the ions parallel to the field (the mobility) will depend on the local environment.

For this ionic conductivity model, it is shown that the behavior is non-ohmic [does not obey Ohm's law, Eq. (1)]. In fact,

$$J = A \sinh(BE) \tag{12}$$

where

$$A = GebN \exp\left[-\frac{(U_i/2) + U_b}{k_B T}\right]$$

$$B = \frac{eb}{2k_B T}$$

and where

> G = constant involving, among other things, the vibrational frequency of the ion
> e = magnitude of the ion charge
> b = "jump distance," the distance traveled by the ion to overcome the potential barrier of height U_b

The quantity U_b is sometimes called the activation energy for mobility. The other parameters have been defined previously.

Theory and experiment are compared in Fig. 3, which gives J versus E for several papers and a regenerated cellulose film. Ohm's law is also shown on the semilogarithmic plot for comparison. The agreement between theory

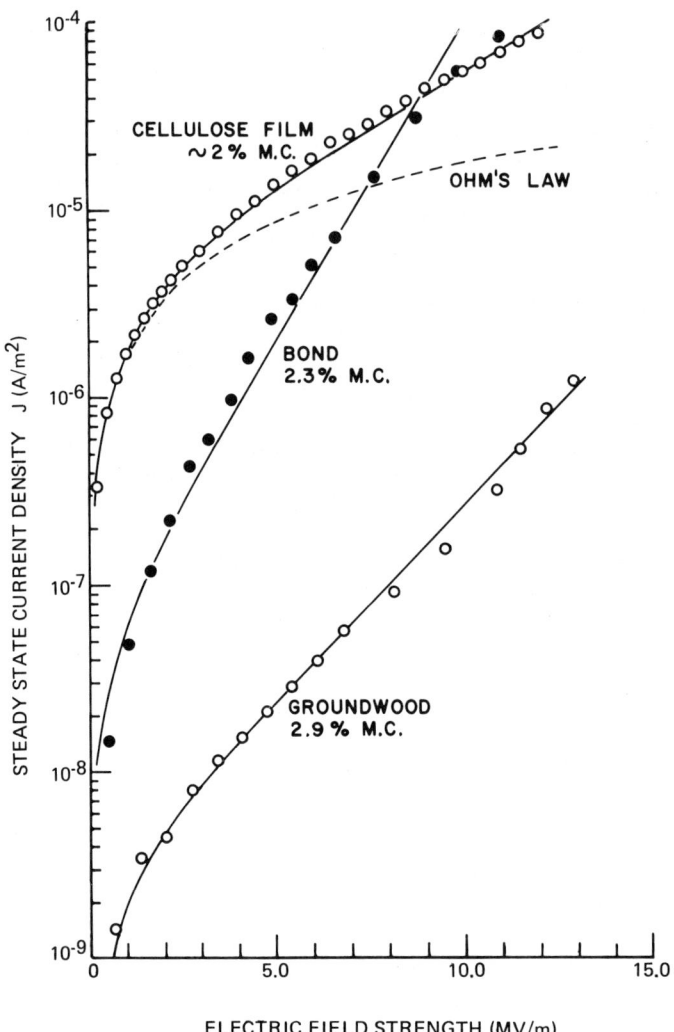

Fig. 3 Steady state current density versus electric field strength for several nondielectric papers and a cellulosic film at low moisture contents (M.C.). The solid lines correspond to the model given by Eq. (12).

and experiment is good, and values can be found for the parameters A and B. From B a value can be estimated for the jump distance b. For the cellulosic film shown in Fig. 3, if one assumes a monovalent ion, this distance turns out to be around 20 nm (200 Å), a surprisingly large value compared to, say, the length of a cellobiose unit (1.03 nm). If a divalent ion were assumed, b would be halved. It is not clear whether values of this magnitude are reasonable or whether the ionic conductivity model is inadequate when it comes to describing a heterogeneous material such as paper.

The samples of Fig. 3 suffered dielectric breakdown after the maximum value given for E for each material. That is, the dielectric strength was about 12×10^6 nt/coul. These samples were nondielectric papers and were not intended for use as insulating materials. Nonimpregnated dielectric papers should be able to withstand much higher field strengths than those in Fig. 3. At very high values of field strength, the hyperbolic sine function will be approximated by an exponential term. That is, J should increase exponentially with increasing E. Such behavior is observed [22,44]. Murphy [44] studied conduction in capacitor tissue as a function of E at field strengths up to 5.0×10^7 nt/coul (500 kv/cm). He found that the conductivity can be expressed as a sum of two exponential terms. The slope of the curve of log resistivity versus E decreases around 200 kv/cm. Murphy concluded that although the paper is an ionic conductor, the behavior is suggestive of a local electronic conduction in the vicinity of transient thermally generated defects.

Equation (12) correctly predicts the temperature dependence of the conductivity over a broad temperature range. If E is held constant, the hyperbolic sine term is relatively insensitive to changing T, compared to the exponential term. Thus for constant E and changing T,

$$\sigma = \frac{J}{E} = \sigma_0 \exp\left(\frac{-U}{k_B T}\right) \tag{13}$$

where the sinh term and constants are contained in the parameter σ_0 and $U = U_i/2 + U_b$. A typical plot of log resistivity ($= 1/\sigma$) versus T^{-1} is shown in Fig. 4 for a regenerated cellulose film. The energy U can be determined from the slope of this line, as 0.92 eV (21.2 kcal/mol). Murphy [43] studied the conductivity of dry capacitor tissue as a function of temperature in the range 25 to 175°C. He found that from room temperature to about 60°C the energy U was 10.6 kcal/mol, which he attributed solely to an activation energy for mobility, U_b, for impurity ion conduction. At higher temperatures, however, the slope of the curve ln σ versus T^{-1} increased, yielding an energy of 30.7 kcal/mol. Murphy argued that this value could be broken into a dissociation energy U_i of 40.2 kcal/mol and a mobility activation energy of 10.6 kcal/mol. That is, $U_i/2 + U_b = 30.7$ kcal/mol. According to Murphy, the 40.2 kcal/mol agrees favorably with the activation energy for thermal decomposition of regenerated cellulose of 39.5 kcal/mol, while 10.6 kcal/mol is approximately equal to the activation energy for conduction in ice.

The ionic conductivity model can predict the effects of moisture on the conductivity only qualitatively. As suggested earlier, the presence of water alters the local dielectric constant, which in turn would affect U (or U_i or U_b). The value of U should decrease with increasing moisture. Data testing this effect in paper appear to be scarce. Lin [36] has discussed the effect in the case of wood.

Other models have been proposed for conduction in paper. Hanneson et al. [22], studying capacitor tissue paper, suggested a model in which ionic space charges accumulate near one or both electrodes and thereby limit a predominantly electronic steady state current. The contribution of the ionic current is presumed to be transient. Although the idea that electrons carry the current in paper (or cellulosic materials) is contrary to all previous thinking, the model is successful in describing some experimental observa-

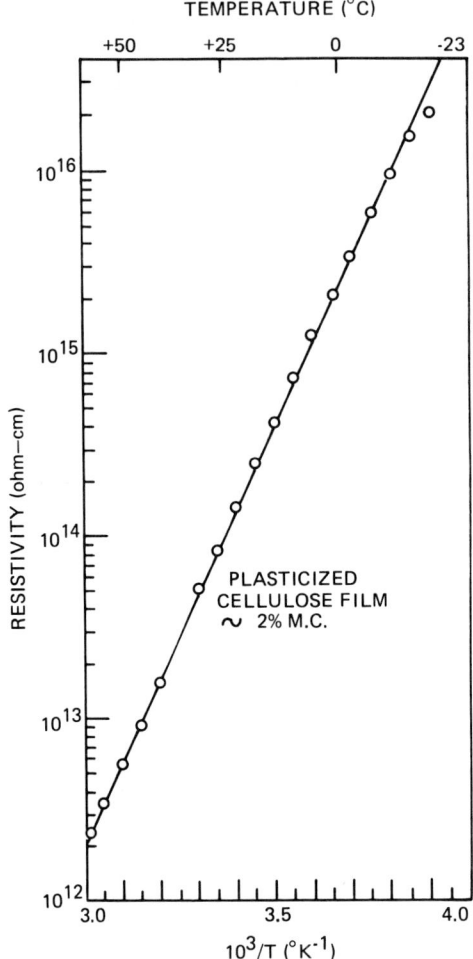

Fig. 4 Log resistivity versus inverse temperature for a plasticized cellulose film.

tions, and it should be given some attention. For example, in the case of the dependence between current density and field strength, the space charge limited electronic model predicts that J should vary exponentially with the square root of E. If the data in Fig. 3 are replotted as log J versus $E^{1/2}$, they do fall in a straight line (except at the lowest values of E). The model supposes that a barrier exists at the (electrode) metal/paper interface. The height of the barrier, in part, depends on the work function of the metal. If different electrode metals were to be used, differences would be expected in the measured currents. Such behavior, however, is not observed [44].

Time Effects If a voltage step function is applied to paper, several distinct time regions are observed in which the behavior of the current with time differs significantly. At very short times, in the order of microseconds, paper behaves as a typical dielectric material, with the current decaying exponentially with time according to

$$I = (V/R) \exp(-t/RC) \tag{14}$$

where

R = the resistance (ohms) of the paper sample
C = capacitance (farads) of the paper sample
V = applied voltage

For typical papers, the product RC, the *relaxation time*, is on the order of a few microseconds [4]. At longer times, greater than about 100 milliseconds, the current tends to approach a "nearly" steady state value. The current may slowly continue to decrease for minutes or hours before a true steady state value is reached, if at all. In the range from about 10 microseconds to 100 milliseconds, however, the current decays according to

$$I = at^{-p} \tag{15}$$

where a and p are constants. Such behavior is observed in many insulating materials.

Figure 5 shows a plot of log current versus log time, from 1 microsecond to 1 min, for a coating base stock paper (solid line). If the steady state current at 1 min is subtracted from the curve, the dashed line is obtained, showing the exponential behavior at short times and the t^{-p} behavior at the intermediate times. For the data shown, the value of p is 0.67. Although studies of this phenomenon in paper are limited, this value appears to be typical for paper and organic polymers [35,71,72]. Several explanations have been offered for the time dependence observed in polymers [70,71]. These involve electrons as charge carriers with trapping sites distributed over a range of energy levels.

At this time it is clear that neither the ionic model presented here nor any of the various electronic models proposed for other polymeric materials [30,55] are completely in accord with the available experimental evidence for paper. Perhaps in the future a definitive theory will be found, but in the interim, there is much additional experimental work that needs to be done.

Wood Pulp Fibers A number of theories have been put forth attempting to relate the mechanical properties of paper to the fiber properties and the fiber orientations in the web. Similar models have not been put forth in regard to the electrical properties of paper and their relationships to the individual fiber characteristics and distribution in the sheet. We will find in the next section that the anisotropy observed in the dielectric properties of machine-made papers can be related to geometry effects and probably anisotropy in the fibers themselves. Norimoto et al. [46] have related the anisotropy of the dielectric constant in the cell wall to the anisotropy of the dielectric constant in coniferous wood. Very little data on individual wood pulp

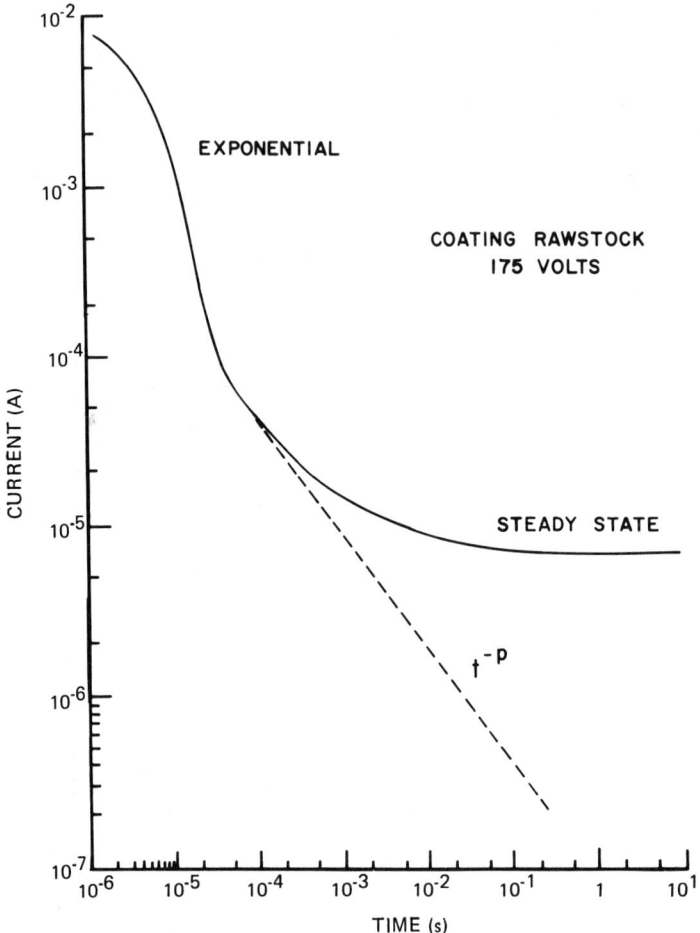

Fig. 5 Time dependence of the current when a constant voltage of 175 V is suddenly applied to a paper sample.

fibers are available. Smith [64,65] studied parallel groups of fibers and attempted to relate sheet conductivities to fiber conductivities. Lowe and Baum [38] made conductivity measurements on individual loblolly pine fibers. They found that the conductivity of both earlywood and latewood fibers varied exponentially with relative humidity. The in-plane conductivity of small handsheets made from the same fibers showed a similar dependence. Over the relative humidity range examined (11 to 84%) the fiber conductivity was 50 to 100 times greater than the paper conductivity.

B. Dielectric Constant

Introduction Historically, the dielectric constants of paper have received more attention than the conductivity, perhaps because the former are more closely related to the actual end-use requirements for most electrical papers.

As expected, the ion and water contents still play important roles. Delevanti and Hansen [12] provided a review of the early literature concerning the dielectric properties of chemical pulps, and Raman and Walker [54] have more recently reviewed the literature on electrical papers.

In order to understand the dielectric properties of paper, it is worthwhile to review the factors that influence the dielectric constants of materials in general. The dielectric constant or permittivity of any material is a measure of the polarizability of its constituents. The polarizability is an atomic property, whereas the dielectric constant also depends on the concentration and spatial distribution of the constituents. The polarizability may usually be separated into several parts. In the presence of an electric field, an electronic contribution arises from a displacement of the electron shell relative to a nucleus and an ionic contribution from the displacement of a charged ion with respect to other ions. In materials that possess molecular groups having permanent dipole moments, such as water or the hydroxyl and carboxyl groups in lignin or cellulose, the molecular dipoles will aslo make a contribution.

At low frequencies all of these parts contribute to the dielectric constant, as will any free ions (space charges) in the material. The space charges and permanent dipoles attempt to oscillate back and forth in phase with the electric field vector. Their ability to do this depends upon the inertia of the group, the nature of the local environment (steric hinderances), temperature, and frequency. Any interactions with the surrounding atomic structure can diminish the contribution of the group to the overall polarizability of the material, and any energy dissipated in the process will be reflected in the magnitude of the loss factor.

As the frequency increases, the space charges and permanent dipoles find it increasingly difficult to keep in phase with the rapidly changing electric field vector, and at some frequency they "relax out." That is, they no longer make a contribution to the polarizability or the real part of the dielectric constant. The relaxation is accompanied by a large increase in the imaginary part of the dielectric constant (loss factor). At still higher frequencies the group does not contribute to either k' or k" [refer to Eq. (6)]. Space charges are usually the first to relax out, followed by the permanent dipole groups (at frequencies in the radio frequency or microwave regions). Ionic and electronic polarizabilities still make contributions at infrared frequencies and above.

Cellulosic materials contain space charges and permanent polar groups, and both are important at low frequencies. If water is present, however, even in relatively small amounts, it is apt to make the major contribution due to its polar nature.

Machine-made papers are anisotropic in their mechanical properties because of fiber alignment with the wire during formation and because of restraints imposed on the web during drying. This anisotropy at the fiber level governs the distribution of polar groups in the material. Because the dielectric properties are sensitive to the manner in which the polar groups are arranged in space, the anisotropy of the paper would be expected to be manifest in the dielectric properties. This has been observed and is discussed in detail later. It is important to understand that the dielectric constant has different values along each of the three principal directions: the machine direction, the cross-machine direction, and the thickness or Z direc-

Table 1 Dielectric Constant (k') for Dry Paper and Other Cellulosic Materials

Dielectric Constant[a] k'	Frequency	Material	Moisture Content[b] (%)	Density (gm/cm³)	Temperature (°C)	Reference
1.33	100 kHz	Low ash filter paper	0	0.496	25	58
1.32	550 kHz					
1.31	1 MHz					
1.4	0.1 to 200 kHz	Printing paper	2.3	—	—	66
1.4	27 MHz	Newsprint	0[c]	0.49		13
1.4	27 MHz	Linerboard	0[c]	0.70		
1.8	27 MHz	Tracing paper	0[c]	1.03		
2.67	1 kHz	Viscose rayon	0	0.627	105	69
2.60	1 kHz	Alpha wood pulp	0	0.721	105	
1.67	10 kHz	Cotton linters	0	0.404		67
1.87	10 kHz	Bleached sulfite	0	0.485		
2.86	200 kHz	Cotton cellulose	0		25	63
2.64	2 MHz					
2.50	5 MHz					
2.42	10 MHz					
1.6	9.6 GHz	Newsprint	0[c]	0.375	25	21
1.8	9.6 GHz	Linerboard	0[c]	0.653	25	

[a]Dielectric constant not corrected for density. See text.
[b]Most measurements carried out in vacuum.
[c]Obtained by extrapolating plots of dielectric constant versus moisture to 0% moisture.

tion, in accordance with Eq. (2). In the following discussion the dielectric constant referred to is that measured in the thickness direction.

Low Moisture and Free Ion Concentrations At low moistures and low free ion concentrations, the polar groups of cellulose should be a major contributor to the dielectric constant at low frequencies. For dry, low ash papers, dielectric constants around 1.3 to 1.8 have been reported [13,21,58,63,66,67, 69]. Table 1 gives some literature values for dry paper and regenerated cellulose samples. The concentration of polar groups has an important bearing on the overall dielectric constant. Since the concentration depends on the density of the material, the dielectric constant would be expected to vary with density. Any process whose effect is to increase paper density, such as refining or wet pressing, should also increase the dielectric constant. Seidman and Mason [58] estimated that for their filter paper samples with a dielectric constant of 1.35 (10 kHz), the dielectric constant of the cellulosic fiber itself was 5.5. They also reported values for transverse and axial dielectric constants of the cellulose crystallite as 5.27 and 7.19, respectively, at 300 kHz.

Delevanti and Hansen [12] and Calkins [8] reported that the relationship between the density ρ and the (real part of the) dielectric constant of paper satisfied a Clausius-Mosotti type relationship:

$$\frac{\varepsilon' - 1}{\varepsilon' + 2} = K\rho$$

where K is a constant for a particular fiber and involves the polarizability of the atoms and their arrangment in space.

The ability of a polar group to respond to an electric field also depends upon the temperature and the local environment of the group in question. In crystalline regions, for example, the groups will not be as labile as similar groups in a noncrystalline region. The differences in steric hindrance between noncrystalline and crystalline regions have been found to result in measurable differences in the dielectric constant. Calkins [8] and Verseput [69] first studied the relationship between dielectric constant and the accessibility of cellulose, with additional work understaken by Kane [29]. Venkateswaran and Van den Akker [68] examined the dielectric properties and crystallinity of ramie, cotton linters, bleached sulfite pulp, and cellophane when the materials were treated with ethylamine in water. They found a linear relationship between the crystallinity and dielectric constant, the latter decreasing as the crystallinity increased.

Frequency Effects As noted earlier, increasing fequencies will tend to decrease k' and increase k". The decrease in k' can be seen in some of the data in Table 1. When relaxation occurs, the rising loss factor will reach a maximum and then decrease to some new low value, while the dielectric constant k' will also decrease to a new value. The polar group in question no longer makes a contribution to the complex dielectric constant. If more than one type of polar group is present, and they relax out at sufficiently different frequencies, separate peaks or bands may be observed in the loss part of the dielectric constant when it is plotted against frequency. The relaxation frequency increases with temperature. Because of the strong dependence upon the conditions of the local environment, such bands are apt to be broad and overlap one another, rather than to be sharp distinct bands, a reflection of the fact that the local conditions are likely to vary on an atomic scale from point to point.

No well-defined bands are observable for paper near room temperatures. Renne [56] observed a loss maximum at −55°C and 1 kHz for condenser paper and attempted to relate it to the cation concentration. Mikhailov et al [41] have reported a similar loss maximum at −67°C and 1 kHz and attributed it to the movement of the primary hydroxyl groups. Seidman and Mason [58] investigated the dielectric relaxation in dry cellulosic material and cellulose-containing sorbed vapors over the temperature range from −60°C to 30°C and a frequency range of 10 to 1000 kHz. For dry material, they found that the dielectric constant ε' increased over this temperature range, whereas the loss component ε'' went through a maximum. The temperature of the maximum increased as the measuring frequency increased.

Klason and Kubát [32] investigated loss factors in cellulosic materials for both the mechanical and the dielectric cases. The samples used were spruce sulfite pulp (containing cellulose I), regenerated cellulose (containing cellulose II) and a cellulose III, prepared by immersing the sulfite pulp in liquid ammonia for 24 h. They used a torsion pendulum for the mechanical measurements and a capacitance bridge for the dielectric measurements. For the latter, on dry samples, after plotting loss factor versus temperature Klason and Kubát found a loss maximum at −60°C and 1 kHz. The peak temperature increased with frequency and drying intensity, reaching about −20°C at 10 kHz for the most intensely dried samples. The dielectric loss peak had a counterpart in the mechanical case if a measuring frequency of 1 Hz was assumed, with a maximum at −73°C. The activation energy in either case was found to be 50 kJ/mol.

Shinouda and Hanna [63] measured the dielectric constants of dry cotton and cellulose derivatives at frequencies from 0.1 to 12 MHz and temperatures from 0 to 70°C. They found that for the cellulose derivatives at a given frequency, k' fell between the values for the cotton (lowest) and regenerated cellulose (highest). The results for the cotton are shown in Fig. 6. The values of k' steadily increase with temperature and decrease with frequency, as expected. Results for the cellulose derivatives behaved in a similar way. Values for the dielectric loss were not reported. The dielectric results, together with infrared and specific volume measurements, were explained in terms of the nature of the side groups and the degree of hydrogen bonding. It was argued that in cotton fibers the OH groups are relatively strongly bonded, whereas in regenerated cellulose they are more weakly bonded, thus resulting in a higher dielectric constant.

In general, for dry, low ash paper at ordinary temperatures and low frequencies (10 Hz to 10 kHz), little dispersion is exhibited. In some instances where effects have been noted, it is likely that water or some other impurity was present, or excessive ions in the system led to the loss [12]. Maxima in the dielectric loss factor have been reported when a DC bias voltage was applied simultaneously with the AC measuring field [48]. These have been attributed to impurities.

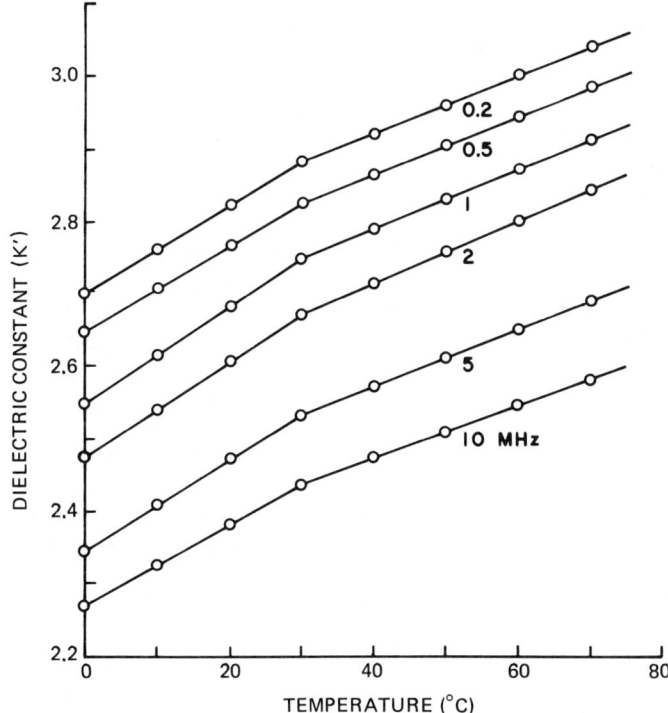

Fig. 6 Variation of the dielectric constant k' of cotton with temperature and frequency. (From Ref. 62.)

Salt Effects The effects of salts on the dielectric properties have been alluded to earlier. Delvanti and Hansen [12] demonstrated that metallic ions were responsible for much of the loss factor in paper at low frequencies. Driscoll [13] measured the loss factor of boxboard samples that had been soaked in various salt solutions. Very high loss factors were reported for samples soaked in tap water or alum in tap water as compared to deionized water. The effect decreased significantly with increasing frequency, suggesting that no salt effect would be observed above several hundred megahertz. Chu and Wyslouzil [11] examined the addition of salt and alum to paper at three microwave frequencies (10, 21, and 33 GHz). They reported a small effect due to the impurities at the lowest frequency but none at the highest frequency. Such a result is to be expected, since any contribution to the dielectric loss of paper due to metallic ions should relax out well before the microwave range.

Effects of Water The effects of moisture in the low frequency region have been studied extensively. References 54 and 28 review the literature for paper and wood and hardboard. At frequencies below the microwave region, the dielectric constant of water is around 80, a value significantly higher than the other normal constituents of paper. Thus if water is present in even relatively small quantities, it can significantly influence the dielectric properties of paper.

Tsuge and Wada [66] found that above a critical water content a dielectric dispersion was observed in their printing paper and cellophane samples, the intensity of which increased linearly with increasing water content. They argued that the first water molecules sorbed in polysaccharide are associated with unbound hydroxyl groups, that is, groups that are not involved in inter- or intramolecular hydrogen bonds. At some critical moisture content, however, the sorbed water begins to break up the inter- and intramolecular bonds (in noncrystalline regions) and in so doing allows more and more polar groups to contribute to the polarizability.

For paper and cellophane, these critical moisture levels were estimated to be about 3% and 6% water, respectively. It was proposed that below the critical value the water had essentially no effect and the polar groups are unable to contribute significantly because of steric hindrances due to hydrogen bonding. At moisture above the critical value, the polar groups are more labile and dielectric dispersion was observed. The higher critical value for cellophane, compared to paper, was assumed to be a result of the greater crystallinity for the cellophane and the presence of other organic materials in the case of the paper.

As more and more water is added to paper, a point may be reached where each additional water molecule added behaves as if it were free. It is quite probable that the bound and free water portions behave dielectrically quite different. In this case, it would be necessary to define different dielectric constants for the bound and free water. Pure water undergoes relaxation at about 22 GHz. When it is sorbed in paper, it is likely that this relaxation would occur at much lower frequencies. That is, the bound and free portions of water also would be expected to behave differently in terms of their frequency dependence.

A number of researchers have measured the dielectric constants of paper at high frequencies [7,9—11,13,14,17,21,34,39,60—62]. The contributions of bound water and free water to the dielectric properties at microwave frequencies have been discussed by Busker [7], Dusoiu [14], and M'Baye and Pellissier [39]. The last-named authors estimated the free and bound water concentrations in paper from microwave measurements of dielectric constants using a form of the Kirkwood equation [31]. These authors applied the equation to relate the dielectric constants of paper to the dielectric constants and volume fractions of its constituents. Assuming values for the dielectric constants of bound and free water, the volume fractions of these two constituents were estimated. The results were dependent upon the relative orientation of the electric field vector with the normal vector to the paper sheet. The inhomogeneity of the mixture was ignored in these calculations but properly should be included [21]. At present, however, the appropriate way to do this is not obvious.

C. Dielectric Anisotropy

Introduction Measurements of dielectric constants in the radio frequency and microwave ranges reveal that paper and board are anisotropic with respect to these parameters. Complex dielectric constants must be defined for each of

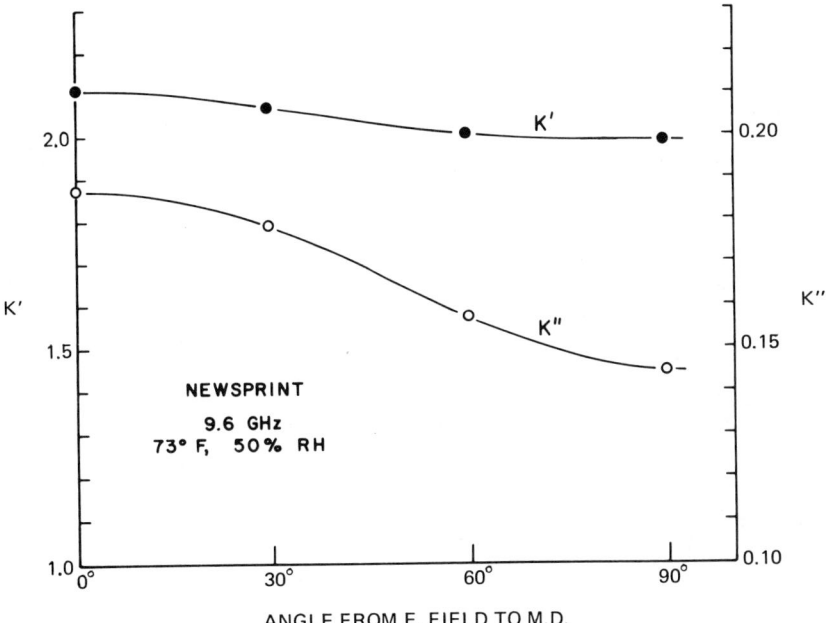

Fig. 7 In-plane dielectric constant for machine-made newsprint as a function of the orientation of the electric field with respect to the machine direction. (From Ref. 21.)

the three principal directions: the machine direction, cross-machine direction, and Z or thickness direction. A number of researchers have studied this anisotropy [13,15,16,18,21,60,62]. Figure 7 shows the real and imaginary parts of the complex dielectric constant of a machine-made newsprint, measured at 9.6 GHz and a constant moisture content of 9%, as a function of the angular displacement between the electric field and the machine direction of the paper. The highest values are obtained when the field is parallel to the machine direction. This in-plane anisotropy is believed to arise from the fiber orientation in the machine direction resulting from formation of the sheet on the wire. Fainberg et al. [18] have used the dielectric anisotropy to estimate the degree of orientation of molecular chains and aggregrates.

Driscoll [13] has examined the three-dimensional dielectric anisotropy in machine-made papers at radio frequencies (13.56 to 100 MHz). Figures 8 and 9 show k' and k", respectively, for a boxboard sample as a function of moisture content. In each figure, the three curves correspond to the electric field vector aligned with each of the three principal directions in the paper. As noted above, the highest values are obtained when the field is parallel to the fibers. On the other hand, the lowest values of the dielectric constant are obtained when E is perpendicular to the sheet, and thus perpendicular to the majority of fibers. The data indicate that the sorbed water apparently reflects the anisotropy of the fibrous structure, although the anisotropy

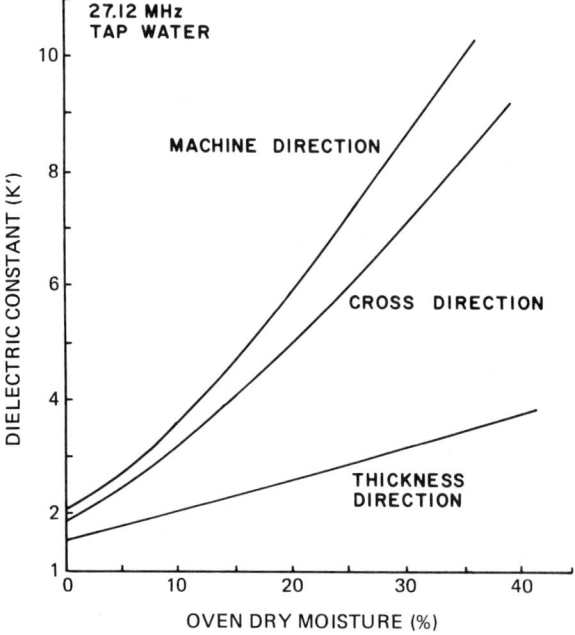

Fig. 8 The real part of the dielectric constant k' in the three principal directions of a boxboard sample as a function of moisture content. (From Ref. 13.)

ratio (the ratio of two dielectric constants at a given moisture) does increase with increasing moisture. Clearly more study of this phenomenon is needed.

Thermally Stimulated Depolarization Currents The ionic thermocurrent or thermal depolarization current technique of Bucci et al. [6] (sometimes also referred to as thermally stimulated currents, although this is a different technique altogether) is a method of investigating the dielectric properties of solids. In particular, it provides a means of studying the response of polar groups in a material to an applied electric field. Since the response will also be sensitive to the local environment, the effect can be examined of any thermal, mechanical, or electrical treatments that alter the number, character, or environment of the polar groups.

The method involves the application of an electric field to a specimen at a constant temperature (Fig. 10). This causes a polarization of the specimen or, if it has permanent dipole groups, the partial alignment of these groups with the applied field. The temperature of the material is then lowered while the electric field is maintained. The temperature is reduced to a value sufficiently low that the polarization due to the permanent dipoles is "frozen in." That is, when the field is removed at the low temperature and the sample electrodes are shorted, the induced (ionic and electronic) polarizations will relax out, whereas the polarization caused by the permanent groups cannot. This is because the time constant for the dipoles to return

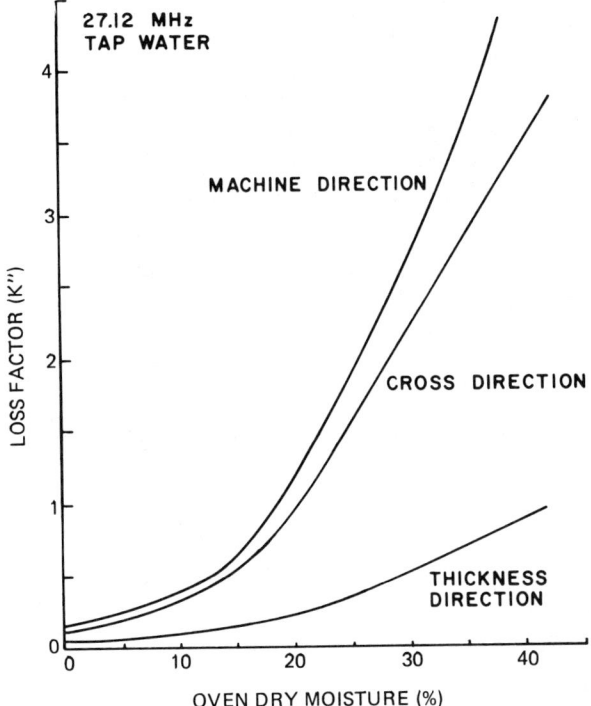

Fig. 9 The imaginary part of the dielectric constant k" in the three principal directions of a boxboard sample as a function of moisture content. (From Ref. 13.)

to a random orientation is very long at the low temperature. If the sample electrodes are connected to an ammeter and the specimen is slowly warmed up, currents will be measured by the meter over certain temperature ranges. Typically, current peaks or bands are observed, the current increasing from zero to a maximum at some temperature and then decaying back to zero.
The shape and area of this current band depends upon the magnitude of the dipole moment, the polar group involved, the concentration of groups present, the local environment of the polar group, the temperature at which the specimen was polarized, and the electric field strength. The temperature of the current maximum depends upon the dipole relaxation time, among other things, which in turn is sensitive to the nature of the local environment. For example, a polar group such as the hydroxyl group in cellulose would be expected to behave quite differently in crystalline and noncrystalline regions, or in the presence of water or some other plasticizer. A number of theoretical treatments of the phenomenon have been published [6,20,45]. McKeever and Hughes [40] have analyzed and described the reverse procedure, which involves cooling the specimen in the absence of an electric field and then warming the material in the presence of the field.

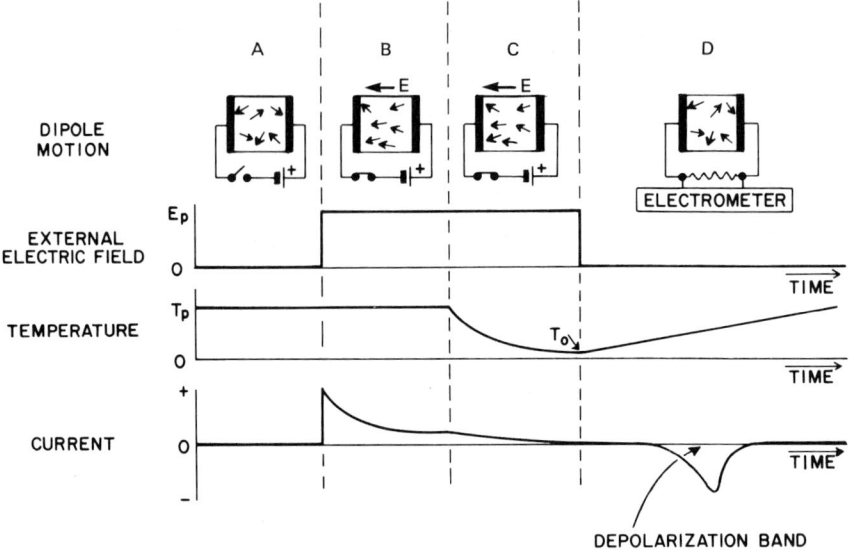

Fig. 10 The thermal depolarization technique. A. At the polarization temperature T_p with no external electric field the dipoles in the material are randomly oriented. B. When the polarization electric field E_p is applied the dipoles will attempt to align with E_p. This results in a polarization current that decays with time to some steady state value. C. With E_p applied the temperature of the material is lowered to temperature T_0, such that the dipoles are frozen in their aligned configuration. In this state, if the field is removed and the sample electrodes are shorted, no current will flow. D. If the sample is heated, eventually the dipoles can relax back to a random orientation. This causes a current in the external circuit, the depolarization bond. A linear heating rate is not essential but makes analysis of the data simpler.

The thermal depolarization current technique is especially suitable for materials, such as cellulose, which have polar side groups. A number of researchers have applied the method to cellulosic fibers and their derivatives. Baum [3] studied dry regenerated cellulose films in the temperature range −100°C to +50°C, and reported depolarization bands at −58, −2, and 25°C. The most intense band was that at −2°C and was shown to be caused by the presence of a glycerol plasticizer in the cellulosic film. This band completely disappeared when the glycerol was extracted. The low-temperature band was not influenced by the presence of the plasticizer and was thought to arise from the rotation of the primary hydroxyl group on the glucose unit. The band with the maximum at 25°C was believed to be related to the transition that is reported to occur in dry cellulose at this temperature. According to Wahba [70] extensive hydrogen bond breakage occurs above the transition temperature. If this is true, the thermal depolarization method should be ideally suited for investigation of such an event. Since the method is sensitive to the density of polar groups and the ease with which they can respond to the external field, the presence of the electric field, which causes changes in the orientation of the dipoles, would prohibit reformation of the hydrogen bond, especially as the specimen is cooled.

Sawatari et al. [57] studied cellulosic powder and some cellulose derivatives in the temperature range from −175 to +120°C using various techniques for drying the specimens. They observed a broad triplet peak at −140°C, which they attributed to the primary hydroxyl group in the amorphous region. The magnitude of a band at −70°C decreased as the drying conditions became more severe and was thus attributed to sorbed moisture. A third peak at 70°C was believed to be related to the motion of the main chain, although in this case the nature of the polar group giving rise to the effect is not clear. Okabe and coworkers [47] used the thermal depolarization method to study interfacial polarization mechanisms in oil-impregnated kraft paper systems. It seems reasonable to assume that in the future this technique will find wider applications with respect to paper systems.

SYMBOLS

a	Constant, Eq. (15)
A	Area (m^2)
A	Constant, Eq. (12)
b	Jump distance
B	Constant, Eq. (12)
C	Capacitance (farads)
C_0	Geometric capacitance (farads) with vacuum dielectric
d	Caliper
D	Electric displacement vector (coulombs/m^2)
D	Dissipation factor (unitless)
δ	Dielectric loss angle
E	Electric field strength (Newtons/coulomb)
ε	Complex permittivity (farads/m)
ε'	Real part of complex permittivity (farads/m)
ε''	Imaginary part of complex permittivity (farads/m)
ε_0	Permittivity of free space (farads/m)
e	Electronic charge (coulombs)
e_i	Electronic charge (coulombs) of ion of species i
f	Frequency (Hz)
G	Constant, Eq. (12)
i	Dummy index
i	$(-1)^{1/2}$
I_c	Charging current (amps)
I_L	Loss current (amps)
I_T	Total current (amps)
J	Steady state current density (amps/m^2)

k   Complex dielectric constant
k'   Real part of complex dielectric constant
k"   Imaginary part of complex dielectric constant
k_B Boltzmann's constant
K   Constant

L   Electrode separation distance (m)

m   Constant, Eq. (10)
m_s Constant, Eq. (11)
M   Moisture content
M_s Moisture content at saturation
μ Carrier mobility, (cm^2/volt · s)
μ_i Carrier mobility of ion species i (cm^2/volt · s)

n   Density of charge carriers (cm^{-3})
n_i Density of charge carriers of ion species i (cm^{-3})
N_i Concentration of generating sites of ion species i (cm^{-3})

p   Constant, Eq. (15)
ϕ Phase angle

Q   Electric charge (coulombs)

R   Electric resistance (ohms)

ρ Density (gm/cm^3)
ρ Resistivity (ohm · cm)
ρ_s Surface resistivity (ohms)

σ Conductivity (mho/cm)
σ_s Surface conductivity (mho)
σ_{sm} Conductivity at moisture saturation (mho/cm)

t   Time (s)
T   Temperature (°K)

U   Activation energy at constant field strength (eV)
U_b Activation energy for mobility (eV)
U_i Ionization energy to liberate ion species i (eV)

w   Angular frequency (sec^{-1})

V   Voltage (volts)
V_0 Voltage at time zero (volts)

REFERENCES

1. Amborski, L. E. (1962). Structural dependence of the electrical conductivity of polyethylene terephthalate. *J. Polymer Sci.* 62:331–346.
2. Barker, R. E., and Thomas, C. R. (1964). Effects of moisture and high electric fields on conductivity in alkali-halide-doped cellulose acetate. *J. Appl. Phys.* 35(11):3203–3215.
3. Baum, G. A. (1973). Thermal depolarization currents of regenerated cellulose films. *J. Appl. Polymer Sci.* 17:2855–2866.

4. Brodie, I., Dahlquist, J. A., and Sher, A. (1968). Measurement of charge transfer in electrographic processes. *J. Appl. Phys.* 39(5): 1618–1624.
5. Brown, J. H., Davidson, R. W., and Skaar, C. (1963). Mechanism of electrical conduction in wood. *Forest Prod. J.* 13(10)455–459.
6. Bucci, C., Fieschi, R., and Guidi, G. (1966). Ionic thermocurrents in dielectrics. *Phys. Rev.* 148(2):816–823.
7. Busker, L. H. (1968). Measurement of water content above 30% by microwave absorption methods. *Tappi* 51(8):348–353.
8. Calkins, C. R. (1950). Studies of dielectric properties of chemical pulps. 3. Dielectric properties of cellulose. *Tappi* 33(6):278–285.
9. Chene, M., Coumes, A., and Lafaye, F. (1965). Cellulose permittivity in the 9000 megahertz wavelength band. *C. R. Acad. Sci. (Paris)* 260:3632–3635. (In French.)
10. Chene, M., Revlo, N., Pellissier, J. P., and Mesnard, G. (1967). Interferometric measurement in band X of the dielectric constant of paper and its application to moisture content determination. *Cellulose Chem. Tech.* 1:597–600. (In French.)
11. Chu, F. Y., and Wyslouzil, W. (1977). Frequency dependence of microwave moisture measurements of paper. *Tappi* 60(10):144-145.
12. Delevanti, C., and Hansen, P. B. (1945). Studies of dielectric properties of chemical pulps. 1. Methods and effects of pulp purity. *Paper Trade J.* 121(26):25–33.
13. Driscoll, J. L. (1976). The dielectric properties of paper and board and moisture profile correction at radio frequency. *Paper Tech. Ind.* (April):T42–46.
14. Dusoiu, N. (1975). Decrease of high-frequency dielectric hysteresis by water desorption at different temperatures. *C. R. Acad. Sci.* B280(24):777–779.
15. Dusoiu, N. (1976). Dielectric anisotropy of stacks of paperboard and paper in the microwave range. *Rev. Roum. Chim.* 21(4):563–570.
16. Dusoiu, N., Balanescu, G., and Liviu, T. (1976). Dependence of dielectric permittivity of paperboard and paper on density. *Celuloza Hirtie* 25(1):34–37.
17. Dusoiu, N., Balanescu, G., and Liviu, T. (1976). Method of measuring the complex permittivity of paperboard and paper in the microwave range. *Celuloza Hirtie* 25(2):57–63.
18. Fainberg, E. Z., Eifer, I. Z., and Mikhailov, N. V. (1966). Electric anisotropy of regenerated cellulose fibers. *Khim. Volokna* 4:38–41. [ABIPC 37:541 (1967).]
19. Friedrich, R. E., and Chiu, T. T. (1970). Comparison of AC and DC methods of measuring conductivities in electrophotographic papers. *Tappi* 53(2):382–384.
20. Gross, B. (1975). On the analysis of thermally stimulated depolarization effects. *J. Phys. D: Appl. Phys.* 8:L127-L128.
21. Habeger, C. C., and Baum, G. A. (1983). The microwave dielectric constant of water-paper mixtures: The role of sheet structure and composition. *J. Appl. Poly. Science* 28:969–981.

22. Hanneson, J. E., Raman, R., and Hart, J. (1971). Electrical conductivity in tissue paper. *Tappi* 54(6):955–958.
23. Hearle, J. W. S. (1952). The electrical resistance of textile materials: A review of the literature. *J. Textile Inst.* 43:P194–P223.
24. Hearle, J. W. S. (1953). The electrical resistance of textile materials: The influence of moisture content. *J. Textile Inst.* 44:T117–T143.
25. Hearle, J. W. S. (1953). The electrical resistance of textile materials: The effect of temperature. *J. Textile Inst.* 44:T144–T154.
26. Hearle, J. W. S. (1953). The electrical resistance of textile materials: Miscellaneous effects. *J. Textile Inst.* 44:T155–T176.
27. Hearle, J. W. S. (1953). The electrical resistance of textile materials: Theory. *J. Textile Inst.* 44:T177–T198.
28. James, W. L. (1975). Dielectric properties of wood and hardboard: Variation with temperature, frequency, moisture content, and grain orientation. USDA Forest Service Research Paper FPL 245, Madison, Wisconsin.
29. Kane, D. E. (1955). The relationship between the dielectric constant and water-vapor accessibility of cellulose. *J. Polymer Sci.* 18:405–410.
30. Karasz, F. E., ed. (1972). *Dielectric Properties of Polymers*. Plenum, New York.
31. Kirkwood, J. G. (1939). The dielectric polarization of polar liquids. *J. Chem. Phys.* 7:911–919.
32. Klason, C., and Kubát, J. (1976). Thermal transitions in cellulose. *Svensk Papperstid.* 79:494–500.
33. Kosaki, M., Sugiyama, K., and Ieda, M. (1971). Ionic jump distance and glass transition of polyvinyl chloride. *J. Appl. Phys.* 42(9):3388–3392.
34. Kumar, A., and Smith, D. G. (1976). The measurement of the complex permittivity of paper at microwave frequencies. *Tappi* 59(1):149–151.
35. Lengyel, G. (1966). Schottky emission and conduction in some organic insulating materials. *J. Appl. Phys.* 37(2):807–810.
36. Lin, R. T. (1965). A study on the electrical conduction in wood. *Forest Products J.* 15(11):506–514.
37. Lin, R. T. (1973). Wood as an orthotropic material. *Wood Fiber* 5(3):226–236.
38. Lowe, G. R., and Baum, G. A. (1979). Electrical conductivity of single wood pulp fibers. *Tappi* 62(6):87–89.
39. M'Baye, K., and Pellissier, J. P. (1975). Determining the nature of water sorbed on cellulosic fibers through microwave measurements of dielectric constants. *Revue ATIP* 29(2):51–54. (In French.)
40. McKeever, S. W. S., and Hughes, D. M. (1975). Thermally stimulated currents in dielectrics. *J. Phys. D: Appl. Phys.* 8:1520–1529.
41. Mikhailov, G. P., Artyukov, A. I., and Borisova, T. I. (1967). Characteristics of relaxation of cellulose hydroxyl groups at low temperatures. *Vysokomol. Soed.* 9(B), no. 2:138–141 (in Russian).

42. Murphy, E. J. (1960). The dependence of the conductivity of cellulose, silk and wool on their water content. *J. Phys. Chem. Solids.* 16:115–122.
43. Murphy, E. J. (1960). The temperature dependence of the conductivity of dry cellulose. *J. Phys. Chem. Solids.* 15:66–71.
44. Murphy, E. J. (1974). High field conduction in native cellulose and its structural implications. *J. Coll. Interface Sci.* 49(3):442–452.
45. Nedetzka, T., Reichle, M., Mayer, A., and Vogel, H. (1970). Thermally stimulated depolarization: A method for measuring the dielectric properties of solid substances. *J. Phys. Chem.* 74(13): 2652–2666.
46. Norimoto, M., Hayashi, S., and Yamada, T. (1978). Anisotropy of dielectric constant in coniferous wood. *Holzforschung* 32(5):167–172.
47. Okabe, Y., Yamashita, H., and Amano, H. (1978). Thermally stimulated current in oil impregnated kraft paper systems. Proceedings of the 11th Symposium on Electrical Insulating Materials in Japan, September, pp. 113–116.
48. Olach, O., and Calderwood, J. H. (1977). The effect of simultaneous application of AC and DC voltages on dielectric loss. *J. Phys. D: Appl. Phys.* 10:L257–L259.
49. O'Sullivan, J. B. (1947). The conduction of electricity through cellulose. 1. The conductance of cellulose sheet impregnated with salts. *J. Textile Inst.* 38:T271–T284.
50. O'Sullivan, J. B. (1947). The conduction of electricity through cellulose. 2. The chemical effects of the current. *J. Textile Inst.* 38:T285–T290.
51. O'Sullivan, J. B. (1947). The conduction of electricity through cellulose. 3. The mobility of hydrogen and hydroxyl ions in cellulose sheet. *J. Textile Inst.* 38:T291–T297.
52. O'Sullivan, J. B. (1947). The conduction of electricity through cellulose. 4. The mobility of various ions in cellulose sheet. *J. Textile Inst.* 38:T298–T306.
53. O'Sullivan, J. B. (1948). The conduction of electricity through cellulose. 5. The effect of temperature. *J. Textile Inst.* 39:T368–T384.
54. Raman, R., and Walker, S. (1972). Electrical papers: A literature review. *Paper Technol.* (April):126–132.
55. Rembaum, A., and Landel, R. F., eds. (1967). Electrical conduction properties of polymers. *J. Polymer Sci., C, Polymer Symposia.*
56. Renne, V. T. (1968). Influence of inorganic impurities on the dielectric properties of electric insulating paper. *Zellstoff Papier* 17:343–345.
57. Sawatari, A., Kurihara, H., and Takashima, T. (1978). Thermally stimulated current of cellulose powder. *J. Japan Wood Res. Soc. (Mokuzai Gakkaishi)* 24(4):224–229 (in Japanese).
58. Seidman, R., and Mason, S. G. (1954). Dielectric relaxation in cellulose containing sorbed vapors. *Can. J. Chem.* 32:744–762.
59. Seitz, F. (1940). *The Modern Theory of Solids.* McGraw-Hill, New York.

60. Servant, R., and Cazayus-Claverie, J. (1957). Dielectric anisotropy of paper at 3400 megahertz. The influence of moisture. *Comptes Rendus* 245:509–511. (In French.)
61. Servant, R., and Gougeon, J. (1960). Birefringence and rectilinear dichroism of paper at 3000 MHz. *Comptes Rendus* 242:2318–2320. (In French.)
62. Servant, R., and Weever, J. W. (1960). Dielectric anisotropy of boards at 3000 megahertz. *J. Phys. Rad.* 21:95S-96S.
63. Shinouda, H. G., and Hanna, A. A. (1977). Dielectric and infrared study of some cellulose derivatives. *J. Appl. Polymer Sci.* 21:1479–1488.
64. Smith, W. E. (1965). Determination of the relative bonded area of handsheets by direct-current electrical conductivity. *Tappi* 59(1):476–480.
65. Smith, W. E. (1968). An investigation of a method for measuring interfiber bonding in pulp handsheets based on sheet and fiber DC electrical conductivities. Ph.D. Dissertation, North Carolina, State University, Raleigh.
66. Tsuge, K., and Wada, Y. (1962). Effect of sorbed water on dielectric dispersion of cellulose at low frequencies. *J. Phys. Soc. (Japan)* 17(1):156–164.
67. Venkateswaran, A. (1965). Formulas for the dielectric constant and dissipation factor of mixtures and their application to the cellulose system. *J. Appl. Polymer Sci.* 9:1127–1138.
68. Venkateswaran, A., and Van den Akker, J. A. (1965). Effect of ethylamine treatment on the dielectric properties and crystallinity of cellulose. *J. Appl. Polymer Sci.* 9:1149–1166.
69. Verseput, H. W. (1951). Studies of dielectric properties of chemical pulps. 4. The relationship between the dielectric constant and crystallinity of cellulose. *Tappi* 34(12):572–576.
70. Wahba, M. (1968). The effect of drying and of temperature on the infrared spectra of regenerated cellulose films, in relation to the second order transition of cellulose around 25°C. *Arkiv Kemi* 29(32):395–413.
71. Wilcox, P. (1972). A dielectric loss model based on interfacial electron tunneling. *Can. J. Phys.* 50:912–924.
72. Wintle, H. J. (1973). Absorption current, dielectric constant, and dielectric loss by the tunneling mechanism. *J. Appl. Phys.* 44(6):2514–2519.

22
ELECTRICAL PROPERTIES: II. PRACTICAL CONSIDERATIONS AND METHODS OF MEASUREMENT OF ELECTRICAL PROPERTIES

SHINJI MATSUDA

Technical Research Laboratory
Tomoegawa Paper Company
Shizuoka-shi, Japan

I.	Practical Considerations	201
	A. Introduction: Requirements of Electrical Grade Papers	201
	B. Interrelationships Between Dielectric Properties and Other Sheet Parameters	204
	C. Relationships Between Dielectric Breakdown Strength and Physical Properties of Paper and Refining Process Parameters	215
	D. Current Insulation Research	220
II.	Methods of Measurement of Electrical Properties	221
	A. Measurement of Dielectric Properties	221
	B. Measurement of Dielectric Breakdown Strength	235
	References	238

I. PRACTICAL CONSIDERATIONS

A. Introduction: Requirements of Electrical Grade Papers

The basic materials for the manufacture of electrical machinery are conductors and insulators. Copper comes first among the conductors, while synthetic insulating materials such as PVC and PE have recently made their debut into the diverse field of insulators. For more than 100 years paper has had unique uses as an insulating material. The reasons for this abundant use of paper as an insulating material may be summarized as follows:

Table 1 Factors Influencing the Dielectric Properties of Paper: Influence of Wood Material, Pulp, and Paper-making conditions.

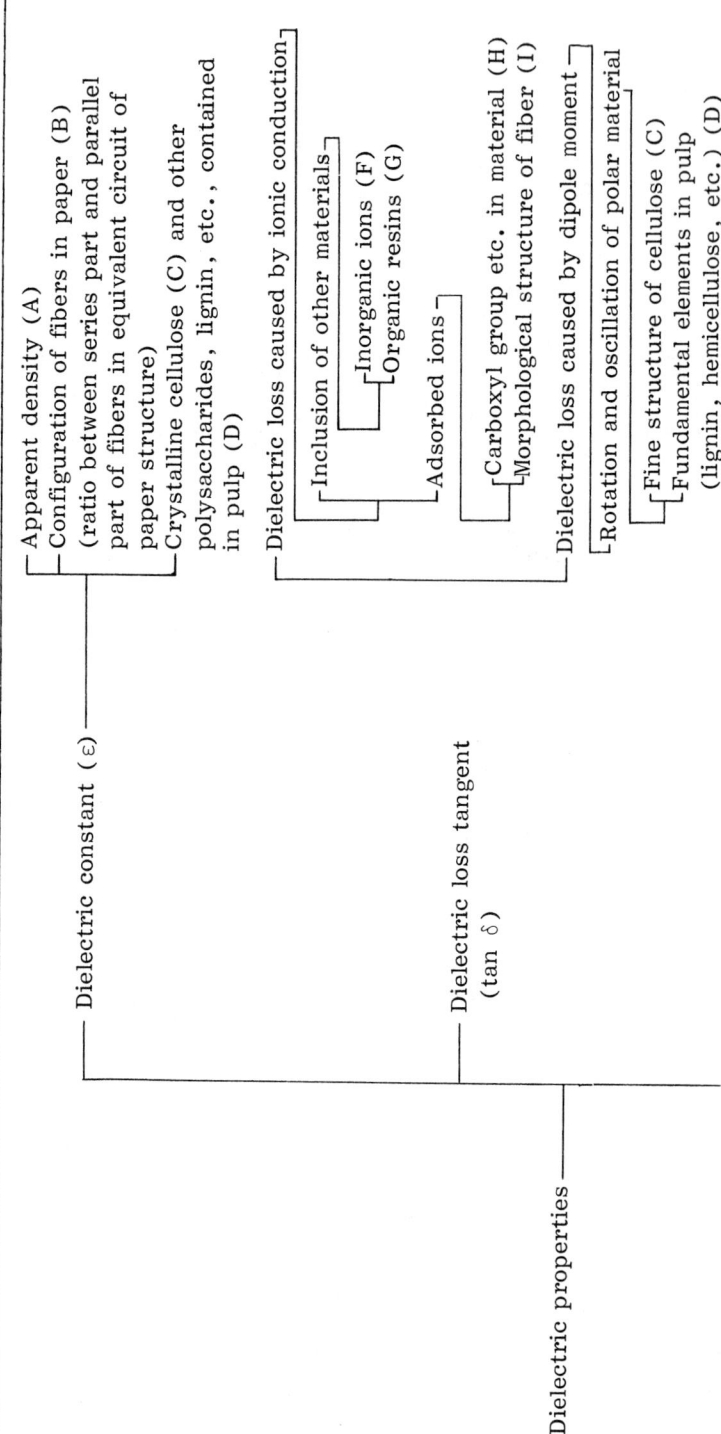

Electrical Properties: II. Practical Considerations

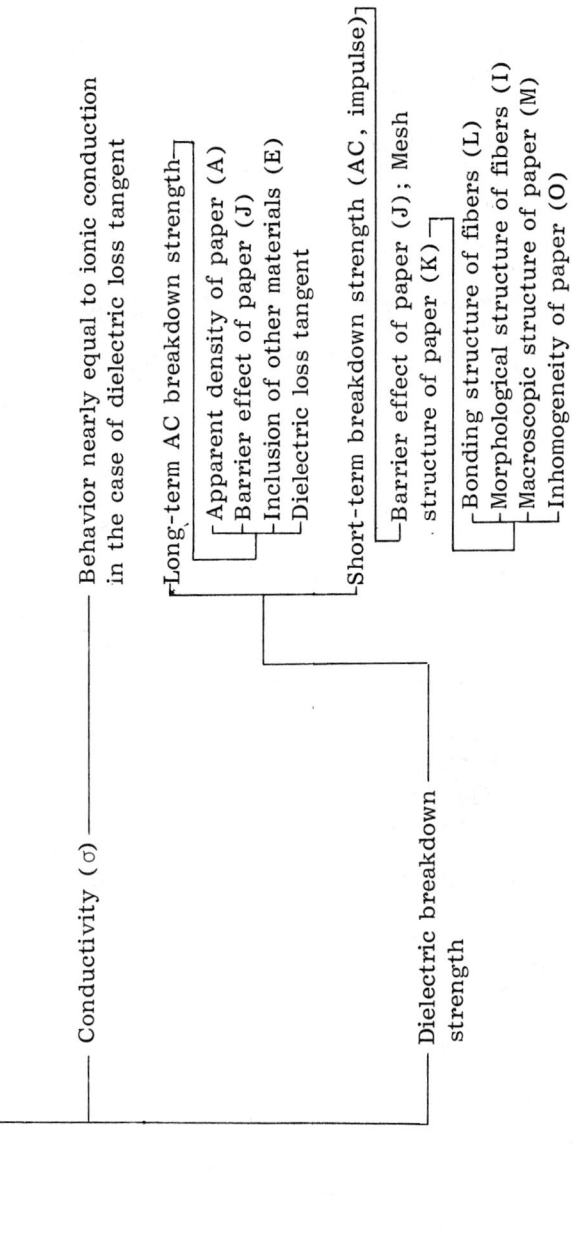

- Conductivity (σ) —— Behavior nearly equal to ionic conduction in the case of dielectric loss tangent
- Dielectric breakdown strength
  - Long-term AC breakdown strength
    - Apparent density of paper (A)
    - Barrier effect of paper (J)
    - Inclusion of other materials (E)
    - Dielectric loss tangent
  - Short-term breakdown strength (AC, impulse)
    - Barrier effect of paper (J); Mesh structure of paper (K)
    - Bonding structure of fibers (L)
    - Morphological structure of fibers (I)
    - Macroscopic structure of paper (M)
    - Inhomogeneity of paper (O)

- Low cost.
- Excellent properties in both mechanical strength and flexibility.
- Thickness can be held rather uniformly over a large area.
- Chemical stability and long life.
- Excellent electrical properties in dry condition.
- High dielectric breakdown strength can be obtained by oil impregnation.

Insulating paper is used for power cables, telephone cables, capacitors, transformers, and so on. For the material to behave properly in these applications, low dielectric loss and high dielectric breakdown strength are required.

The dielectric loss Q is expressed by the following equation:

$$Q = 2\pi f C V^2 \tan \delta \tag{1}$$

where

f = frequency
C = capacity
V = applied voltage
$\tan \delta$ = dielectric loss tangent (the dissipation factor)

The dielectric loss tangent was defined in Eq. (3) of Chapter 21. In order to decrease the dielectric loss Q, it is necessary to lower $\tan \delta$, which is a property peculiar to insulating materials. The $\tan \delta$ (and therefore the dielectric loss) of the paper has much to do with the kinds and quantities of metallic ions, lignin content, hemicellulose content, and carboxyl group content in the paper. However, the dielectric breakdown strength of the paper depends not only on its physical properties and sheet formation, but also on various conditions in the paper manufacturing process. Various factors that reflect on the electrical properties of the paper as mentioned above are shown in Tables 1 and 2. Table 1 shows the factors that wield significant influence on the electrical properties of paper, while Table 2 shows in which stages of the paper manufacturing process these factors are decided. For example, the dielectric constant of the paper is influenced by apparent density (Table 1), which in turn depends on the conditions of refining, pressing, and calendering in the paper manufacturing process (Table 2).

B. Interrelationships Between Dielectric Properties and Other Sheet Parameters

Dielectric Loss Tangent (Dissipation Factor) This chapter deals with the relationships between $\tan \delta$ and lignin content, hemicellulose content, carboxyl group content, cellulose crystallinity, and ash content (effect of metallic ions).

Effect of Lignin Content Although it is sometimes stated that the removal of lignin has little effect on $\tan \delta$ for insulating paper [23], some researchers claim that the removal of lignin tends to lower $\tan \delta$ because lignin is typically polar in nature [5,15] and because lignin removal results in the removal of its constituent phenolic hydroxyl groups [17]. Some evidence on this question was produced by Take et al. [32], who prepared paper samples

Table 2 Relation Between Various Factors and Conditions in the Papermaking Process

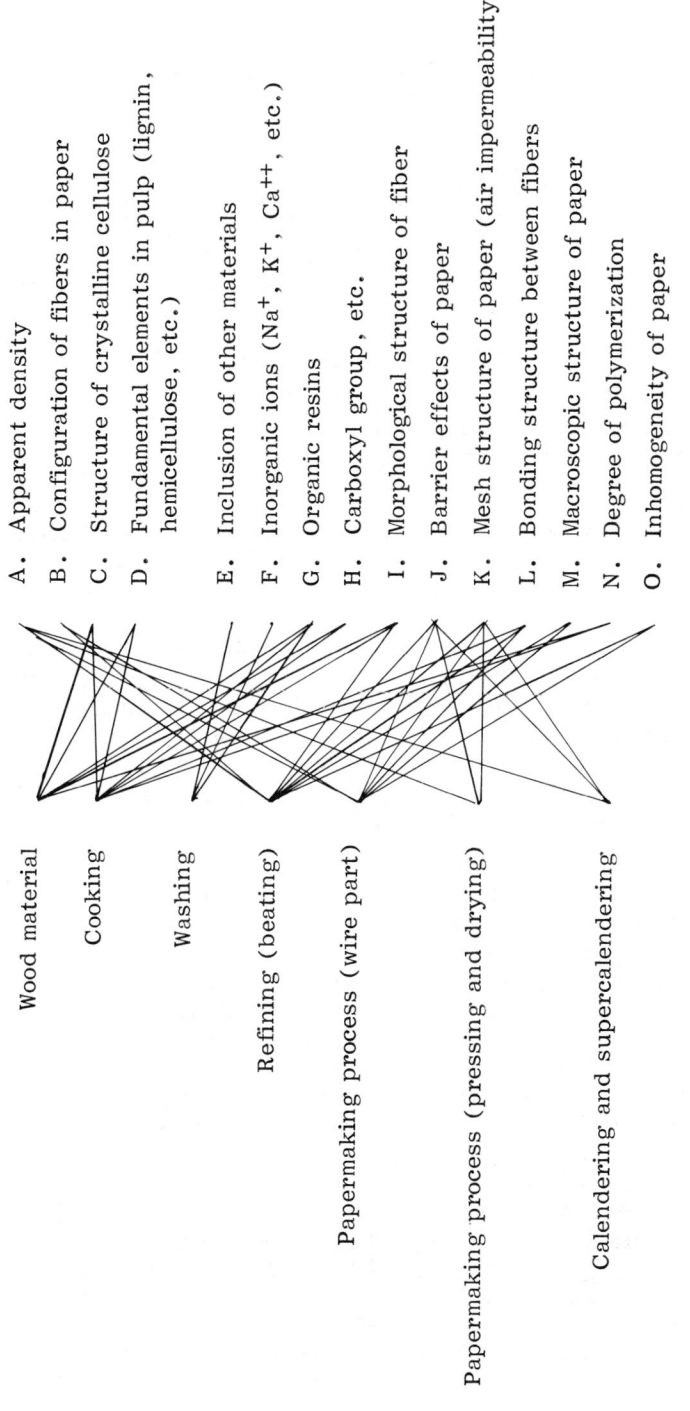

Wood material
Cooking
Washing
Refining (beating)
Papermaking process (wire part)
Papermaking process (pressing and drying)
Calendering and supercalendering

A. Apparent density
B. Configuration of fibers in paper
C. Structure of crystalline cellulose
D. Fundamental elements in pulp (lignin, hemicellulose, etc.)
E. Inclusion of other materials
F. Inorganic ions (Na^+, K^+, Ca^{++}, etc.)
G. Organic resins
H. Carboxyl group, etc.
I. Morphological structure of fiber
J. Barrier effects of paper
K. Mesh structure of paper (air impermeability)
L. Bonding structure between fibers
M. Macroscopic structure of paper
N. Degree of polymerization
O. Inhomogeneity of paper

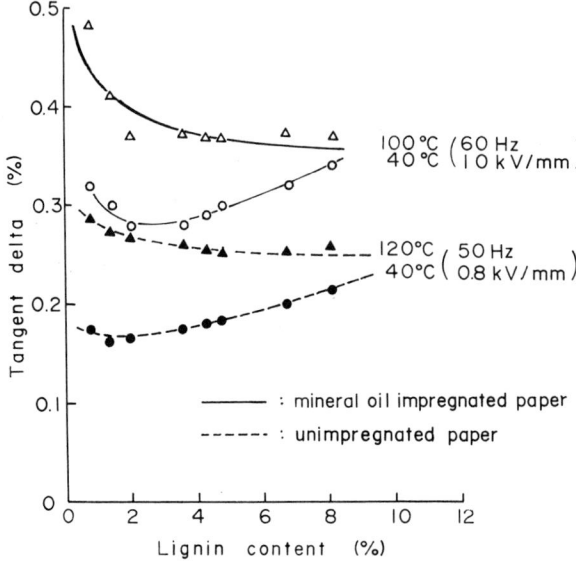

Fig. 1 Relation between lignin content and dielectric loss tangent of paper. (From Ref. 32.)

having various lignin contents as a result of treatment of the (kraft) pulp with sodium chlorite. The results are shown in Fig. 1. At lower (40°C) temperatures, both dry (unimpregnated) paper and mineral oil-impregnated paper generally showed tan δ increases as lignin content increased, whereas tan δ decreased with the increase of lignin content under higher temperatures of 100 to 120°C. When the lignin content is below 2%, a tendency for tan δ to increase is observable even at lower temperatures. Perhaps this effect is caused by excessive lignin removal.

Effect of Hemicellulose Content A number of studies have been made of the effect of hemicellulose content upon tan δ of insulating paper. Miller and Hopkins [23] asserted that the removal of 20% hemicellulose is beneficial, but that the removal of more than this worsens the tan δ of the paper. Renne and coworkers [26] claimed that the removal of hemicellulose increases tan δ While Calkins [3] reported that a decrease in hemicellulose content improves tan δ.

Take and coworkers [33] prepared sulfate pulp having various hemicellulose contents by a prehydrolysis cooking method, keeping lignin content as constant as possible. Measurements of tan δ were made on handsheets prepared from these pulps. Figure 2 shows the results. At low temperature (40°C), tan δ increases as the hemicellulose (pentosan) content increases. This is in contrast to the decrease of tan δ with pentosan content in the higher temperature region (100 to 120°C). Tan δ is independent of hemicellulose content in the medium temperature region (80°C).

Figure 3 shows the relation between tan δ and pentosan content for paper made from kraft pulps prepared from several kinds of softwoods and

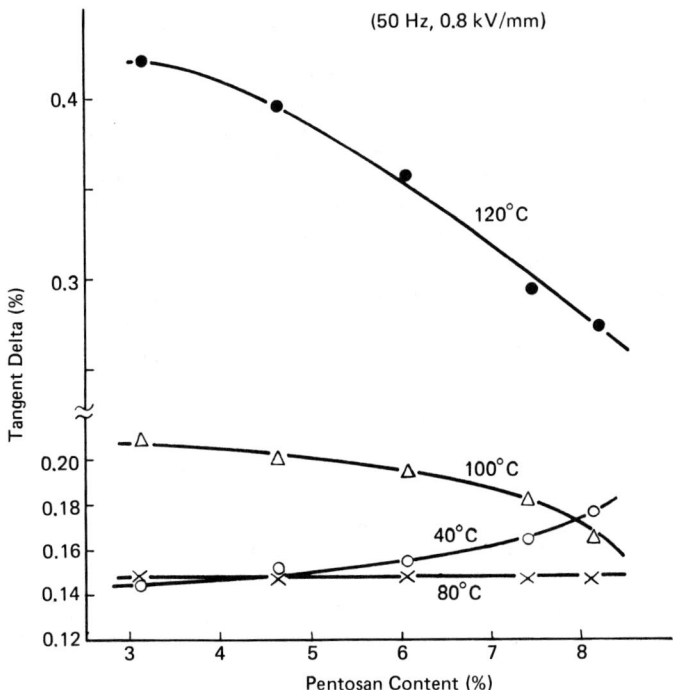

Fig. 2 Relation between pentosan content of prehydrolyzed sulfate pulp and dielectric loss tangent of unimpregnated paper. (From Ref. 33.)

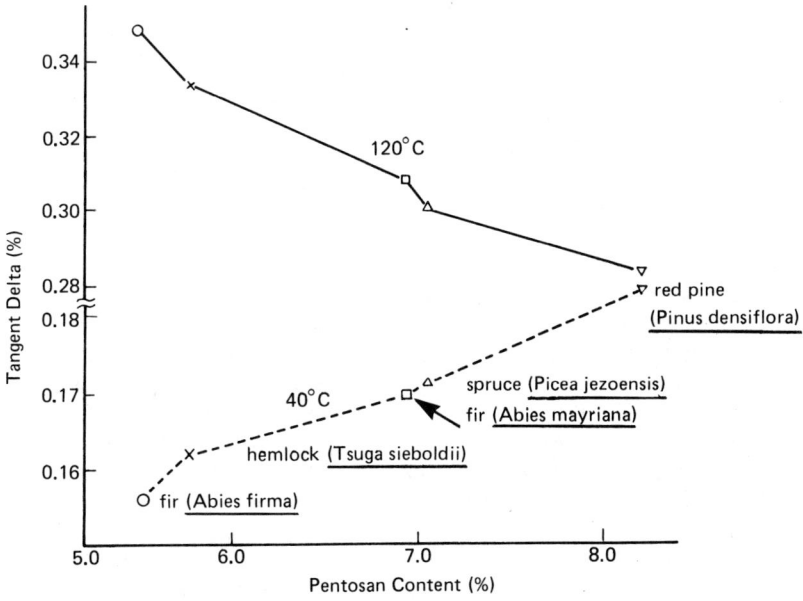

Fig. 3 Relation between dielectric loss tangent and pentosan content of paper made of pulp obtained from various species of woods. (From Ref. 33.)

hemicellulose. These results are consistent with those of Fig. 2 in that the increase of hemicellulose content causes an increase in tan δ in the lower temperature region and a decrease in tan δ in the higher temperature region.

Effect of Carboxyl Groups Carboxyl groups do not exist in pure cellulose, but they are present in large quantities in wood pulps, which are the principal raw materials of insulating papers. Because of their ion exchange ability, such carboxyl groups tend to be powerful ion adsorption centers and can have a large influence on the tan δ of paper. Calkins [3] reported that at high temperatures carboxyl groups wield a great influence and that this influence becomes larger at lower frequency. On the basis of their experiments, Shimoyamada and Satoh [30] stated that carboxyl groups have a bad effect on the tan δ of paper.

Results that differed from those described in the preceding paragraph were obtained by Take and coworkers [31,34]. They found that when carboxyl groups are increased in quantity, the tan δ of paper increases at higher temperatures, even though the total number of metallic ions is decreased. Their explanation was as follows: as carboxyl groups are reduced, metallic ions bound to them are liberated and ionic conduction is increased substantially, resulting in a higher tan δ in the higher temperature region. Since a decrease in the lignin and hemicellulose contents causes an increase in tan δ at high temperatures, this may be also due to an increase of the free metallic ions liberated from the carboxyl groups contained in the lignin and hemicellulose.

Influence of the Crystalline Fraction At low temperatures, dielectric loss is principally related to the rotations and oscillations of polar groups in the molecular constituents of the fibers. The three OH groups that exist per glucose unit of cellulose, as well as the comparable OH groups that are present in the repeat units of other polysaccharides, are considered to play the main role. Within crystalline regions the OH groups are in a comparatively stable state, but large dipole moments may be ascribed to those that exist in noncrystalline regions of the fiber.

Verseput [39] and Kane [16] have pointed out that there exists a straight-line relation between accessibility, which is closely related to the noncrystalline fraction, and dielectric constant. It may be deduced from the relation between dielectric constant and tan δ represented by Eqs. (4), (5), and (6) in Chap. 21 that tan δ is dependent on the crystalline-noncrystalline ratio just as dielectric constant is related to this ratio. Few reports, however, have been made concerning the relation with tan δ so far.

Take [31] prepared handmade sheets from various pulps having different proportions of crystalline fractions, and he measured tan δ for each type to obtain the results shown in Fig. 4. This figure shows that an increase in the amount of crystalline area leads to a decrease of tan δ at lower temperatures. Accordingly, the increase of tan δ shown at lower temperature, which is concomitant with increased lignin and hemicellulose contents, might be accounted for by the increase of dipole moments resulting from the increase of the noncrystalline fraction. On the other hand, tan δ at higher temperature becomes higher when the crystalline fraction increases. This is mainly related to the correlation of that fraction with the number of carboxyl groups. That is, the number of carboxyl groups decreases with an increase

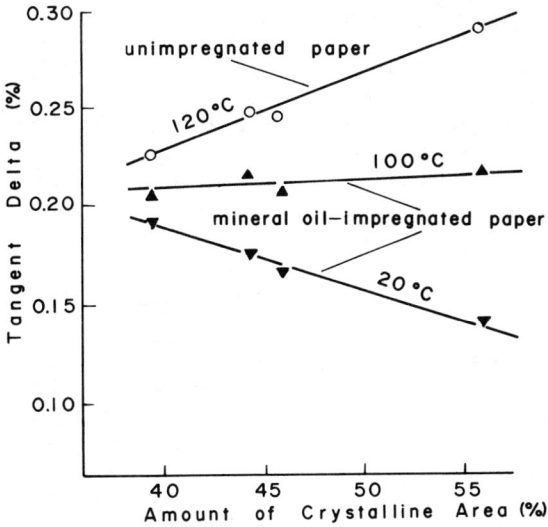

Fig. 4 Relation between the amount of crystalline area and dielectric loss tangent of paper. (From Ref. 31.)

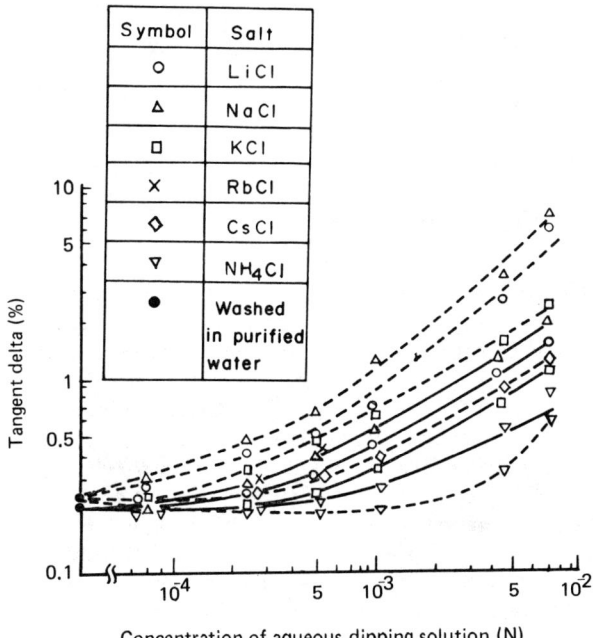

Fig. 5 Influence of monovalent metallic ion concentration on dielectric loss tangent of paper. (From Ref. 31.)

in the crystalline fraction, and tan δ at higher temperature becomes higher, as mentioned under "Effect of Carboxyl Groups" on p. 208.

Influence of Ash Content (Effect of Metallic Ions) At high temperature and at commercial frequency, the dielectric loss tangent of paper depends mainly upon the conduction current. This conduction current is greatly affected by the very small amounts of metallic ions contained in the paper. The total quantity of metallic ions may be estimated from the ash content of the paper.

Some examples of the effects that certain ions have on tan δ of paper are shown in Fig. 5. These results show the influence of univalent ions. These experiments were conducted on many paper samples dippled in aqueous salt solutions and dried in hot air. The values of tan δ determined for these samples are plotted against the concentrations of the solutions.

Insulating papers ordinarily contain many more divalent cations than univalent ones, but the latter group wields a greater influence on tan δ than the former. The presence of univalent cations makes for a greater tan δ of paper, in the order $Na > Li > K > Rb > Cs > NH_4$. Although anions have less influence on tan δ than cations, the influence of anions can become greater as the influence of cations diminishes. In the case of cations with little influence, such as NH_4 (see Fig. 5), the influence of the anion Cl^- is pronounced. It should be noted that the effects of these metallic ions on tan δ at lower temperature are negligibly small.

Insulation Resistance Although insulation resistance involves both volume and surface resistances, this section will deal only with the former. The intrinsic resistance of a material is often expressed by the volume resistivity ρ. As defined in Chap. 21, the volume resistivity ρ is expressed by

$$\rho = \frac{RA}{L} \text{ (ohm-cm)}$$

where

R = volume resistance
A = cross-sectional area of the specimen between electrodes
L = thickness of the specimen

But, in many cases, the thickness of the specimen L cannot be measured precisely (see Chap. 26). For this reason, the product of capacity C and the volume resistance R, CR, is frequently used in electrical engineering in place of the volume resistivity ρ. The relation between ρ and CR is $CR = (\varepsilon A/L)(\rho L/A) = \varepsilon \rho$ (ohm-farads). In the case of paper, dielectric constant ε does not change very much with temperature, so that the change of CR with temperature approximately represents the change of ρ with temperature. This CR value is in practice very convenient, because it is independent of the specimen and electrode geometries.

Insulation resistance has long been used as an index for the insulating properties of paper. Few reports, however, have been published that relate insulation resistance to the various parameters of paper described in the preceding section. Some useful experimental work has been done by Take [31] in this regard. Some results will be discussed here. As mentioned before, it is reasonable to assume that a decrease of lignin or hemicellulose leads not only to a decreased presence of carboxyl groups in the pulp but

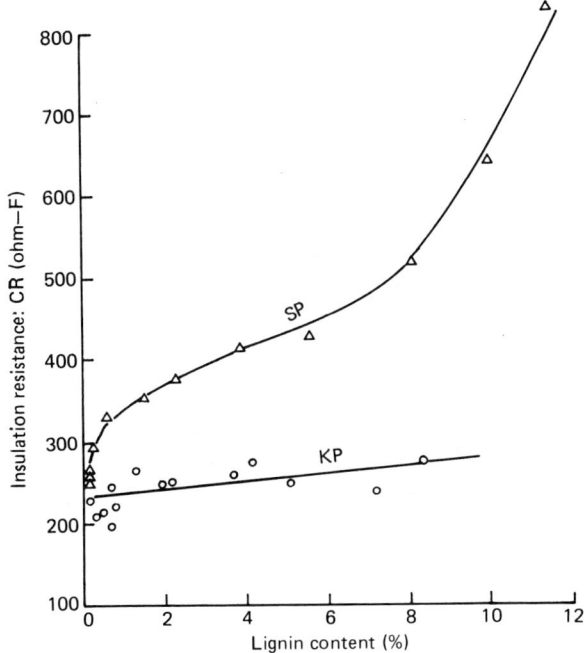

Fig. 6 Relation between lignin content and insulation resistance (CR) of unimpregnated paper at 100°C. (From Ref. 32.)

also to an increase in free metallic ions, resulting in a decrease of the insulation resistance of paper made from the pulp, other things being equal. Figures 6 and 7 show the influence of the lignin and hemicellulose contents, respectively, on the insulation resistance of paper. As anticipated above, the decrease in lignin and hemicellulose contents is accompanied by an increase in ionic conduction and a decrease in the insulation resistance of paper. The difference in this property shown by sulfite pulp (SP) and sulfate pulp (KP) might be accounted for by the presence of smaller quantities of Na^+ ions in SP as a result of the differences in cooking conditions between the two.

From the foregoing, it is clear that variation in insulating resistance—in which ionic conduction by metallic ions in paper would play a main role—is considered to be analogous to that of tan δ in the higher temperature region. In other words, the factors that raise tan δ at higher temperatures would function in such a way as to lower almost all insulation resistances.

Dielectric Constant Using the model shown in Fig. 8 as representative of a small element of paper or paperboard between two electrodes, Sakamoto and Yoshida [28] presented the following equation for the dielectric constant of paper (ε_0):

$$\varepsilon_0 = \varepsilon_f \theta_{pf} + \varepsilon_i \theta_{pi} + \frac{\varepsilon_f \varepsilon_i (\theta_{sf} + \theta_{si})^2}{\varepsilon_f \theta_{si} + \varepsilon_i \theta_{sf}} \tag{2}$$

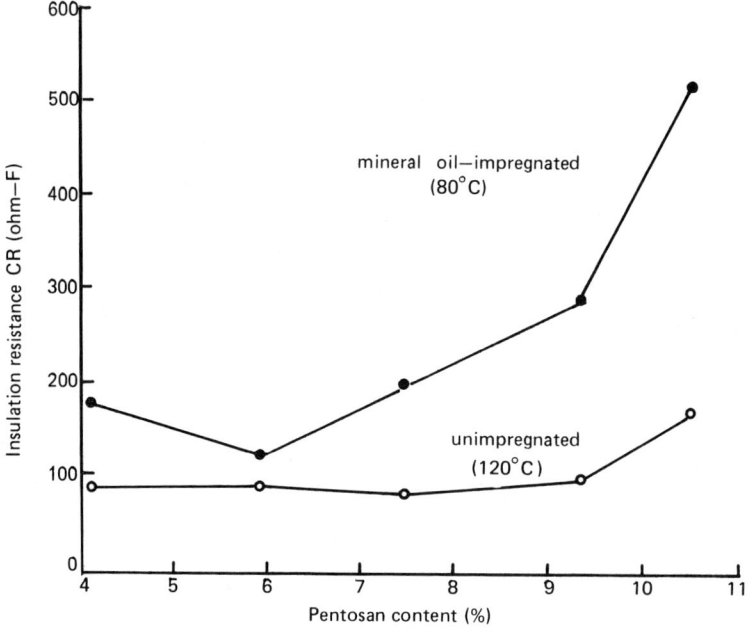

Fig. 7 Relation between pentosan content and insulation resistance of paper made from pulp obtained under various cooking conditions. (From Ref. 33.)

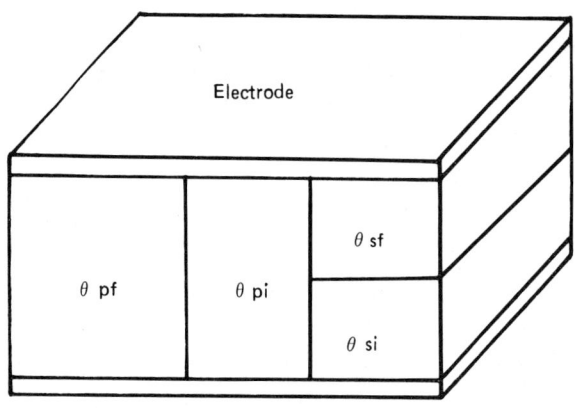

θ pf: parallel fiber portion

θ pi: parallel vacant portion

θ sf: series fiber portion

θ si: series vacant portion

Fig. 8 A model of paper structure. (From Ref. 28.)

where

ε_f = dielectric constant of paper fiber
ε_i = dielectric constant of impregnant, or vacant portion in the case of unimpregnated paper
$\theta_{pf}, \theta_{pi}, \theta_{sf}, \theta_{si}$ = fractional volumes of each portion of the model paper structure shown in Fig. 8

The following relations flow from this model:

$$\theta_{sf} + \theta_{si} + \theta_{pf} + \theta_{pi} = 1 \tag{3}$$

$$\theta_{sf} + \theta_{pf} = \frac{d}{d_f} \tag{4}$$

where d and d_f stand for the densities of paper and paper fiber, respectively. From a study of the dielectric constant of a cellulosic film (cellophane), it is possible to estimate that the dielectric constant of paper fiber is about 6 and has positive temperature dependence.

The dielectric constant of paper fiber may be obtained experimentally from measurements of the temperature dependence of the dielectric constant of a paper impregnated with a liquid dielectric having a known dielectric constant of approximately 6 and a negative temperature dependence. The crossing point of the temperature-dependence curve of the impregnated paper with that of the impregnating liquid itself yields the dielectric constant of paper fiber at that temperature. By measuring several crossing points, using various liquid dielectrics having different dielectric constants, and then connecting these crossing points, the temperature dependence of the dielectric constant of paper fiber is obtained. Sakamoto and Yoshida [28] employed the liquid dielectrics trichlorodiphenyl (TCD), pentachlorodiphenyl (PCD), trichlorobenzene (TCB), and mixtures of these in their experiments.

Figure 9 demonstrates the use of the above-mentioned method to obtain the dielectric constant of one type of insulating paper. From this figure, the temperature dependence of paper fiber itself is given by the following equation:

$$\varepsilon_f = 5.98 + 0.005t \tag{5}$$

where t is temperature (°C). Sakamoto and Yoshida also measured the resistivities of paper impregnated with various deteriorated mineral oils having very low resistivities, in order to obtain the value of θ_{pi}; they obtained results expressed by the following relation:

$$\theta_{pi} = \frac{1}{3 \times 10^{2d}} \tag{6}$$

where d is paper density. Setting $\theta_{sf}/\theta_{pf} = n$, the relation between density and dielectric constant with n as a parameter can be obtained from Eqs. (2), (3), (4), (5), and (6) (see Fig. 10).

In Fig. 10 the dielectric constants of insulating papers made from red pine sulfate pulps are plotted; it appears that the n value for these insulating papers should be within the range $0 - 1/3$. As shown in Eq. (5), it is ex-

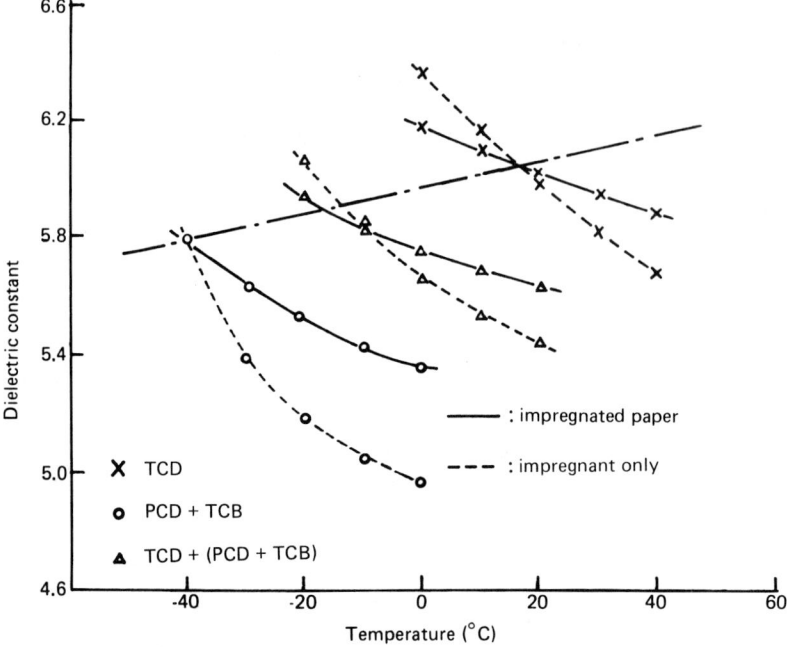

Fig. 9 Experimental determination of the dielectric constant of paper fiber. TCD: trichlorodiphenyl; PCD: pentachlorodiphenyl; TCB: trichlorobenzene. (From Ref. 29.)

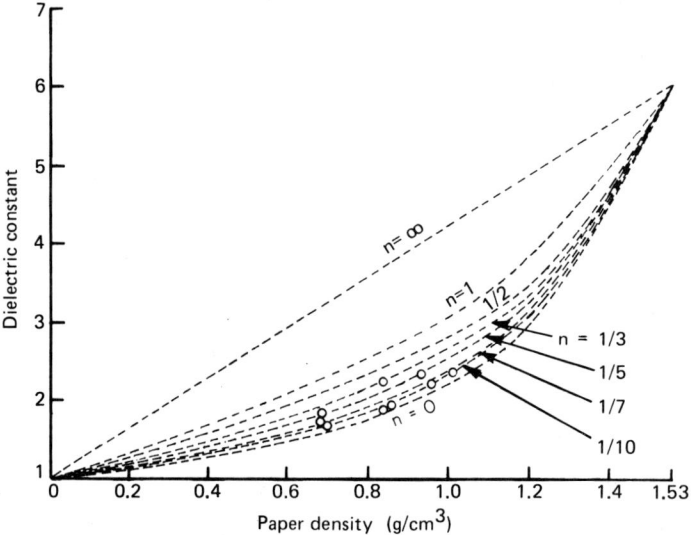

Fig. 10 Calculated and observed values for dielectric constant of unimpregnated insulating paper. Calculated curves are carried to a presumed maximum specific gravity of 1.53, which is the density of cellulose. (From Ref. 29.)

Electrical Properties: II. Practical Considerations 215

pected that the paper dielectric constant increases to some extent with temperature rise, a relationship also supported by the studies of Race and co-workers [25], Veith [38], and Calkins [3].

C. Relationships Between Dielectric Breakdown Strength and Physical Properties of Paper and Refinining Process Parameters

General Phenomena As explained before, the dielectric breakdown of an insulating material is a phenomenon in which the material becomes electroconductive; once this breakdown occurs, the insulating material no longer serves its purpose. The dielectric breakdown strength of paper is a property of great importance when the paper is to be used for high-voltage machinery. Generally, dielectric breakdown strength is expressed as a breakdown voltage per unit thickness, that is, the value of the breakdown voltage divided by the thickness. (It is also called dielctric breakdown stress, and its units are kV/mm, MV/m, etc.)

The breakdown mechanism of paper may roughly be summarized as follows. Paper possesses internal pores, and partial electrical discharge takes place in the voids existing between fibers when the electric field is intensified. Following such an electrical discharge, a stream of ions or electrons begins to flow and propagate among the paper fibers, creating a short cir-

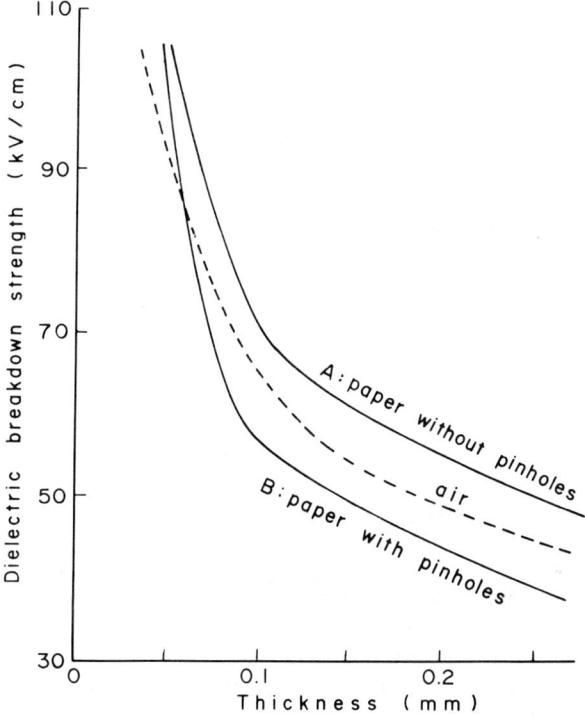

Fig. 11 A comparison between dielectric breakdown strength of paper with artificial pinholes and that of paper without pinholes. (From Ref. 27.)

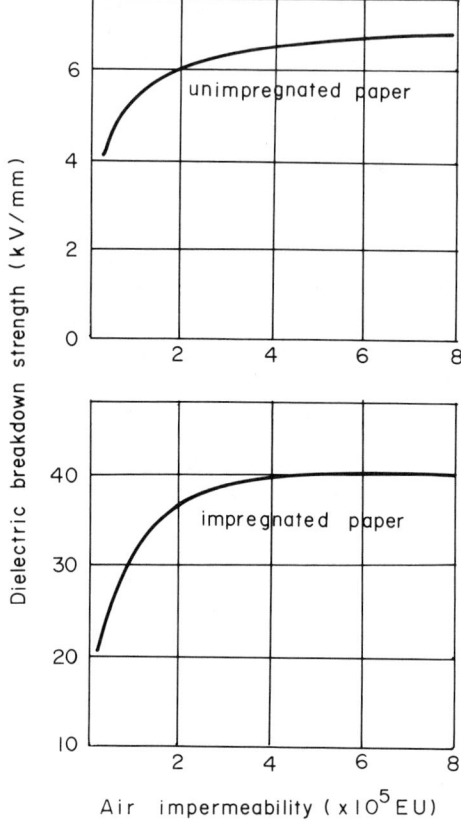

Fig. 12 Effect of oil impregnation on dielectric breakdown strength of cable insulating paper. (From Ref. 8.)

cuit between the electrodes and, ultimately, dielectric breakdown. The propagation of such a "streamer" is accelerated by the electric field but decelerated by collision against the fibers. These charges may accumulate on fiber surfaces, giving rise to high field strengths within the fiber. This accumulation may cause breakdown of the fibers and thus permanent degradation of the paper [1]. The experimental results by Riley and Scott [27] are introduced to prove this mechanism. Figure 11 shows the relation between thickness and dielectric breakdown strength of an intact, unimpregnated paper (A) and a perforated paper (B); the strength of paper B with artificial pinholes in it is not only lower than that of ordinary paper A, but it is also lower than that of air in the same electrode gap. As the stream of ions could move readily into the pinholes, the dielectric breakdown strength of paper B was lowered relative to A. That B had lower strength than air was ascribed to leakage along the side surfaces of the pinholes.

Generally, insulating paper is impregnated with insulating oils or varnishes when it is used for high voltage apparatus and machinery. The intent

Table 3 Summary of Experimental Conclusions from Published Data Concerning Dielectric Breakdown Strength of Paper

Air impermeability	With the rise of air impermeability, AC short-term, DC, and impulse breakdown strength will all increase. AC long-term breakdown strength is not affected.
Density	Generally, AC short-term, DC, and impulse strength increase with density. (However, there are some reports contrary to this conclusion.)
Thickness	As thickness decreases, there is a rise in AC short-term and impulse strength. But according to the author's experiments, thickness does not affect the breakdown strength of paper (see Fig. 13).
Uniformity	As sheet formation becomes more uniform, impulse breakdown strength becomes higher.

is to fill the voids existing among the fibers with impregnants possessing dielectric strengths higher than air or vacuum in order to prevent partial breakdown. An example of how the dielectric strengths of unimpregnated versus impregnated paper compare is shown in Fig. 12. Tests by Emanueli show that the latter, impregnated with a kind of heavy oil, has about 6 times the strength of the former [8].

The barrier effect against streamer propagation is in proportion to how densely the paper fibers are piled up in the paper structure. In general, it may be readily assumed that paper having higher density and greater impermeability is also higher in dielectric breakdown strength, other things being equal.

Influence of Macroscopic Properties on Dielectric Breakdown Strength of Paper Many reports are available on the relation between the macroscopic properties of paper—thickness, density, and air impermeability—and dielectric breakdown strength. The results of these reports are summarized in Table 3. Generally, in the manufacture of paper it is difficult to change thickness, density, and air impermeability independently, which in turn makes it difficult to distinguish the individual effects of these factors on dielectric breakdown strength. Figure 13 shows AC and impulse breakdown strength of papers with these factors varied independently using a small Fourdrinier machine for test use. These measurements were conducted using mineral insulating oil-impregnated samples; from the results obtained it may be said that paper thickness has almost no effect on the dielectric strength.

Hall and Kelk [12] stated that besides thickness, density, and air impermeability, the uniformity of paper also exerts considerable influence on

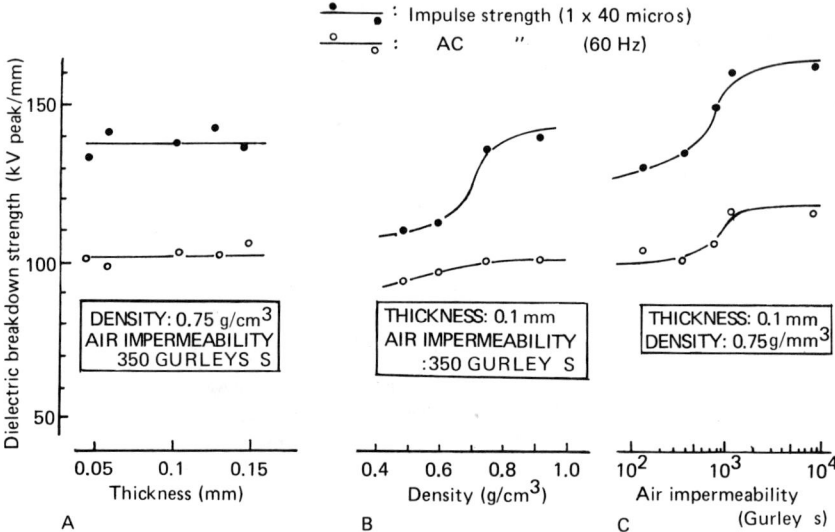

Fig. 13 Effects of physical properties in relation to dielectric breakdown strength of mineral oil-impregnated paper. (A) Effect of thickness. (B) Effect of apparent density. (C) Effect of air impermeability.

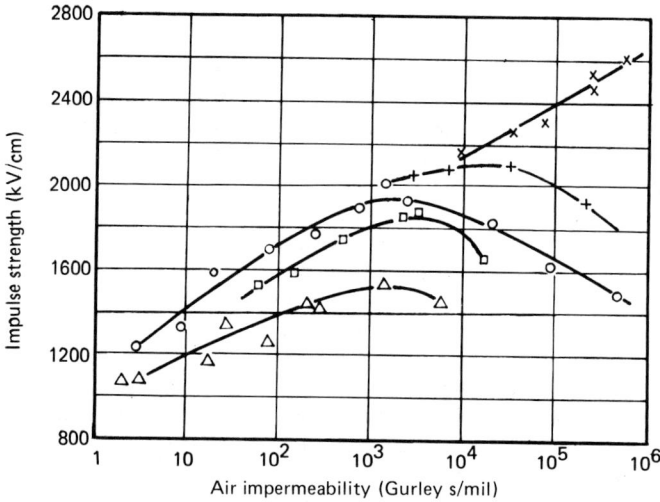

△ 3 mil papers, density 0.8 g/cm^3
□ 3 mil papers, density 0.9 g/cm^3
+ 2 mil papers, density 1.0 g/cm^3
× 1 mil papers, density 1.0 g/cm^3
○ 3 mil papers, density 1.0 g/cm^3

Fig. 14 Experimental relationship of air impermeability to impulse breakdown strength of oil-impregnated paper. (From Ref. 18.)

the dielectric strength. Using an electromicrometer with a ball-point sensor and photometer, they evaluated uniformity concurrently with the change in air impermeability in a small area, and they reported that the dielectric strength of paper measured by impulse voltage is affected the most by air impermeability plus uniformity. Later, Kelk and Wilson [18] investigated the relation between impulse strength and air impermeability of paper again and obtained the results shown in Fig. 14. From these results, they concluded that papers of high impermeability made from heavily beaten pulp (which results in slow drainage and nonuniform deposition of the fibers), tend to show declines in dielectric strength.

Influence of Refining Process Parameters Refining (beating) is one of the most important processes in papermaking. Generally, when pulp is beaten to a greater degree, fiber fibrillation increases and the quantity of fines also increases. The production of longer fiber fibrillations at higher beating levels results in the creation of a dense mesh structure in which fines are entrapped. The blocking, or filling up, of many of the sites in the mesh by fines enhances the barrier effect and acts to improve the dielectric breakdown strength accordingly. Thus, refining conditions have much to do with the dielectric strength of a given paper.

Matsuda [21] investigated the influence of refining conditions on dielectric breakdown strength of paper. Figure 15 shows the relationships of the impulse breakdown strength to paper density and to beating methods. As is

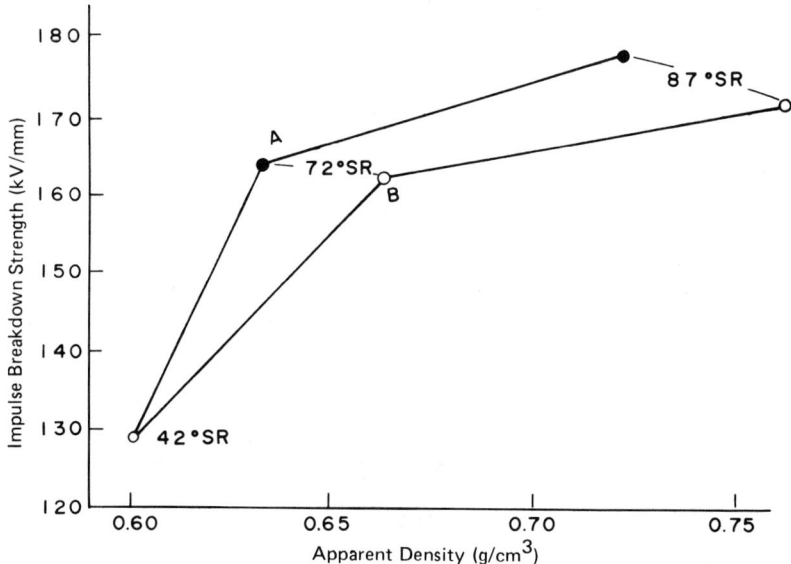

Fig. 15 Impulse strength versus apparent density of paper (influence of fines on the impulse breakdown strength of mineral oil-impregnated paper). (A) Paper containing larger quantities of fine. (B) Paper containing smaller quantities of fine. (From Ref. 21.)

clear from this figure, papers from specially beaten pulps having larger quantities of fines have the same or higher levels of dielectric strength in spite of their lower densities, than papers from normally beaten pulps.

D. Current Insulation Research

In recent years there has been a general rise in the voltage levels used in transmission. This rise has been accompanied by demands for improvement of the properties of insulating paper for high-voltage apparatus and machinery, particularly in regard to reduction of dielectric loss and increase in dielectric breakdown strength. The field of power cable manufacture is one important example. Also, in many applications, insulating paper made from ordinary wood pulp is not able to meet the more severe operating requirements. For instance, polypropylene film has been substituted for conventional insulating paper in some types of capacitors. Various synthetic papers have been tried in power cables, but none of them have achieved commercial success because of high price and inferior mechanical as well as electrical strength when compared with conventional insulating papers.

Also recently, combinations of conventional insulating paper and synthetic polymer film have been employed to obtain the advantageous mechanical properties of the former and the lower dielectric loss of the latter. A typical combination is polypropylene-laminated paper [6,20]; the development of EHV (extra high voltage) and UHV (ultra high voltage) cables using this material is now under way [1a,7,19]. Another proposal has been made for lamination of conventional insulating paper with fluoroethylene propylene, silicone-grafted high density polyethylene, or polymethyl pentene-1 films. It is reported that they have without exception low tan δ and high dielectric strength values not found in conventional insulating papers. In 1976 there was announced a novel type of insulating paper that has the porosity characteristics of insulating paper but whose structure consists of polypropylene nonwoven fabric sandwiched between two sheets of polypropylene fiber and wood pulp mixed to form a composite paper material [11]. This material has desirable mechanical and electrical characteristics, including the advantage of low dielectric loss; it also possesses the feature of being as easy as conventional insulating paper to impregnate with oil.

On the other hand, Fujita and Ishitobi [9,10] have proposed a new method to improve the tan δ of conventional insulating papers. In this method, the paper is passed between a pair of electrodes, one of which DC high voltage is applied to in order to expel ionic impurities present in the paper. According to their reports, the ash content of the paper was reduced to about one-half that of untreated paper by this method, and divalent ions—which are resistant to removal by ordinary methods—were also reduced considerably.

Regarding the dielectric breakdown strength of paper, Murata and Nakata [24] have commented on the influence of sheet formation on the dielectric strength of insulating paper in relation to pulp refining. While agreeing that fines have an effect on dielectric strength (as discussed in the previous section), they point out that the effect is related not only to the quantity of fines but also to their quality.

II. METHODS OF MEASUREMENT OF ELECTRICAL PROPERTIES

A. Measurement of Dielectric Properties

Dielectric Constant, Dielectric Loss Tangent (tan δ) Measurements of capacitance C and tan δ are conducted in a broad range of frequencies, from as low as 0.1 Hz up to the ultra high frequency of some 10 GHz. Measurements at such ultra low and ultra high frequencies, however, are limited to theoretical interest only. Useful measurements are generally conducted in the frequency range between about 50 Hz and 100 MHz. The bridge method is used at comparatively low frequencies. It has an advantage in that its measurement is precise; the properties can be measured precisely since this method is not affected by edge capacity because of the use of a guard electrode. However, at higher frequencies the bridge method is subject to great errors, so in such cases the resonance method is substituted. Generally speaking, the resonance method is inferior to the bridge method in precision, and it has a weak point in that it cannot make use of a guard. Since the tan δ of insulating papers increases considerably at high frequencies, these materials are generally unable to serve the purpose in the high region. Accordingly, for most practical purposes, the bridge method is the more important.

Schering Bridge This type of bridge is the most widely used. It has a circuit of the type shown in Fig. 16. In this diagram the impedance is expressed as Z_1, Z_2, Z_3, and Z_4 for each arm respectively. When the galvanometer G deflects to zero, that is, when equilibrium is established,

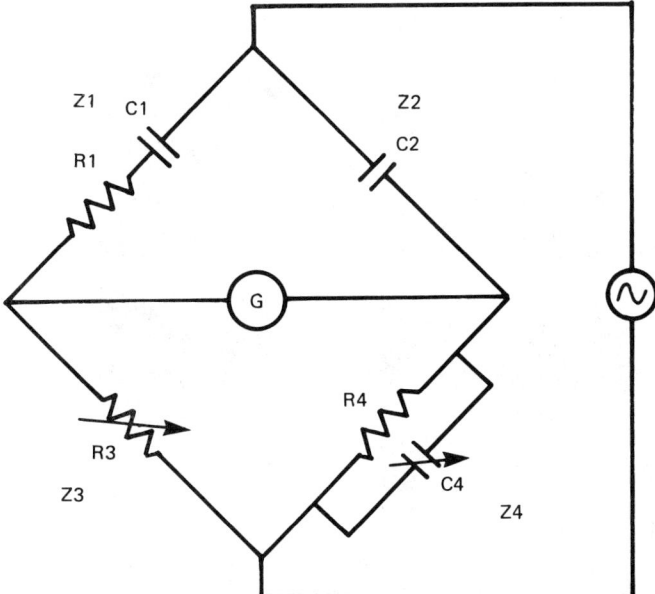

Fig. 16 Fundamental circuit of the Schering bridge.

$$Z_1 \cdot Z_4 = Z_2 \cdot Z_3 \tag{7}$$

Z_1 is the equivalent circuit of a specimen and

$$Z_1 = R_1 + \frac{1}{j\omega C_1} \tag{8}$$

$$Z_2 = \frac{1}{j\omega C_2} \tag{9}$$

$$Z_3 = R_3 \tag{10}$$

$$Z_4 = \frac{R_4}{1 + j\omega C_4 R_4} \tag{11}$$

where ω is angular frequency ($\omega = 2\pi f$, f is frequency). From these equations,

$$R_1 = \frac{C_4 R_4}{C_2} \quad \text{and} \quad C_1 = \frac{C_2 R_2}{R_3} \tag{12}$$

Accordingly, tan δ of the specimen is given as follows:

$$\tan \delta = \omega C_1 R_1 = \omega C_4 R_4 \tag{13}$$

The value of ωR_4 is selected so that direct reading of the value of tan δ can be made in the Schering bridge. An actual Schering bridge is more complicated than that shown in Fig. 17 since a Wagner ground or guard circuit is normally used and an electronic null detector would replace the galvanometer.

Fig. 17 Schering bridge. (Courtesy of Nisshin Electric Co., Ltd.)

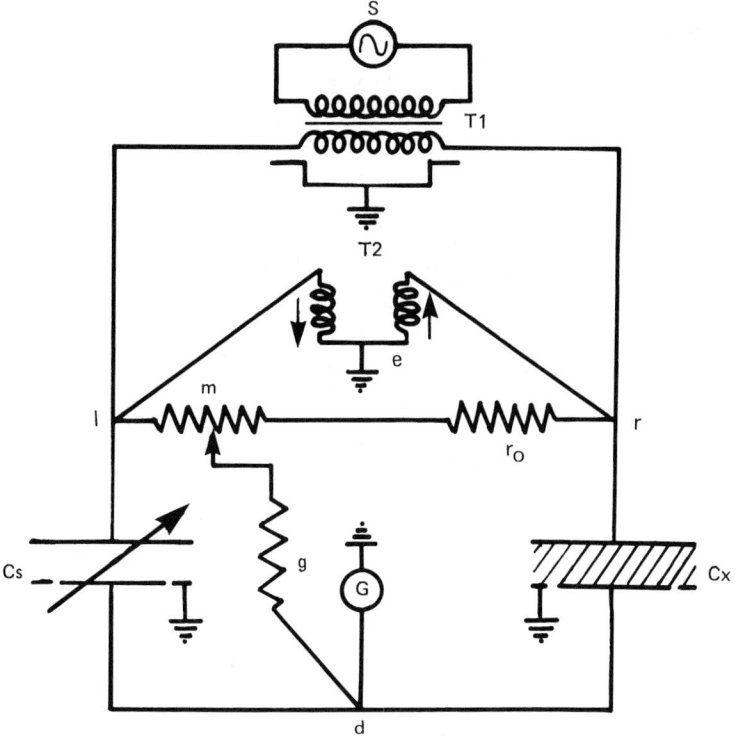

Fig. 18 Fundamental circuit of the transformer bridge.

The frequency range in which measurements can be conducted using a standard Schering bridge is from about 10 Hz to 10 kHz, but measurements up to 1 MHz are possible in a conjugate Schering bridge.

Transformer Bridge This bridge has a basic circuit of the type shown in Fig. 18. Precise measurement is possible using this bridge in the frequency range from about 30 Hz to 3 MHz. The circuit consists of a transformer with a 1:1 turn ratio, in which leakage inductance and winding resistance are minimized; a conductance shifter composed of two each of r_0 and g resistors; and a standard variable capacitor Cs. If the specimen Cx is connected, g represents the conductance between m and d of the conductance shifter, $2r_0$ represents the resistance between l and r, $r_0(1 - s/100)$ is the resistance between l and m, and $r_0(1 + s/100)$ is the resistance between r and m, then

$$Cx = Cs \tag{14}$$

and the conductance of the specimen Gx is

$$Gx = \frac{g(s/100)}{1 + (1/2)gr_0[1 - (s/100)^2]} \tag{15}$$

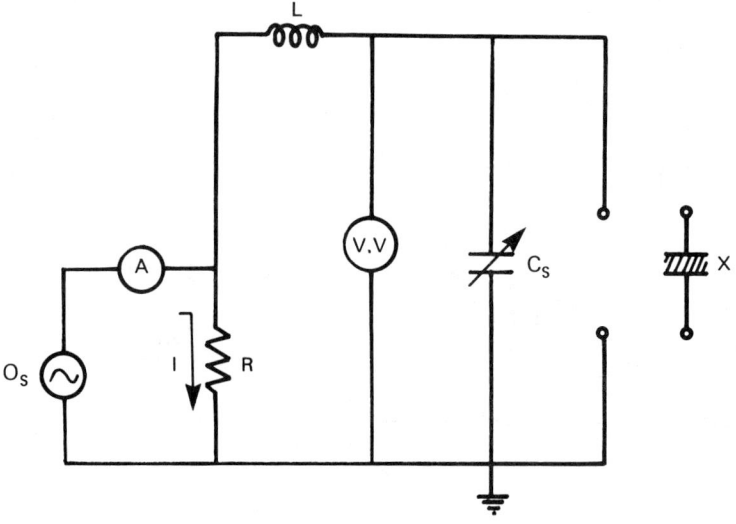

X: Specimen
Cs: Standard variable capacitor
V.V: Valve voltmeter
L: Inductance
R: Low resistance
A: Ammeter
Os: Oscillator

Fig. 19 Fundamental circuit of the Q meter.

if $gr_0 \ll 1$,

$$Gx \fallingdotseq g(s/100) \tag{16}$$

and tan δ of the specimen is given as follows:

$$\tan \delta = \frac{Gx}{\omega Cx} \tag{17}$$

Resonance Methods The resonance method is generally used when frequencies higher than those for which the bridge method is suited are encountered. Two typical resonance methods are the Q meter and the variable capacitance methods.

The circuit of the Q meter method is shown in Fig. 19. The resistance R is supplied with the constant current I and the resonance circuit is fed with the constant voltage V = RI. A specimen is connected to this circuit, and the capacitor Cs is adjusted to make a resonant circuit. At this point, readings are taken of the resonance voltage and capacitance values with the specimen connected. The readings are designated V_1 and Cs_1, respectively. The specimen is then removed from the circuit and Cs is adjusted to make a

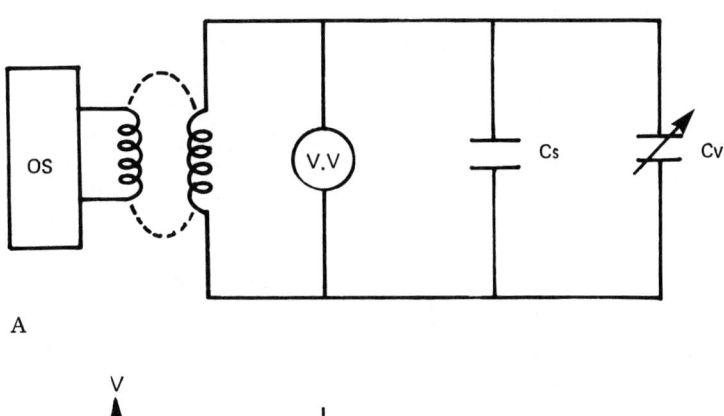

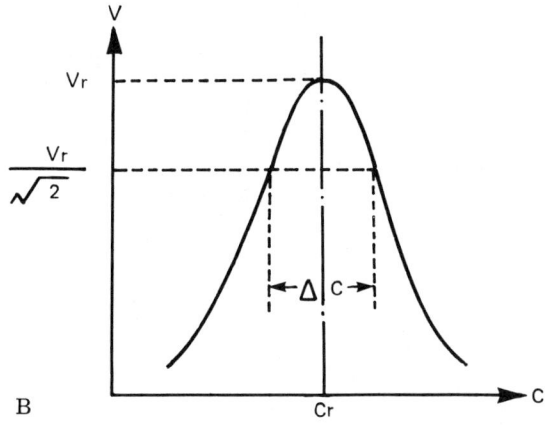

Fig. 20 Fundamental circuit of the variable capacitance method. (A) Circuit diagrams. (B) Resonance curve.

resonant circuit again. The resonance voltage V_0 is then read. Also, the corresponding capacitance value Cs_0 is read. From these values, the equivalent capacity Cx and $\tan \delta$ of the specimen are given by these equations:

$$Cx = Cs_0 - Cs_1$$

$$\tan \delta = Cs_0 \frac{V/V_1 - V/V_0}{Cs_0 - Cs_1} \qquad (18)$$

A circuit diagram for the variable capacitance method is shown in Fig. 20A. The specimen is first inserted between the surfaces of the capacitor Cs. A resonance curve as shown in Fig. 20B is obtained by changing Cv and noting the vacuum tube voltmeter (V.V.). The curve width ΔC_1 at the point of $1/\sqrt{2}$ of the resonance voltage Vr is also obtained. After returning Cv to the resonance point, the specimen is then removed from the circuit and the electrode gap of Cs is adjusted to cause resonance again. After so adjusting the gap at the resonance point, Cv is changed again to obtain the half-value width ΔC_0. The capacitance Cx and loss tangent $\tan \delta$ of the specimen are calculated by the following equations:

$$Cx = Cs_0 - Cs_1 + C_0$$

$$\tan \delta = \frac{\Delta C_1 - \Delta C_0}{2C} \tag{19}$$

where

Cs_1 = capacitance at the time of resonance when the specimen is inserted
Cs_0 = capacitance when the specimen is removed

C_0 represents the geometric capacitance of the electrode calculated from its area A (cm^2) and the thickness of the specimen d (cm) as follows:

$$C_0 = \frac{1}{3.6\pi} \frac{A}{d} \text{ in picofarads (pF)} \tag{20}$$

The term in the bracket has units of pF/cm.

The Q meter method, although simple, is somewhat inferior in precision. The frequencies for which it may be used lie in the range of several hundred kHz to 10 MHz. The frequency range for the variable capacitance method is 10 kHz to 100 MHz.

Dielectric Constant The dielectric constant ε can be calculated from the measured capacitance Cx [$\equiv C_1$ of Eq. (12) for Schering bridge] by the following equation:

$$\varepsilon = \frac{Cx}{C_0} = \frac{3.6\pi d}{A} Cx \tag{21}$$

The units of Cx are pF, and A and d are given in cm^2 and cm, respectively.

When a guard electrode cannot be used, as in the resonance methods, a correction as shown in the following equation is made:

$$\varepsilon = \frac{Cx - Ce}{C_0} \tag{22}$$

Correction factors for Ce are given in Table 4 for the types of concentric circular electrodes shown in Fig. 21.

Measurement of DC Conductivity Generally, DC conductivity σ is given by reciprocal of the specific volume resistivity ρ, and is represented by the following equation:

$$\sigma = \frac{t}{RA} \text{ which is expressed in units of } S \cdot cm^{-1} \tag{23}$$

where

R = resistance (ohms)
A = electrode area (cm^2)
t = thickness of the specimen (cm)

Table 4 Edge Correction Ce in Eq. (22)

Type of electrode shown in Fig. 21	Ce (pF)
Diameter of the electrodes = Diameter of the specimen	$(0.045 \log_{10} \frac{P}{d} - 0.037)P$
Equal electrodes smaller than the specimen	$(0.060 \log_{10} \frac{P}{d} - 0.0885 + 0.020\, \varepsilon')P$
Unequal electrodes	$(0.084 \log_{10} \frac{P}{d} - 0.090 + 0.041\, \varepsilon')P$

Where

P = circumference of the electrode (cm)
d = thickness of the specimen (cm)
ε' = an approximate value of the specimen dielectric constant

In all cases, the thickness of the electrodes is much less than that of the specimen and the diameter of the specimen is above 30 mm.

The measurement of resistance R is explained in the next paragraph.

Direct Deflection Method This method uses a circuit as shown in Fig. 22. Rx represents the resistance of the specimen, Rs a standard resistance, G a galvanometer, and S a shunt resistance. Let θx represent the deflection of G when Rx is inserted in the circuit with the switch K, and let Fx represent the magnification of the shunt resistance. Next, let θs represent the deflection of G when Rs is inserted in the circuit with K switched, and let Fs represent the magnification of the shunt. Then

$$\frac{Rx}{Rs} = \frac{Fs\,\theta s}{Fx\,\theta x} \qquad (24)$$

Since Rs is known,

$$Rx = Rs\, \frac{Fs\,\theta s}{Fx\,\theta x} \qquad (25)$$

From this equation, the resistance of the specimen Rx is obtained; substitution of this value into Eq. (23) yields the value of conductivity σ.

Generally, a reflecting galvanometer is used for G, but it is very difficult to read precisely deflections smaller than ±0.5 mm. Since the sensitivity

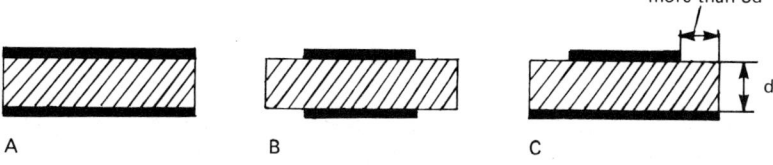

Fig. 21 Construction of plane-to-plane electrodes.

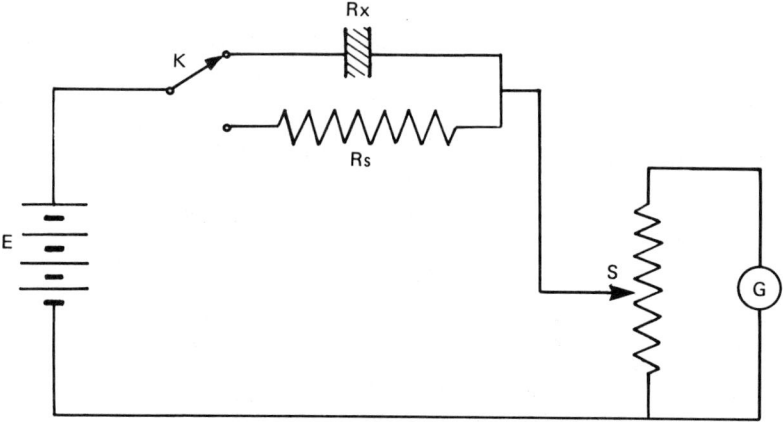

Fig. 22 Fundamental circuit of the direct deflection method.

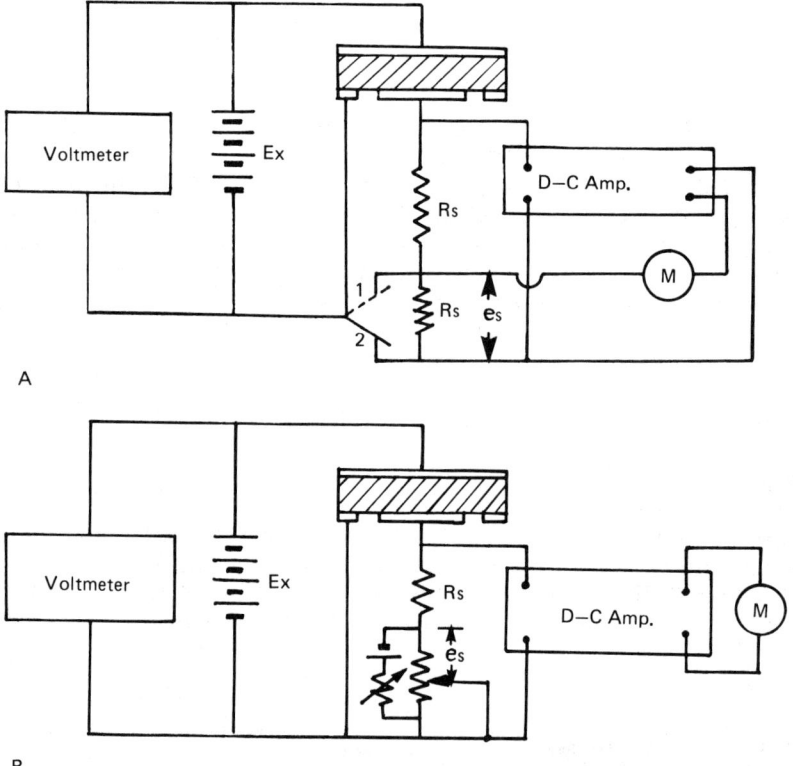

Fig. 23 Fundamental circuits for voltmeter-ammeter methods. (A) Normal use of amplifier and indicating meter. (B) Amplifier and indicator as null detector.

of the most excellent galvanometer is somewhat near 1×10^{-11} A/mm, the maximum value of Rx that can be measured with this method, at a measuring voltage of 500 V, is Rx = $500/0.5 \times 10^{-11} = 1 \times 10^{14}$ ohms at best. In our experience, it is almost impossible for a small-area electrode, such as shown in Fig. 24, to measure the resistance of insulating paper precisely at a temperature below 20°C.

Voltmeter-Ammeter Methods Using DC Amplification This method makes use of DC amplification to increase the sensitivity of the current-measuring device for measurement of a specimen with high resistance (such as insulating paper of high quality). Typical measuring circuits are shown in Fig. 23A and B [2].

In the case of Fig. 23A, a DC amplifier amplifies the voltage drop produced in the standard resistance Rs by the current Ix flowing through the specimen, and the voltage drop is indicated on the meter. In the case shown in Fig. 23B, the voltage drop produced in Rs due to Ix is balanced by adjusting the opposing voltage supplied from a calibrated potentiometer. In this case, the DC amplifier and the indicating meter are so sensitive that they are used as a null detector of high input resistance. Either a DC coupling type or a modulation type of DC amplifier can be used. Ordinarily, all or part of the output voltage is fed back negatively through the feedback resistance Rs in order to stabilize the net gain. Also, a resistance-capacitor filter is usually mounted in front of the amplifier to protect it from fluctuating voltage induced by the specimen.

The resistance of the specimen Rx can be calculated by the following equation:

$$Rx = \frac{Ex}{Ix} = \frac{Ex}{e_s Rs} \quad \text{(ohms)} \tag{26}$$

where

Ex = applied voltage (V)
Ix = current through the specimen (A)
Rs = standard resistance (ohms)
e_s = voltage drop across Rs, which is indicated by either the amplifier output meter or the calibrated potentiometer (V)

Electrode and Specimen Holder for Measurements of Dielectric Properties
In this section the term *dielectric properties* refers to dielectric loss tangent, capacity, and resistance of paper. The electrode for measurement of dielectric breakdown strength will be discussed on pp. 236–238. The dielectric properties of paper are greatly influenced by moisture content. It therefore follows that experiments should be conducted with the specimen in a completely dry state in order to evaluate its intrinsic dielectric properties. An effective way to achieve the completely dry state is to dry the paper in a suitable electrode system in vacuum at high temperature. Also, since impregnation with insulating oil is a common practice for insulation purposes, the sample holder should also allow for oil impregnation. The electrode system should therefore be completely airtight. Many suitable electrode systems have been developed, but only the typical one widely used in Japan is introduced here [33].

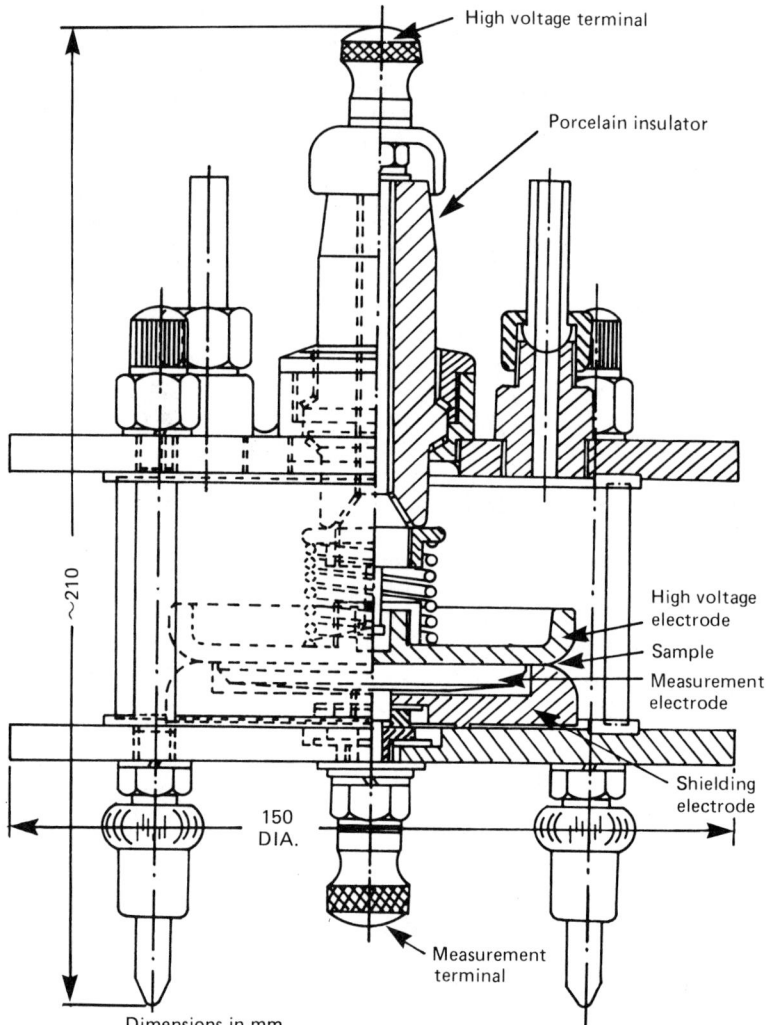

Fig. 24 Typical electrode system for measurment of dielectric properties of paper. (Courtesy of Nisshin Electric Co., Ltd.)

Figure 24 shows an electrode system within a container that can be evacuated to produce a high vacuum. In this container are located (1) a measuring electrode with a guard and (2) a high voltage electrode. The high voltage electrode is pushed downward with a spring to bring the specimen, inserted between the pair of electrodes, into close contact with these electrodes. The side wall of the container is a glass tube, which enables observation within. The interior cannot only be evacuated to conduct a test, but it can also allow impregnation of the specimen with insulating oil.

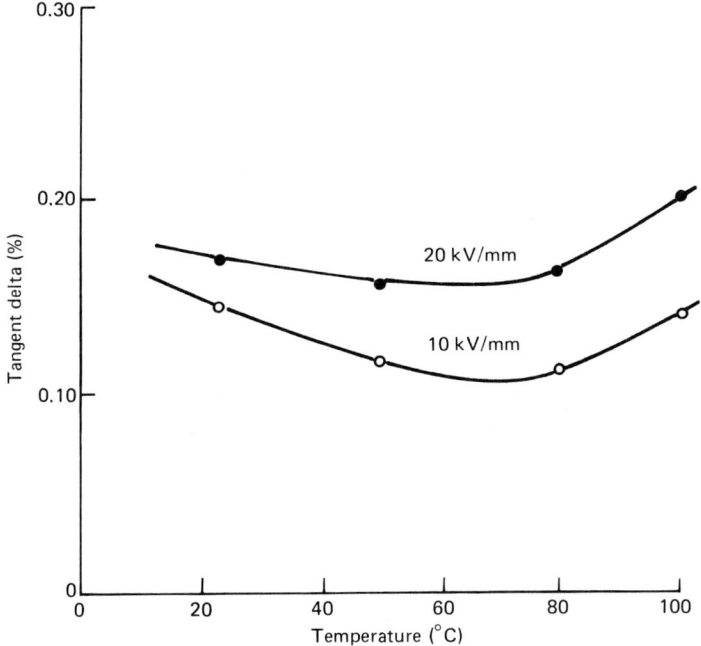

Fig. 25 Temperature dependence of tan δ of a hard type alkylbenzene-impregnated cable insulating paper using the electrode system shown in Fig. 24.

Figure 25 shows an example of the temperature dependence of tan δ for a hard-type alkylbenzene-impregnated cable insulating paper. These results were obtained using the electrode system shown in Fig. 24 and a Schering bridge (as illustrated in Fig. 17). As is clear from this figure, tan δ tends to be minimal in the range of around 60 to 70°C. At higher temperatures there is a notable rise in tan δ resulting from ionic conduction, while at lower temperatures there is an increase in the loss tangent related to large dipole moments. OH groups in noncrystalline areas of the fiber cell walls are considered chiefly responsible for the latter effect.

Figure 26 illustrates the temperature dependence of conductivity as measured by the electrode system of Fig. 24 [36]. These results are for 0.125 mm cable insulating paper impregnated with hard-type alkylbenzene. As this figure clearly shows, the conductivity increases in proportion to the temperature rise. In explaining this phenomenon, it is presumed that ionic conduction in paper becomes larger in the higher temperature range.

Generally, when DC voltage is applied to an insulating material, an electric current flow that diminishes with time can be observed. This current I is expressed by the following equation:

$$I = I_{sp} + I_a + I_d \tag{27}$$

I_{sp} is a transient current including the current charging the geometrical capacitance of the electrode as well as inducing electronic and atomic polari-

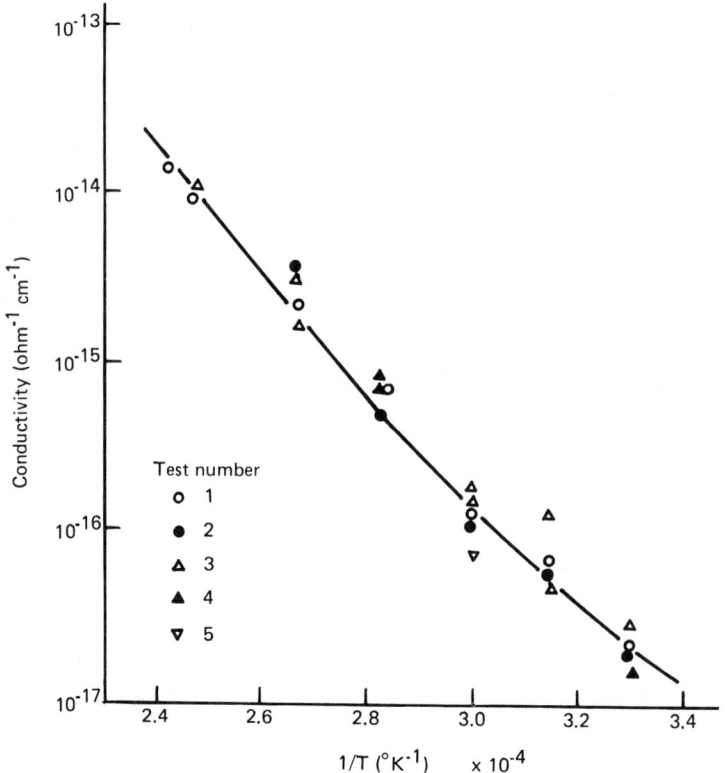

Fig. 26 Temperature dependence of conductivity for hard type alkylbenzene-impregnated cable insulating paper using the electrode system shown in Fig. 24. (From Ref. 36.)

zations. Ia is a current caused by comparatively slow polarization involving oriented and interfacial polarizations, and it is called the absorption current. Id is the steady state leakage current. The DC conductivity is to be calculated from the leakage current. However, in the case of dried and impregnated good quality insulating papers, this steady state leakage current may not be reached even after a long time. The conductivity, therefore, is obtained by measurement of the current after a fixed time for convenience. This fixed time can be 1, 5, 10, or 60 min; the results shown in Fig. 26 are calculated from the current at 60 min after voltage was applied.

In recent years there have been extensive developments in reprographic papers for use in office copy machines, communications systems, data storage devices, computer readout printers, and so on. Many of these papers have surfaces rendered conductive by the application of special chemicals. Accordingly, it may be useful to consider some measurements to characterize the surface conductivity (resistivity) of paper. Cooprider [4] studied the effects of environmental, geometrical and electrical variables on the surface resistivity of conductive base paper and found that the humidity

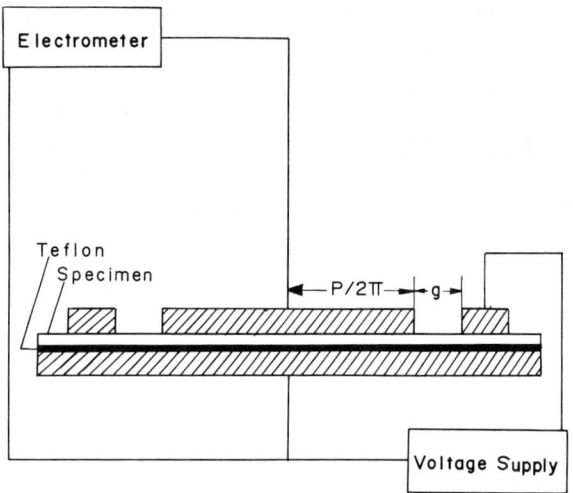

Fig. 27 Surface resistivity test circuit. (From Ref. 4.)

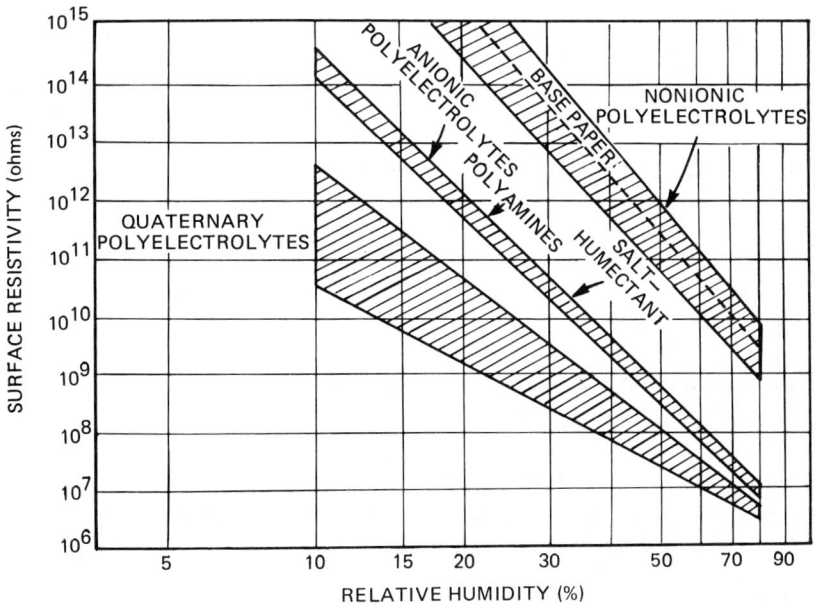

Fig. 28 Surface resistivity versus relative humidity for polyelectrolyte-coated papers. (From Ref. 13.)

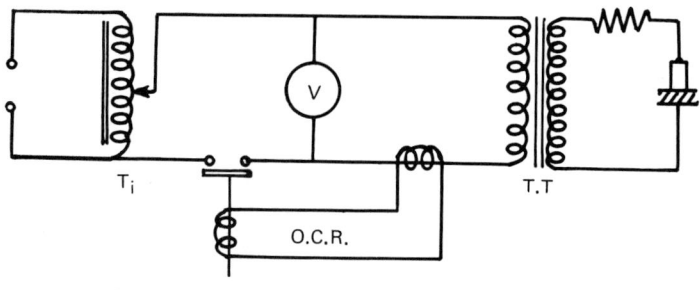

A

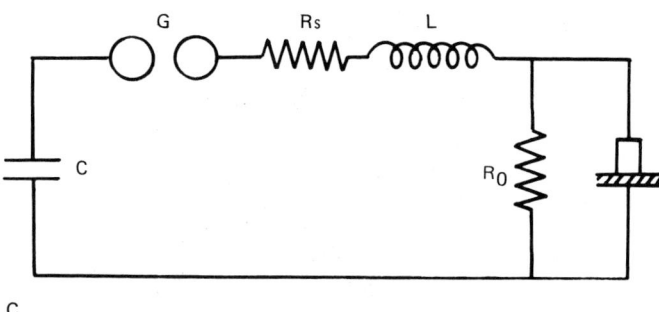

B

C

Fig. 29 Fundamental circuits of high voltage generators for measurement of dielectric breakdown strength of insulating materials. (A) AC high voltage source. (B) DC high voltage source. (C) Impulse generator.

and electrode configuration had the greatest effects on the resistivity. He concluded that to maintain a satisfactory test procedure, the humidity must be controlled to within ±0.5% RH, and the circular electrode system shown in Fig. 27 should be specified, with about 500 V potential. The electrode configuration should have a ratio P/g of 1.0 or less (where P is the effective perimeter of the electrode and g is the distance between electrodes (Fig. 27).

Hoover and Carr [13] used such a circular electrode system to study the relationships between surface resistivities of three kinds of polyelectrolyte-coated papers and relative humidity. Their results are shown in Fig. 28.

B. Measurement of Dielectric Breakdown Strength

Power Source The dielectric breakdown strength of insulating paper is generally measured for AC voltage, DC voltate, or impulse voltage, but in special cases it can be measured using a damped oscillation voltage, simulating a surge. Here, however, circuits that measure breakdown voltage using typical AC, DC, and impulse voltages are described.

Figure 29A depicts an AC high voltage generating circuit in which the secondary winding of a voltage regulator T_1 is connected to the primary winding of a testing transformer (T.T.) whose voltage is controlled by regulating the voltage of the secondary winding of T_1. Accordingly, the output voltage of T.T. is controlled by the voltage of the secondary winding of T_1. Figure 29B shows an example of a generator in which AC high voltage is rectified to DC high voltage with a kenotron or a silicone rectifier; this DC is then smoothed by appropriate circuitry. And Fig. 29C is the fundamental circuit of an impulse generator. After charging the capacitor C with a DC high voltage, the gap G is short circuited with a spark to generate an impulse voltage, which in turn is fed to the inductance L and the resistor R_0 through the damping resistance Rs. The wave-front length of this impulse is adjusted by controlling the value of L, while the wave-tail length is adjusted by R_0.

Methods of Voltage Application As the dielectric breakdown strength is affected by the duration of the applied voltage, it is necessary to provide a method to increase the voltage and the duration time of the voltage application. The methods are generally the following three.

Short-Term Test Applied voltage is increased from zero at a constant rate, say 10 kV/mm/s. The voltage is sometimes increased in such a way as to cause dielectric breakdown in a specimen in about 10 s.

Step-Up Test Applied voltage is increased as quickly as possible to about one-half of the breakdown voltage measured by the above-mentioned method and then is increased stepwise. The holding time of the voltage at each voltage level is constant, that is, 1, 10, or 30 min. This method is recommended in cases where it takes a rather long time before potential distribution is achieved in an insulating paper, such as when DC voltage is applied.

Long-Term Test Both of the above tests are relatively short term tests. A characteristic test, known as the V-t test, is often performed on insulating materials to obtain a relation between applied voltage and time duration; this

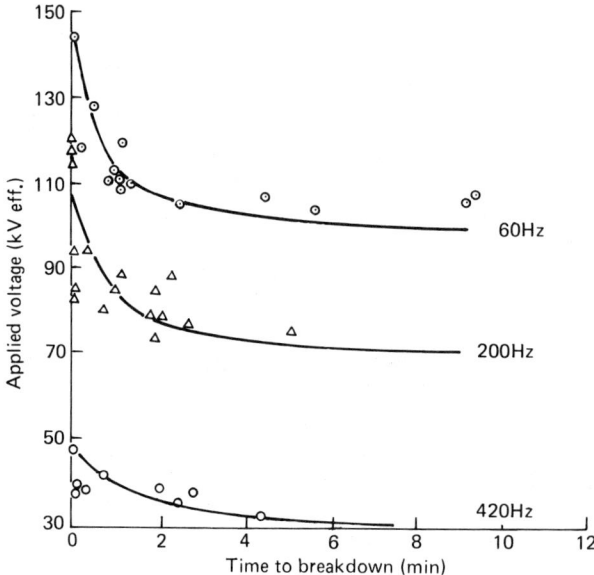

Fig. 30 V-t characteristics of oil-impregnated pressboard. (From Ref. 22.)

test is also conducted on insulating paper. The V-t test may be thought of as a life endurance test, in which a certain voltage is applied to the specimen for a time sufficient to cause breakdown. The voltage level is then changed several times in order to obtain a relationship between applied voltage and time to breakdown. Figure 30 shows an example of the V-t characteristics of an oil-impregnated pressboard obtained by Montsinger [22].

The V-t characteristic is usually measured using AC voltage, and a circuit is employed in which a timer and a circuit breaker are incorporated in the circuit of Fig. 29A.

Electrode for Measurements of Dielectric Breakdown Strength The size and form of the electrode used exert significant influence on the measurement of dielectric breakdown strength. As the electrode area becomes larger, the probability that the weak points in a sample of insulating paper will lie within the electrode area is enhanced, with the result that the measured dielectric breakdown strength becomes lower and approaches a constant value. Figure 31 shows the relation between the dielectric breakdown strength and the area of the electrode obtained by Toriyama and Inada [37].

A plate-to-plate electrode system is generally used for proving the dielectric breakdown strength of insulating paper. In many cases, the paper breaks down at seemingly low voltage because of concentration of the electric field at the edge of the electrode (edge effect) unless certain precautions are taken. In order to prevent the edge effect, the edge is rounded after the pattern of the Rogowski electrode [14]. For example, Japan Industrial Standard (JIS) C-2111 stipulates that for measurement of the dielectric

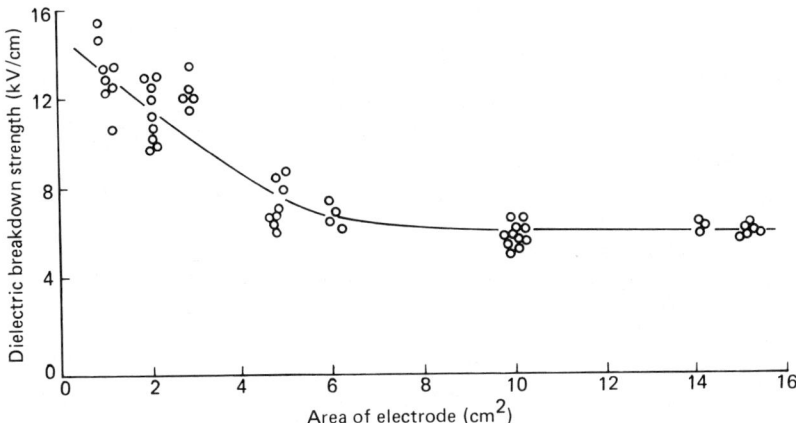

Fig. 31 Influence of electrode area on dielectric breakdown strength of insulating paper. (From Ref. 37.)

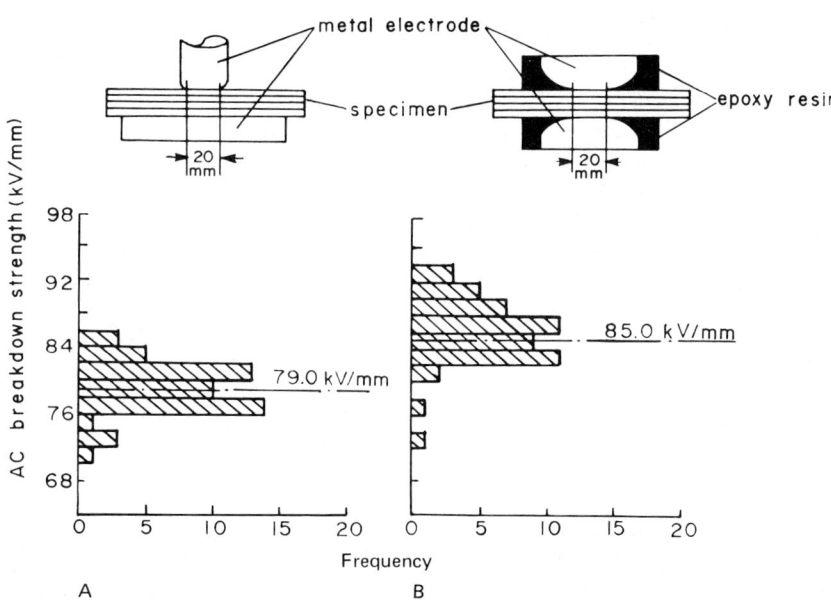

Fig. 32 Comparison of AC breakdown strength of mineral oil-impregnated paper between an electrode specified by JIS C-2111 and that having an epoxy-reinforced edge. (A) Electrode specified by JIS C-2111. (B) Electrode with epoxy-reinforced edge.

breakdown strength of insulating paper, a 25 mm diameter brass electrode with an edge rounded to a radius of 1.25 mm should be used. However, edge rounding alone is not enough to remove the edge effect completely. Figure 32 shows a comparison of tests on samples of mineral oil-impregnated cable insulating paper. For the results shown in Fig. 32A, an electrode of the type stipulated by JIS was used; Fig. 32B depicts the results when an electrode having an edge reinforced with epoxy resin was used. As this figure shows, the edge effect is considerably reduced and high breakdown strength is obtained when the measurement is done with an electrode having an epoxy-reinforced edge. It is also essential that the electrode system be isolated from ambient air because of the influence of atmospheric moisture content on dielectric breakdown strength.

REFERENCES

1. Adams, D. O. (1946). Studies of dielectric properties of chemical pulps. 2. Study of certain factors influencing the dielectric strength of paper. *Paper Trade J.* 122(7):43–52.
1a. Allam, E. M., Cortelyou, W. H., and Doepkin, H. C. (1978). Low-loss 765 kV pipe-type power cable. IEEE Trans. on Power Apparatus and Systems PAS-97, No. 6, pp. 2019–2030.
2. ASTM D 257 (1961). Electrical Resistance of Insulating Materials.
3. Calkins, C. R. (1950). Studies of dielectric properties of chemical pulps. 3. Dielectric properties of cellulose. *Tappi* 33:278–285.
4. Cooprider, T. E. (1968). Resistivity testing methods for conductive base paper. *Tappi* 51(11):520–527.
5. Delevanti, C. and Hansen, P. B. (1945). Studies of dielectric properties of chemical pulps. 1. Method and effects of pulp purity. *Paper Trade J.* 121(26):25–33.
6. Edwards, D. R., Council, J. A. H., Gibbons, J. A. M., and Scarisbrick, R. M. (1972). Polymer and polymer/paper laminated tapes for EHV oil-filled cables. International Conference on Large High Voltate Systems, paper no. 15-05.
7. Edwards, D. R., and Melville, D. R. G. (1974). An assessment of the potential of EHV polypropylene/paper laminated self-contained oil-filled cables. IEEE Underground Transmission and Distribution Conference, pp. 529–535.
8. Emanueli, L. (1959). High voltage cables, Chapman and Hall, London.
9. Fujita, H., and Ishitobi, M. (1976). A novel technique for improving dielectric loss of power cable insulating kraft paper. 1. Improvement of dielectric properties of insulating paper by applying DC potential in aqueous solution. *J. Inst. Elec. Eng. (Japan)* 96-A(4):181–188 (in Japanese).
10. Fujita, H., and Ishitobi, M. (1976). A novel technique for improving dielectric loss of power cable insulating kraft paper. 2. Properties of insulating paper treated by DC potential in aqueous solution. *J. Inst. Elec. Eng. (Japan)* 96-A(4):189–196.

11. Fujita, H., Itoh, H., and Matsuda, S. (1976). A novel type of synthetic paper for use in EHV underground cable insulation. NAS-NRC Conference on Electrical Insulation and Dielectric Phenomena, paper no. C-11.
12. Hall, H. C., and Kelk, E. (1956). Physical properties and impulse strength of paper. *Proc. Inst. Elec. Eng.* 103A:564–567.
13. Hoover, M. F., and Carr, H. E. (1968). Performance-structure relationships of electroconductive polymers. *Tappi* 51(12):552–559.
14. Institute of Electrical Engineers of Japan, representative ed., Y. Omoto (1964). Discharge Phenomena. Denki Gakkai, Tokyo, p. 49 (in Japanese).
15. Kalyazina, N. N. (1957). Investigation of dielectric losses in insulating papers. *Soviet Phys. Tech. Phys.* 2(12):2524–2528 (in Russian).
16. Kane, D. E. (1955). The relationship between the dielectric constant and water-vapor accessibility of cellulose. *J. Polymer Sci.* 18(89): 405–410.
17. Katsuura, K., and Suzuki, T. (1962). Some observations on electrical insulating paper. *Japan Tappi* 16(5):440–443 (in Japanese).
18. Kelk, E., and Wilson, I. O. (1965). Constitution and properties of paper for high-voltage dielectrics. *Proc. Inst. Elec. Eng.* 112(3):602–612.
19. Kubo, H., and Miyazaki, T. (1978). Long-term stability of impregnated polypropylene laminated paper insulation under high electric stress. IEEE International Symposium on Electrical Insulation, paper no. B4 (in Japanese).
20. Kubo, H., Sasajima, Y., Miyazaki, T., Matsuda, S., Kuwabara, H., and Fukamachi, T. (1977). A study of polypropylene laminated paper for EHV and UHV cable. *J. Inst. Elec. Eng. (Japan)* 97-A(8):403–410.
21. Matsuda, S. (1979). Electrical properties of paper. *Japan Tappi* 33(4):271–280 (in Japanese).
22. Montsinger, V. M. (1925). *Elec. World* (Oct.). Summarized in Handbook on Electric Discharge (1963). Denki Gakkai, Tokyo, p. 277 (in Japanese).
23. Miller, H. F., and Hopkins, R. J. (1947). A new kraft capacitor paper. *General Electric Rev.* 50(2):20–24.
24. Murata, M., and Nakata, K. (1977). Influence of the fines on the dielectric and tensile breakdown of oil-impregnated paper. *IEEE Trans. Elec. Insul.* EI-12, no. 1, pp. 90–96.
25. Race, H. H., Hempell, R. G., and Endicott, H. S. (1940). Important properties of electrical insulating papers. *General Electric Rev.* 43(12):492–499.
26. Renne, V. T., Kalyazina, N. N., and Morozova, M. N. (1958). Dielectric loss in capacitor paper. *Electrichestvo* 9:47–52 (in Russian).
27. Riley, T. N., and Scott, T. R. (1929). Electrical insulating papers for the manufacture of power cable. *J. IEE* 67:946–967.

28. Sakamoto, T., and Yoshida, Y. (1955). Research on dielectric properties of impregnated paper as composite dielectrics. *J. Inst. Elec. Eng. (Japan)* 75(800):504–514 (in Japanese).
29. Sakamoto, T., Ohshima, K., Yoshida, Y., Shinoda, R., and Take, Y. (1962). Improvement of dielectric properties of capacitor paper. International Conference on Large High Voltage Systems report no. 133.
30. Shimoyamada, T., and Satoh, H. (1958). Thermal deterioration of insulating paper for power cable (interrelationships between dielectric properties and drying conditions). *Hitachi Rev.* 40:1235–1242 (in Japanese).
31. Take, Y. (1960). Investigation of dielectric losses in insulating paper. Ph.D. thesis, Tokyo Institute of Technology (in Japanese).
32. Take, Y., Suzuki, Y., and Itohara, F. (1964). The effect of lignin contents on the dielectric properties of insulating paper. *J. Inst. Elec. Eng. (Japan)* 84-4(907):609–614 (in Japanese).
33. Take, Y., Suzuki, Y., and Itohara, F. (1964). Effects of hemicellulose on the dielectric properties of insulating paper. *J. Inst. Elec. Eng. (Japan)* 84-4(907):601–608 (in Japanese).
34. Take, Y., Suzuki, Y., Itohara, F., and Matsunaga, R. (1964). Influence of carboxyl group on dielectric properties of insulating paper. Conference Papers of Electrical and Electronic Engineers in Japan, no. 495 (in Japanese).
35. Take, Y., Suzuki, Y., Itohara, F., and Matsunaga, R. (1964). Influence of the amount of crystalline region on dielectric properties of insulating paper. Conference papers of Electrical and Electronic Engineers in Japan no. 555 (in Japanese).
36. Tanaka, T. (1969). Electrical conduction in oil-imprenated insulating papers. Report on Central Research Institute of Electric Power Industry paper no. 68094 (in Japanese).
37. Toriyama, Y., and Inada, K. (1957). High voltage engineering. Corona Co., Tokyo, p. 92 (in Japanese).
38. Veith, H. (1949). The dependence of DC resistance and dielectric loss of paper upon drying conditions and temperature. *Frequenz* 3(6):165–223 (in German).
39. Verseput, H. W. (1951). Studies of dielectric properties of chemical pulps. 4. The relationship between the dielectric constant and crystallinity of cellulose. *Tappi* 34(12):572–576.

23
THERMAL PROPERTIES

SABURO NAKAGAWA

Scientific Research Laboratory
Tomoegawa Paper Company
Shizuoka-shi, Japan

FRED SHAFIZADEH [†]

Wood Chemistry Laboratory
University of Montana
Missoula, Montana

I. Introduction	242
II. Thermal Conduction (S. Nakagawa)	242
A. Introduction	242
B. Specific Heat	243
C. Thermal Conductivity	245
III. Thermal Expansion (S. Nakagawa)	251
A. Introduction	251
B. Definitions	251
C. Measurements	252
D. Typical Results	253
IV. Combustion (S. Nakagawa)	255
A. Introduction	255
B. Definitions	255
C. Flammability	256
D. Ignition Temperature	257
E. Flash Point	257
F. Oxygen Index	258
G. Heat of Combustion	258
V. Thermal Decomposition of Cellulosic Materials (F. Shafizadeh)	259
A. Introduction	259
B. Lower Temperature Reactions	261
C. High-Temperature Reactions	266
D. Significance	271
E. Methods of Measurement	273
References	277

†Now deceased.

I. INTRODUCTION

The importance of thermal properties in relation to the general field of polymer properties is well known. However, it seems that these thermal properties are often lightly regarded in the field of paper and paperboard. One reason for this neglect is that since the major constituent of paper and paperboard is usually wood pulp, it is considered that the thermal properties of paper and paperboard can readily be estimated if only the thermal properties of cellulose, hemicellulose, and lignin, which are the principal ingredients of pulp, are known. Another reason is that since the characteristics of paper and paperboard are affected greatly by moisture content, it is often thought that measurement of the thermal properties of paper and paperboard under moisture-free conditions is not important from a practical point of view.

However, as more varied applications of paper and paperboard are developed, the use of many synthetic polymers and various chemicals in combination with conventional paper materials has increased, with the result that some characteristics of the composite papers are markedly different from those of papers heretofore used. Accordingly, it is to be expected that a greater emphasis will gradually be placed on the importance of the thermal properties of paper and paperboard in the future.

It is highly desirable that the level of research in this area be accelerated, as there are relatively few reports available in the present literature. The thermal properties covered in this chapter include thermal conduction, thermal expansion, combustion, and thermal decomposition.

II. THERMAL CONDUCTION (S. Nakagawa)

A. Introduction

Specific heat and thermal conductivity are important among the parameters of thermal conduction. Since specific heat is strongly related to material structure in the field of the physical properties of polymers, the importance of its measurement is well recognized in that field.

In the paper field, however, such measurements are seldom made. There is relatively little useful information presently obtained from measured values of specific heat. Although the heat conductivity of the web is an important parameter in paper drying, most of the energy consumed in the process is related to the evaporation of water, and in many cases, little consideration is given to the specific heat or the thermal conductivity of the paper itself.

As new and varied applications have developed for paper and board materials, the importance of the measurement of various physical and mechanical characteristics of paper has increased. However, measurements of specific heat and thermal conductivity are made less frequently, and the importance of these measurements is generally less well recognized than those of other properties of paper at present.

B. Specific Heat

Definitions In SI units, specific heat is expressed in terms of energy (joules) divided by the product of mass in kilograms times the absolute (Kelvin) change in temperature; the units are therefore J/(kg · K) or kJ/(kg · K). Specific heat is defined as the heat energy required per unit mass of material to raise the temperature of the mass by 1°C. Accordingly, 1kJ/(kg · K) = 0.23889 cal/(g · °C). In some cases, as in ASTM D2766-71 [4], specific heat is a unitless number. It is classified in two ways: when the volume is kept constant (Cv) and, alternatively, when the pressure is kept constant (Cp). In the case of a solid such as paper or paperboard, one measures Cp. The measurement of Cv is not practical for these materials.

In dealing with heat conduction, thermal capacity is used along with specific heat. It is expressed as the ratio Q/dT, where a temperature rise dT is observed if Q calories are added to the material; it can be calculated by multiplying specific heat by the mass (number of grams) of the material.

Measurement For measurement of specific heat, several methods are available, for example, adiabatic method, thermodifferential method (heat conductive method), and mixing method. The adiabatic method is recommended for measurement of the specific heat of a cellulosic material such as paper [11]. This method was devised by Sykes et al. in 1936; substantial improvements have been made since then. It is often used for the measurement of specific heat because of its convenience and high measurement accuracy. For the measurement of cellulose or cellulosic materials, an improved measuring device invented by Götze and Winkler [12] is available (see Fig. 1). The Götze-Winkler device consists of a calorimeter, switchboard, regulator with attachment, galvanometer, Weston standard cell, and 6 V accumulator. The main measuring part is the calorimeter, in the center of which is a 50 cm^3 copper container with a platinum resistance thermometer and a wound filament for continuous heating of a sample placed in the container. In order to prevent heat exchange with its surroundings, the container is encircled with a shield, which is automatically adjusted by means of a heater so as not to have any temperature difference with the container. The shield is double walled and the outer shield is also fitted with a heater for automatic temperature control to prevent any temperature difference.

Furthermore, the shield is put in an alumite pot, which in turn is put in a Dewar's vessel containing a freezing mixture (alcohol-dry ice) for cooling. The shield is kept at a temperature slightly lower than that of the sample with the heater. The temperatures are detected with copper-constantan thermocouples placed between the container and the shield and between the shields. The upper part of the Dewar's vessel and the pot are fitted with a polyvinyl chloride (PVC) cap. To prevent loss of insulation due to water condensate settling, silica gel or another desiccant is placed between the pot and the shield.

First, the heat capacity of the container is measured without a sample, and then a fully dried sample whose mass is already known is put in the container. The calorimeter is assembled and put in the Dewar's vessel, which contains alcohol, and the vessel is capped with the PVC cap. Dry ice is put in through a hole in this cap. The galvanometer is adjusted so that the reading is zero in the absence of current flow (after the electrical connections are

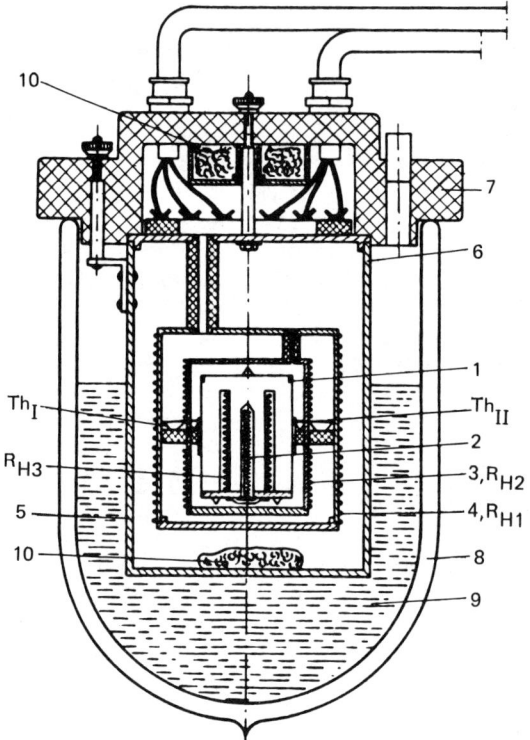

Fig. 1 Main measuring part of Götze-Winkler's calorimeter. (1) Copper container; (2) platinum resistance thermometer; (3,4) double-walled shield, (5,6) alumite pot; (7) PVC cap; (8) Dewar's vessel; (9) Freezing mixture; (10) Desiccant; (Th) copper-constantan thermocouple; (R_H) heater. (From Ref. 12.)

made). The temperature is raised by applying constant voltage and current to the heater of the container. Next, adjustment is made by applying currents to the heaters of the two shields respectively, so that there will be no heat flow between the container and the shield. This operation is repeated so as to raise the temperatures of the sample and the container gradually in order to obtain the rate of temperature rise for the sample under the particular heating schedule employed.

From this value, the specific heat Cp can be calculated according to the following equation:

$$mCp + Ce = \frac{E}{\Delta t} \tag{1}$$

where

    Cp = specific heat of the sample [kJ/(kg · K)]
    Ce = heat capacity of the container (kJ/K)

Table 1 Measured Specific Heats for Paper and Cellulosic Materials

Material	Data[a] [kJ/(kg · K)]	References
Paper	1.17–1.34	[39]
Ground pulp	1.369	[33]
Sulfite pulp	1.336	[33]
Soda pulp	1.352	[33]
Cotton	1.214–1.357	[6,12]
Flax	1.344–1.348	[6,12]
Hemp	1.327–1.353	[6,12]
Jute	1.357	[6,12]
Ramie	1.365	[21]
Viscose fiber	1.357	[10]
Isotopic rayon	1.776	[21]
Oriented rayon	1.403	[21]
Oriented rayon (Chatillon)	1.595	[21]
Oriented rayon (Meryl)	1.449	[21]
Bemberg	1.357	[10]

[a]All data are converted from cal/(g · °C) to kJ/(kg · K).

E = calories (product of voltage by current) (kJ) applied to the container heater
m = mass of the sample (kg)

Typical results Although very few measurement results are available regarding the specific heats of paper and paperboard, a number of Cp measurements for wood pulp, wood fibers, cotton, bast fibers, viscose fibers, and so on, have appeared in the literature. Since these results are reported in various measurement units, they have been converted to kJ/(kg · K) [= J/(g · K)] and are shown in Table 1.

The specific heat of paper is typically very close to that of natural cellulosic fiber. However, one can expect variation from this norm in cases where the paper contains materials other than native fibers. It must be taken into consideration also that if the paper has a high moisture content, the apparent specific heat will become larger because of the influence of the water. This is especially true if specific heat measurements are made on paper machine wet webs.

Recently, the measurement of specific heat has come to be conducted by differential scanning calorimeter (DSC) in a simple manner, but the influence of adsorbed water should be taken into consideration in this case as well.

C. Thermal Conductivity

In a practical sense, the thermal conductivity of paper and paperboard is a very important characteristic; reported data are abundant in comparison with those on other thermal properties. Since paper is porous and generally very hygroscopic, its thermal conductivity is influenced by apparent density

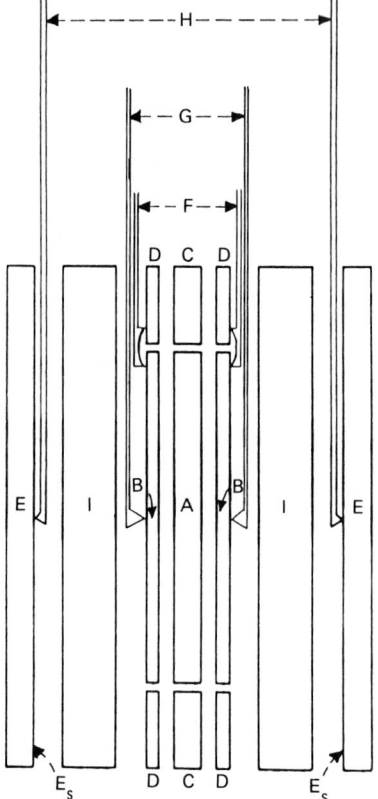

Fig. 2 General features of the metal-surface hot plate in ASTMC 177 method. (A) Central heater; (B) central surface plates; (C) guard heater; (D) guard surface plates; (E) cooling units; (E_s) cooling unit surface plates; (F) differential thermocouples; (G) heating unit surface thermocouples; (H) cooling unit surface thermocouples; (I) test specimens. (From Ref. 1.)

and adsorbed moisture. It is to be noted, however, that in many cases the values for thermal conductivity given in the literature make no mention of these influential factors.

Definition Thermal conductivity is a measure of the quantity of heat that passes through a unit area in a unit of time, as expressed by the relation shown in the following equation:

$$\frac{dQ}{d\theta} = -kB\frac{\partial t}{\partial L} \tag{2}$$

where

k = thermal conductivity
Q = heat quantity
$dQ/d\theta$ = time rate of heat flow

t = temperature
B = sample area perpendicular to the heat flow
L = distance
∂t/∂L = temperature gradient in the direction of the heat flow

Measurement When a specimen of paper or paperboard is in an absolutely dry state or contains little adsorbed moisture, its thermal conductivity can be measured in a steady state condition. However, if it contains much water, measurement in the steady state becomes difficult, and measurement in an unsteady state is required because of the release of water caused by heating at the time of measurement.

Steady State

ASTM Method This method, described in ASTM C 177-63 [1], yields the value for k that is employed in the above equation (see Fig. 2). In the center of the hot plate device is a heater, on both sides of which specimens are placed. Thermal conductivity can be obtained from the quantities of heat generated by the heater and the temperatures attained by the specimens on both sides. The heater is controlled electrically and is fitted with a guard to prevent heat loss. The size of the specimen and the guard are stipulated. The specimen for measurement should be homogeneous, with good flatness.

At the outset, the temperature difference between the hot plate and cold plate is set greater than 12°C, but the temperature on the cold plate side should not be below the ambient dew point. The thickness of the specimen is decided by the distance between the hot and cold plates. Temperature measurement is conducted by reading the electromotive force up to 0.5 μV on a potentiometer in the steady state. Here steady state means that the variation of the temperature difference between the hot and cold plates is below 0.5% of the measured temperature difference and that temperature drop through the two specimens does not differ by more than 1%.

Thermal conductivity is calculated by the following equation:

$$k = \frac{qL}{A(t_1 - t_2)} \tag{3}$$

where

  k = thermal conductivity (W/mK)
  q = time rate of heat flow (W)
  L = specimen thickness (m)
  A = two-side area of heating plates (m^2)
  t_1 = temperature of hot plate (K)
  t_2 = temperature of cold plate (K)

Terasaki's Method. This is a method for measuring the thermal conductivity of paper under conditions of constant temperature and humidity [37]. It is illustrated in Fig. 3. The specimen is wound 3 times around a copper tube containing an electric heater. Thermostatically regulated air is in contact with the specimen on the outer surface. The temperature difference between the copper tube surface and the thermostat air is adjusted at around 2 to 3°C. A copper-constantan thermocouple is used for the actual temperature measurement. In addition to the above-mentioned method, Terasaki

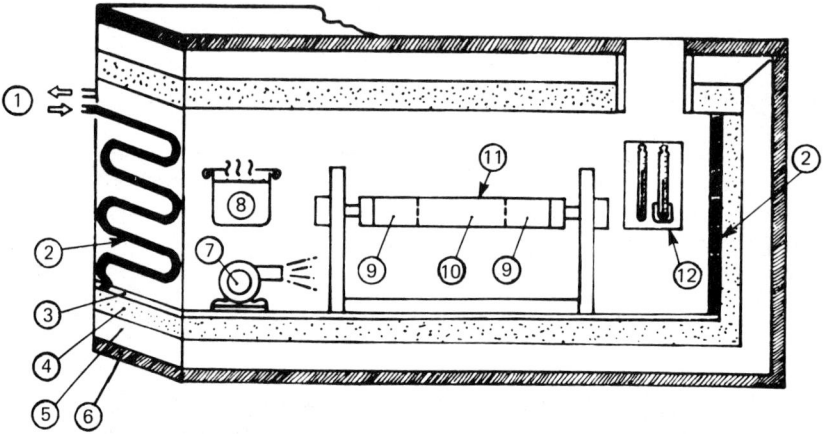

Fig. 3 Main measuring part of Terasaki's device. (1,2) Heating medium; (3) panel heater; (4) foamed styrol panel; (5) glass wool; (6) wood panel; (7) blower; (8) humidifier; (9) guard heater; (10) copper tube with main heater; (11) test specimen; (12) wet and dry bulb thermometer. (From Ref. 37.)

et al. devised two other methods, specifically, a method to measure heat transfer coefficients by passing hot and cold currents inside and outside a copper tube, respectively [36], and a method to measure effective thermal conductivity of paper with a plate heater [38].

Terada's Method. Terada et al. [35] devised a method to measure the thermal conductivity of electrical insulating papers (Fig. 4). In this method are used, as a high-temperature source, copper stranded cable wires on which electrical insulating paper as thick as 13 mm and 23 mm is wound. The specimens are dried in a vacuum for at least 14 h. The thermal conductivity of electrical insulating paper is generally measured in both a vacuum state and an atmosphere of dry nitrogen gas with three stages of pressure (100, 380, 760 mm Hg).

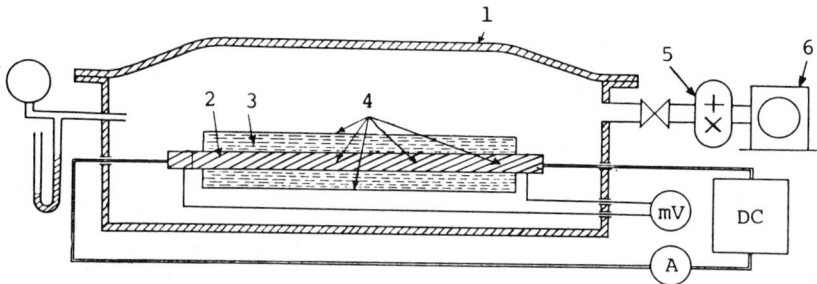

Fig. 4 Schematic of Terada's apparatus for determining the thermal conductivity of lapped paper layer. (1) Drying tank; (2) conductor (heat source); (3) test specimen layer; (4) thermocouples; (5) roots pump; (6) rotary pump. (From Ref. 35.)

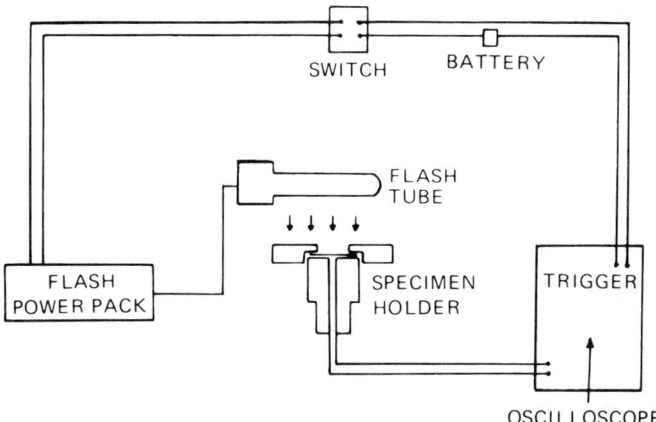

Fig. 5 General experimental layout of heat pulse method for determining the thermal conductivity of paper. (From Ref. 17.)

Unsteady state Whereas in the steady state methods heat is furnished continuously at one side of the specimen and removed at the other, the rate of heat supply is in general variable in the unsteady state method. The time and temperature gradients and rates of heat supply and removal are measured and used in appropriate forms of the general heat transfer equation.

Method of Kirk and Tatlicibasi. These authors [17] devised the following method, which offers the advantages of speed and special applicability to paper and other porous sheet materials containing volatiles such as water. The device shown in Fig. 5 employs a flash power pack and a flash tube. The energy supplied by the former is radiated from the latter as a heat pulse to the specimen. The specimen, held in a sample holder, receives heat on one side. On the other side of the specimen is a thermocouple connected to an oscilloscope for rapid detection of temperature change. From the data measured, thermal diffusivity is obtained. Thermal conductivity k is determined from the density ρ and the specific heat c of the specimen as well as diffusivity α by the relation $k = \alpha \rho c$. The measurements are conducted after the specimens have been conditioned at 20°C and 50% RH.

Typical Results Values obtained for the thermal conductivity of paper and paperboard differ, dependent on the measuring methods and conditions. Under normal conditions, these values fall within 5 to 20×10^{-2} W/(m · K). Paper being very porous, the influence of the air within the sheet is great. According to Terasaki et al. [38], 90% of the effective thermal conductivity of paper derives from the thermal conductivity of the air present inside the paper, with the solid contents of the sheet contributing only 10%. The experimental data of Kirk and Tatlicibasi [17] reveal that in the case of the sulfite pulp handsheet, the thermal conductivity increases slightly in proportion to the increase in degree of beating as well as in thickness. And the results of Han and Ulmanen [13] show that the influence of water is great; the thermal conductivity is 1.6×10^{-4} cal/cm · s. C [= 6.7×10^{-2} W/(m

Table 2 Thermal Conductivity of Paper and Paperboard

Material	Data[a] [W/(m·K)] (all entries × 10^{-2})	Condition	References
Paper	6	Room temperature	[39]
Cardboard	21	Room temperature	[39]
Rice paper	4.6	40°C	[26]
Blotting paper	6.3	20°C	[26]
Corrugated cardboard	6.3	20°C	[26]
Cement paper, plain 14 layers, each 0.38 mm, 0.62 g/cm³	12.7	20°C	[26]
Fish paper, 21 layers,	17.2	20°C	[26]
Handsheet, bleached sulfite pulp	8.4–11.7	20°C, 50% RH	[17]
Electrical insulating paper, 0.152 mm, 0.86 g/cm³	4.6–7.9	100°C, vacuum	[35]
Sulfite pulp sheet, 1500 g/m²	6.7	moisture content 0%	[13]
Sulfite pulp sheet, 1500 g/m²	19	moisture content 110%	[13]
Fine paper, 0.074 mm, 0.858 g/cm³	1.3	30°C, 40% RH	[37]
Fine paper, 0.074 mm, 0.858 g/cm³	3.1	30°C, 60% RH	[37]
Fine paper, 0.074 mm, 0.858 g/cm³	3.7	60°C, 40% RH	[37]
Fine paper, 0.074 mm, 0.858 g/cm³	5.8	60°C, 60% RH	[37]
Copy paper, 0.060 mm, 0.840 g/cm³	14.6	30°C, 40% RH	[37]
Copy paper, 0.060 mm, 0.840 g/cm³	15.1	30°C, 60% RH	[37]
Copy paper, 0.060 mm, 0.840 g/cm³	16.3	60°C, 40% RH	[37]
Copy paper, 0.060 mm, 0.840 g/cm³	17.0	60°C, 60% RH	[37]

[a] All data re converted from ca./(cm·s·°C) or kcal/(m·h·°C) to W/(m·K).

· K)], for a sulfite pulp sheet of 0% moisture content in contrast to 19×10^{-2} W/(m · K) for the same material at 110% moisture content. Table 2 shows some typical examples of reported values for the thermal conductivity of paper and paperboard.

III. THERMAL EXPANSION (S. Nakagawa)

A. Introduction

Paper and paperboard expand and shrink with changes in temperature and moisture. However, because of the greater change in expansion and shrinkage caused by moisture than by temperature, thermal expansion behavior has often been ignored. As paper and board find their way into more diverse applications, cases are encountered in which these materials have technical uses in moisture-free environments. Thermal expansion may represent a significant problem in such situations. However, there is relatively little data in the published literature on this subject.

The degree of expansion of a material is generally expressed in terms of the coefficient of linear thermal expansion. In the case of a solid, coefficients of linear thermal expansion indicate dimensional changes; a coefficient of volume thermal expansion is used to show a volumetric change. For paper or paperboard, dimensional change is more important than volumetric change. In the normal state, however, dimensional change due to absorption or adsorption of moisture is much bigger than that due to heat, and the dimensional change due to heat may fall within the tolerance of measuring error.

Since paper and paperboard, whose main constituents are cellulosic materials, are very hygroscopic, a small amount of moisture may remain in a specimen to be measured. Accordingly, the influence of even a slight amount of moisture should be given careful consideration if thermal transition points are to be obtained from thermal expansion measurements.

B. Definitions

The coefficient of linear thermal expansion α_l and the coefficient of volume thermal expansion α_v represent the percentages of change in length and volume, respectively, when the temperature rises 1°C. They are defined by the following equations:

$$\alpha_l = \frac{\partial l/\partial t}{l} \tag{4}$$

$$\alpha_v = \frac{\partial v/\partial t}{v} \tag{5}$$

l = specimen length
v = specimen volume
t = temperature

In the vicinity of room temperature, the following equation may be used as an approximation:

$$\alpha_l = \frac{l_2 - l_1}{(t_2 - t_1)l_0} \tag{6}$$

where

l_0 = specimen length at 273.15 K (0°C)
l_1 = specimen length at t_1
l_2 = specimen length at t_2

$$\alpha_v = \frac{v_2 - v_1}{(t_2 - t_1)v_0} \tag{7}$$

v_0 = specimen volume at 273.15 K (0°C)
v_1 = specimen volume at t_1
v_2 = specimen volume at t_2

Measurement is conducted under constant pressure, and the unit of the coefficient of expansion is K^{-1}.

C. Measurements

Coefficient of Linear Thermal Expansion Generally, the absolute value of the change in length caused by thermal expansion is measured. The following methods of measurement of length are available:

- A direct reading of the change in length
- •• Light interference
- ••• Measurement of the movement of an electrode plate with changes in the capacity of an electrode

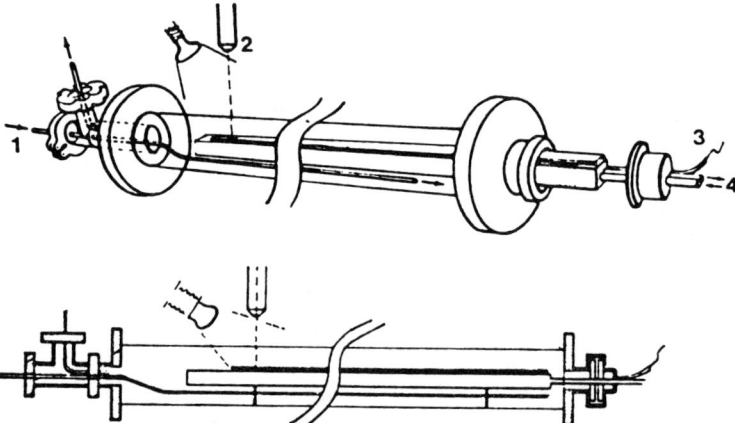

Fig. 6 Apparatus of Kubát et al. for determining the thermal expansion of paper. (1) feed of gas (nitrogen); (2) traveling microscope; (3) thermocouple; (4) feed of thermostating liquid. (From Ref. 20.)

For measurement of paper, the direct reading method by Kubát et al. [20] is the best available. It is drawn in Fig. 6 and may be described as follows. The measuring device consists of a glass tube containing a hollow stainless steel bar upon which the specimen to be measured is placed. Dry nitrogen gas flows through the glass tube to maintain an absolutely dry state in order to prevent any influence of moisture. The cross section of the stainless steel bar is square. Within it is circulated a thermostatting liquid (glycerol or ethanol), whose temperature is controlled by a thermostat. A specimen for measurement placed on the stainless steel bar is then covered by a small U-section copper bar in order to eliminate temperature gradients through the thickness of the specimen. The copper bar is slit so that the position of one end of the specimen can be followed and measured with an overhead traveling microscope. By means of a clamp, the other end of the specimen is fixed at a point to the stainless steel bar. The specimen is dried at 105°C for 10 h in a stream of the dry nitrogen gas passed through the glass tube. An initial length is determined, and thereafter the increase or decrease in length due to temperature change is measured with the traveling microscope. A thermocouple is used for temperature measurement. In the experiment of Kubát et al. [20] the speed of temperature rise is 2.5°C/10 min. The thermal expansion for both ascending and descending temperatures between 0 and 90°C is measured. From the rate of change in length and the temperature variation, the linear expansion coefficient can be obtained. In the case of paper and paperboard, the precise temperature at the time of each measurement should be recorded without fail, together with the measured values of the coefficient of linear expansion, because of the presence of several transition points between 0 and 100°C.

Klason and Kubát [18] used a dilatometer and a torsion pendulum for measurements of the coefficient of linear expansion of spruce sulfite pulp sheets, cellulose II prepared by the viscose method, and sheets formed with ammonia-treated pulp fibers. In these experiments, the speed of temperature rise was 0.2 to 0.5°C/min, and the measurement was conducted between −163 and 197°C (110 to 470 K).

Coefficient of Volume Thermal Expansion Measurements of the coefficient of volume thermal expansion of pulp and regenerated cellulosic materials are often conducted, but no data have been reported so far in the literature regarding paper or paperboard. For measurment of α_V a specimen has to be formed into pellets or some similar form; the specimen pellets are quite different from the original. Accordingly, it is considered that the values obtained are of little significance as data appropriate for paper.

In those instances where pulp or cellulose was measured by Ramiah and Goring [24] or Kubát et al. [19], a dilatometer was used as a measuring device. This device is often used, as per ASTM D 684, to measure the volume thermal expansion of plastics [3]. Measurement with the dilatometer requires great skill, but the automated device recently put on the market has made the measurement comparatively easy.

D. Typical Results

Kubát et al. [20] and Ruvo et al. [25] have published the results of measurements of the coefficient of linear thermal expansion. According to them,

Table 3 Coefficients of Linear Thermal Expansion of Paper and Paperboard (at 25°C)

Material	Data (K^{-1}) (all entries × 10^{-6})	
	Machine direction	Cross direction
Greaseproof paper, 40 g/m², 0.038 mm	7.5	15.5
MG kraft paper, 50 g/m², 0.056 mm	6.4	13.4
Kraft sack paper, 70 g/m², 0.099 mm	6.1	16.2
Newsprint paper, 52 g/m², 0.080 mm	5.6	13.6
Groundwood printing, 60 g/m², 0.054 mm	3.7	10.1
Fluting, 127 g/m², 0.234 mm	7.4	12.1
Whitelined duplex, 400 g/m², 0.581 mm	5.9	12.3
Solid bleached board, 240 g/m², 0.246 mm	4.2	8.7
Whitelined chip board, 400 g/m², 0.552 mm	3.6	15.2
Line chip board, 400 g/m², 0.574 mm	2.0	13.7
Bleached sulfate pulp, 944 g/m², 0.920 mm	6.1	7.9

Source: From Ref. 20.

the coefficient of linear thermal expansion of various paper and board grades is always smaller in the machine direction (MD) than in the cross direction (CD). Thus thermal expansion, like the modulus of elasticity, shows strong directional dependence, but of an opposite trend. That is, thermal expansion coefficients and elastic properties are inversely related. When one takes the product of the coefficient of linear thermal expansion and the modulus of elasticity, the numerical values are shown to be approximately equal for MD and CD samples of the same paper grade. This numerical value varies from grade to grade, however.

Dimensional changes caused by moisture in the ambient air (hygroexpansion) follow the same trend as thermal changes, being less in the MD than in the CD. Nevertheless, the absolute values for hygroexpansion are an order of magnitude greater than linear thermal expansion if one compares a relative humidity increment of 0 to 95% with a temperature rise of 100°C [25].

Thermal expansion coefficients do not seem to be dependent on sheet density, although certain processing variables such as the degree of beating do exert an effect [25]. Thermal transition points can be obtained from the change of the coefficient of thermal expansion with temperature. Kubát et al. discovered that the thermal transition points are near 35°C and 65°C by both the direct reading method [20] and the dilatometer method [19]. It is supposed, however, that this transition phenomenon may be caused by the influence of a slight amount of moisture remaining in the paper, and similar results were obtained by Klason and Kubát [18]. Ramiah and Goring [24] determined the volume expansion of pulp fiber components in water swollen states and found that the values measured in the case of sulfite pulp are two to three times larger than those in the dry state.

Table 4 Coefficient of Volume Thermal Expansion of Cellulosic Materials

Material	Data (ml/g/K) (all entries × 10^{-5})	
	Below T^a	Above T^a
Sulfite pulp, dry	5.5	7.7
Sulfite pulp, water-swollen	18.1	12.7
Cellulose, Avory B, dry	6.0	8.9
Cellulose, Avory B, water-swollen	14.2	11.8
Cellulose, Avory C, dry	5.0	6.5
Cellulose, Avory C, water-swollen	16.0	11.1

[a]Transition point.
Source: From Ref. 23.

The coefficients of linear thermal expansion for several grades of paper and paperboard are shown in Table 3, and the coefficients of volume thermal expansion of some cellulosics are shown in Table 4.

IV. COMBUSTION (S. Nakagawa)

A. Introduction

The flammability of paper and paperboard is well known. This property is advantageous in that paper materials can be readily disposed of by burning, whereas the disadvantage in their use as interior decoration materials (wallpapers, and so on) and other products where there is a significant fire hazard is obvious.

For better protection against fire, much work is needed on the flame resistance of paper, and several evaluation methods to measure flammability have been devised. The results obtained from these methods serve well as an index of the flammability of paper and paperboard, but these data should be evaluated with the understanding that they are obtained under specific conditions.

B. Definitions

The reader should note that certain words related to combustion are often used erroneously. Several terms are defined here:

Combustion: An oxidizing reaction in which heat and light are evolved.
Ignition temperature (autoignition temperature): The lowest temperature at which a material, when heated in air, will spontaneously ignite from the heat of oxidation in the absence of any other source of ignition energy [15].
Flash point: The lowest temperature at which flammable gas given off from a material heated in air is ignited with external ignition energy when the density of the evolved flammable gas has reached the combustion limit.

Burning point: The lowest temperature at which the flammable gas evolved from a material at a temperature slightly higher than flash point is kept burning after it has been ignited (usually about 10°C higher than flash point).

Oxygen index: The minimum concentration of oxygen, expressed as volume percent in a mixture of oxygen and nitrogen, that will just support combustion of a material [2].

C. Flammability

Measurement

TAPPI Method A flammability test is designated in TAPPI T 461 su-72 [34]; it is identical with the method of ASTM D 777-74 [2]. The test is of the vertical burning type. The testing apparatus consists of a metal cabinet, a holder, and a burner. A specimen 210 mm in length and 70 mm in width is inserted in the holder and is suspended vertically on the knob in the center of the cabinet. A Bunsen or Tirrill gas burner 10 mm in inner diameter is used. The distance between the top of the burner tube and the lower edge of the specimen is adjusted to 19 mm, and the flame height is adjusted to 40 mm. After the cabinet door is closed, the burner is fired in such a way that the lower edge of the specimen attached to the holder is exposed directly to the yellow flame of the burner for 12 s; then the flame is removed. Flaming time, the time the specimen glows after it has ceased flaming, and char length are recorded. According to the TAPPI method, it is further stipulated that a water immersion test be conducted for waterproof flame-resistant paper [34].

UL Method Another flammability test is designated in Underwriter's Laboratory standard UL 94 [41], which was originally devised for plastics. In this test, a vertical burning method is used for measurement of the flammability of laminated board consisting of resin-saturated papers. The testing device is similar to that of the above-mentioned TAPPI method, and the TAPPI device can be used instead.

The specimen size is 127 mm in length and 13 mm in width. Each set consists of 5 specimens. A ring stand with clamps that adjust for vertical positioning is used for a specimen holder. The longer axis of the specimen is aligned vertically. The specimen is held by a clamp at a point 6 cm from the upper end. The height of the blue flame of the burner is adjusted to 19 mm, and the center of the lower edge of the specimen is exposed to the flame continuously for 10 s. Then the time the specimen continues to burn after removal of the flame is recorded.

After the flame is extinguished, the specimen is exposed again for 10 s to the burner flame, and the duration of flaming and glowing after the flame is removed is again recorded. The flame resistance of the sample is graded by the total flaming combustion time for the total of 10 flame exposures for each set of 5 specimens.

Typical Results Ordinary paper and paperboard are consumed in a burning test. In the case of papers treated with flame retardants, the degree of burning differs, dependent on the degree of flame resistance. When 13 kinds of commercially available flame-resistant wall-backing papers were tested according to the TAPPI method, the average values of 10 points of char length were 52 to 104 mm. Generally, if the char length exceeds 120 mm, the probability that the specimen will burn completely is high. Since the values obtained differ a great deal, attention should be paid to the measuring method. In the UL method as well, the results of measurement differ, and there is an observable tendency that the difference in the values obtained becomes greater as the degree of flame-retardant treatment is lower. Results of measurements on laminated board for electrical insulation show that the total time of burning is 10 to 15 s for material with a high degree of flame-retardant treatment and 50 to 100 s for material with a low degree of treatment. Since the reproducibility of flammability test results is generally low, it is desirable to compare the results of treated samples with untreated controls.

D. Ignition Temperature

Measurement For the measurement of the ignition temperature of paper and paperboard, several common devices can be employed, but here the UL method [42] is introduced. This method is stipulated in UL 94. A combustion chamber to be used for measurement is put in a metal-melting bath and is heated with a special electric furnace with a temperature controller. The combustion chamber is a flat-bottomed cone flask made of heat-resistant glass. The diameter of the bottom is 60 mm, that of the top, 28 mm, and the height, 114 mm. A thermocouple protected with quartz is used for temperature measurement. Specimen pieces 6.4 mm square are dropped into the flask at intervals of 5°C and are observed to see whether they ignite. The lowest temperature at which ignition occurs, which results in flaming or glowing combustion within 2 min, is taken to be the ignition temperature.

Typical Results The ignition temperature of paper is about 450°C [15], but this value varies somewhat with the type of paper. The ignition temperatures of cellulosic fibers are 475°C for cotton, 550°C for flame-resistant cotton (treated with N-methyl-dimethyl-phosphonopropionamide), and 450°C for rayon [16]. It is probable that the ignition temperature for a paper treated with flame retardants is 100°C higher than that for an untreated paper.

E. Flash Point

Measurement The method devised by Ishii et al. [16] is explained here. An electric furnace is used for this method. At the lower end of the electric furnace a chromel-alumel thermocouple is fixed; aluminum foil, in which a specimen is put, is placed on the thermocouple. A nichrome wire is used for an ignition source. A 25 mg pulverized specimen is employed, and the rate of rise in the furnace temperature is 30°C/min. The flash point is the temperature at which a flame begins to appear. This device can be also used for measurement of ignition temperature if the nichrome wire is removed as an ignition source.

Typical Results Although no measurements of the flash point of paper and paperboard are available, the flash point of cotton is 361°C, that of rayon is 327°C, and that of flame-resistant cotton is more than 650°C, according to measurements by Ishii et al. [16]. Based on these results, the flash point of paper and paperboard is estimated to be around 350°C.

F. Oxygen Index

Measurement A method for measuring oxygen index (OI) is specified in ASTM D 2863-70 [5]. The testing device consists of a heat-resistant glass tube 75 mm in inside diameter and 450 mm in height, a specimen holder, a gas supplier, a flow measurement and control device, and an ignition source. The size of specimen is 70 to 150 mm long, 6.5 ± 0.5 mm wide, and 3.0 ± 0.5 mm thick. At least 10 pieces of the specimen are measured. Since the ASTM test method was devised for plastics of moderate ($\sim$ 3 mm) thickness, a special device is needed for a thin sheet material such as paper in order to properly affix a specimen in the specimen holder. The test proceeds by the use of an ignition source with various oxygen-nitrogen mixtures to provide several levels of oxygen concentration. Gas of known mixing ratio is passed into the glass cylinder at a speed of 4 ± 1 cm/s for at least 30 s. Then the upper edge of the specimen is ignited by means of a gas burner. The minimum amount of oxygen that either will keep the test piece burning for more than 3 min or will consume more than 50 mm of the specimen length must be measured together with the amount present in order to obtain an oxygen index by the following equation:

$$OI(\%) = \frac{100 \times [O_2]}{[O_2] + [N_2]} \tag{8}$$

where

$[O_2]$ = minimum flow quantity of oxygen required for combustion (cm^3/s)
$[N_2]$ = flow quantity of nitrogen corresponding with the above (cm^3/s)

Typical Results Although measurements of the oxygen index for paper and paperboard do not appear in the literature, there are some results for other cellulosic materials. The index for cotton is 18%, that for rayon 19%, and that for flame-retardant cotton 35%, according to Ishii et al. [16]. From these figures, the oxygen index of ordinary paper and paperboard is assumed to be near 20%. Generally, the oxygen index is below 20% in the case of flammable materials and above 30% for fire-retardant materials.

G. Heat of Combustion

The heat of combustion is measured by calories produced when a material burns. It is expressed in terms of combustion energy per unit mass. For measurement of paper and paperboard, a bomb calorimeter of the type used for the heat of combustion of a solid can be employed. A specific method of measurement for paper is omitted here because there is no reported procedure. However, there are available some measured values of the heat of combustion of pulps [6,33] which are shown as follows:

Wood pulp	17,460 J/g (4173 cal/g)
Linter	17,426 J/g (4165 cal/g)
Bleached sulfite pulp	17,924 J/g (4284 cal/g)
Bleached rag stock	17,962 J/g (4293 cal/g)

Paper and paperboard are present in large quantities in municipal refuse. As such they can be an important heat source. In the future it can be expected that greater emphasis will be placed on the combustion energies of paper and paperboard for this reason.

V. THERMAL DECOMPOSITION OF CELLULOSIC MATERIALS
(F. Shafizadeh)

A. Introduction

As paper is heated it undergoes a series of physical and chemical changes that affect the physical properties and ultimately lead to charring and pyrolysis. The rate of deterioration and degradation is enhanced not only by the increasing temperature but also by the partial pressure of oxygen and water and the presence of reactive compounds—particularly acidic and alkaline materials and decomposition products, which produce an inductive or autocatalytic effect. Conversely, the thermal stability of paper may be improved by the careful removal of inorganic materials or by the introduction of neutral oxides to prevent development of incipient acidity. At lower temperatures it is difficult to draw a line of demarcation between normal aging and oxidative thermal degradation, which accelerates by heating. When rag paper is heated at 38°C (311°K) for about 6 months, the accelerated aging results in a 19% reduction in the folding strength.

Thermal degradation precedes combustion of cellulosic materials and affects not only the flammability and rate of combustion but also how the ma-

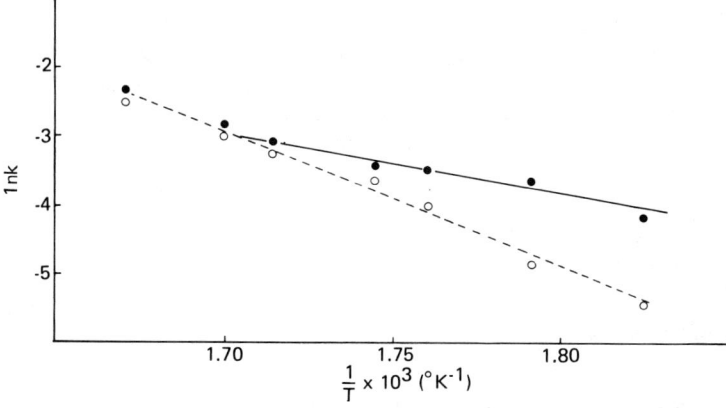

Fig. 7 Arrhenius plot for the first-order reaction in the isothermal degradation of cellulose in air (—) and nitrogen (---). (From Ref. 31.)

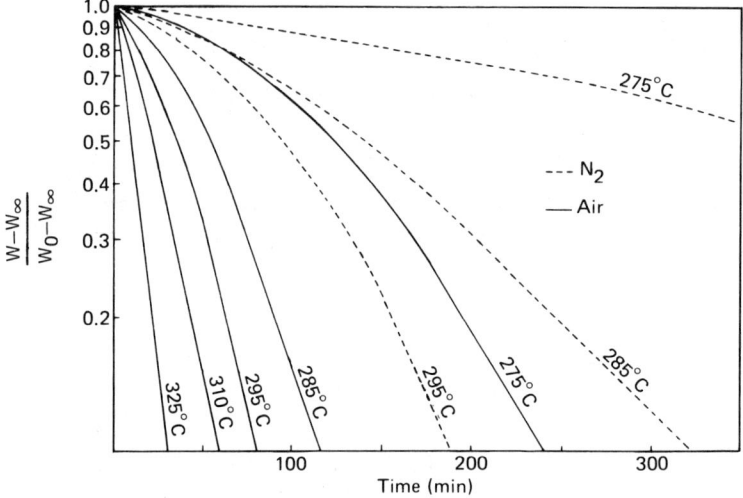

Fig. 8 First order plot for the residual cellulose weight (normalized) versus time. Plots at 310°C and 325°C for air and nitrogen are similar. (From Ref. 31.)

terial burns and how the process may be controlled. Finally, thermal degradation provides the basis for analytical pyrolysis, which is developing as a powerful tool for the analysis of polymeric materials, including cellulose and paper.

In view of the above considerations and because of the current interest in the conversion of biomass and waste paper to fuel and chemicals, the thermal degradation of cellulosic materials has been extensively investigated and discussed in numerous publications. The following summary may provide a general understanding of the related reactions and phenomena [28–30].

The complexity of the thermal reactions and the inhomogeneity of the substrates presents a major problem in generalizing on this subject and accounts for the variation of the quantitative data. This problem is handled by working with better defined substrates and unraveling the sequence of the thermal reaction under different conditions. Since paper contains mainly cellulose and smaller amounts of hemicelluloses and lignin, the basic studies on this subject have been focused on the thermal degradation of cellulose.

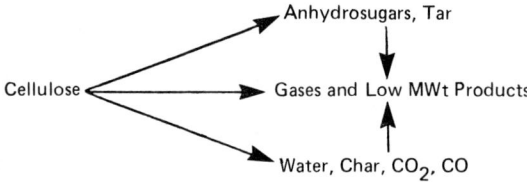

Fig. 9 Competing pathways for cellulose pyrolysis.

As cellulose is heated, the availability of sufficient energy of activation for different reactions results in production of char and a variety of volatile products that can be measured as weight loss by thermogravimetry (TG). An Arrhenius plot for rates of weight loss of cellulose in air and nitrogen is shown in Fig. 7. These data indicate a transition at ∿ 300°C (∿ 573 K) which reflects the existence of two different pathways. As shown in Fig. 8, the rate of pyrolysis followed by weight loss under isothermal conditions shows an initial period of acceleration and proceeds much faster in air than in an inert atmosphere. As the pyrolysis temperature is increased, the initiation period and the difference between pyrolysis under nitrogen and air gradually diminish and disappear at 310°C, when pyrolysis by the second pathway takes over. The general pathways for pyrolysis of cellulose [5a,31] are shown in Fig. 9.

B. Lower Temperature Reactions

As shown in Fig. 9, the reactions in the first pathway, which dominates at lower temperatures, involve reduction in molecular weight or DP by bond scission; appearance of free radicals, elimination of water, formation of car-

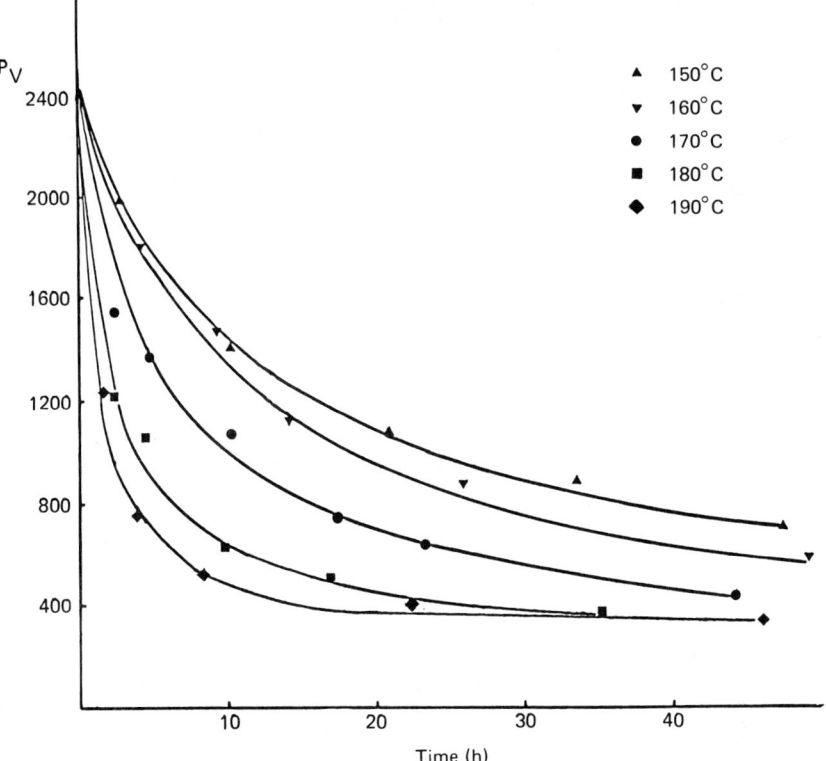

Fig. 10 Viscosity average degrees of polymerization (P_V) of cellulose heated in air at 150 to 190°C.

Table 5 Rate Constants for the Depolymerization of Cellulose in Air and Nitrogen

Temperature (°C)	$k_V \times 10^7$ in N_2 mol/162 g · min[a]	$k_V \times 10^7$ in air mol/162 g · min[a]
150	1.1	6.0
160	2.8	8.1
170	4.4	15.0
180	9.8	29.8
190	17.0	48.9

[a] 162 g represents 1 mol of monomer unit.

bonyl, carboxyl, and hydroperoxide groups (especially in air); evolution of carbon monoxide and carbon dioxide, and finally production of a charred residue. The mechanism and kinetics of these reactions, which contribute to the overall rates of pyrolysis of cellulosic materials, have been individually investigated. Reduction in the degree of polymerization of cellulose on isothermal heating in air or nitrogen at a temperature within the range of 150 to 190°C (423 to 463 K) has been measured by the viscosity method as shown in Fig. 10. These data have been correlated with rates of bond scission as given in Table 5 and used for calculating the Arrhenius plot shown in Fig. 11.

Figure 12 shows the rate of production of carbon monoxide and carbon dioxide at 170°C (443 K) in air and in nitrogen. The rate of evolution of these gases is much faster in air than in nitrogen and, furthermore, accelerates on continued heating. It is instructive to compare the initial linear

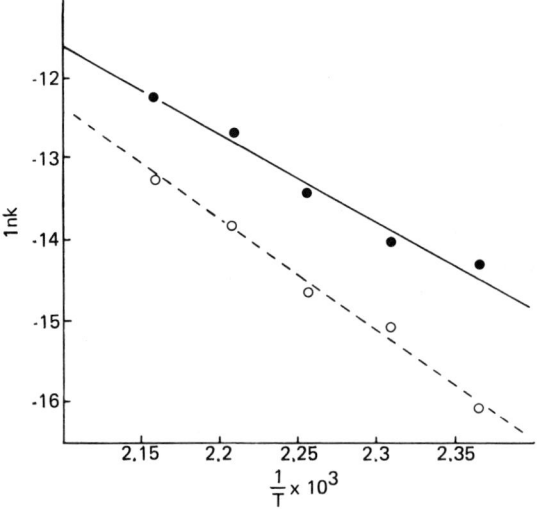

Fig. 11 Arrhenius plot for the rate of bond scission in air (—) and nitrogen (- - -). (From Ref. 31.)

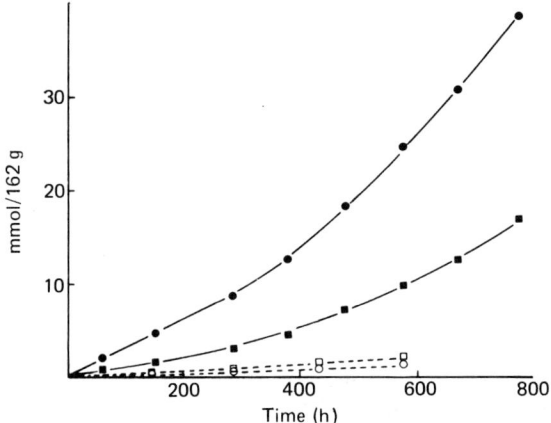

Fig. 12 Yields of CO and CO_2 from heating cellulose at 160°C: (○)CO_2 in N_2; (□)CO in N_2; (●)CO_2 in air; (▫)CO in air. (From Ref. 31.)

rates for the emission of these gases with the rates of bond scission obtained for depolymerization at 170°C, discussed before. As can be seen in Table 6, for air the rate of bond scission approximately equals the rate of production of carbon dioxide plus carbon monoxide in moles per glucose unit. In nitrogen, however, the rate of bond scission is greater than the combined rates of carbon monoxide and carbon dioxide evolution.

The thermal degradation of cellulose in air, similar to that of synthetic polymers, apparently involves a free radical mechanism and formation of hydroperoxide groups. The hydroperoxide functions are simultaneously formed and decomposed, and their concentration rapidly climbs until a steady state is reached. Figure 13 shows the development of a steady state concentration in air at 170°C during a period of about 100 min. It also shows the rate of decay in nitrogen over a similar period. This decay of the hydroperoxide function appears to follow first-order kinetics with a rate constant of 2.5×10^{-2} min^{-1} at 170°C. From the steady state concentration of 3.0×10^{-5} mol/162 g, the rate of hydroperoxide decomposition is thus 7.5×10^{-7} mol/162 g · min. When compared with the initial rate of bond scission in air of

Table 6 Initial Rates of Glycosidic Bond Scission and Carbon Monoxide and Carbon Dioxide Formation at 170°C (443 K)

Reaction	Rate $\times 10^5$ in N_2 mol/162 g · hr[a]	Rate $\times 10^5$ in air mol/162 g · hr[a]
Bond scission	2.7	9.0
CO evolution	0.6	2.1
CO_2 evolution	0.4	6.4

[a]162 g represents 1 mol of monomer unit.

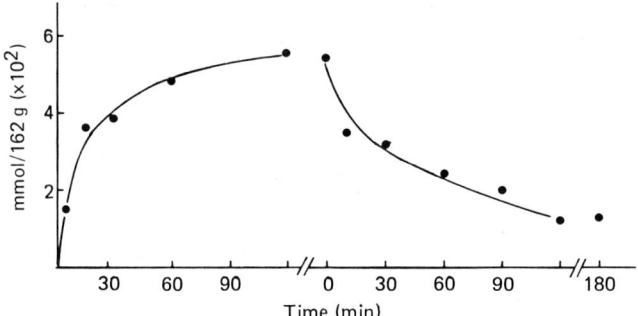

Fig. 13 Rate of formation and decay of hydroperoxide groups in cellulose at 170°C (443 K). Formation is in air and decay is in nitrogen. (From Ref. 31.)

1.5×10^{-6} mol/162 g min at 170°C (Table 5), it is apparent that the hydroperoxide formation could make a significant contribution to bond scission.

Based on the above considerations and in analogy with studies on radiation of carbohydrates, it is proposed that three stages are involved in the low-temperature pathway: initiation of pyrolysis, propagation, and product formation. As shown in Fig. 14, the initiation period apparently involves the formation of free radicals facilitated by the presence of oxygen or inorganic impurities. Subsequent reactions of the free radicals could lead to bond scission, oxidation, and decomposition of the molecule, producing char, water, carbon monoxide and carbon dioxide. Figure 15 shows the nature of the initiation, propagation, and decomposition reactions involved in the thermal decomposition of cellulose by this pathway [31].

Initiation

$$\text{Initiator (I)} \xrightarrow{\text{Heat}} \text{I·}$$

$$\text{Cell-H} + O_2 \rightarrow \text{Cell·} + HO_2·$$
$$\text{I·} + O_2 \rightarrow IO_2·$$
$$\text{Cell-H} + \text{I·} \rightarrow \text{Cell·} + IH$$
$$\text{Cell-H} + IO_2· \rightarrow \text{Cell·} + IO_2H$$

Propagation

$$\text{Cell·} + O_2 \rightarrow \text{Cell } O_2·$$
$$\text{Cell } O_2· + \text{Cell-H} \rightarrow \text{Cell } O_2H + \text{Cell·}$$

Formation of Products

$$\text{Cell } O_2H \rightarrow \text{Cell } O·, \text{Cell } O_2·$$
$$\text{Cell } O_2H \rightarrow \text{Products}$$
$$\text{Cell } O_2·, \text{Cell } O· \rightarrow \text{Products}$$

Fig. 14 Thermal autooxidation of cellulose in air. (From Ref. 31.)

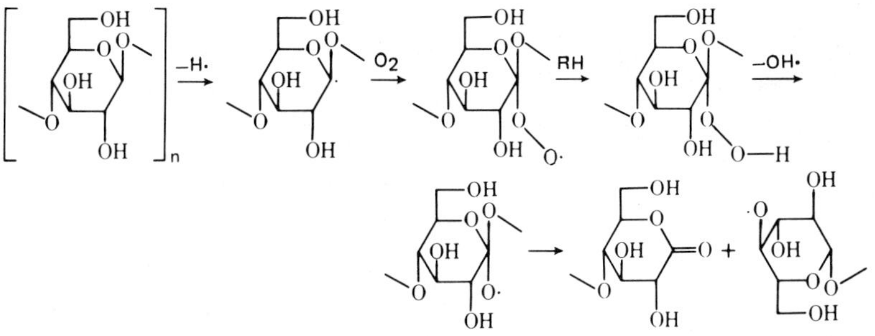

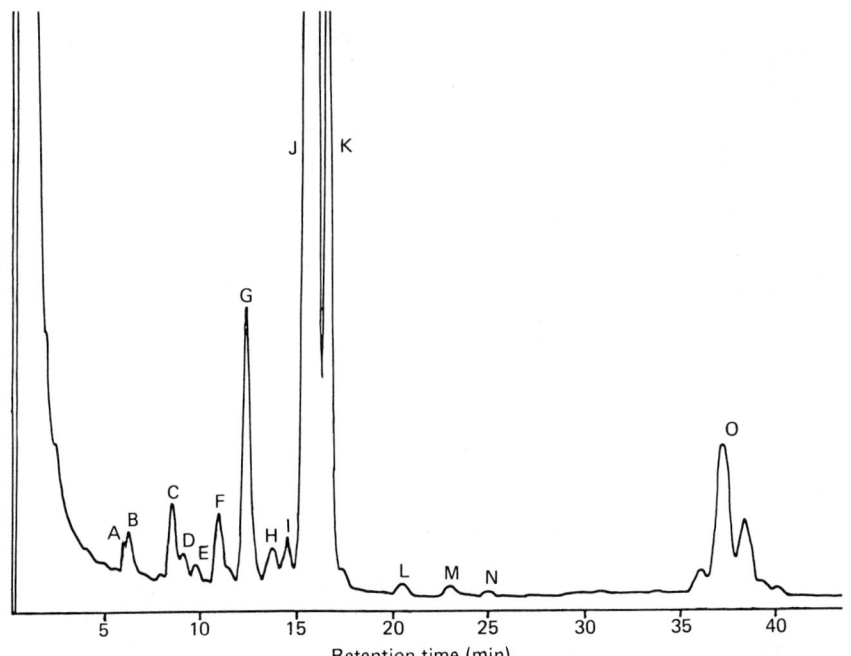

Fig. 15 Possible mechanism of formation and decomposition of cellulose hydroperoxide formed thermally in air. (From Ref. 31.)

Fig. 16 Gas liquid chromatograph of tar from pyrolysis of (trimethylsilyl)ated cellulose (column 2, 110 to 275°C at 4°/min). (a) 5-(Hydroxymethyl)-2-furaldehyde; (B) 1,4:3,6-dianhydro-α-D-glucpyranose; (C) 2,3-dihydro-3,5-dihydroxy-6-methyl-4H-pyran-4-one; (D) unknown; (E) 3,5-dihydroxy-2-methyl-4H-pyran-4-one; (F) unknown; (G) 1,5-anhydro-4-deoxy-D-glycero-hex-1-en-3-ulose; (H,I) unknown; (J) levoglucosan; (K) 1,6-anhydro-β-D-glucofuranose; (L) α-D-glucose; (M) β-D-glucose; (N) 3-deoxy-D-erythro-hexosolose; (O) O-D-glucosyl-levoglucosans.

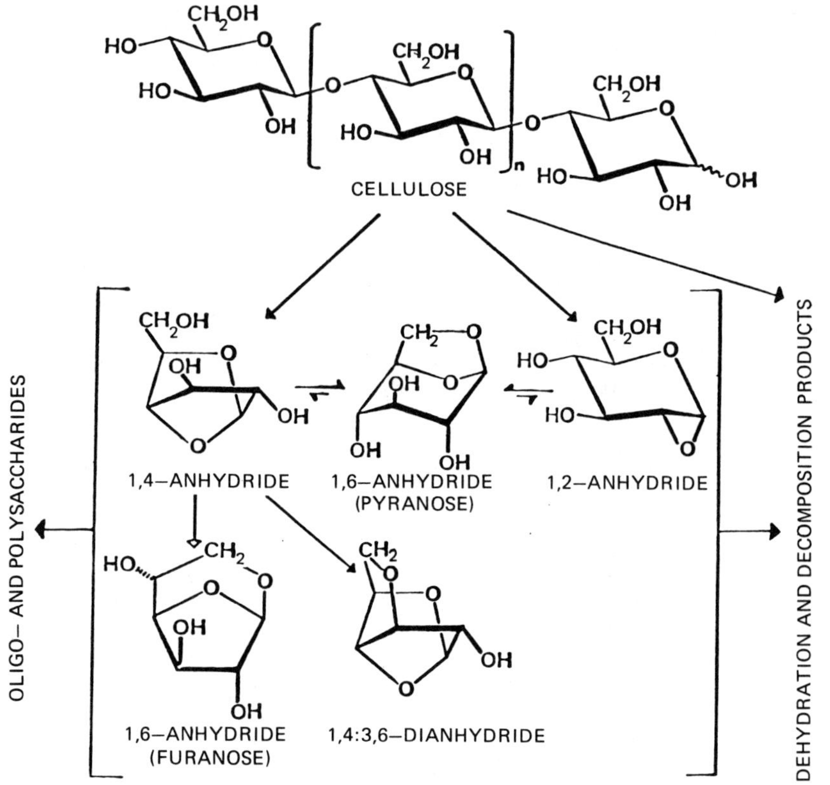

Fig. 17 Pyrolysis of cellulose to anhydrosugars and other compounds by transglycosylation reactions.

C. High-Temperature Reactions

At temperatures above 300°C (573 K), cellulose is decomposed by an alternative pathway that provides a tarry pyrolyzate containing levoglucosan (1,6-anhydro-β-D-glucopyranose), other anhydroglucose compounds, randomly linked oligosaccharides, and glucose decomposition products (see top line of[2] Fig. 9). Figure 16 shows the gas liquid chromatography (GLC) analysis of the tarry pyrolyzate after trimethylsilylation of the free hydroxyl groups. The mechanism for the formation of these compounds has been established by extensive investigation of the pyrolytic reactions of phenyl glucosides and other related model compounds. This mechanism, shown in Fig. 17, involves intramolecular substitution of the glycosidic linkage in cellulose by one of the free hydroxyl groups (transglycoslylation). Subsequent inter- and intramolecular transglycosylations provide several anhydrosugars and randomly linked oligosaccharides, which can dehydrate and decompose on further heating to form a tar. The initial substitution requires changes in the conformation of the sugar units and increase flexibility of the molecule. This may be achieved at elevated temperatures by reduction of molecular weight,

Table 7 Effect of Temperature on the Products from Pyrolysis of Cellulose Powder Under Vacuum

Oven temp (°C)	Pyrolysis time (min)	Percent yield from cellulose				
		Char	Tar	Levo-glucosan	1,6-Anhydro-D-glucofuranose	Reducing sugar
300	180	21	60	34	4	47
325	60	10	70	38	–	54
350	30	8	70	38	4	52
375	10	6	70	38	–	59
400	5	5	77	39	4	60
425	4	4	78	40	4	59
450	3	4	78	39	4	57
475	3	3	80	38	4	58
500	3	3	81	38	4	57

breaking of hydrogen bonds, and glass transition, all of which could be expected to activate the molecule.

Kinetic studies have shown that at higher temperatures the tar-forming reactions accelerate rapidly and overshadow the production of char and gases. The data in Table 7 show the production of diminishing amounts of char and increasing amounts of tar, the anhydrosugars, and other compounds (that could be hydrolyzed to reducing sugar) as the oven temperature is raised from 300 to 500°C (573 to 773 K).

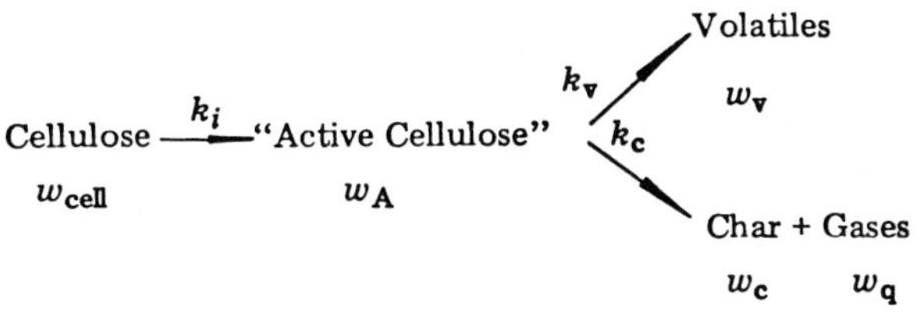

$$\text{Cellulose} \xrightarrow{k_i} \text{"Active Cellulose"} \begin{array}{c} \xrightarrow{k_v} \text{Volatiles } w_v \\ \xrightarrow{k_c} \text{Char + Gases } w_c \; w_q \end{array}$$

w_{cell} w_A

where

$$\frac{-d(w_{cell})}{dt} = k_i [w_{cell}]$$

$$\frac{d(w_A)}{dt} = k_i [w_{cell}] - (k_v + k_c)[w_A]$$

$$\frac{d(w_c)}{dt} = 0.35 k_c [w_A]$$

For pyrolysis of pure cellulose under vacuum the rate constants k_i, k_v and k_c were found to correspond with $k_i = 1.7 \times 10^{21} e^{-(58.000/RT)}$ min^{-1} $k_v = 1.9 \times 10^{16} e^{-(47.300/RT)}$ min^{-1} and $k_c = 7.9 \times 10^{11} e^{-(36.000/RT)}$ min^{-1}

Fig. 18 Pyrolysis model for cellulose.

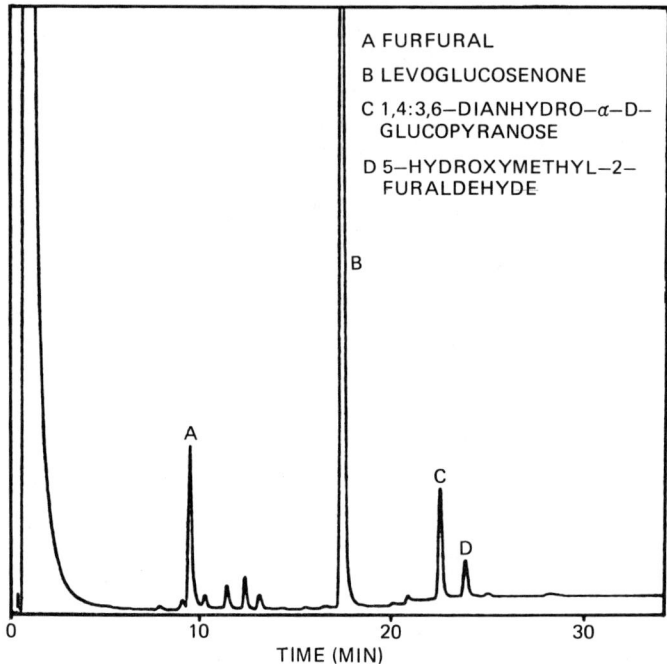

Fig. 19 Gas liquid chromatograph analysis of the pyrolyzate from cellulose +2% H_4PO_4 at 350°C.

The chemical kinetics of cellulose pyrolysis can be represented by the three-reaction model shown in Fig. 18. In this model, it is assumed that the initiation reactions discussed before lead to the formation of an active cellulose, which subsequently decomposes by two competitive first-order reactions, one yielding anhydrosugars (transglycosylation products) and the other char and a gaseous fraction [5a].

As discussed before, detailed analysis of the pyrolysis tar shows the presence of levoglucosan, its furanose isomer (1,6-anhydro-β-D-glucofuranose), and their transglycosylation products as the main components. In addition to these compounds, the pyrolyzate contains minor amounts of a variety of products formed from dehydration of the glucose units.

As in aqueous reactions, the dehydration and charring reactions are strongly catalyzed by presence of acidic reagents. GLC analysis (see Fig. 19) has shown that acid-catalyzed pyrolysis of cellulose at 350°C (623 K) produces a pyrolyzate containing levoglucosenone (instead of levoglucosan) as the major component and 1,4:3,6-dianhydro-α-D-glucopyranose, 2-furaldehyde, and 5-(hydroxymethyl)-2-furaldehyde as minor components [32]. Levoglucosenone, formed by dehydration reactions shown in Fig. 20, can be separated by fractional distillation and is a highly reactive compound that can be obtained by pyrolysis of waste paper treated with mineral acids.

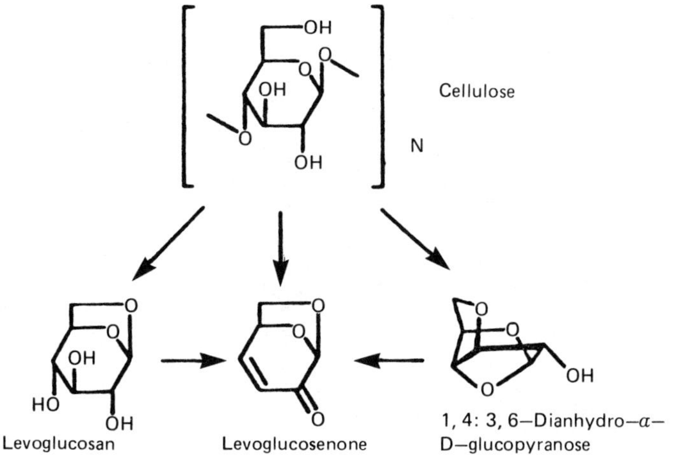

Fig. 20 Dehydration of cellulose and glucose derivatives to levoglucosenone.

Table 8 Pyrolysis Products of Cellulose and Treated Cellulose at 550°[a]

Product	Neat	+5% H_3PO_4	+5% $(NH_4)_2HPO_4$	+5% $ZnCl_2$
Acetaldehyde	1.5	0.9	0.4	1.0
Furan	0.7	0.7	0.5	3.2
Propenal	0.8	0.4	0.1	T
Methanol	1.1	0.7	0.9	0.5
2-Methylfuran	T[b]	0.5	0.5	2.1
2,3-Butanedione	2.0	2.0	1.6	1.2
1-Hydroxy-2-propanone / Glyoxal	2.8	0.2	T	0.4
Acetic acid	1.0	1.0	0.9	0.8
2-Furaldehyde	1.3	1.3	1.3	2.1
5-Methyl-2-furaldehyde	0.5	1.1	1.0	0.3
Carbon dioxide	6	5	6	3
Water	11	21	26	23
Char	5	24	35	31
Tar	66	16	7	31

[a]Percentage, yield based on the weight of the sample.
[b]T - trace amounts.

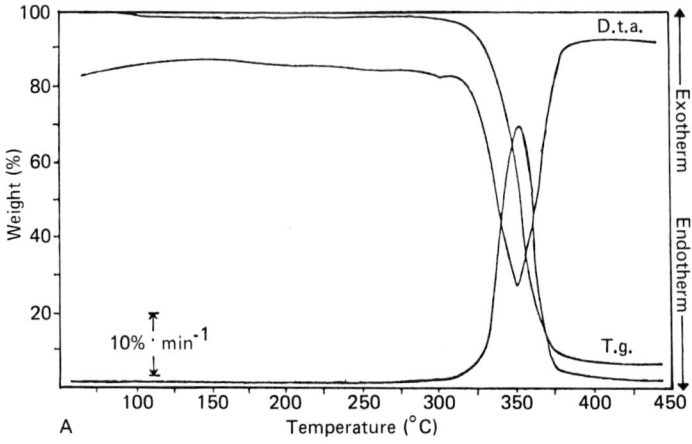

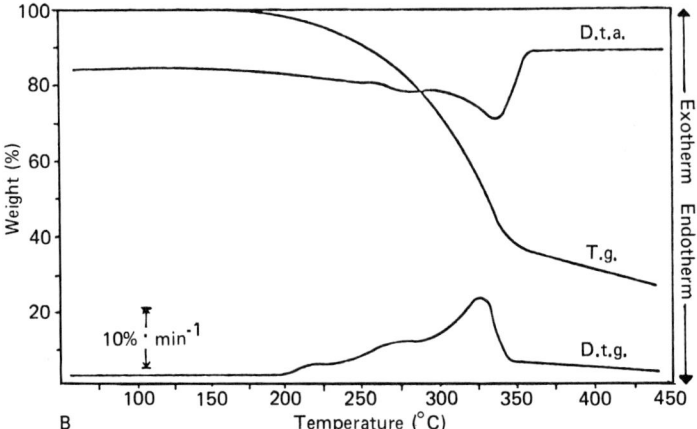

Fig. 21 Thermal analysis curves of (A) pure cellulose and (B) cellulose containing 5% $ZnCl_2$.

On further heating, fission of the sugar units at higher temperatures accompanied by dehydration, disproportionation, decarboxylation, and decarbonylation provides a variety of carbonyl, carboxyl and olefinic compounds as well as water, carbon dioxide, carbon monoxide, and char. The analyses of these products are closely similar to those obtained from pyrolysis of levoglucosan (see Table 8). These data also show the effect of acidic salts in producing char and water at the expense of the tar and combustible volatile pyrolysis products.

Ionic inorganic materials in general and flame retardants in particular suppress the gasification of cellulose by catalyzing the dehydration reactions, thus augmenting char production. This effect can be readily detected by thermal analysis of cellulose, particularly thermogravimetry (TG), that shows the amount of volatiles and residue. Figure 21A and B shows that pyrolysis of pure cellulose proceeds rapidly at 325 to 375°C (598 to 648 K) and leaves little char, whereas in the presence of $ZnCl_2$, cellulose is gradually decomposed at a wider and lower temperature range of 150 to 350°C (423 to 623 K) and leaves substantial amounts of char.

D. Significance

Although, for the reasons stated before, the above discussions deal with cellulose, their implications for and applications to paper and other cellulosic materials are significant. The thermal oxidation and depolymerization of cellulose fibrils and the accompanied inductive effect and catalysis, particularly by acidic materials, account for thermal deterioration of paper under different conditions.

At higher temperatures the rapid breakdown of cellulose to combustible volatiles accounts for the flammability of paper. The hemicelluloses behave in a similar manner, while lignin provides more char and fewer volatiles. A quantitative description of these properties may be observed on TG of wood and its components, such as the example shown in Fig. 22.

The existence of the alternative pathways for the pyrolysis of cellulosic materials also leads to two different modes of combustion, as shown graphically in Fig. 23. In the first pathway, which operates at lower temperatures, pyrolysis gives mainly a reactive, carbonaceous char and a gas mixture containing water and carbon dioxide which is not very flammable. Oxidation of the resulting char then provides glowing or smoldering combustion, which is

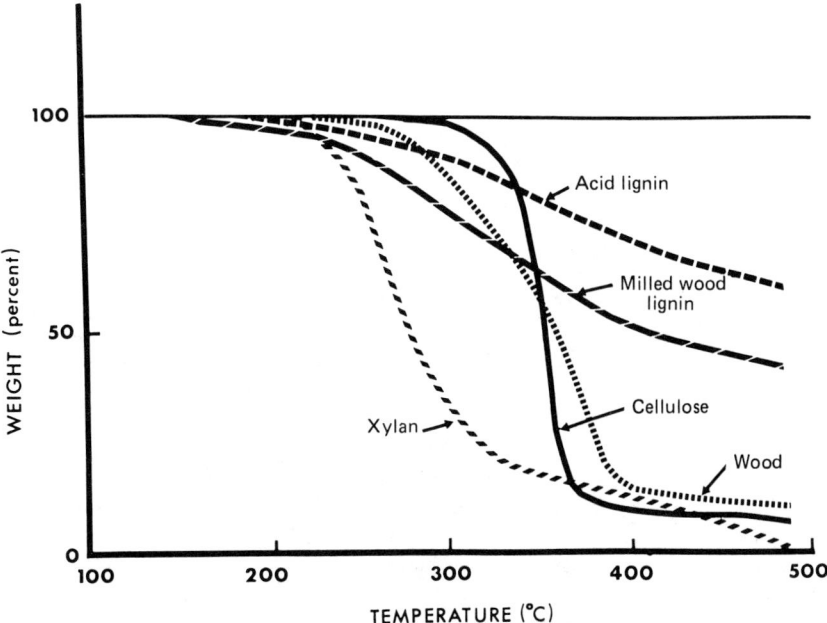

Fig. 22 Thermogravimetry of cottonwood and its components.

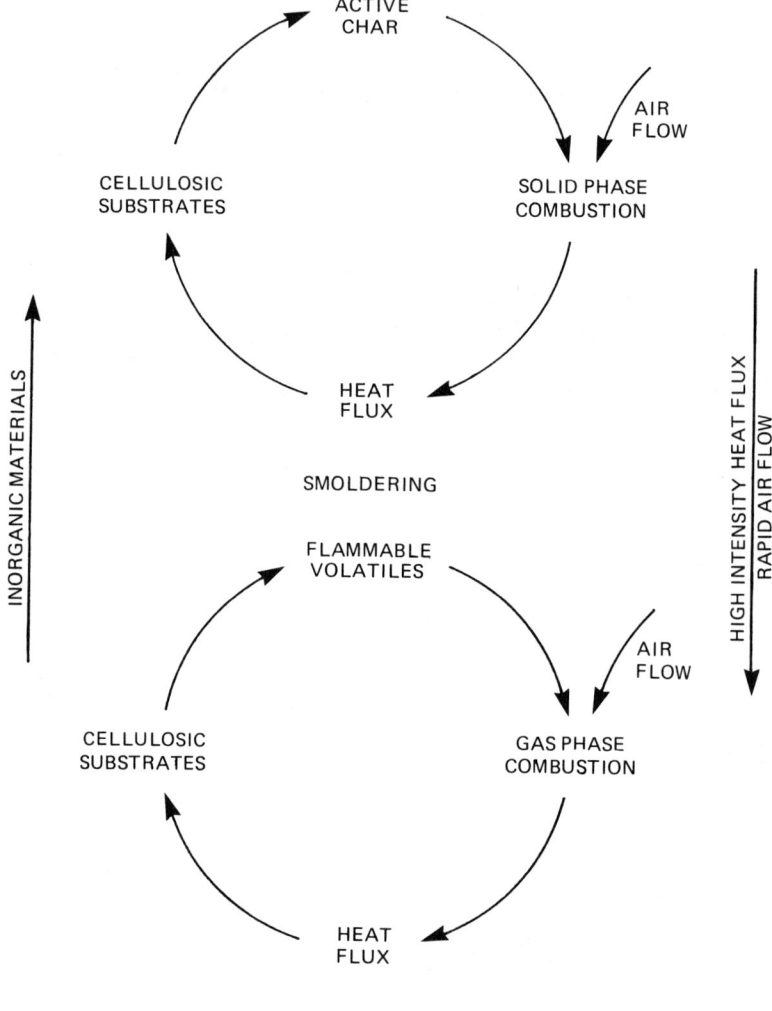

Fig. 23 Graphic description of flaming and smoldering combustion.

a localized and slower process, proceeding as a front in the solid phase as oxygen becomes available.

In the second pathway, which operates at higher temperatures, pyrolysis or thermal decomposition of the cellulosic materials yields a mixture of combustible gases. These gases mix with air and fuel and the flaming combustion that can rapidly spread in the gas phase. The flaming combustion is inhibited by materials that reduce the production of combustible volatiles by enhancing the dehydration and char-forming reactions. The smoldering combustion is inhibited by materials such as boric acid that prevent complete oxidation of char to CO_2 and produce more CO with a heat combustion of -22.9 kcal/mol (-95.9 kJ/mol), as compared to -88.5 kcal/mol (-370.5 kJ/mol) for CO_2.

E. Methods of Measurement

The physical changes taking place on thermal decomposition of paper can be measured by thermal analysis methods, including differential thermal analysis (DTA) and thermogravimetry (TG), and the decomposition products can be analyzed by spectroscopic or chromatographic methods, including mass spectroscopy and gas liquid chromatography (GLC). These methods are often combined to provide more comprehensive measurements of the same phenomena. For instance, DTA and TG are combined to measure the heat and mass transfer occurring during pyrolysis, or a pyrolysis unit is combined with GLC and mass spectroscopy to achieve fragmentation of the substrate by the pyrolysis unit, separation of the pyrolysis products by GLC, and identification of the individual products by mass spectroscopy. The characterization or identification of the substrate or its pyrolysis products may be achieved in a simpler combination of pyrolysis with mass spectroscopy or GLC, as described below. Several methods of analtyical pyrolysis are being developed for rapid characterization or fingerprinting of small amounts of complex and polymeric compounds such as cellulosic materials, which cannot be handled by existing powerful analytical tools unless the specimen material is reduced to smaller, soluble or volatile fragments. A detailed description of such equipment and methods is beyond the scope of this handbook, however, information is available in related monographs and periodicals [15a, 16a, 43]. The following introduction may serve to acquaint the reader with the basic principles involved.

Pyrolysis Unit The selection or design of the pyrolysis unit depends on the purpose or nature of the experiment. It may consist of a simple aluminum boat (containing the substrate), which is heated in a quartz tube surrounded by a thermally controlled furnace. In this unit, the pyrolysis products are swept by vacuum or an inert gas into a series of condensers where they are recovered as tar or liquid condensate and analyzed separately. Alternatively, the pyrolysis products may be swept directly into the injection port of a GLC for analysis of individual products. To avoid overlapping, it is essential for the pyrolysis products to reach the injection port and start the separation process at the same time. In analytical experiments, the pyrolysis should take place very rapidly to achieve this purpose and also to prevent excessive fragmentation by secondary reactions. Rapid pyrolysis is particularly desirable for pyrolysis-mass spectroscopy combinations. In these experiments, rapid fragmentation can be achieved by Curie-point, Pyroprobe, ribbon probe, or laser beam pyrolyzers. In the Curie point pyrolyzer, the sample is heated by filaments of certain ferromagnetic alloys to the pyrolysis temperature in nanoseconds. At this temperature, called the Curie point (which is characteristic for the alloy), a change in conductivity of the alloy automatically prevents further heating. In the Pyroprobe, a platinum filament that is electronically programmed for a rapid and linear heating rate serves as the heating source and sample holder. These probes are directly attached to the mass spectrometer or GLC [15a, 15b, 16a].

Mass Spectroscopy A mass spectrometer is a highly sensitive and powerful tool that can be used for analysis of as little as nanogram quantities of various compounds, provided that they can be vaporized. In this instrument,

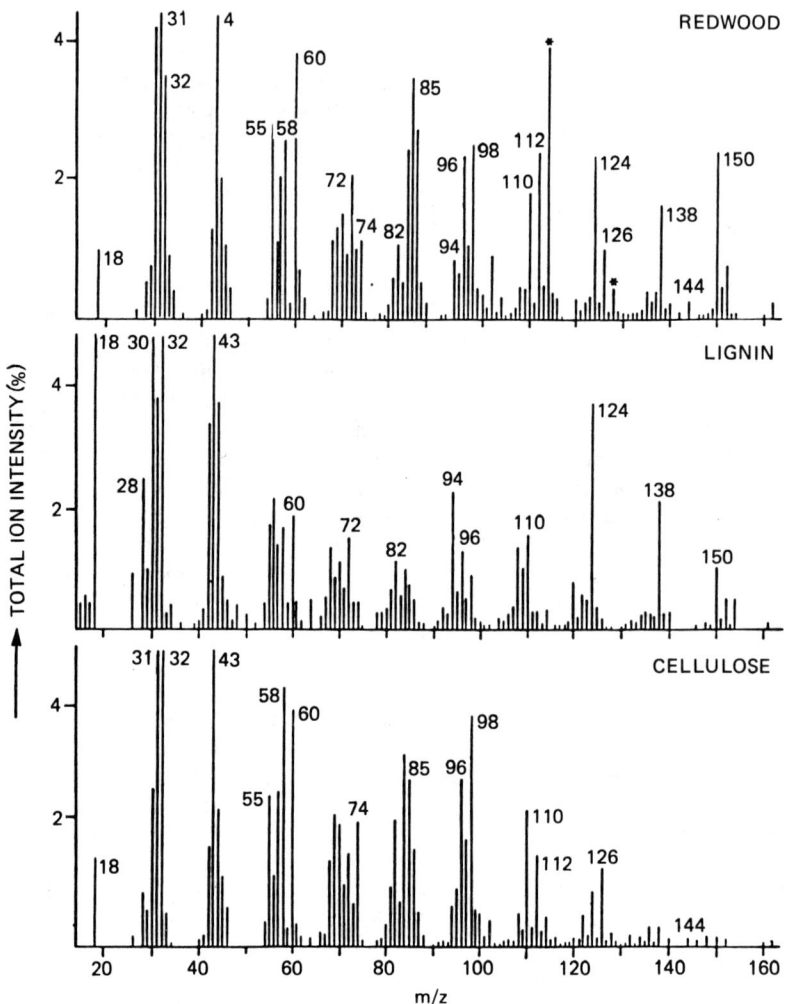

Fig. 24 Pyrolysis and field ionization mass spectroscopy of cellulose, lignin, and redwood.

the sample is evaporated by vacuum pumps and fragmented and ionized by an ionization source—for example, an electron beam (impact ionization). The positive ions formed are then separated according to their mass/charge (m/z) ratios. The quantitative detection and recording of these mass ions provides a spectrum that is quite characteristic for the individual compounds. Different combinations of pyrolysis and mass spectroscopy can be used for analytical pyrolysis [15a,16a,21a,27] of wood and cellulosic materials, including Curie-point pyrolysis and field ionization or field desorption mass spectroscopy. These spectroscopic methods are preferred because they minimize further fragmentation of the pyrolysis products. They are based on removal of electrons at a high electric field, rather than on electron impact ionization. Integrated Curie-point pyrolysis and field ionization of hexosans yields characteristic molecular ions for polymeric units such as $C_6H_{10}O_5$ (m/z 162), $C_6H_8O_4$ (m/z 144), and $C_6H_6O_3$ (m/z 126) that correspond to the anhydro sugar derivaties and other pyrolysis products discussed before. Cellulose is distinguished from other hexosans by the ratio of the major peaks and by its relative thermal stability, which necessitates a high energy transfer to give satisfactory amounts of volatile materials [27]. Xylan yields significant mass ions for $C_5H_8O_4$ (m/z 132), $C_5H_6O_3$ (m/z 114), and $C_5H_4O_2$ (m/z 96). Lignin gives characteristic peaks at m/z 124, 138, and 150 for the substituted phenolic fragmentation products [21a]. Analytical pyrolysis is particularly suited for fingerprinting of wood and paper products because it provides the characteristic peaks for the carbohydrates, lignin, and other components to the extent that they may be present (see Fig. 24).

Gas Liquid Chromatography Pyrolysis products can be fractionated to gases, liquids, and tars according to their volatility. These mixtures in turn can be separated to individual compounds by gas liquid chromatography [43]. In this process, the mixture is usually introduced into a heated block, where it is evaporated rapidly and swept by an inert carrier gas (nitrogen or helium) into a thin and long stainless steel column containing a liquid adsorbant that is supported on solid particles to act as a stationary phase. The volatiles adsorbed on the stationary phase are desorbed by the inert carrier gas moving toward the end of the column. Thus, the products are partitioned between the two phases and move down the column according to their partition coefficient, which is a function of the adsorbent as well as of the individual compounds. In this manner, different components of the mixture separate and exit from the column as separate bands. These bands are detected by special devices such as flame ionization detectors and are electronically recorded as a series of peaks, each peak corresponding to individual components of the mixture. Location of the peak or the time that it takes to travel through the column is indicative of the composition; the peak area is proportional to the concentration. To achieve better separation, different types of columns are used for the gases and liquids, and the tarry products are rendered more volatile by converting them to their (trimethylsilyl) ether derivatives through chemical reaction. The volatility of the mixtures can also be increased by heating the injection port and the column. The results obtains from GLC analysis of tars from pyrolysis of pure cellulose and cellulose-containing phosphoric acid as catalyst are shown in Figs. 16 and 19, respectively. In the former case, the tarry pyrolyzate, containing a variety of sugar derivatives, has been treated with trimethylsilyl chloride, which reacts with the free hydroxyl groups, before GLC analysis. This treatment helps in separation of the less volatile compounds containing free hydroxyl groups and excludes interference from other compounds. It is also used for analysis of the sugars obtained from hydrolysis of paper and cellulosic materials. Etherification is not used for GLC analysis of the pyrolyzate from acid-catalyzed pyrolysis of cellulose (see Figs. 13 and 20), because the compounds involved are sufficiently volatile; also, some of them lack free hydroxyl groups. As mentioned before, when composition of the components is unknown, pyrolysis and GLC can be interfaced with mass spectroscopy (PY-GC-MS) to provide quantitative and qualitative analysis of the pyrolysis products [15a,16a].

Differential Thermal Analysis Differential thermal analysis (DTA) is a method of measuring the thermal properties of a sample by measuring the temperature difference between the sample and a reference material when the two materials are heated or cooled at a constant rate at the same time. The sample and a thermally inactive material are put in a furnace to form a comparative test. When the furnace is heated or cooled, the test sample undergoes a thermal change due to some transition or decomposition different from that of the reference material, and this change is measured with a thermocouple to be recorded automatically. The measuring devices are generally available commercially, and some reports on paper and paperboard have been published [7,8,14,22,23,40].

The study of the process of thermal decomposition of paper by DTA reveals that there are usually two endothermic peaks; one is due to the evaporation of adsorbed moisture at 100 to 150°C, and the other is due to the decomposition of paper at 300 to 370°C. According to an experiment performed in a nitrogen atmosphere by Herbert et al. [14], the endothermic peak of decomposition had no definite relation with the species of pulp used, but it had a close relation with the pH of the paper. As the pH is lowered, the endothermic peak is shifted to the low-temperature side. It is notable that the endothermic peak appears at 340 to 370°C for cellulose whereas the exothermic peak is observed at 306°C for xylan and 430°C for lignin. Similar results were obtained in an experiment by Ramiah [23], but the DTA curve in the stationary air or under the flow of oxygen differs from that under the flow of nitrogen. According to the results of thermal analysis of modified cellulose by Parks [22], oxidation of cellulose leads to a slight destabilization of the modified cellulose, resulting in a slightly higher endothermic peak temperature. The oxycellulose can be stabilized when treated with calcium.

Thermogravimetric Analysis Thermogravimetric analysis (TGA) is a method of measuring change in mass with a thermobalance when the temperature of a material is raised at a constant rate. With most devices on the market, both DTA and TGA can be measured. Important information on the behavior of thermal decomposition of the material in question can be derived from a comparison of the DTA curve with the TGA curve.

If the initial temperature at which the weight begins to decrease can be obtained from the TGA curve, then this temperature can be regarded as the initial temperature of thermal decomposition. Results published by Ramiah [23] showed that in an atmosphere of nitrogen the weight of a sample of softwood sulfite pulp began to decrease at 280°C, that of microcrystalline cellulose at 295°C, that of xylan at 195°C, and that of Klason lignin at 320°C. According to the results on cellulose by Parks [22], the carboxyl group content is in inverse proportion to the initial temperature of weight decrease.

An apparent activation energy can be obtained from the results of TGA [8,22,23]. According to the results of Ramiah [23], the activation energies of cellulosic materials are in the range of 36 to 60 kcal/mol, those of hemicellulose 15 to 26 kcal/mol, and those of lignin 13 to 19 kcal/mol. Cardwell and Luner [8] reported the following activation energies for pulps and related materials (all units in kcal/mol): cotton linters, 34.5; bleached pine kraft pulp, 43.9; bleached birch kraft pulp, 43.1; bleached pine sulfite pulp, 41.6; unbleached mixed hardwood sulfite pulp, 33.3; unbleached birch semichemical pulp, 33.6; and rayon fiber, 32.8.

REFERENCES

1. ASTM C 177-63 (reapproved 1968). Thermal conductivity of materials by means of the guarded hot plate.
2. ASTM D 777-74. Flammability of treated paper and paperboard.
3. ASTM D 864-52. Coefficient of cubical thermal expansion of plastics.
4. ASTM D 2766-71. Specific heat of liquids and solids.
5. ASTM D 2863-70. Flammability of plastics using the oxygen index method.
5a. Bradbury, A. G. W., Sakai, Y., and Shafizadeh, F. (1979). A kinetic model for pyrolysis of cellulose. *J. Appl. Polym. Sci.* 23:3271-3280.
6. Brandrup, J., and Immergut, E. H., eds. (1966). Properties of cellulose materials. In *Polymer Handbook*, vol. 1. John Wiley and Sons, New York, p. 36.
7. Browning, B. L. (1969). 3. Laboratory Techniques, C. Gas Chromatography. In *Analysis of paper*. Marcel Dekker, Inc., New York, pp. 205-206.
8. Cardwell, R. D., and Luner, P. (1978). Thermogravimetric analysis of pulp: Kinetic treatment of dynamic pyrolysis of papermaking pulps. *Tappi* 61(8):81—84.
9. Forsythe, W. E., and Powell, R. L. (1957). Thermal conductivity. In *American Institute of Physics Handbook*, vol. 4. McGraw-Hill, New York, pp. 65—69.
10. Götze, W., and Winkler, F. (1967). Calorimetric studies on textile fibrous materials. 1. Specific heat: Literature review. *Faserforsch. Textiletech.* 18(3):119—123.
11. Götze, W., and Winkler, F. (1967). Calorimetric studies on textile fibrous materials. 2. Measuring methods and apparatus. *Faserforsch. Textiltech.* 18(5):222—227.
12. Götze, W., and Winkler, F. (1967). Calorimetric studies on textile fibrous materials. 3. Adiabatic calorimeter AKM 1. *Faserforsch. Textiltech.* 18(5):292—295.
13. Han, S. T., and Ulmanen, T. (1958). Heat transfer in hot-surface drying of paper. *Tappi* 41(4):185—189.
14. Herbert, R. L., Tyron, M., and Wilson, W. K. (1969). Differential thermal analysis of some papers and carbohydrate materials. *Tappi* 52(6):1183—1188.
15. Iimure, N. (1958). Ignition temperature. In *Manual for Dangerous Article Treating (Kikenbutsu Toriatsukai Hikkei)*, Sangyo Tosho Co., Tokyo, pp. 62-63.
15a. Irwin, W. J. (1979). Analytical pyrolysis: An overview. *J. Anal. Appl. Pyrolysis* 1:3—25.

15b. Irwin, W. J. (1982). *Analytical Pyrolysis: A Comprehensive Guide*, chromatographic Science Series, vol. 22, Marcel Dekker, Inc., New York.

16. Ishii, K., Sekigushi, T., and Takaya, T. (1972). Thermal properties and flammability of various fibers. *J. Soc. Fiber Sci. Technol. Japan (Senii Gakkaishi)* 28(9):359–367.

16a. Jones, C. E. R. and Cramer, C. A., eds. (1977). *Analytical Pyrolysis*, Elsevier, Amsterdam.

17. Kirk, L. A., and Tatlicibasi, C. (1972). Measurement of thermal conductivity of paper by a heat pulse method. *Tappi* 55(12):1697–1700.

18. Klason, G., and Kubát, J. (1976). Thermal transitions in cellulose. *Svensk Papperstidn.* 79(15):494–500.

19. Kubát, J., Martin-Löf, S., and Söremark, C. (1969). A dilatometric study of secondary transition in cellulose between − 5°C and + 70°C. *Svensk Papperstidn.* 72(22):731–734.

20. Kubát, J., Martin-Löf, S., and Söremark, C. (1969). Thermal expansivity and elasticity of paper and board. *Svensk Papperstidn.* 72(23):763–767.

21. Mikhailov, N. V., and Fainberg, E. Z. (1962). The heat capacity and phase composition of cellulose fibers of various structures. *Vysokomol. Soedin.* 4(2):230–236.

21a. Muezelaar, H. A. C., Haverkamp, M., and Hileman, F. D. (1982). *Pyrolysis Mass Spectrometry of Recent and Fossil Biomaterials: Compendium and Atlas*, Elsevier, Amsterdam.

22. Parks, E. J. (1971). Thermal analysis of modified pulp. *Tappi* 54(4):537–544.

23. Ramiah, M. V. (1970). Thermogravimetric and differential thermal analysis of cellulose, hemicellulose and lignin. *J. Appl. Polymer. Sci.* 14(5):1323–1337.

24. Ramiah, M. V., and Goring, D. A. I. (1965). Thermal expansion of cellulose, hemicellulose, and lignin. *J. Polymer Sci., C: Polymer Symposia* 11:27–48.

25. de Ruvo, A., Lunberg, R., Martin-Löf, S., and Söremark, C. (1976). Influence of temperature and humidity on elastic and expansional properties of paper and the constituent fiber. In *Fundamental Properties of Paper Related to Its Uses* (F. Bolam, ed.), British Paper and Board Industry Foundation, London, pp. 785–810.

26. Schofield, F. H., and Hall, J. A. (1927). Thermal insulating materials for moderate and low temperatures. In *International Critical Tables of Numerical Data, Physics, Chemistry and Technology*, vol. 2, McGraw-Hill, New York, pp. 312–316.

27. Schulten, H. R., Bahr, U., and Görtz, W. (1982). Pyrolysis field ionization and mass spectrometry of carbohydrates. *J. Anal. Appl. Pyrolysis* 3:229–241.

28. Shafizadeh, F. (1968). Pyrolysis and combustion of cellulosic materials. *Adv. Carbohy. Chem.* 23:419–474.

29. Shafizadeh, F. (1975). Industrial pyrolysis of cellulosic materials. *Appl. Polymer. Symp.* 28:153–174.

30. Shafizadeh, F. (1982). Introduction to pyrolysis of biomass. *J. Anal. Appl. Pyrolysis* 3:283-305.
31. Shafizadeh, F., and Bradbury, A. G. W. (1979). Thermal degradation of cellulose in air and nitrogen at low temperatures. *J. Appl. Polymer Sci.* 23:1431–1442.
32. Shafizadeh, F., Furneaux, R. H., Stevenson, T. T., and Cochran, T. G. (1978). 1,5-Anhydro-4-Deoxy-D-glycero-hex-1-En-3-Ulose and other pyrolysis products of cellulose. *Carbohy. Res.* 67:433–447.
33. Sutermeister, E. (1948). The physical properties of cellulose. In *Chemistry of Pulp and Paper Making*, John Wiley and Sons, New York, pp. 31–34.
34. TAPPI T 461 su-72. Flame resistance of treated paper and paperboard.
35. Terada, T., Ito, N., and Goto, Y. (1969). Effective thermal conductivity of insulating paper. *Japan TAPPI* 23(5):191–197.
36. Terasaki, K., and Matsuura, K. (1972). The study of heat properties of papers and consideration to use for heat exchangers. *Japan TAPPI* 26(4):173–178.
37. Terasaki, K., and Matsuura, K. (1972). The study of the effective thermal conductivities of papers for temperature and humidity, 2nd report. *Japan TAPPI* 26(10):511–515.
38. Terasaki, K., Matsuura, K., and Okada, M. (1973). The study of the effective thermal conductivity of papers, 3rd report. *Japan TAPPI* 27(11):525–529.
39. Tokyo Astronomical Observatory (1979). Physical table 63. In *Annual Scientific Tables (Rikanenpyo*, Maruzen Co., Tokyo.
40. Tsuchiya, Y., and Sumi, K. (1970). Thermal decomposition products of cellulose. *J. Appl. Polymer Sci.* 14(8):2003–2013.
41. UL 94 (1972). Tests for flammability of plastic materials. 3. Vertical burning test for classifying materials.
42. UL 94 (1972). Tests for flammability of plastic materials. 7. Ignition temperature test.
43. Willard, H. H., Merrit, L. L., Jr., Dean, J. A., and Settle, J. A., Jr. (1981). *Instrumental Methods of Analysis*, 6th ed., D. Van Nostrand Co., New York.

OTHER PHYSICAL PARAMETERS

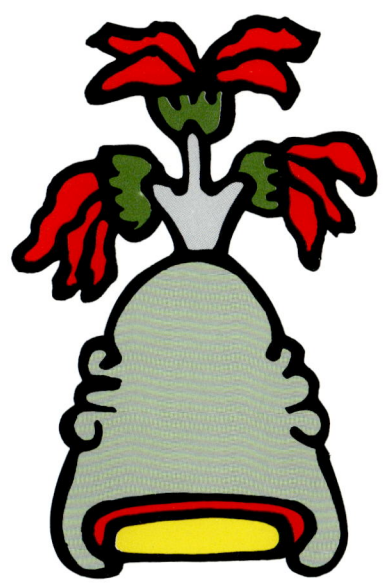

Emblem of Huaxtepec

Located in the "Hill of Huaxin Trees," this tribute town provided a fruit similar to carob derived from the Huaxin (*Leucaena esculenta* Benth.). The emblem clearly shows a green hill surmounted by a stylized *Leucaena* tree with bananalike red flowers. The Aztec word for hill was "tepec." The modern city of Oaxaca is also thought to derive its name from the leguminous Huaxin tree.

24
STRUCTURE AND STRUCTURAL ANISOTROPY

RICHARD E. MARK*

Empire State Paper Research Institute
State University of New York
College of Environmental Science and Forestry
Syracuse, New York

I.	Introduction	284
II.	Fiber Orientation Distribution	285
	A. Methods for Fiber Orientation Distribution	285
	B. First Principles (Illustrated for a Direct Method)	287
	C. Selection of a Representative Function	295
	D. Curve-Fitting Method and Goodness of Fit Test	297
	E. Relations Between Orientation Functions	299
	F. Fiber Orientation by Scattering (Contributed by Jens Borch)	300
	G. Silvy's Orientation and Segment Length Distribution Method ("The Equivalent Pore")	306
	H. Mean Orientation Angle	311
	I. Orientation in the Z Direction	312
III.	Dimensional Characterization of Fibers	312
	A. Fiber Length	313
	B. Fiber Curl	334
IV.	Other Dimensional Parameters Used to Characterize Fiber Raw Materials and Fiber Networks	338
	Notes on Procedures for Various Structural Parameter Determinations	339

---

*Section II.F is authored by Jens Borch, IBM Corporation, General Products Division, Tucson, Arizona.

V. Dimensional Characterization of Shives, Slivers, and Fines   358
   A. Shives and Slivers   358
   B. Fines   359

References   367

I. INTRODUCTION

A quantitative knowledge of the geometric structure of paper and board is fundamental to the understanding of why these materials possess the properties they exhibit and how those properties are changed when the structure changes. In this chapter, some of the methods used to determine several important structural parameters are described, and the principles behind these testing methods are discussed.

The internal structural parameters that are the most important depend on whether one considers a relatively dense, highly bonded material such as linerboard or a low-density material such as tissue, whether one is primarily interested in mechanical or other physical properties and, within either of these last 2 categories, what type of physical (e.g., opacity or absorbency) or mechanical (e.g., elastic properties or properties at failure) characteristics are of particular interest.

In Chap. 2 of this book and in Refs. 107 and 108, Perkins has identified the network parameters that are considered in theories relating to mechanical properties as follows:

1. Fiber length (including length/width and length/thickness ratios) and curliness
2. Fiber segment length (distance between bond centroids along a fiber)
3. Fiber cross-sectional shape (includes area, width, wall thickness, degree of collapse, aspect ratio); characteristics of the distribution of these quantities
4. Mechanical properties of fibers
5. Size and mechanical nature of the fiber-to-fiber bond area
6. Percentage of the fiber surface bonded to other fibers (relative bonded area)
7. Fiber orientation distribution
8. Sheet density and uniformity; distribution of mass
9. Sheet shrinkage strains

For testing related to other (physical or mechanical) properties, one may need information regarding some or all of the above parameters, plus additional information on other properties, such as:

10. Basis weight (grammage)
11. Apparent fiber density
12. Proportions of cell wall layers (compound middle lamella, S1, S2, S3)
13. Filament winding angles (microfibril angles)
14. Conformation of molecular components
15. Surface charges
16. Internal surface volume (including pore size distribution)
17. Electrical conductivity

18. Acoustical conductivity
19. Indices of refraction

The above list, by no means complete, includes many items that are considered in detail in other chapters. In this chapter, we concentrate on methods for ascertaining some important parameters related to internal structural geometry. Special attention will be given to item 7, fiber orientation distribution, because the amount of previously published information available in the paper testing literature is quite limited, despite the importance of this parameter.

II. FIBER ORIENTATION DISTRIBUTION

The influence of in-plane fiber orientation on the physical and mechanical properties of paper sheets has been well documented in the literature [24,42, 60,62,66,67,108,111,114,122,127,151,165]. Whether one is evaluating the effects of fiber alignment or processes that affect it or attempting to predict properties by analytical methods, it is important to obtain accurate data regarding the distribution of these alignments or orientations in the sheet.

In this section we consider both direct and indirect methods of making such determinations. Both have advantages and drawbacks. Both require some statistical manipulation to give the researcher a complete picture of the orientation parameters, which may or may not be coupled to other parameters, such as fiber length distribution. In addition, it is important for the investigator to consider such questions as what actually needs to be measured and how representative samples are to be obtained, which are illustrated in this chapter for the case of a direct (digitizer) method. Certainly it can be well demonstrated (Chap. 2) that the choice of distribution function (as discussed in this chapter) used to represent fiber orientation has some serious implications relative to our ability to understand and predict elastic constants and other mechanical properties for sheets of different fiber alignment configurations.

A. Methods for Fiber Orientation Distribution

The initial work done in this field was by Danielsen and Steenberg [29], who developed a sort of rotatable protractor for directly measuring the orientations of dyed fibers that are added to the furnish when the sheet is made. The dyed fibers, amounting to less than 1%, are typical of the pulp stock used and are measured as a representative sample of the whole. Subsequently, other workers [42,62,63,111,127,151] modified the Danielsen-Steenberg mehtod. For example, Glynn et al. [42] constructed a circular turntable fitted with 48 equal sheet-metal bins on its periphery, each bin thus comprising 7.5° of arc. The paper sample, containing dyed fibers, was placed at the center of the turntable under fixed hairlines and a magnifying glass. A hopper at the edge of the turntable dispensed small metal balls, one at a time, into the bin whose arc corresponded with the orientation of a given fiber as observed by the hairlines. It was determined that 2000 measurements should be made to obtain reproducible results; thus, 2000 balls were required for one set of measurements. The number of fibers in each 7.5° of arc was then calculated from the weight of the balls in each bin. It required an operator about 2.5 hr to make 2000 counts on both sides of a sample in this way.

The alternative (indirect) methods that have been developed include mechanical testing [42,64,127,144], light diffraction [119,120,166], small angle light scattering [10,17,82,85,93], X-ray diffraction [110,118], and line intersection methods. In the last-mentioned methods, used in various forms by Corte and Kallmes [24], Forgacs and Strelis [41], and Silvy [128, 129], counts are made of the numbers of dyed fibers falling on straight reference lines. These counts provide indirect data that are converted to orientation distribution functions via mathematical treatment of the data.

The indirect method most often used is by calculation of the strength anisotropy of the sheet as determined from zero span tensile tests. This method is embodied in the former TAPPI Suggested Method T 481 [144] and has been extensively discussed by Kallmes [64]. It has the virtues of being relatively simple, inexpensive, and rapid, and it employs equipment often found in paper testing facilities. On the other hand, the agreement between this method and the classic manual direct method (dyed fiber orientations) is not very good (Fig. 1) and great care must be employed to obtain accurate, reproducible results [25,49]. The most serious drawback of the zero span

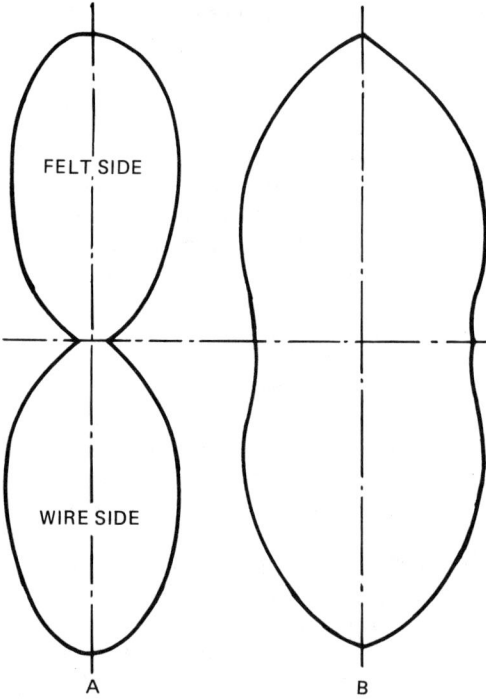

Fig. 1 Polar diagrams of fiber orientation (A) determined directly by protractor with turnable pointer and (B) a strength versus orientation relation determined from tensile tests on sheets conducted in various directions. The machine direction is vertical in these diagrams. (From Ref. 151.)

method is that it is a measure not of orientation but of mechanical anisotropy. Anisotropy is related strongly to orientation, but it is also related to the draw or stretch imposed by the paper machine [127] and other aspects of drying, such as the solids content of the paper web at the time of application of drying stress [54b].

The methods of determining fiber orientation by scattering, which are discussed by Borch in a contributed section of this chapter (Sec. II.F) are under development; some show considerable promise for the future, although at present most of them have special procedural requirements that render them relatively difficult to perform routinely. Fortunately, devices designed to extract quantitative information from slides, photographs, and other images have been developed within the past few years and enable much more rapid collection of data on geometric parameters. These *image analysis* devices are discussed on pp. 319–334 in this chapter. One type of semi-automatic device is the graphic digitizer, which is basically an instrument for the rapid and accurate reading of coordinates. It is especially well adapted for accumulation of fiber orientation information in paper.

The digitizing system used in the author's laboratory is shown in Figs. 2 and 3. A description of its operation is given in Refs. 27 and 109. Some uses of digitizers and other image analyzers to obtain structural information will be discussed in this chapter.

B. First Principles (Illustrated for a Direct Method)

Determination of the orientation of fibers by any direct method requires some preliminary decisions as to what types of information are needed.

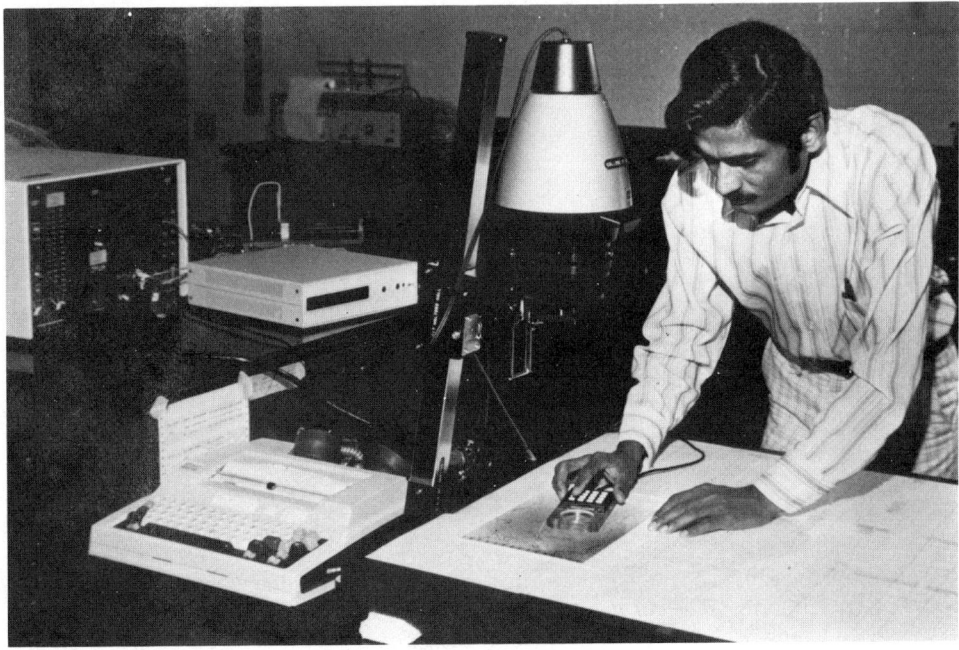

Fig. 2 Operation of a graphic digitizer.

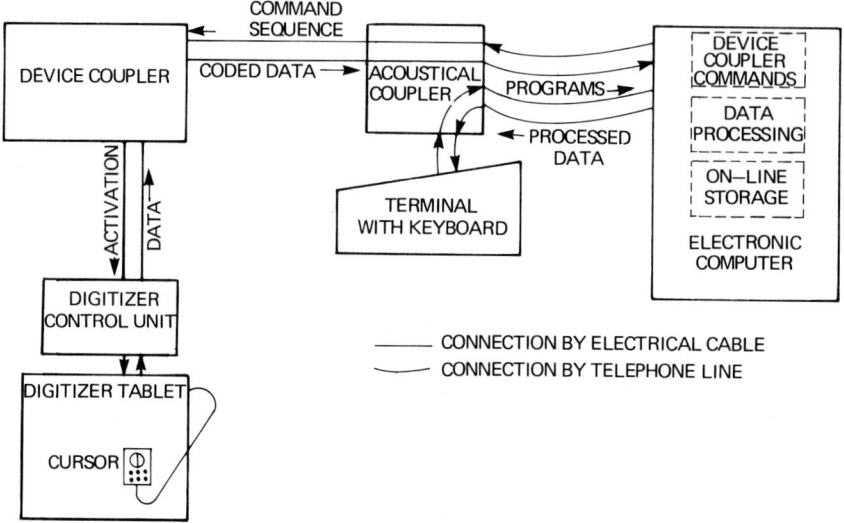

Fig. 3 Schematic of graphic digitizing system, including device coupler, terminal, and computer. (Modified from Ref. 28.)

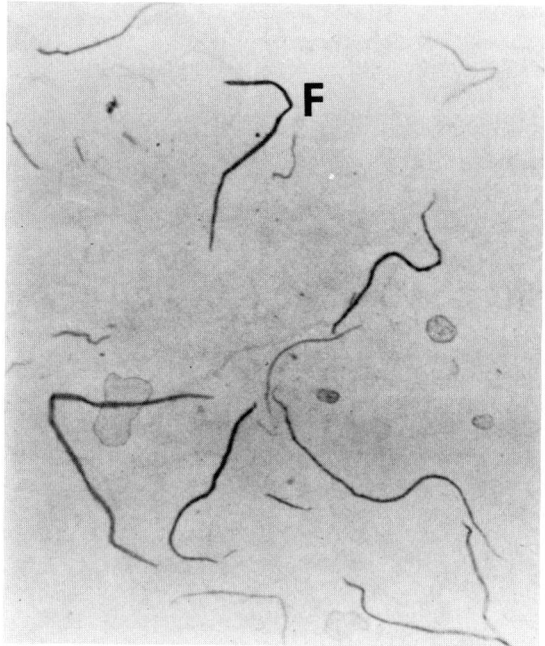

Fig. 4 Photograph of sheet containing dyed fibers. The sheet is impregnated with silicone oil. Its image, projected by the lamp of a photographic enlarger, clearly delineates the black fibers regardless of their depth in the sheet. (Photo by A. Eusufzai.) Approximately 10 ×.

Measurements Needed Since fibers in a sheet of paper or paperboard are of finite (short) length and usually contain bent, curled, or broken sections, one has to make a decision as to what constitutes a fiber for purposes of determining fiber orientation. The same statement can be made regarding determination of fiber length, and it should be emphasized that there is no basis a priori for assuming that fiber length and orientation are independent of each other. These parameters may, in fact, be highly correlated in some cases [18].

With reference to Figs. 4 and 5, one may observe several possibilities for defining the fiber in terms of orientation. Each possibility has different implications for length categorization. Figure 4 shows a sheet in which a small fraction (0.25%) of black dyed fibers show contrast against the remaining 99.75%, which are bleached. The paper samples are impregnated with silicone oil (with or without the assistance of vacuum) to enhance visibility of the dyed fibers by making the rest of the sheet almost transparent. The orientation of the fiber in Fig. 4 labeled F, for example, may be described in any of the following ways:

- With respect to a fixed axis (e.g., the machine direction, MD), the orientation of a straight line joining the ends of the fiber (Fig. 5A).
- • With respect to a fixed axis, the orientation of the best fit of all the segments of the fiber together. Essentially, a regression line is drawn that

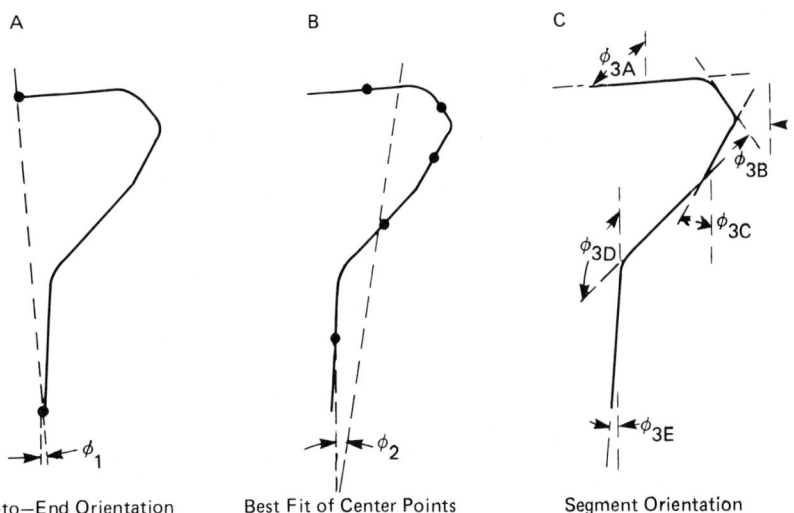

Fig. 5 Diagrammatic representation of dyed fiber F in Fig. 4.

A. Black circles represent endpoints; dashed line represents end-to-end distance; ϕ_1 equals orientation of fiber in sheet.
B. For computational purposes, fiber is divided into 5 segments. Black circles are segment midpoints; dashed line represents regression line of midpoints; ϕ_2 equals orientation of fiber in sheet by best fit of center lines.
C. Orientation ϕ_{3i} determined for each segment.

best fits a set of points taken to lie at the midpoint of each segment of the fiber (Fig. 5B).
- ••• With respect to a fixed axis, the orientation of each segment of the fiber taken individually (Fig. 5C).
- •••• With respect to a fixed axis, the orientations of segments according to length category. In such cases, criteria have to be set as to how many segment length categories are needed. For example, a preliminary study can be made to determine the most probable (statistically) segment length one encounters in a given type of sheet material. One can establish the length categories according to some system based on most probable length (mpl). For example:

Segment group no. 1: $0 < \ell \leq \ell_{mpl}$

Segment group no. 2: $\ell_{mpl} < \ell \leq 2\ell_{mpl}$

Segment group no. 3: $2\ell_{mpl} < \ell \leq 3\ell_{mpl}$

Segment group no. 4: $3\ell_{mpl} < \ell$

The four modes of description itemized above are listed in increasing order of reliability as to the results obtained, assuming an adequate sample size is taken. However, the difficulty of measurement also increases in the same order.

Fiber (or fiber segment) length data can be obtained simultaneously with orientation. The acquisition of data on length distribution will usually be highly desirable, and plans should be made to record them at the start. As noted previously, length and orientation may or may not be independent variables. The probability of finding certain lengths in conjunction with certain orientations may be large for some materials. Another use for length data is to make possible the determination of fiber curl (difference between total fiber length and straight line end-to-end length divided by total length),* which varies greatly according to the fiber processing methods used in making the sheet.

In some cases, such as when sheet curl is being studied, it may be necessary or desirable to examine the variation of fiber orientation (and/or length) distribution with respect to the Z direction of the sheet. To accomplish this, one can employ a sheet-splitting technique, using either a cryostatic-type apparatus or one that employs pressure-sensitive tape, to examine the fiber pattern at various depths (see Chap. 27 and Ref. 88). An apparatus that uses pressure-sensitive tape is generally able to split a sheet more finely. Alternatively, one can adopt a simplified (and much less accurate) assumption that there are only two orientations—those corresponding to the patterns observed on the top- and wire-side surfaces. However, if only the surfaces are to be examined, the sheet should not be impregnated with oil. Oil impregnation makes all the dyed fibers stand out with about equal intensity at all depths in the sheet, so that an observer cannot tell if a dyed fiber lies near the surface or near the center.

---

*This is one definition of fiber curl; other definitions are discussed on pp. 334-336.

Choice of Sector Size The frequency with which fibers or fiber segments are found with alignments falling in a given range (radians or degrees) is usually the immediate objective of the experiment. Preselection of an angular interval that makes it possible to obtain a comprehensive, accurate picture of fiber orientation distribution is essential. A very commonly selected interval is 5°.

Obtaining Representative Samples If possible, the incorporation of a small percentage (0.10 to 0.25%) of dyed fibers into the sheet is desirable. For many purposes, a chlorazol black E dye is excellent. These dyed fibers stand out among the other fibers, as illustrated in Fig. 4. The use of dyed fibers makes it possible to accurately cross check against other methods of determination. However, the presence of scattered dyed fibers may be unacceptable. In such cases, it is sometimes possible to incorporate dyed fibers into a sheet only during the last few minutes of a machine run. Alternatively, a colorless additive that fluoresces in light of an appropriate wavelength may be acceptable. A third possibility, developed by Glynn et al. [42], is to add tannic acid as a mordanting agent to a small fraction of the pulp stock. The mordant is then reacted with a metallic salt, such as lead acetate. With approximately 1% of these fibers in the stock, machine-made paper is unaltered in appearance. Samples are cut from the roll and dipped in a bath containing basic red dye, which is taken up preferentially by the treated fibers. The contrast may be enhanced by dipping in chlorine water, then washing and dipping in a yellow dye bath. After a final washing and drying, the mordanted fibers show up dark red on a buff yellow background. Samples should be taken across the width of the sheet, with sufficient replication to provide reproducibility of results. The dyed fibers should be representative of all the fibers in the sheet.

If it is not possible to incorporate any type of individually identifiable fiber into the sheets to be tested, one of the indirect methods will have to be used. However, it is essential that when any indirect method is considered for use on a particular material, it is independently verified that the indirect method will yield results in agreement with a direct method, using similar test material that does contain dyed fibers.

Compilation of Data

Angular Distribution Frequency Usually, fiber orientation data have to be compared or fitted to a generally bimodal (symmetrical with respect to the machine direction) mathematical function of some form to be useful. Distribution functions that have been used or suggested for use with fiber networks include the cosine [24], elliptical [41], and von Mises [86].

The goodness with which any of these functions will fit a set of experimental data for fiber orientation depends on the degree of anisotropy of the sheet, the scatter in the experimental data, and the shape parameter(s) for the function. Some explanation of these functions and parameters will illustrate this point.

Elliptical The elliptical distribution function has the form

$$f(\theta) = \frac{\lambda}{\pi} \frac{1}{\cos^2\theta + \lambda^2 \sin^2\theta} \qquad (1)$$

The degree of ellipticity of this function is controlled by the shape parameter λ, which is equal to the ratio of the major and minor semiaxes. Selection or determination of an appropriate value for λ enables one to fit the curve of the probability density of finding a fiber within a given (say, 5°) sector of orientation. The determination of λ usually is done by the least squares error method (see p. 297) when the elliptical function is fitted to the observed data. Allowable values of λ are never less than unity. The symbol θ refers to the angle of orientation. It varies from $-90°$ to $90°$ when the machine direction is taken to be $0°$. Plotted in Cartesian form, the elliptical function will generate a smooth, symmetrical, rounded-peak curve. When plotted in polar coordinates, the function generates an ellipse, of course.

Cosine As shown by Cox [26], there is a useful distribution function consisting of a series expansion of cosine terms of the form

$$f(\theta) = \frac{1}{\pi}(1 + \eta_1 \cos 2\theta + \eta_2 \cos 4\theta + \cdots + \eta_n \cos 2n\theta) \qquad (2)$$

In this expression there are a series of shape parameters η that modify the basic curve form. Again, this makes it possible to fit the function to the experimental probability distributions (i.e., data points) sector by sector. In the work of Corte and Kallmes [24], the series expansion was truncated to

$$f(\theta) = \frac{1}{\pi}(1 + \eta \cos 2\theta), \quad 0 \leq \eta \leq 1 \qquad (3)$$

for ease of mathematical manipulation. The form of Eq. (3) is designated as *single cosine term* in this chapter. Plotted in polar coordinates, the form is bimodal cardioid. As Perkins has noted in his discussion of Eq. (118) in Chap. 2 (p. 67), the truncation to only one cosine term may have undesirable consequences from the standpoint of determining or predicting certain mechanical properties. It is better to retain a second cosine term, that is, the expression

$$f(\theta) = \frac{1}{\pi}(1 + \eta_1 \cos 2\theta + \eta_2 \cos 4\theta) \qquad (4)$$

is generally superior to Eq. (3) for purposes of analyzing experimental data. When plotted in polar coordinates, the two cosine term form is cusped differently than the single cosine term. The use of Eqs. (3) and (4) also generates rounded-peak curves when plotted in Cartesian coordinates.

The determination of shape parameters η_i in Eqs. (2) and (4) can also be done using the least squares error method (p. 297). In Eqs. (2) and (4), these shape parameters are subject to limits in allowable values that ensure that no negative probabilities are generated. In Eq. (3) the limits are $0 \leq \eta \leq 1$.

Von Mises Another powerful function suited to the handling of fiber orientation distribution data is known as the von Mises distribution. More specifically, the function used here is a multimodal distribution of the von Mises type (see Ref. 86, pp. 72–74). Here, the probability density function is given by

$$f(\theta) = \frac{1}{\pi I_0(\kappa)} e^{\kappa \cos 2(\theta - \mu_0)} \tag{5}$$

where $I_0(\kappa)$ is a modified Bessel function of the first kind and order zero, that is,

$$I_0(\kappa) = \sum_{n=0}^{\infty} \frac{1}{(n!)^2} \left(\frac{\kappa}{2}\right)^{2n} \tag{6}$$

Two parameters are present in Eq. (5). The parameter μ_0, which is discussed further in the ensuing paragraphs, establishes the mean direction, while κ is a shape (concentration) parameter. The determination of parameters κ and μ_0 is discussed by Mardia [86]. Allowable values of κ must always be nonnegative. Since an increase in the anisotropy of the test material will result in a set of experimental points with a relatively high, narrow peak and, conversely, a more random sheet will have less variation between the MD and other directions, it can be inferred from Fig. 6 that the value of the concentration parameter κ will be larger for the distribution curve that approximates the points generated from the more oriented material.

Establishment of Mean Directionality It can often be assumed that the mean fiber orientation* will lie in a sector around the machine direction, for example, in the interval $-2.5°$ to $2.5°$ when MD lies at $0°$. If this is a valid assumption, then the mean direction parameter that is included in Eq. (5), μ_0, has the value of $0°$, as in Fig. 6. If the fibers are more oriented in the cross-machine direction (CD), then data plotted in the same way will approximate a curve whose peak lies at $90°$ (see Fig. 7) and $\mu_0 = 90°$. In such cases θ will be taken for the interval $0°$ to $180°$. For sheets formed under conditions wherein the greatest fiber alignment is neither MD nor CD, μ_0 will have to be determined according to the method described by Mardia [86].

It should not be too readily assumed that $\mu_0 = 0°$ or that $\mu_0 = 90°$ (see Fig. 8). Some paper machines, in fact, operate in a manner that results in an offset mean direction [42]. One of the advantages in fitting Eq. (5) (for the von Mises distribution) to the experimental data is that the determination of μ_0 can serve as an indication whether the sample edges are, in fact, aligned with the MD and CD directions in cases where that is the intention. The elliptical and cosine functions can be similarly modified by replacing θ with $(\theta - \mu_0)$ in Eqs. (1) through (4); this was done, for example, by Corte and Kallmes for the single cosine function [24]. The determination of μ_0 can at times provide insights that make it possible to correct faulty experimental or analytical procedures. The use of μ_0 is not limited to orientation distributions; other functions with angular dependence may be shifted by use of such a parameter.

---

*This refers to the interval $-90°$ to $90°$ [86]; the term is also (differently) used with reference to a single quadrant [129]. See also pp. 310–312.

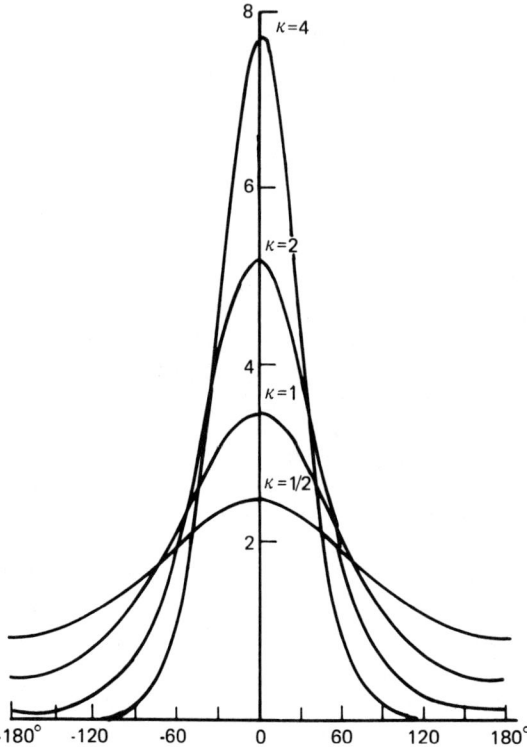

Fig. 6 Cartesian density of the von Mises distribution for $\mu_0 = 0°$ and $\kappa = 1/2, 1, 2, 4$. (From Ref. 86, with permission from K. V. Mardia, Statistics of Directional Data, 1972. Copyright Academic Press Inc. Ltd., London.)

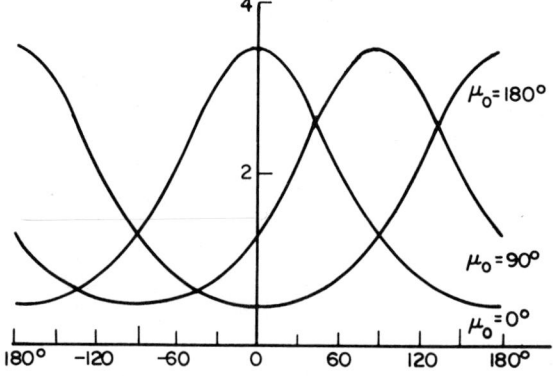

Fig. 7 Cartesian density of the von Mises distribution for $\kappa = 1$ and $\mu_0 = 0°, 90°, 180°$. (From Ref. 86, with permission from K. V. Mardia, Statistics of Directional Data, 1972. Copyright Academic Press Inc. Ltd., London.)

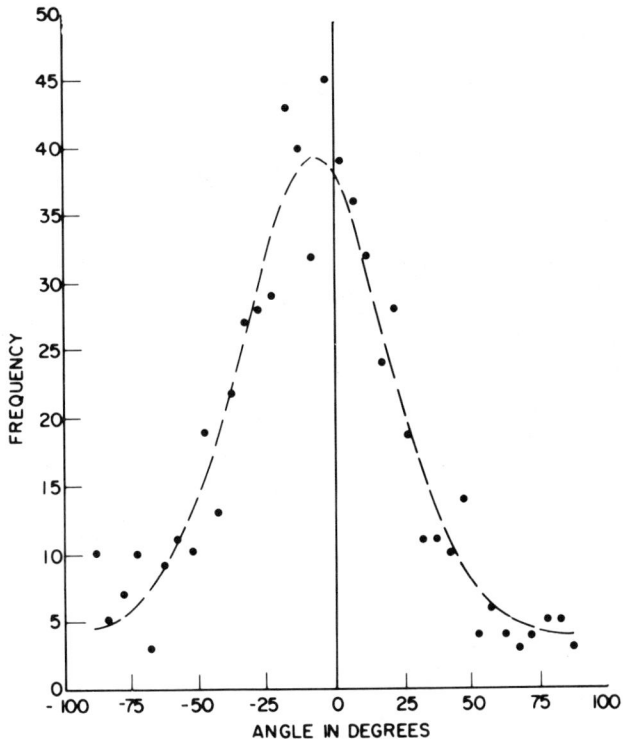

Fig. 8 Frequency data for fiber end-to-end angular orientation measurements in 5° increments compared with a von Mises distribution. This sheet has a high degree of fiber orientation. Note that the distribution is not centered around 0° but is displaced $-7.1°$ (-0.12 radians); therefore $\mu_0 = 0.12$. The shape factor κ is 1.12. With these parametric values, P (probability) = 64.2%.

C. Selection of a Representative Function

Once the true axes of orientation have been determined, the appropriate distribution function(s) (elliptical, cosine, or von Mises) can be used as analytical tools to develop structure-property relationships with respect to different directions in the sheet. One important example is the precise determination and prediction of anisotropic elastic constants. Given the true orientation axes, the term μ_0 is set equal to zero from that point on. Accordingly, the Fourier expansion forms for the above mentioned distributions are very helpful. A Fourier expansion for the von Mises distribution has been given by Mardia [86]:

$$f(\theta) = \frac{1}{\pi I_0(\kappa)} e^{\kappa \cos 2\theta} = \frac{1}{\pi}\left[1 + 2 \sum_{n=1}^{\infty} \frac{I_n(\kappa)}{I_0(\kappa)} \cos 2n\theta\right] \tag{7}$$

where $I_n(\kappa)$ is a modified Bessel function of the first kind and order n. A polar diagram of the von Mises-type distribution shows that it is bimodal when plotted in that manner.

For the elliptical distribution, the Fourier series can be written as

$$f(\theta) = \frac{\lambda}{\pi} \frac{1}{\cos^2\theta + \lambda^2 \sin^2\theta} = \frac{1}{\pi}\left[1 + 2\sum_{n=1}^{\infty}\left(\frac{\lambda-1}{\lambda+1}\right)^n \cos 2n\theta\right] \qquad (8)$$

A comparison of Eqs. (2), (7), and (8) shows that the n-term cosine function has n degrees of freedom, whereas the von Mises and elliptical functions each have a single degree only. As Perkins has noted in Chap. 2, the fact that the last two functions each have only one shape parameter makes them especially suitable in evaluating fiber orientation data (see his Table 1 and pp. 67–69).

When the von Mises distribution is displayed in Cartesian coordinates it is a highly useful function for fitting the curve of a plot of probability density versus orientation angle, as shown in Fig. 8. A cumulative frequency distribution for the same data as are used in Fig. 8 is shown in Fig. 9.

For other possible distribution functions, refer to Mardia [86].

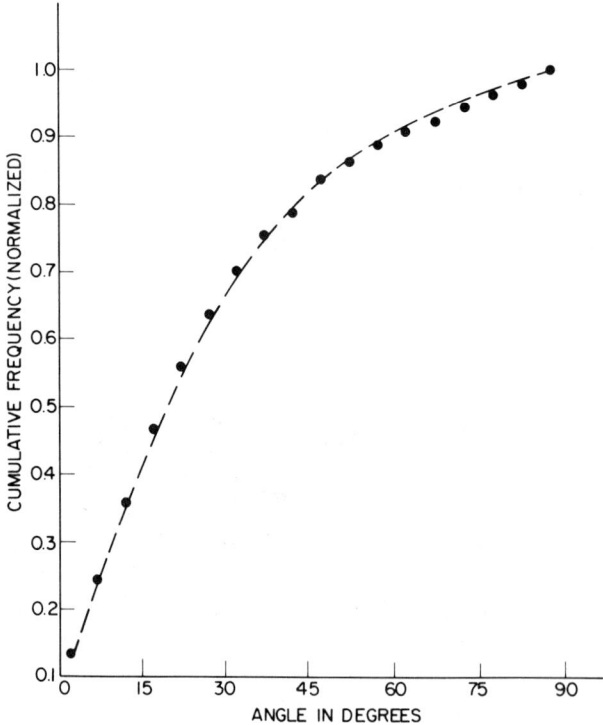

Fig. 9 Cumulative frequency data for the end-to-end angular orientation measurements used to plot Fig. 8. The cumulative von Mises distribution frequency probability in this case exceeds 99.9%.

In general, angular frequency distribution functions can be used with results generated by indirect as well as direct methods. For example, use was made of the single cosine term function to interpret results obtained by optical diffraction in the study described on pp. 300–303.

D. Curve-Fitting Method and Goodness of Fit Test

Given a need to match a curve to a set of observed data, one inevitably demands that the assumed mathematical model fit the data as closely as possible. Therefore, some measure of goodness of fit is needed. Probably the most popular evaluation of fit involves application of the least squares principle.

According to the least squares method [47], the parameters of the assumed model are adjusted so as to minimize the sum of the squares of the differences between the observed and assumed distributions. It is usually necessary to solve a set of nonlinear simultaneous equations with the parameters as unknowns; however, where there is only one parameter involved [e.g., λ in Eq. (1), η in Eq. (3)] one nonlinear equation suffices. A numerical method is required to solve these equations [47].

An important method for making statistical inferences is known as the chi-square test; it is often used to justify whether a given distribution model is sufficiently acceptable, in the sense of goodness of fit [17,53]. The principle of the chi-square test is based on the following rationale.

It can be shown that the variable chi-square

$$\chi^2 = \sum_i \frac{(\text{observed} - \text{expected})^2}{\text{expected}} = \sum_i \frac{(\chi_i - f_i)^2}{f_i} \tag{9}$$

obeys the chi-square distribution

$$f(\chi^2) = \frac{e^{-\chi^2} (\chi^2)^{(\nu/2)-1}}{2^{\nu/2} \Gamma(\nu/2)} \tag{10}$$

Here χ_i and f_i represent the observed and expected frequency, respectively, ν is the degree of freedom (number of classes subtracted from number of restrictions on expected frequency), and

$$\Gamma(\nu/2) = \int_0^\infty e^{-y} y^{(\nu/2)-1} \, dy \tag{11}$$

For a particular set of data, the chi-square χ_p^2 is calculated according to Eq. (9). One can then compute the probability P in accordance with the relation

$$P = 1 - \int_0^{\chi_p^2} f(\chi^2) \, d\chi^2 \tag{12}$$

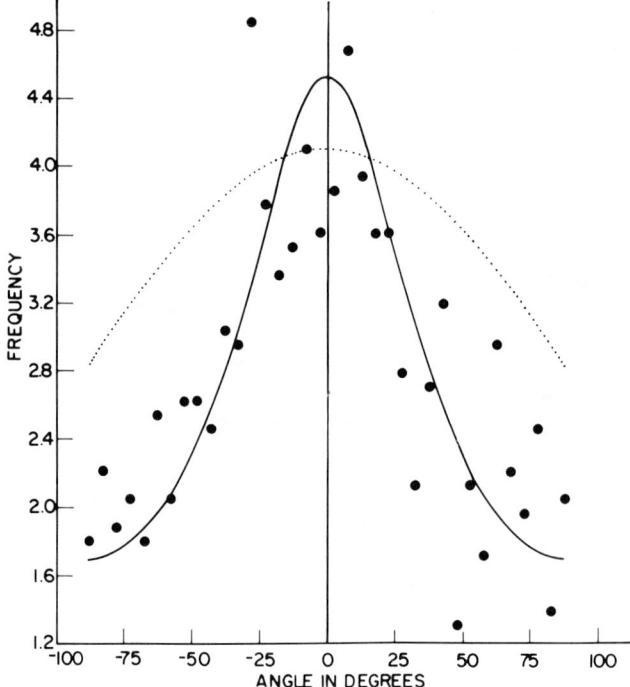

Fig. 10 Frequency data for fiber segment angular orientations measured in 5° increments compared with elliptical (solid line) and single cosine term (dotted line) distributions with shape parameters $\lambda = 1.459$ and $\eta = 0.368$, respectively. This sheet has a fairly low degree of fiber orientation. The mean direction parameter μ_0 has been assumed to equal 0°, so the distributions are symmetrical about 0°. The elliptical function shows the better fit in this case, with $P = 32.5\%$.

P can be explained as the percent probability that the expected distribution (that is, the assumed mathematical distribution model) can describe the observed distribution. If the value of P is found to be equal to one (100%), it means that the assumed distribution model is actually a true representation of the observed distribution. For the other extreme case, the assumed distribution cannot describe the observed distribution at all if the probability is found to be zero. In general, one accepts that the assumed model is a good representation of the observed frequency when the probability P is greater than 0.05, that is, greater than 5%. In the examples shown in Figs. 8 to 10 this requirement is satisfied for appropriate distribution functions.

    An important application of the chi-square test comes from the requirement that P be large to consider that a curve is well fitted to a set of data points. However, one can continue to adjust the shape parameters of an assumed model obtained by the least squares method until the probability is the largest one possible for that particular model.

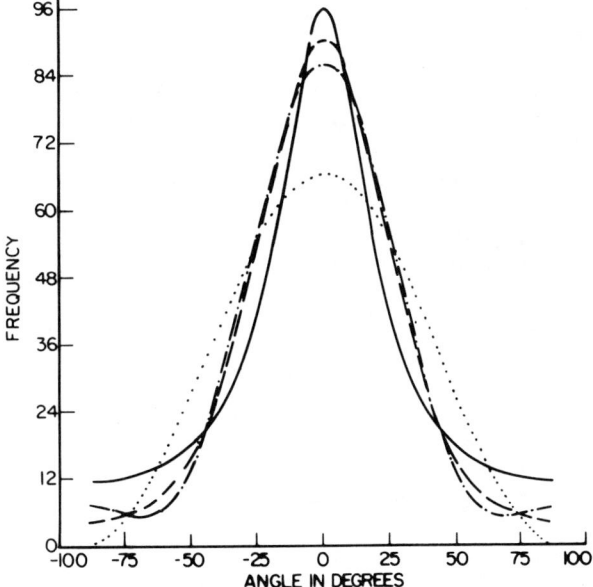

Fig. 11 Elliptical (solid line), single cosine term (dotted line), and two cosine term (dot-dashed line) distributions adjusted for best fit to a von Mises distribution (dashed line) with shape parameter $\kappa = 1.5$.

The least squares method is a well-established method for selecting parameters in a way that adequately fits the data, but the following alternate procedure could be suggested:

1. Use the least squares method to find parameters (e.g., η_1, η_2) that will match the function to the data reasonably well.
2. Find P by the chi-square test.
3. Adjust parameters η_i so as to maximize P.

E. Relations Between Orientation Functions

In Fig. 11, a graphic comparison is shown of the four distributions that have been discussed in this chapter. The graph may enable the reader to visualize the mathematical interrelationships between the functions. In this figure, a plot is made of a von Mises distribution for $\mu_0 = 0$ and arbitrarily assigned $\kappa = 1.5$ (corresponding to a highly anisotropic sheet). An attempt is made to adjust the other 3 functions to fit the von Mises curve, using the least squares method. It is seen that no congruence occurs; the best-fitting elliptical function generates a curve between $-50°$ and $50°$ that is narrower and higher, while the cosine function curves are wider and lower. In the 50° to 90° range, other disparities occur. This illustration is included to emphasize that the choice of distribution function may be different for each data set; there is neither a "right" function nor a universal correspondence between data and function or function and function.

300 Mark

F. Fiber Orientation by Scattering (Contributed by Jens Borch)

In addition to the indirect methods that have been discussed on pp. 286—287, efforts have been made to apply visible light and X-ray scattering techniques to the quantitative determination of fiber orientation in paper sheets [19,82,110,119,166]. Since scattering techniques are nondestructive and in their mode of transmission provide integrated average values of fiber orientation through the thickness of the sheet, there is potential application in routine and on-line control of commercially made sheet structures. Radiation methods for the on-line monitoring of moisture, basis weight, and optical properties are already well established. Accurate measurement techniques for structural properties such as fiber orientation are less well developed but will undoubtedly increase with the development and application of proper theories, measuring techniques, and instrumentation.

Light Diffraction The multiple scattering and absorption processes of visible light in paper structures create their opaque or translucent, light-diffusing appearance, as described in Chap. 16, pp. 32—40. In order to examine scattering details from individual fibers, it is necessary to isolate individual fibers and apply a suitable experimental setup. The laser light source is ideally suited for illuminating single fibers of small (1 to 100 µm) diameter as well as providing coherent illumination of larger objects through beam expansion. Low intensity helium-neon (He-Ne) and ruby lasers have been used for both light diffraction and small angle light scattering (SALS) analysis (Fig. 12) [119,131,166].

The diffraction characteristics of single pulp fibers are similar to those of textile fibers where the scattered light intensity is broken up into a series of periodic maxima in the plane perpendicular to the fiber direction

Fig. 12 Laser scan setup for light diffraction analysis. (From Ref. 166.)

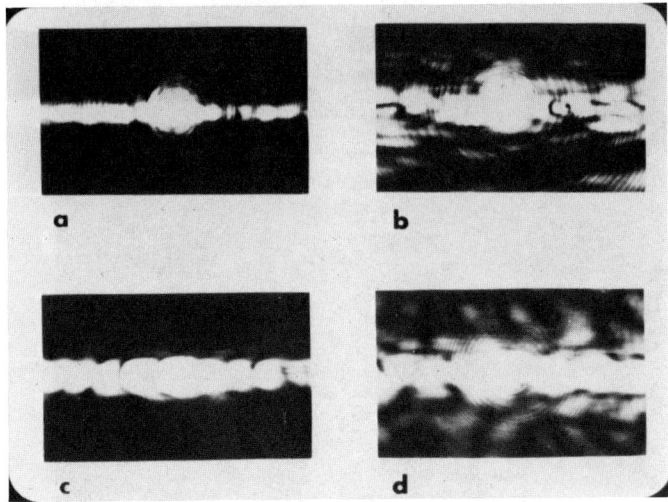

Fig. 13 Scattering from single paper fibers viewed from the forward direction (each fiber a to d aligned perpendicularly to scattering streak).

[84]. Shape distortions and internal structural heterogeneities create imperfections in the symmetry and periodicity of the intensity peaks (Fig. 13). The possibility of measuring fiber orientation by light diffraction depends on the spatial distribution of scattering streaks from individual fibers. The scattered light is collected in the back focal plane of a suitable lens system and analyzed through a rotating aperture in a photomultiplier [119] or photographically via microdensitometry [166]. The individual scattering streaks are lost, but the diffraction patterns (power spectra) of sufficiently well oriented fiber assemblies still indicate preferential orientation direction, as shown in Fig. 14. The data may be further treated by scanning the angular intensity distribution in the plane of the photograph and relating the variation to the probability that the constituent fiber is inclined to a preferential orientation direction. By including a parameter D_1 that depends on the number of (equal) sectors selected for scanning over a specified circular arc and an orientation distribution parameter D_2 determined by least squares fitting of the intensity data, the drop-off in probability density $F(\theta)$ can be expressed by the relation

$$F(\theta) = D_1(\pi^{-1} - D_2 \cos 2\theta) \qquad (13)$$

as shown in Fig. 15 [166]. Note that Eq. (13) is a variant form of Eq. (3). So far, this technique has been most successful in analyzing split paper sheets and synthetic nonwoven sheets due to the opaque nature of most paper structures made of pulp fibers.

Small Angle Light Scattering Experimentally, the small angle light scattering (SALS) technique is only slightly more complex than the visible light dif-

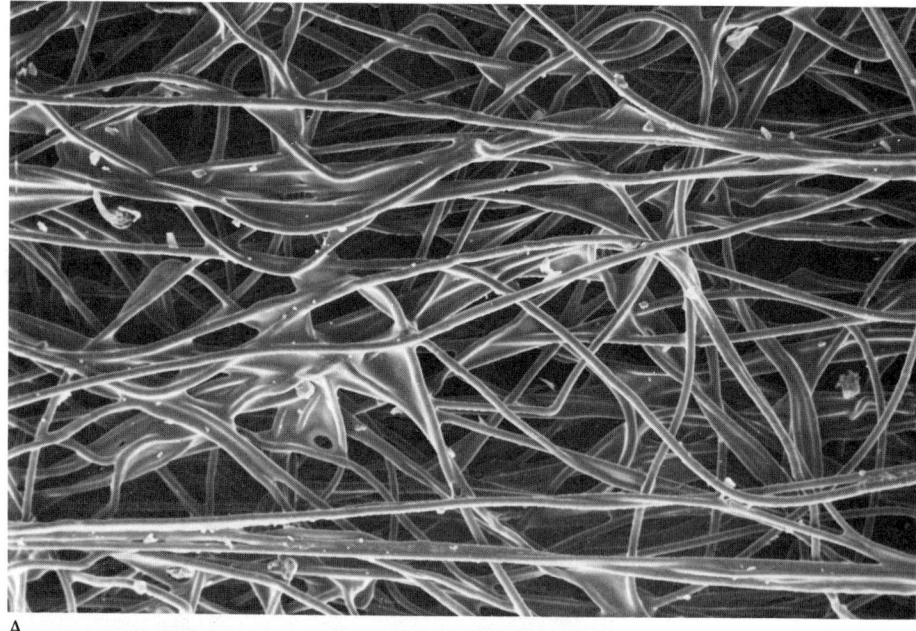

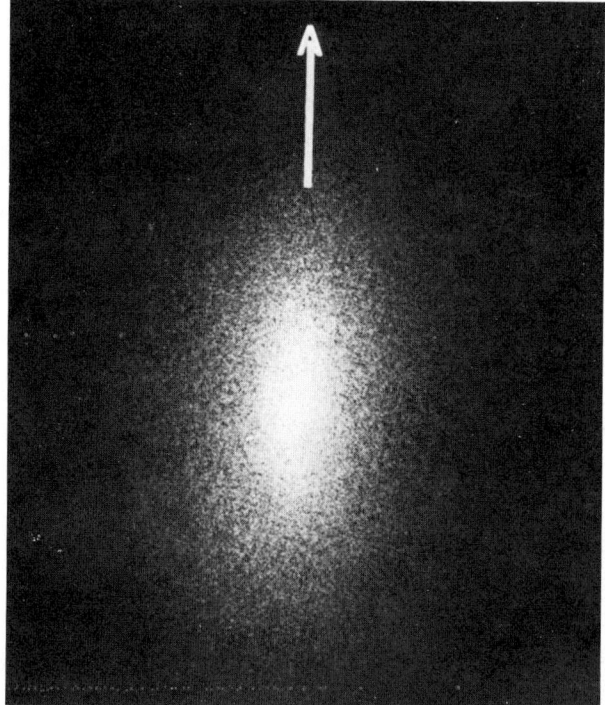

Fig. 14 A highly oriented synthetic fiber network (A) photographed at 60 × magnification, and (B) its diffraction pattern (Fourier transform). The arrow in (B) indicates the plane of maximum scattered light perpendicular to orientation direction. (From Ref. 166.)

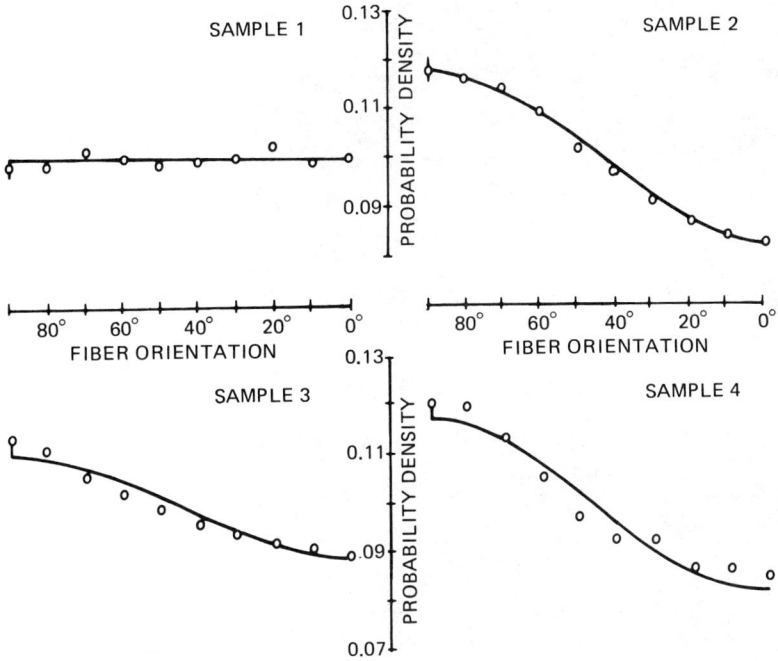

Fig. 15 Fiber orientation distributions of fiber assemblies of varying orientations. Values for D_2 [Eq. (13)] varied from 0 to 0.055, while $D_1 = 0.314$ for all cases. (From Ref. 166.)

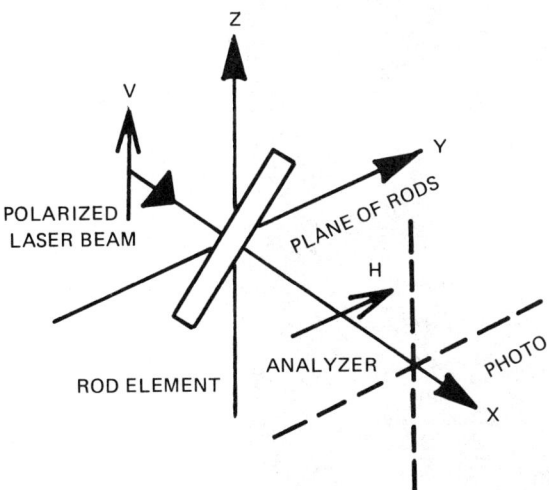

Fig. 16 Light scattering geometry. The incident laser beam is vertically polarized. The scattered light from elements of a rod assembly is filtered through a horizontally oriented polar (the analyzer).

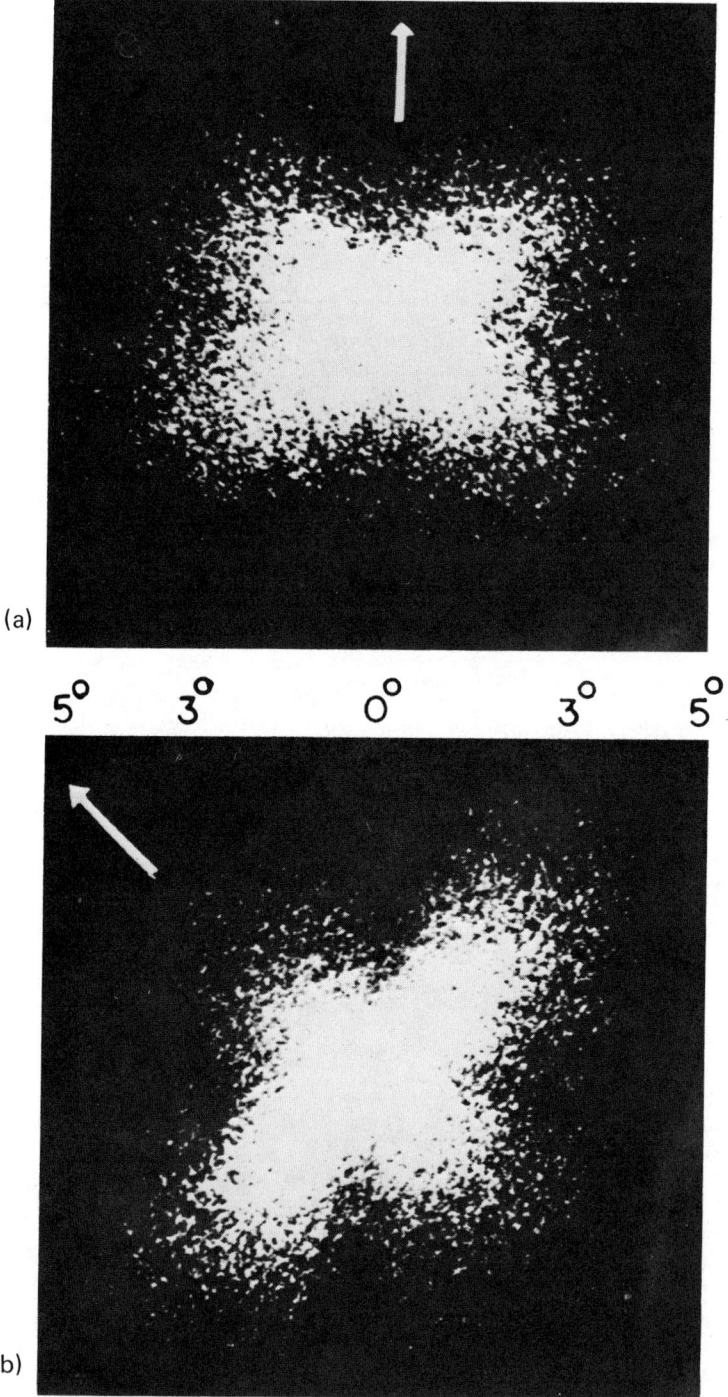

Fig. 17 Photographic H_v scattering patterns of condenser paper. Machine direction is indicated by arrows. (From Ref. 85.)

fraction procedure. One procedure is to imbed the anisotropic scattering object in a liquid or resin of closely matching refractive index and to filter the scattered light through a polarizer with its optic axis oriented as shown in Fig. 16 (H_v scattering) [10,19,82,85,93,110,131,158]. The scattered light intensity that is recorded is the result of optical anistropy (birefringence) rather than of structural anisotropy. The interpretation of scattering patterns relies heavily on theories developed for visible light scattering from anisotropic polymer films [131]. For "rodlike" scattering (Fig. 16), the rod is visualized as consisting of scattering elements situated along the rod axis and for which scattered intensity will depend on the angular inclination of the rod axis toward the XZ plane. Consequently, a rodlike assembly oriented around a preferred angular direction will induce asymmetry to the distribution of scattered light (scattering envelope), as discussed by Charrier and Marchessault [19]. The H_v scattering envelope of condenser paper shows the machine-direction orientation created during manufacture [10,19,82,85]. X-shaped scattering streaks are flattened around the horizontal axis when rod orientation (machine direction) is vertical (Fig. 17A). When rod orientation is 45° to the horizontal/vertical directions, the scattering streaks are orthogonal but of unequal intensity variation (Fig. 17B). The intensity variation in both patterns can be computer-calculated and related to orientation parameters and rod dimensions [19,82]. Additional V_v scattering envelopes (vertically oriented polarizer) provide additional structural information at the expense of added intensity contributions from imperfections in the sample and optical components of the scattering fixture (density fluctuations) [19,131].

The SALS technique provides a powerful method of analyzing cellulosic material that is strongly broken up (condenser paper) [10,19,82,85], regenerated (solvent cast films) [10], or elucidated for internal structure (single fibers) [110,158]. The most serious obstacles preventing the more universal use of this technique in paper science are those that pertain to fiber size and distribution and to the thickness of most commercially made papers. The radiation wavelengths of the visible light spectrum tend to favor the asymmetry of micron-size substructures rather than complete paper fibers in the intensity distributions, even at small scattering angles. A significant sample thickness requires corrections for multiple scattering effects similar to those needed for multiple scattering corrections in light diffraction analysis [19].

X-Ray Diffraction The orientation of crystalline cellulosic microfibrils in the middle secondary layer of the pulp fiber (S2 layer, see p. 424 of Vol. 1 and Sec. V below) affords a means of measuring fiber orientation by X-ray scattering [110]. The method is indirect in that the short-wavelength X-ray radiation interacts with the fibrillar subunits inside the collapsed fiber that are oriented at angles $\pm \beta$ to the fiber axis, itself inclined at angle ϵ to the XZ plane (Fig. 18). Consequently, it is necessary to measure S2 fibrillar orientation using one of the techniques described on pp. 464—480.

Both the SALS technique [19] and X-ray diffraction analysis [110] have been treated through orientation distribution functions where ϵ and C, an orientation parameter, provide calculation parameters for matching theoretically assumed distributions with experimentally measured intensity variations. For X-ray diffraction analysis, an elliptical-type orientation distribution has been employed.

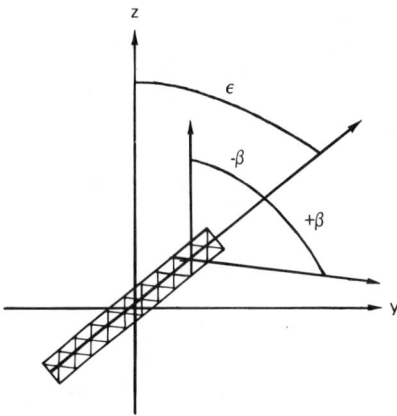

Fig. 18 Geometric representation of fiber in the YZ plane tilted at an angle ε. (From Ref. 110.)

$$N(\varepsilon) = \frac{C}{C^2 \sin^2\varepsilon + \cos^2\varepsilon} \tag{14}$$

The intensity distribution is both a function of fiber orientation distribution $N(\varepsilon)$ and the fibril angle distribution $D(\beta)$, as described by Prud'homme et al. [110].

Experimentally, the X-ray method relies on the azimuthal intensity drop-off of equatorial reflections in the X-ray pattern (Fig. 19A and B). Better mutual fiber orientation produces a stronger intensity decrease (A compared with B in Fig. 19). The decrease is measured quantitatively and matched to a best fit value of the orientation parameter C for the fibril orientation in question (Fig. 20) [110].

Compared with the methods that utilize visible light, this method is less ambiguous as to what creates the scattering. However, the added complexity in instrumentation and the need for knowledge of the fibril angle distribution create a procedure that is generally better suited for research and development than for routine and on-line control. On the other hand, the X-ray procedure may provide the only method available for paper embedded in plastics. Other X-ray methods suffer from limitations in evaluating scattering patterns—with associated complicated analysis procedures—as discussed by Rudström and Sjölin [119].

G. Silvy's Orientation and Segment Length Distribution Method ("The Equivalent Pore")

The Concept The orientation of a fibrous network may be examined in terms of its void configuration as well as of its solid component. Up to this point, the discussion of orientation has focused on the arrangment of the solid frac-

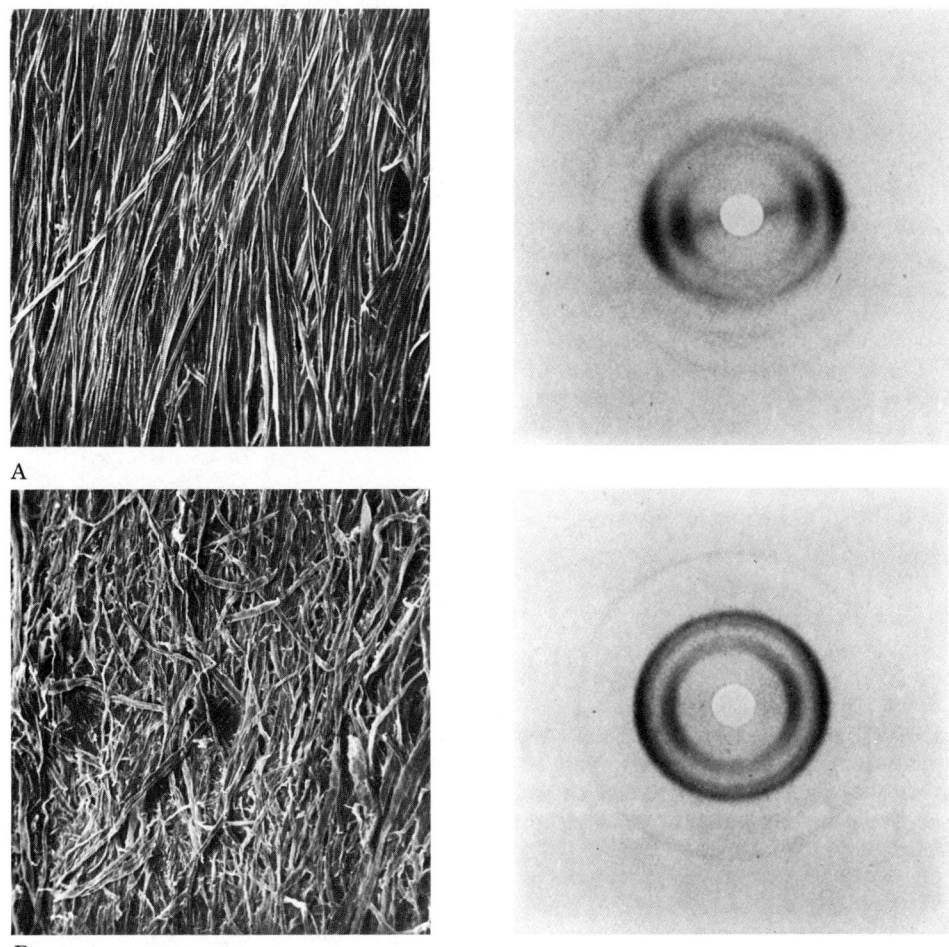

Fig. 19 Scanning electron micrographs of paper samples with different levels of orientation in the fiber structure at ∿ 30 ×. Better mutual orientation (A compared with B) produces stronger azimuthal intensity drop-off of equatorial X-ray reflections. (From Ref. 110.)

tion, but Silvy* [128,129] has developed an approach that examines the effects of the geometry of pore space on the mechanical and physical behavior of paper. Void configuration is correlated with air permeability, moisture removal, strength, and optical properties.

---

*Readers are advised to use reference 129, as some of the material in reference 128 has been subsequently reformulated by Silvy.

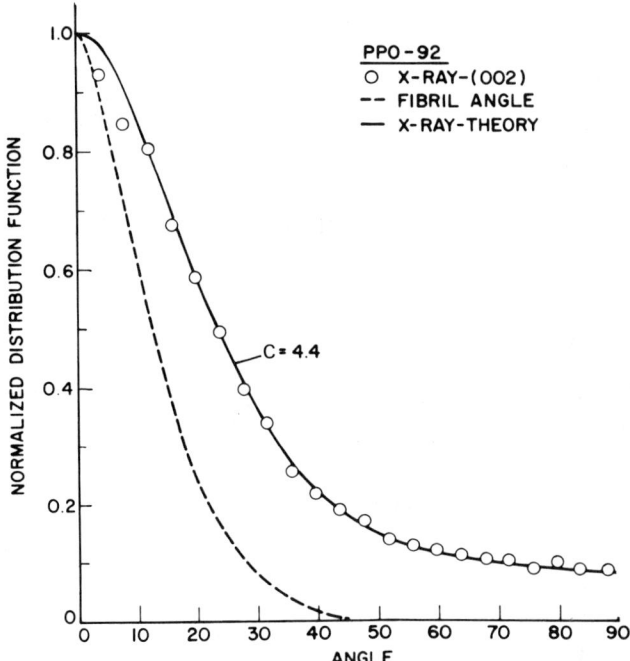

Fig. 20 Fibril angle (S2 filament winding angle) distribution function and X-ray intensity distribution (experimental points) for the sample shown in Fig. 19A. The continuous line through the experimental points is obtained theoretically with the use of an orientation parameter C = 4.4. (From Ref. 110.)

Silvy notes that the pore spaces in an anisotropic network will by necessity be elongated and can be approximately described as ellipsoidal in shape. One can also describe a mean elliptical cross-section in the plane of the sheet by considering successive contour lines enveloped by surrounding fibers in parallel planes at various depths in the sheet. From a fiber orientation distribution plot such as the one appearing in Fig. 21, one may calculate an elliptical pore that is "equivalent" to the shape of the void space in the sheet by postulating that the radius of curvature of the ellipse at any point M is proportional to the total length of fibers or fiber segments $n_{\ell\theta}$ that are tangentially aligned in an interval $d\theta$ centered on point M. Silvy's method normalizes this length with reference to the cumulative length of fibers per unit surface area of the sheet. Thus, $n_{\ell\theta}$ represents a length-weighted orientation density of the fibers. Accordingly, the cumulative length L_θ of fibers or segments per unit surface area whose orientations lie between θ and $\theta + \Delta\theta$ can be represented by an arc of a curve s_θ formed by placing all the fibers aligned within that interval end to end. In Fig. 21, one sees that the cumulative length of fibers oriented at θ is greater than the corresponding cumulative length of fibers oriented at θ'. Thus, the arc set off by the

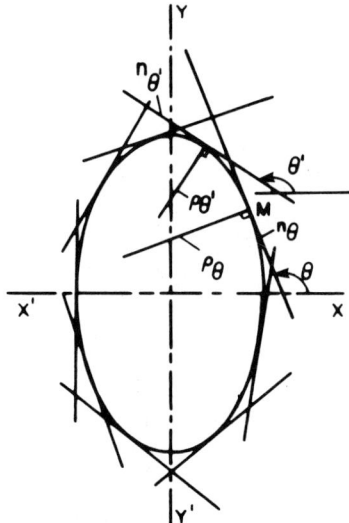

Fig. 21 Diagram of equivalent pore, showing that radius of curvature is directly proportional to orientation density. (From Ref. 128.)

former is longer than the latter, therefore the radius of curvature of the former is greater. A relation exists such that

$$\frac{dL_\theta}{d\theta} = \lim_{\Delta\theta \to 0} \left(\frac{s_{\theta+\Delta\theta} - s_\theta}{\Delta\theta} \right) \tag{15}$$

represents the radius of curvature at each point on the curve of the equivalent pore ellipse (see Fig. 21). Any point on the equivalent pore can thus be defined from the radius of curvature r_θ of the pore at angle θ and/or the appropriate density by the following relations:

$$n_{\ell\theta} = \frac{\bar{\ell}_\theta n_\theta}{\langle\lambda\rangle} = \frac{2}{\int_0^{2\pi} r \, d\theta} r_\theta \tag{16}$$

where

$\bar{\ell}_\theta$ = mean length of the fibers of orientation θ, determined by a summation technique in which numbers of fibers of a given length are weighted by that length
n_θ = orientation density (number/radian) of the fibers
$\langle\lambda\rangle$ = mean length of all fibers at all orientations

Properties of Equivalent Pores of Elliptical Form The orientation density of fibers weighted by length may be derived from an elliptical curve's major and minor semi-axes, a and b, respectively. In polar coordinates, the orientation density is proportional to the cube of the radius vector OM_θ of the equivalent pore. Thus,

$$n_{\ell\theta} = \frac{2}{abN\ell}(OM_\theta)^3 \qquad (17)$$

Here N = the number of fibers per unit area.

It follows that the anisotropy of the pore is defined by

$$\frac{a}{b} = \sqrt[3]{\frac{n_{\ell\theta}MD}{n_{\ell\theta}CD}} \qquad (18)$$

where

$n_{\ell\theta}MD$ = weighted orientation density in the machine direction
$n_{\ell\theta}CD$ = weighted orientation density in the cross machine direction

In both cases, the numbers of fibers are those counted in the angular interval selected (usually an arc of between 5° and 10°).

If the distribution is symmetrical, the mean angle of orientation θ^* is defined (with respect to the cross machine direction) as

$$\theta^* = \frac{\int_0^{\pi/2} n_{\ell\theta}\theta\, d\theta}{\int_0^{\pi/2} n_{\ell\theta}\, d\theta} = 2\int_0^{\pi/2} n_{\ell\theta}\theta\, d\theta = 2\int_0^{\pi/2} \theta\, ds_\theta \qquad (19)$$

Note that the integration interval is 0 to $\pi/2$. Calculated over the length of the perimeter of the pore, this integral has the value $\theta^* \cong \tan^{-1}(a/b)$. Thus, for the assumption of an elliptical shape for the equivalent pore, the mean orientation angle θ^* is sufficient to describe the ellipticity of the pore and accordingly, the fiber orientation distribution in the plane of the sheet for a wide range of papers.

Silvy employs the calculation on papers produced on industrial paper machines as well as on laboratory centrifugal formers by (1) using both line crossing and direct measurement methods to count dyed fibers and then (2)

Table 1 Mean Orientation Angle θ^* Determined According to Three Methods

Stock speed / Wire speed	Mean angle of orientation		
	From measurements of dyed fibers	From equivalent pore	From measurements of zero-span tensile strength
0.9	52°21'	52°26'	51°
0.7	58°15'	58°30'	58°
0.5	61°45'	62°07'	64°
0.1	67°00'	67°42'	67°

Source: From Ref. 128.

integrating the number of fibers per radian in steps of 0.175 rad (10°) to obtain the proportion of fibers in each 10° interval. The calculation of the ellipticity of the equivalent pore is then fitted to the data points by a least squares method. Table 1 shows the mean angle of orientation θ^* for different types of papers calculated by various methods.

H. Mean Orientation Angle

Several investigators have developed orientation indices or other orientation parameters for the purpose of simplifying the relationship between mechanical properties and structural anisotropy in the plane of the sheet. Such an index is usually based upon the mean angle of orientation θ^* in the first quadrant, defined as

$$\theta^* = \frac{\sum_{i=1}^{\tilde{n}} n_i \bar{\theta}_i}{\sum_{i=1}^{\tilde{n}} n_i} \quad (20)$$

where $\bar{\theta}_i$ is the mean angle of the fibers in an angular increment (sector) containing n_i fibers and $\tilde{n}$ equals the number of sectors lying between orientations 0° and 90°.

The above expression, similar to the first form of Eq. (19), is limited in that

1. Either the angular increment θ_i must be small in order to justify an assumption that $\bar{\theta}_i$ lies at the center of the increment, or else $\bar{\theta}_i$ must be computed for each sector.
2. No weight is given to such factors as length of fibers ℓ or linear density of fibers ω.

To account for item 2, Eq. (20) would have to be modified by changing n_i terms to $n_i \ell_i \omega_i$ terms. It may be noted that Silvy's method [129] given in Eq. (19) does take into account ℓ_i as well as n_i terms in the development of the equivalent pore.

The grouping of fiber orientation data in increments θ_i lends itself to weighting according to the number n_i or frequency f_i of observations in the ith sector of one or more quadrants. Taking fiber mass $\ell\omega$ as an example of a weighted grouping, one can select M class intervals representing different $\ell\omega$ fractions and determine how frequently the various $\ell\omega$ fractions occur in each increment. Now each $\bar{\theta}_i$ is the angle of M vectors whose directions all lie at the center of the ith increment but whose lengths vary according to the magnitude of $\ell\omega$. The mean orientation θ^* is now defined as the direction of the resultant of these vectors, whose magnitude $\bar{R}$ is determined by

$$\bar{C} = \frac{1}{N} \sum_{i=1}^{N} \sum_{j=1}^{M} (\ell\omega)_j f_i \cos \bar{\theta}_i$$

$$\bar{S} = \frac{1}{N} \sum_{i=1}^{N} \sum_{j=1}^{M} (\ell\omega)_{ji} f_i \sin \bar{\theta}_i \qquad (21)$$

$$\bar{R} = (\bar{C}^2 + \bar{S}^2)^{1/2}$$

where

N = sum of the frequencies or numbers of observations
$\bar{R}$ = resultant magnitude of vectors $\bar{S}$ and $\bar{C}$.

The mean orientation is found by

$$\theta^* = \tan^{-1}(\bar{S}/\bar{C}) \qquad (22)$$

For examples of mean orientation determinations for grouped data, the reader is referred to Chap. 2 of Ref. 86.

The mean orientation angle can be related to physical or mechanical properties (e.g., curl, MD versus CD breaking length), when other processing variables are held constant; experimental reports often show linear relationships [cf. 41,42,111,118,128,129]. In general, the elastic and inelastic (e.g., strength) mechanical properties of paper in the plane of the sheet should vary with mean fiber orientation angle in a nonlinear manner [127]. This variation follows the exponential laws in good agreement with theoretical prediction (cf. Chaps. 2 to 6).

I. Orientation in the Z Direction

The thickness-direction orientation of fibers in paper may be detected by taking advantage of the fact that inorganic materials such as clay tend to be adsorbed along the cellulosic fibers with their crystallographic axes at a definite disposition with respect to the fiber axis [46]. Thus, Sundararajan [133] noted the preferential orientation of clay filler in bond paper, which is influenced strongly by the shape of the inorganic constituent, and characterized the orientation by the use of X-ray patterns taken in the edge and end view of sheets stacked 1 to 2 mm thick.

Earlier, Aaltio et al. [1] had shown that orientation in the Z direction can be qualitatively derived from X-ray diagrams that reveal the density distribution of the paratropic reflections of cellulose crystallites in the fibers, especially the (002) reflection. Such diagrams are obtained when the incident beam makes an angle with respect to the normal on the sheet.

The Z-direction orientation is important with respect to pick resistance in printing grade papers, for abrasive papers, and for papers used as a backing for peeling tapes, as examples.

III. DIMENSIONAL CHARACTERIZATION OF FIBERS

The term *dimensional characterization* includes such parameters as the length, width, cross-sectional area, surface area, degree of flattening (collapse),

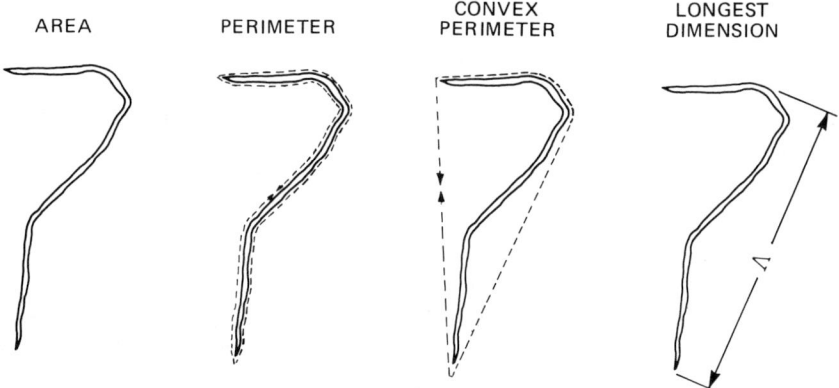

Fig. 22 Four of the basic measurements that can be made using fast logic circuitry in image analysis.

and aspect ratio of fibers and fines, as well as the kinking (segmentation),* curving, and twisting of fibers and the resultant curl that these deformed fibers exhibit. Fiber curl may be considered from the standpoint of in-plane curvature or from combined in-plane and out-of-plane curvature and is quantitatively assessed as a ratio of the total convoluted length to the straight-line length between the ends of the fiber [74] or by several other criteria (see pp. 334–336).

A great deal of work has been done on dimensional characterization by data-gathering observations and measurements with light and electron microscopes. These excellent instruments have been joined by powerful electronic allies—the graphic digitizer and the automatic image analyzer—with data transmitted directly to computers for rapid processing (refer to Chap. 13). There are also indirect methods for quantitatively assessing certain structural parameters, some of which have become standard test procedures.

A. Fiber Length

The length of a typical fiber that is ultimately incorporated into a sheet of paper or paperboard undergoes two kinds of change from the time it is part of a tree or plant stem until it is part of the sheet. First, the chipping or grinding and subsequent chemical, physical, and mechanical processes shorten it, and second, these same processes change the rather straight, spindle-shaped structure that is the original fiber (Chap. 28) to a generally convoluted (curled), roughened, and sometimes bent or twisted object. Thus, length is changed both absolutely and in longest dimension (Fig. 22). For some purposes, it is important to ascertain fiber lengths at various stages.

---

*That is, segmentation by abrupt bends (changes in direction); segmentation in this sense does not mean a physical separation of the segments.

In the living stem, fiber length is often regarded as a prime indicator of quality in evaluations of tree improvement trials and other genetic programs. Fiber length, together with fiber strength and fiber flexibility, are generally highly correlated to the development of good mechanical properties in paper, with tear strength and folding endurance being especially dependent on length* [4,20,22,30,31,54,80,123a,160–162].

Fiber length is important in the development of strength, because although fibers transmit load axially, the load transmission from fiber to fiber takes place in the fiber-to-fiber bonds, which are stressed in shear. The actual shear transfer length is not great; thus, concentrated stress transfer by shear occurs over a shorter percentage of the total fiber length as undamaged fiber length increases and bond density remains constant. Reduced fiber length, reduced fiber moduli, and reduced bonding result in lower paper elastic moduli [98a,98c]. These concepts are embodied in the literature of paper physics and will not be expanded here; the interested reader can refer to Chap. 2 and the outstanding series of Fundamental Research Symposium Proceedings of the British Paper and Board Industry Federation Technical Division and the paper physics conferences organized by the Technical Association of the Pulp and Paper Industry (TAPPI) and the Canadian Pulp and Paper Association (CPPA).

Classical Methods of Length Determination

Microscopy The determination of fiber length in the stem or other tissue of a plant is generally accomplished with the instruments available in a good light microscopy laboratory equipped to kill, fix, embed, section, stain, and mount tissues. In sections of wood, the longitudinal-tangential (L-T) plane (Chap. 28, Fig. 1) is normally used to measure fiber length, since there is less ambiguity as to the locations of the fiber ends than in any other orientation of section. Detailed descriptions of the procedures in obtaining satisfactory sections for observation are available in a good reference on botanical microtechnique, such as that of Berlyn and Miksche [9].

The light microscope has also been the classic tool for determination of fiber length in pulp. A well-known reference source is Isenberg's book [56], which includes 18 pages on methods of making length and width determinations microscopically.

Projection Most of the early microscopic procedures proved to be too time consuming. A substantial developmental effort over approximately 25 years [20,21,43,163] produced projection methods such as those described in the book by Isenberg [56] and TAPPI suggested method T 232 [140]. In these methods, a slide containing a dispersion of pulp fibers is placed on the stage of a projection microscope. An image of the fibers is projected onto a screen that incorporates a ground plate-glass grid so that the line spacing on the grid bears a known quantitative relation to the actual length dimensions on the microscope stage as determined by a transparent micrometer scale.

---

*Unlike most mechanical properties assays, tensile tests of paper and board often show little or no correlation or even a negative correlation with fiber length [89,135,157,161], except in the case of wet web tensile strength [75].

Measurements of fiber length, width, and coarseness are all possible in the projection apparatus. For additional details, the reader is referred to the TAPPI suggested method for coarseness T 234 [142] and the description and references given in Chap. 10, pp. 443−445.

Screen Classification During roughly the same time period that witnessed the development of projection methods, screening methods were also developed for pulp fiber length determination [20,115]. Wilson [163] has provided a brief but well-referenced developmental history. The devices that have been developed for separation or fractionation by length depend on a series of screens of different mesh sizes through which the fibers pass, at a dilute consistency of about 0.15 to 0.4%, starting with the coarsest screen. Most commonly, 4 screen opening sizes are used, although more screen gradations can provide greater fractionation of the sample. Preferably, the series of screens used should be selected so that approximately one-fourth of the fibers are held on the coarsest screen. A given mesh opening size generally retains fibers whose lengths are over twice the mesh opening, except in the case of very flexible fibers.

Two of the most popular types of these screening devices (classifiers) are (1) a horizontal type wherein the stock passes through a series of compartments separated by removable screens, and (2) a series of vertical tanks arranged in a cascade series so that the stock not retained on the first tank's screen overflows to the next tank, and so on. These classifiers form the basis of such industry standards as JIS P-8207 [58], TAPPI T 233 [141], and SCAN-M6 [123].

Number Average Length and Weighted Average Length There are basically three kinds of length calculations made using fiber length data. The number average (or arithmetic average) can be found for ungrouped fibers by dividing the total length of the fibers measured by the number of fibers measured,

$$<L> = \frac{\sum_{i=1}^{n} L_i}{n} \qquad (23)$$

where

L_i = actual measured length of the ith fiber
n = number of fiber length measurements

The same calculation made for fibers grouped into certain class intervals according to length has the form

$$<L> = \frac{\sum_{i=1}^{n} F_i L_i}{\sum_{i=1}^{n} F_i} \qquad (24)$$

where

F_i = frequency (number) of fibers in a particular length group or class interval
L_i = specified or measured length of the individual fibers in the ith class interval
n = number of fiber length groups or class intervals

There is little practical advantage in ascertaining <L> as in Eq. (23) or (24), for in such a calculation the smallest fiber fragments or other fines count for as much numerically as full-length fibers in the denominator even though they contribute very little to the sum in the numerator. Also, the lower size limit of the fines counted is inherently arbitrary or subjective.

A second, more useful value is obtained by weighting the fiber lengths to reflect the fact that longer fibers and thicker (coarser) fibers exert a disproportionate influence on most sheet properties. The weighted average length by length is found, for ungrouped fibers, by

$$<L>_\ell = \frac{\sum_{i=1}^{n} L_i^2}{\sum_{i=1}^{n} L_i} \tag{25}$$

and for grouped fibers, by

$$<L>_\ell = \frac{\sum_{i=1}^{n} F_i L_i^2}{\sum_{i=1}^{n} F_i L_i} \tag{26}$$

The third kind of calculation refers to the distribution of mass and is referred to as the weighted average fiber length by weight. Since fiber mass represents a fundamental property of pulp and paper sheets, the calculation of weighted average fiber length by weight is generally the most useful of length calculations. Although one could theoretically obtain this value by summing the products of individual fiber lengths and linear densities, multiplying by lengths and dividing by the total weight of the sample, there is no way to ascertain coarseness (linear density) of an individual fiber; accordingly, the weighted average length by weight computation is generally restricted to grouped or classified fibers. The following formula applies:

$$<L>_w = \frac{\sum_{i=1}^{n} F_i (Lc)_i L_i}{W} \tag{27}$$

where

$(Lc)_i$ = length of fiber times fiber coarseness for the ith fraction
W = total weight of sample

L_i is defined as for either Eq. (23) or (usually) Eq. (24). For the particular case of fiber length by classification [141], Eq. (27) takes the form

$$<L>_w = \frac{W_1 L_1 + W_2 L_2 + W_3 L_3 + W_4 L_4 + W_5 L_5}{W} \tag{28}$$

where

W_i = ovendry weight of the fiber fraction retained on each of 4 screens in the classifier for i = 1, 2, 3 and 4
W_5 = ovendry weight of the pulp lost through the finest screen
L_i = average length of each fraction i in the pulp sample
W = ovendry weight of the material supplied to the classifier

It is seen that in Eq. (28) the number of F_i terms is 5 and the W_i terms are the equivalent of the $(Lc)_i$ terms in Eq. (27). An alternative way to write Eq. (27) is therefore

$$\langle L \rangle_w = \frac{\sum_{i=1}^{n} F_i (Lc)_i L_i}{\sum_{i=1}^{n} F_i (Lc)_i} \tag{29}$$

Quite commonly, an assumption is made that the mass of a given fiber is proportional to its length. When this *constant coarseness* approximation is applied to Eq. (29) for grouped fibers, the result is

$$\langle L \rangle_w = \frac{\sum_{i=1}^{n} F_i L_i^3}{\sum_{i=1}^{n} F_i L_i^2} \tag{30}$$

and the corresponding relation for ungrouped fibers is expressed at times in the form

$$\langle L \rangle_w = \frac{\sum_{i=1}^{n} L_i^3}{\sum_{i=1}^{n} L_i^2} \tag{31}$$

The investigator should carefully consider the purposes for which fiber length data is to be acquired, for the different methods of weighting the fiber length data will yield results that differ from each other as well as from the number average. Tasman [136] observed that the average fiber lengths of fractions retained on the respective screens of a cascade-type classifier are relatively constant* and that the number average distributions of the fractions are approximately normal (Gaussian). This cannot be said for the weighted averages, which tend to have a log-normal distribution. Yan [164] has provided a method for transforming number distributions into weight distributions from classifier data and a procedure for determining the variance or standard deviation of the curve fit when the log-normal distribution is plotted. The microscopic and projection techniques yield both averages and standard deviations from data in a more direct but tedious procedure [56].

The constant coarseness approximation may be usefully exploited for a number of operations involving rapid means of data acquisition, provided it is valid for the particular pulp in question. The approximation may be quite reasonable for a kraft pulp of a species having fairly uniform fibers and be invalid for a blend of kraft and groundwood, for example.

A manual trial should always be made to provide baseline data against which the assumptions can be checked. Coarseness in pulp should be determined by one of the methods described in Chap. 10, pp. 443–445 or TAPPI T 234 [142]. A method for determining fiber coarseness in wood has been

---

*Corson and Uprichard [23] reported similar results for a horizontal classifier.

described by Britt [12]. The fiber lengths should be determined, along with widths, for at least 800 fibers, fiber fragments, or other cells at a magnification of approximately 75×, which will permit measurements accurate to the nearest 0.01 mm. It should be verified that 800 fibers are a sufficient sample size for reproducible data. From the coarseness data and the percentage frequency by which fibers, fragments, and other cells (e.g., ray cells) are encountered, one can determine the relative amount and frequency distribution by number and by weight of each component of the pulp. The overall percentage of probable error for 800 measurements to determine weighted average fiber length should be around 3%, although the frequency distribution results will generally show larger deviations for the individual points used to construct a curve or histogram. Other things being equal, the number average fiber length is subject to a higher percentage of probable error [56].

Newer Methods

In the last 25 years, a series of improvements have been made as a result of the development of new instruments, adaptations of older instruments to improve speed and accuracy, and utilization of technology developed for other branches of science. While some of the descriptions below discuss such developments principally as they apply to the measurement of fiber length and other fiber dimensions in pulp, the material on digitizing (pp. 320-321 and 331-334) emphasizes technological advances that permit these measurements to be made in the paper sheet.

Coulter Counter and Other Particle Size Analyzers An unmodified Coulter particle counter is an electronic device that measures the sizes of particles in the diameter range of 1 to 1000 μm by displacement of electrolyte as particles pass through an aperture. At any moment there is a voltage drop between the 2 electrodes in the counter, and this drop is increased by the interruption of ion flow as (approximately spherical) particles pass through the aperture. The magnitude of the interruption is primarily governed by the relation of particle volume to aperture size. The counter has analysis circuits that generate a cumulative distribution curve of particle size.

Valley and Morse [156] modified such an instrument to accept slender objects such as fibers by changing the counter from a volume-sensing to a length-of-time-sensing device. Although the fiber lengths can be determined to within ±0.1 mm, the instrument is not well adapted for extremely curled or bent fibers or for determining other fiber dimensions. However, it has been adapted quite successfully to the measurement of floc size and floc strength [37] and characterization of fines [8]. Other instrument makers, for example, Zeiss GmbH, have developed particle size analyzers. The Zeiss TGZ-3 has been used to obtain fiber length and coarseness measurements on long fiber fraction chemical pulps [39].

The most recent development in this field is the Kajaani fiber size analyzer, which senses the length of fibers drawn in dilute (0.01%) solution through a narrow capillary under suction. Fibers moving singly through the capillary pass through a beam of polarized light. Since cellulosic fibers are birefringent in polarized light, the passage of the fiber creates a discrete interval in which the birefringence occurs. In this manner the passage of several thousand fibers may be recorded in a short time span. The optical fiber length data is then treated statistically in ancillary computing equipment to provide a frequency distribution for 35 length categories.

Planimetric and Related Developments The search for more rapid and less tedious methods was concentrated, from 1955 to 1970, on improved electromechanical instruments for measuring projected fiber images semiautomatically and on electronic equipment for storing, transferring, and processing the electrical impulses into useful data.

The measuring gages that were developed in this era, based on the principle of a planimeter or map curvimeter (map reader) consist essentially of a hand-held probe with a measuring wheel at the lower end [11,55,95, 132]. The operator traces the fibers. The rotation of the wheel actuates switchgear that transforms the traced distances at the wheel surface into electrical pulses, which can then be interpreted in an analog computer [154] or other signal processor to yield fiber length distribution data, usually according to predetermined length classes. In the electromechanical type of probe switchgear, a microswitch is closed momentarily at preset intervals by cam action. If, for example, the fiber lengths are to be classified in increments of 0.5 mm, the switch will close each 0.5 mm of traced length (the circumference of the wheel is of course matched to or calibrated for the projection magnification) [132]. Another system incorporated small gears and clutch plates for 28 fiber length classes in increments of 0.2 mm [55]. Öhrn's modification [95] was to transfer the measuring wheel movement along fiber lengths to electrical pulses through a photoelectric device. A spindle with slitted drum attached is geared to the measuring wheel. A small light bulb is located within the slitted drum. Outside the drum is a photocell that receives a series of light pulses from the drum. The corresponding electrical pulses are then amplified and recorded. This development enabled faster acquisition of the fiber data.

The processing equipment used with such measuring probes varies. For example, it can consist of a relatively simple system in which the probe microswitch activates a stepping switch connected to electromechanical counters and a reset relay [132] or it may be a laboratory computer system of the type described in Chap. 13 or by Unger and Unger [154]. The system illustrated in Fig. 3, in which a device coupler is used to encode and transmit coordinate data from a digitizer to a large computer for processing, could be readily adapted to any of the planimetric measuring systems described in this section. The output should tell the operator the number of fibers measured and all other data needed to make fiber length calculations of the type described on pp. 315-317 or perform those calculations via suitable programming (see flowchart in Fig. 23).

Automatic Image Analysis The electromechanical instruments for photo-optical planimetry and data processing described above collectively represented a major advance in our ability to acquire and process fiber dimensional and related data. The next generation is represented by the semiautomatic image analyzer or digitizer, for which some description has already been given in this chapter, and the automatic image analyzer in its several configurations.

Analysis of an image can mean many things. The eye analyzes images constantly; it is most adept at pattern recognition (e.g., a signature, the

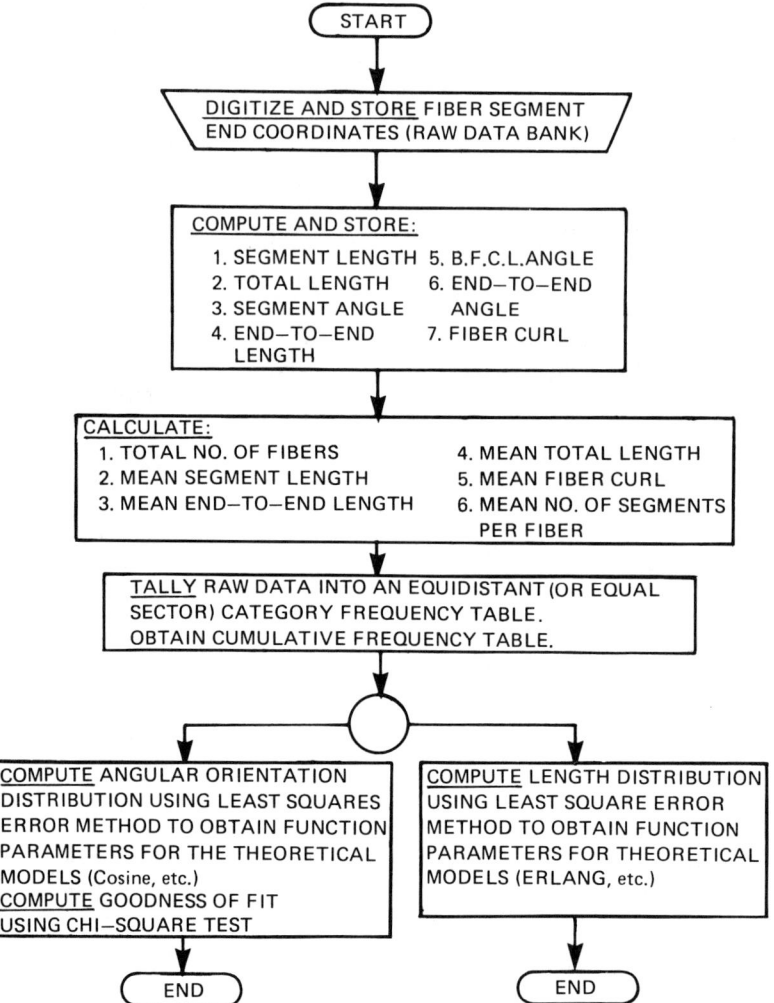

Fig. 23 Flowchart for acquisition and processing of fiber orientation and length data using a digitizing system.

facial features of an acquaintance, the cover of a familiar book) and spatial perception (e.g., distances and relative positions of objects). It is not well suited to the production of numerical information, particularly when large counts must be made. It suffers from fatigue, low productivity, and inaccuracy.

A digitizer interfaced to a computer or calculator may be thought of as a semiautomatic image analyzer. It requires manual operation of a light pen, stylus, or cursor to define the boundaries of the features to be measured on a map, photograph, or projected image, but the X,Y coordinate positions of these features are recorded and may be processed into useful data automatically by the computer or calculator. Digitizing systems have the advantages of accurate and flexible boundary detection under the guidance of the human operator, detection and separation of true versus spurious elements in the image, and low cost. Their major disadvantage is that they are considerably slower than the fully automatic image analyzers. Some commercially available systems combine digitizing and automatic image analysis equipment.

Within the past 10 to 15 years, instruments designed to make rapid, fully automatic measurements on images have been developed to provide data that would otherwise require vast expenditures of human effort. In its typical configuration, an automatic image analysis system comprises a programmable minicomputer interfaced to a producer of images, usually a microscope. The advantages include speed, convenience, and accuracy. There are also some ancillary benefits. Because of the speed, a larger area of sample may be analyzed than would be practical using manual methods. A greater variety of information can often be obtained. With the calculator or computer interface, results can be automatically formatted and processed, since the direct connection between the measurement section and the data processing section eliminates the necessity of manually copying and keypunching the data. Some image analysis systems include refinements such as automatic microscope stage motion or focusing.

Types of Automatic Image Analyzers The major types of automatic image analyzers can be categorized as (a) television (TV) based and (b) scanning electron microscope (SEM) based.

Television Based. Television-based image analysis systems are equipped with a TV camera to transform the image into electrical information from which measurement information can be obtained. Therefore, the instrument itself has the responsibility for defining the boundaries of the objects to be measured as well as for producing the desired counts and measurements.

Television-based systems can be divided (somewhat) according to their capabilities and, thus, the types of measurements they are best suited to perform.

- Systems that perform basic stereological measurements. These are systems designed to perform the basic measurements of area (or point count), projected length (or count per unit length), and feature count (Fig. 24). In addition, the distribution of intercept lengths and the distribution of maximum horizontal chord of the features are usually obtainable from these systems, which may have some manual interaction capability. These are generally the least expensive of the automatic systems; they can be very useful for obtaining data such as average particle diameters and particle concentrations, but they are generally quite limited as to their data processing capabilities, as they rarely include an integral computer or calculator.

- • Systems that perform feature-specific measurements. These systems can provide measurements of area, perimeter, longest dimension, breadth, and convex perimeter (see Figs. 22 and 24) automatically. Furthermore, such systems have the capability to make the measurements on each feature individually as well as cumulatively, so that the individual results are available for later data processing.

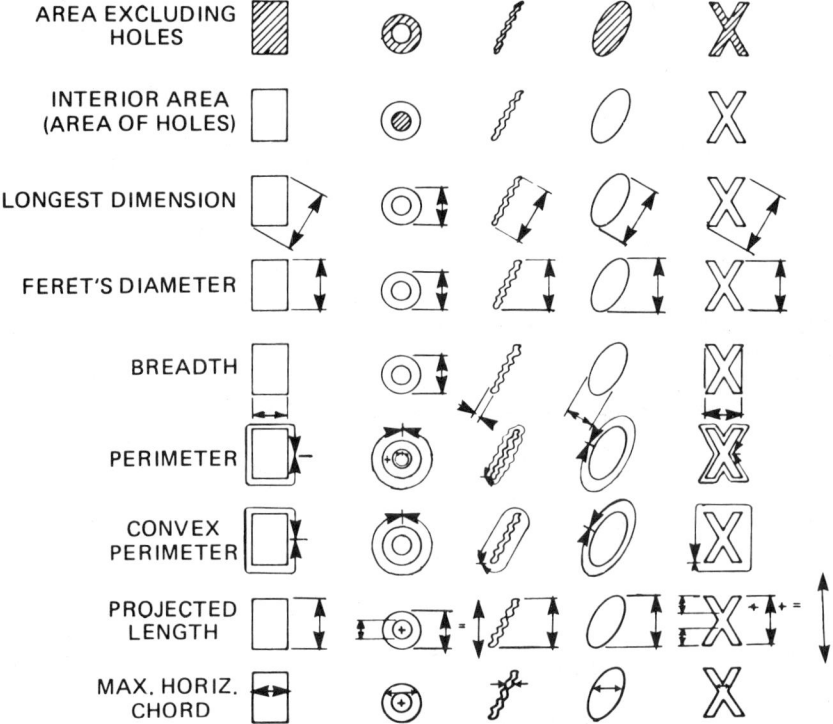

Fig. 24 Examples of commonly used measurements in image analysis produced by algorithms executed in the digital analysis function. (From Ref. 91.)

Because a large amount of data can be acquired in such a system, a computer or advanced-type desk calculator is normally included in order to handle processing and reduction of the data. Reduction operations include

Sorting results into a histogram or cumulative size distribution
Extracting statistics concerning the measurement of features
Determining those features that exceed predetermined sizes
Determining the shape factors of features in the field of view
Selecting features based on specific shape factors

The data processing capabilities of such systems may often include packaged software or operator-generated programs written in a suitable language such as BASIC or FORTRAN. Usually, the standard equipment includes an alphanumeric display and hard copy (with plotting) features (Fig. 25). Their cost is substantially greater than the basic TV-based units.

••• *Systems for size and shape identification.* In these systems, algorithms are employed that subdivide the features in the image according to shape and size either by selection of a size parameter such as longest dimension (Figs. 22 and 24) or a shape factor derived from a computation. Examples of the latter type are (1) longest dimension divided by

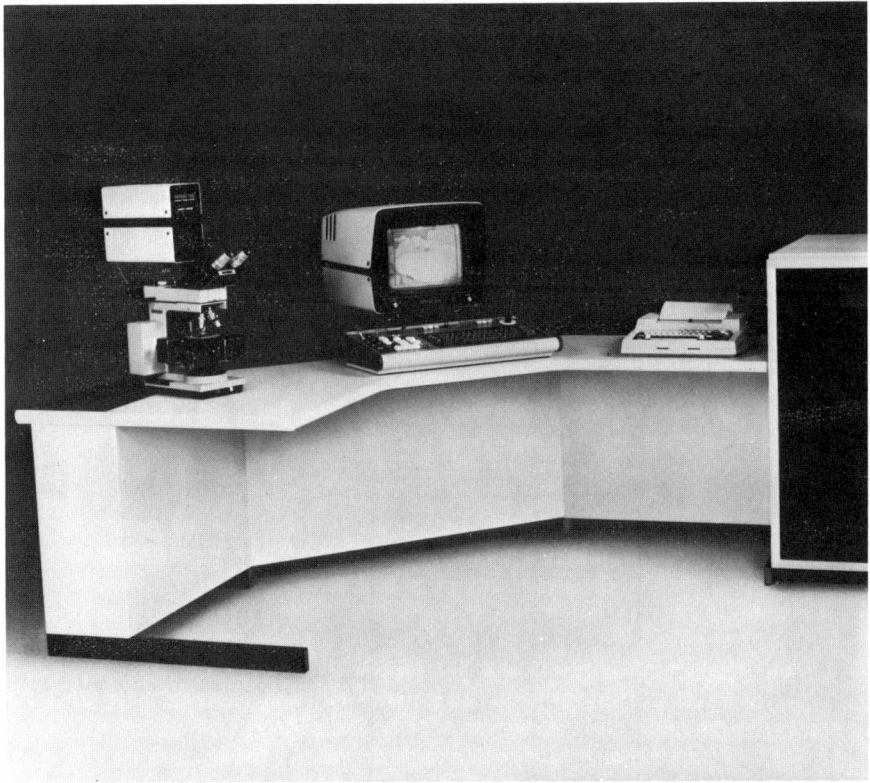

Fig. 25 An automatic image analyzer with (from left) a light microscope and television camera, a video processor and display, an output terminal and dedicated computer. (Photo courtesy Bausch and Lomb Analytical Systems Division, Rochester, New York.)

breadth (a measure of elongation) and (2) convex perimeter squared divided by area, to differentiate between two classes of features.

Image analysis systems in this group are frequently interactive, so that the operator is given a choice as to which measurements best fit the selection requirements and can change the criteria when the need arises. These systems invariably include a computer to perform the selection operations. In addition to the programming capabilities for sequencing and control that are frequently part of the preceding group, this type of TV-based system may have a special storage facility such as a disk or extended memory.

SEM Based. In these systems, the electronic video signals that represent the image are generated from a scanning electron microscope rather than a TV camera. Except for a generally slower scanning rate, appropriate to the SEM, the configuration of image analyzers of this type is the same as that required for the operation with a TV scanner.

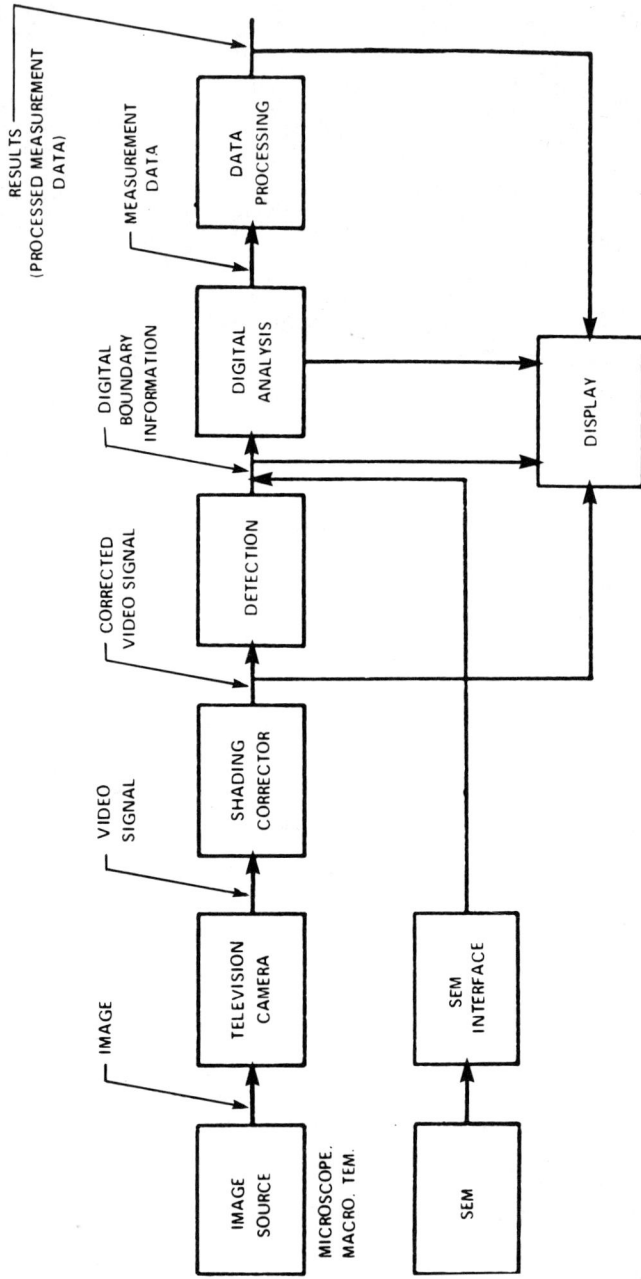

Fig. 26 Block diagram of a typical TV-type image analysis system. (From Ref. 91.)

Systems are now available that can produce from the SEM the same range of measurements and operations as would be obtained from an optical microscope but that provide the additional capability of locating the positions of inorganic and other particles in the field of view. These systems can switch the SEM to X-ray mode and direct the electron beam sequentially to the centroid of each of the particles of interest. When the centroid is targeted, the X-ray analyzer is activated and structural data for the particle are produced. The subsequent output from such systems provides not only size and shape information about the particles, but also X-ray data concerning their physicochemical makeup as well. A related development is the EDXA (energy dispersive X-ray analysis) system such as employed at the Center for Ultrastructure Research at the College of Environmental Science and Forestry in Syracuse, The Institute of Paper Chemistry, North Carolina State University [121], McCrone Laboratories, and elsewhere. The EDXA system in combination with scanning electron microscopy has the capability of providing information on the topochemical distribution of fiber cell wall constituents, on contaminants in pulp or in the air, and many other areas of interest in the field of pulp and paper research.

Basic Functions of Image Analyzers An understanding of how an image analyzer derives numerical information from an image can perhaps be best explained in conjunction with the block diagram shown in Fig. 26.

The image source for a typical system is a microscope of good laboratory quality. Other specialized optical image sources include petri-dish viewers to view macerated fibers, holographic viewers to view reconstructed holograms, and ciné viewers for movie films. Another type of image source is the transmission electron microscope (TEM), which usually requires a special type of interfacing. Also, when a TEM is used, it is employed either by utilization of micrographs or by operation in the scanning mode.

The illumination of the field is extremely important in image analysis. The most commonly used microscope illumination techniques are bright field transmitted illumination and bright field incident illumination. The use of dark field incident, which normally requires a powerful external light source, is less common. As seen in Fig. 27, the direction of the illumination largely determines the type of imaging that is achieved. When illumination is from above the sample and lies within the collection cone of the objective, the light is reflected from the specimen directly into the objective; this is bright field incident. When illumunation is outside the collection cone of the objective, it is dark field incident; and when illumination comes from below and passes through the specimen directly into the collection cone of the objective, it is called bright field transmitted.

The purpose of the television camera is to pick up the image and convert the visual information to video signals. TV cameras specifically designed for image analysis are used. They are generally referred to as scanners. Most scanners accommodate several types of pick-up tubes. The most common type is the vidicon tube, which is relatively inexpensive, can tolerate a wide range of light levels and provides good sensitivity and resolution. However, other specialized pick-up tubes are available for more demanding uses.

Although shading correction is an option in many systems, it is very desirable for fiber measurements. It ensures a uniformity of contrast

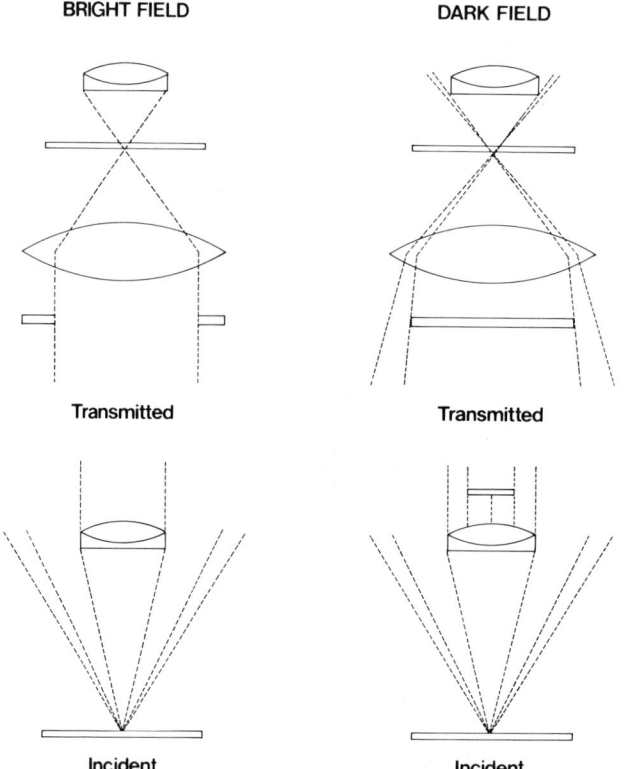

Fig. 27 Light paths for incident and transmitted illumination. (From Ref. 92.)

throughout the field of view. The system used by the Pulp and Paper Research Institute of Canada possesses this feature.

The fourth block in the diagram in Fig. 26 is labeled Detection, which refers specifically to the process of electronically determining the boundaries of the features of interest from the video signal. This is a critical step, since the objects are represented quantitatively by the list of position coordinates at which the edges of each object of interest intersect the scan lines (rasters) generated in the scanner. There are many detection techniques. The simplest is manual, wherein the boundary (of a fiber, for instance) is defined as the point where the video signal crosses some threshold as determined by the operator. Another threshold technique, called semiautomatic, defines the position of the boundary as being the point midway between the maximum and minimum values of the video transition on the boundary. There are a variety of other methods for thresholding, including thresholding based on the sharpness of a transition, sometimes called contrast or automatic thresholding. Other approaches, also related to sharpness, use differential detection techniques.

The responsibility for determining which discrete points (pixels) lie within, or without, the object boundary thus lies with the detection function. Another aspect of detection is that of distinguishing intermediate shades of gray, which requires two or more threshold levels in some applications.

The output of the detection electronics in a typical TV-based system is a set of digital (black and white) signals defining the boundaries of the fibers or other objects to be measured. The image from this set of signals is called the binary video image, and it is passed to the digital analysis function, which then determines the desired parameters of the detected features.

Whether implemented by a programmable processor such as a computer or specially designed electronics, the digital analysis function measures the size and shape factors of each feature of the binary image from its boundary information electronically. In the more advanced digital analysis systems, a variety of algorithms are used to extract measurements such as those shown in Fig. 24, and others, such as the X,Y coordinates of the centroid of an object, the moment of inertia of the object, and so on. The digital analysis function also corrects for features that partially intercept the border of the frame or field of view. All of this information is passed to the data processing function in numerical form.

Not all image analysis systems include the data processing function. In some, the data processing function is combined with the digital analysis function in a single high-speed processor or minicomputer. In others, the data processing function is implemented using a desk calculator or external computer.

In fully programmable systems, the hardwired detector is replaced with an analog to digial converter (ADC). The ADC converts the brightness of each individual pixel into a number—usually a 6-bit number—that results in assigned brightness values from 0 to n. It takes a large amount of computer memory to store brightness values for each pixel; thus intermediate storage capacity must be available. In such systems the process of detection can be programmed; it is thus possible to supersede gray level thresholding by performing different measurements on and around the detected image by program changes. In such systems the detection, digital analysis, and data processing functions are combined; processors of advanced design and large memories are required.

The need for data processing arises because many measurements are not in their most useful form. The output of the digital analysis is typically the list of measurements of each feature; the data processing function extracts the length or other size distribution from the list of numbers. Other data processing operations include calibration for magnification, performing statistical calculations such as the mean and the standard deviation, testing for homogeneity in the raw data base, and so forth.

The purpose of the TV display is to present the various images generated within the system—that is, the image scanned by the scanner, the detected image, the results of measurements, and the outputs of the digital analysis function. The operator selectively interacts with this part of the image analysis system to obtain the desired visual presentation.

328　Mark

Derived Measurements	Significance	
Area excluding holes + area of holes	Total area of feature	
Convex perimeter/π	Average Feret's diameter	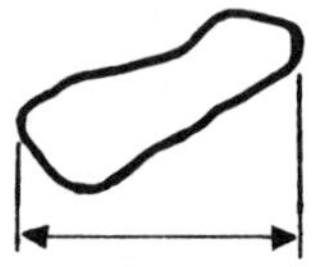
Area/longest dimension	Average feature width at right angles to the longest dimension	
$\pi \cdot$ area/perimeter	Average chord length	
$\dfrac{\pi \cdot (\text{Area})}{4 \cdot \text{longest dimension}}$	Equivalent cylindrical volume	

Fig. 28 Examples of derived size measurements from automatic image analysis. (From Ref. 92.)

Image Analyzer Operations in Fiber Length Determinations In Fig. 24, some examples of image analysis measurements are given. Examples of derived size measurements from automatic image analysis are shown in Fig. 28. When fibers are to be measured, it is advantageous to dye them intensely and use very dilute suspensions in the slide mounts on the microscope stage. The intense staining enhances accuracy in locating the fiber boundaries on each raster by providing a strong contrast between neighboring pixels in the detection function. The dilution will minimize occurrences of crossed fibers, which the automatic image analyzer would, in general, recognize only as an X-shaped single object as in the right column of Fig. 24.

Jordan and Page [61,61a] used four basic measurements—fiber area, perimeter, convex perimeter, and longest dimension (Fig. 22)—and derived from them the length, width, coarseness, and curl of the fibers. From fiber boundary coordinates sent to the digital analysis function by the detection function, their image analyzer generates the area of the fiber, its perimeter including concavities, its perimeter excluding concavities (the convex perimeter), and the length of the fiber, with its typical kinks and twists, projected onto a line. The orientation of the line is scanned at 2° intervals over 180°. The maximum projected length is specified as the longest dimension (see Fig. 22).

From these basic measurements, fiber length can be taken equal to one-half the perimeter (approximately). For greater accuracy, especially if the fibers tend to be short or square-ended, length may be computed as half the perimeter minus the width. Mean fiber width can be taken as the area divided by the length of the fiber. Because the method of preparation used by Jordan and Page transfers known weights of fibers to each slide, they are able to calculate coarseness from the total length of fiber divided by the fiber weight on each slide. Their measurement of fiber curl will be discussed in Sec. III.B.

According to Jordan and Page [61,61a], excellent agreement can be obtained between fiber lengths measured manually with a map reader from enlarged micrographs and lengths obtained by automatic image analysis. Figure 29 shows the length distribution obtained by image analysis for a commercial bleached kraft pulp that had previously been manually characterized. Very close agreement between the two methods was reported. From statistical considerations, they also found an empirical relationship between weighted average fiber length and number average length, specifically,

$$\langle L \rangle_w = (1 + a^2)\langle L \rangle \tag{32}$$

where a is a constant whose value, for their (Bauer-McNett classified) samples and from results reported in the literature, is in the range of 0.32 to 0.40.

The problem of crossed fibers has received special attention. A method called skeletonization has been employed by Taylor and Dixon [137] and Ericsson and Rudgård [36]. A computational algorithm reduces the detected widths of the overlapped fibers progressively until they are represented only by lines of unit width. At this point the crossed structure is decomposed into four segments, each with two free ends. A reconstruction of the segments that belong together is accomplished by matching these segments to form best fits on geometric grounds. Once the path of each individual fiber

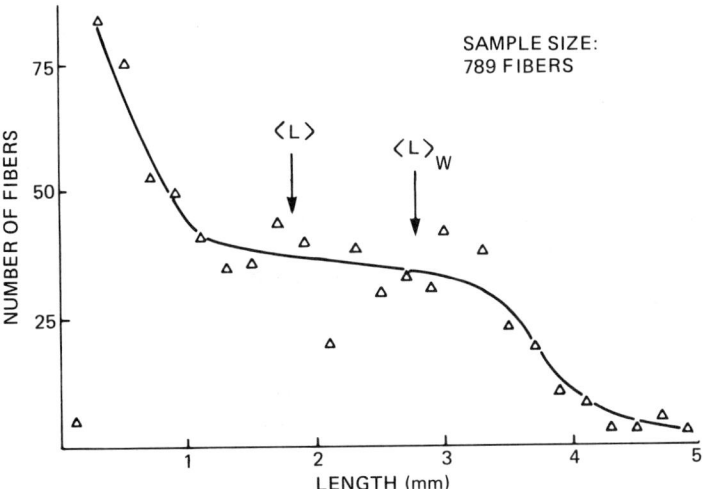

Fig. 29 The length distribution for a commercial bleached kraft pulp. The weighted average $\langle L \rangle_w$ and the number average $\langle L \rangle$ are shown by arrows. (From Refs. 5,61a.)

has been located, fiber length can be defined as simply the distance traced by a path along its successive constituent segments. Skeletonization does introduce some artifacts in measurement that have to be accounted for when other parameters—width, for example—are to be determined.

The measurement of fiber width, or average diameter, which is discussed further on pp. 357-358, is performed in automatic image analysis systems generally by either a *diameter analysis* or a *grid analysis* method. The software programs that are used in the two methods differ significantly. In diameter analysis, the program instructs the scanner to find a particle, then finds the center of gravity of the particle and performs a set of diameter determinations from the center. This method works fairly well for straight fibers and other more or less regularly shaped particles. However, a detection problem may arise in the case of kinked or highly curled fibers.

The grid analysis method is based upon the construction of a grid or network of lines that cover all solid parts of an object regardless of shape or whether internal holes lie within it. The grid encounters boundaries of the object at points, and uses an approximation to the Pythagorean theorem to calculate distances. A recent version of such a grid analysis program developed by LeMont Scientific uses 64 points* on the periphery of a fiber. A large amount of information is generated in such a system, so that mass storage of approximately 40,000 bytes is needed. The types of output that can be obtained include length, width, length/width ratio, surface area, surface/volume ratio, area and volume equivalent diameters, and estimated ratio of perimeter to area.

Advantages and Limitations of Image Analyzers In their present (evolving) stage, automatic image analyzers are useful where large numbers of pulp fibers must be detected and measured. Several stringent criteria must be met. Among these are the following:

- Fibers are well defined, by appropriate staining and illumination, to provide high optical density and a large signal-to-noise ratio.
- • The instrument is calibrated so that light intensity and other factors are fully reproducible.
- • • The instrument is not required to detect and measure at or below the limits of resolution of any of its optical components.

The automatic image analyzer can perform some functions, for example, width and coarseness determinations, with greater speed and accuracy than other available methods. Length measurements can certainly be made faster, but there is an out-of-plane dimensional component that may be significant in pulp suspensions as a three-dimensional system is reduced to two-dimensional analysis. The third dimension may, in some cases, create detection as well as measurement difficulties. Detection of fibers in the paper sheet is extremely difficult if not impossible for an automatic image analyzer, since the embedded fiber—even though stained intensely—appears as an object (or objects) of greatly varying optical density as crossing fibers partly obscure it. Furthermore, the appearance of identically stained fibers is vastly different at different depths in the sheet. For related reasons (of poor con-

---

*This number can be increased or decrease programmatically.

trast between objects of interest in the field), automatic image analysis does not lend itself well to quantitative fiber bonding studies (see Chap. 25).

In summary, automatic image analysis lends itself best to situations in which fibers and other objects are clearly identifiable, as in pulp suspension. Already there is progress in developing these systems for on-line analysis (and feedback) of fiber measurements as part of paper pulp production control [36,44]. Automatic image analysis is currently deficient in making such measurements in paper sheets, where the detection function is best done with the human eye. In such cases, a graphic digitizer interfaced with a high-speed programmable computer offers the fastest system for accurate, reproducible results.

Digitizing A short discussion has already been given, on pp. 319–321, between the fully automatic image analysis systems and a semiautomatic system such as digitizing. A digitizing system has been shown photographically and schematically in Figs. 2 and 3, and an example of flowcharting for fiber length and orientation measurements has been presented in Fig. 23. Digitizing systems, with or without associated storage and computing facilities, graphics display screens, printers, video interfaces to allow tracing directly from video images, and so on, are commercially available from several manufacturers. Also available is interactive computer software to provide users with necessary data acquisition, analysis, and storage capabilities.

Many early digitizers operated only in point mode, which required the operator to locate discrete points on maps, photographs, and other images with the pen or cursor. Newer systems can operate in either point or continuous line mode. Both have advantageous uses. The point mode type enables (with appropriate electronics) the calculation and presentation of:

Coordinate locations
Distances between points
Area enclosed by discrete points
Total number of points entered

In continuous line mode, which is in reality a dense spacing of discrete points, the cursor or pen is moved steadily along lines or other features of interest in the image, with the result that continuous coordinate information is acquired in the data base. This information enables the calculation of:

Individual point coordinates
Line length
Area and perimeter of closed curves
Form factors of closed curves
Centroids and moments of inertia of plane figures
Other parameters of the types shown in Figs. 24 and 28.

It must be borne in mind, however, that these operations are slower (generally, much slower) than with automatic image analysis. On the other hand, the detection function is under the operator's personal control, thus reducing spurious or amibguous detection and ensuring that no features of interest are missed. The relative merits, in regard to accuracy and precision, of digitizers and automatic image analyzers will vary greatly according to the type of image source and the nature of the features to be measured.

A graphic digitizer lends itself well to the detection and measurement of dyed fibers in a sheet of paper, as illustrated in Fig. 4 and discussed on pp. 289–291. The measurement of fiber or fiber segment length *in the sheet*— which is difficult, tedious, or impossible with various other methods—can be efficiently accomplished even in fairly thick sheets. For light-to-medium basis weight sheets composed of mostly bleached fibers, the method of oil impregnation described on pp. 288–290 works very well to highlight the scattered dyed fibers to be measured. For sheets composed of unbleached or colored fibers, sheets with coatings or fillers, sheets of high basis weight, and any other type of sheet that does not permit observation by transmitted light after oil impregnation, it is necessary to employ one of the sheet-splitting techniques [88,105,149] in order to provide thin enough layers for observation. Refer to Chap. 27 for additional information on sheet splitting.

The fiber F in Fig. 4, shown diagrammatically in Fig. 5, can be measured in several ways. Based on the digitized coordinates of its endpoints and the intermediate points acquired in continuous line mode operation, it is possible to calculate fiber straight line end-to-end length, fiber segment length, and total length based on summing of segment lengths. Fiber curl can also be calculated from the data so accumulated [28,109].

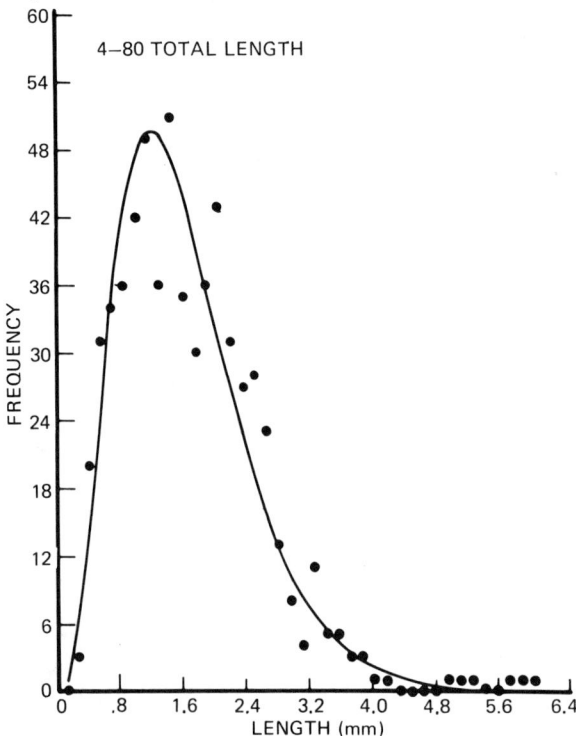

Fig. 30 Frequency data for fiber total length measurements in 0.2 mm increments compared with an Erlang distribution.

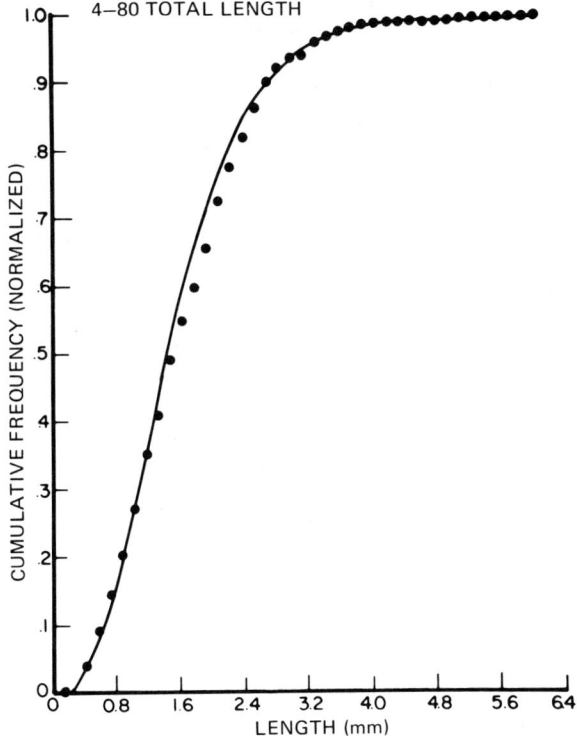

Fig. 31 Cumulative frequency data for fiber total length measurements compared with cumulative Erlang distribution.

Typically, the type of length distribution data that are generated is in the form of a skewed curve when lengths are plotted versus frequency of occurrence. The data will generally match quite well to a type I or type III frequency curve in the Pearson system [35]. Fiber length distributions can in most cases be fitted readily to an Erlang distribution function [51], for example, using the least squares error method.

The Erlang probability density function is

$$f(x) = \frac{(x/b)^{c-1} \exp(-x/b)}{b[(c-1)!]} \tag{33}$$

where

b = scale parameter, always > 0
c = shape parameter, an integer > 0
bc = mean value of x

The range of this function is determined by $0 \leq x \leq +\infty$.

Figures 30 and 31 show the data points for total length distribution and cumulative length distribution for a sample size of approximately 600 dyed fibers in a bleached softwood kraft sheet of 80 g/m² basis weight. The Erlang function (solid line) fits the data according to the chi-square test [Eqs.

(9—11)] with a probability of 0.87% (Fig. 30); for the cumulative frequency, the probability that the curve fits the data is 94.7% by a chi-square test. Other probability functions, for example, Rayleigh [51], do not generally match fiber length distributions as well as the Erlang function.

The above results indicate that the sample size should be increased somewhat, in this case probably to approximately 800 fibers, to yield a better fit of data to function. In many cases, a sample size of 600 fibers will be sufficient with this method.

The procedure used for such samples in the author's laboratory is to cut samples 32 mm × 19 mm (1.25 in. × 0.75 in.), impregnate them with silicone oil, and insert them in the film holder of a standard 35 mm photographic enlarger. The illuminating lamp of the enlarger projects the image of the dyed fibers directly onto the digitizer tablet (Fig. 2) for coordinate data acquisition by cursor. It should be emphasized that although a film holder is used, no film is involved. What is projected onto the digitizer tablet is the image of all the dyed fibers *in the sheet itself*, since the oil impregnant renders them visible regardless of the depth at which they are located within the sheet, as shown in Fig. 4. A rear projection digitizer may be useful for measurement of slide-mounted pulp fibers or fibers in thin sheets [57,155].

B. Fiber Curl

Fiber curl, also called curliness, may exhibit itself by directional changes in the plane of the sheet, as in the case of the dyed fibers shown in Fig. 4. Any out-of-plane deviation observed as the fiber passes over and under crossing fibers, or as it folds and kinks in drying or mechanical straining, may also be considered as a component of fiber curl. Out-of-plane curl is quite evident in the synthetic fiber network shown in Fig. 14. The role of fiber curvature and curl in paper properties has been examined theoretically and experimentally by several investigators with somewhat divergent conclusions [61,68,79,98b,106,109]. Its importance is highly variable, as an examination of the electron micrographs in the atlas of Parham and Kaustinen [104] will readily reveal. Fiber curl in a well-refined softwood kraft formed into a dense sheet is far less prominent than in nonwoven lens tissue made essentially of highly curlated rayon fiber, for example. These and other paper structures of wood fibers, other plant fibers, synthetic fibers, mineral fibers and blends are all illustrated in the aforementioned reference.

In-plane curl has been given much greater quantitative assessment in the recent literature; considerable future effort needs to be applied to the measurement of out-of-plane curl, in order to provide better data for the evaluation of Z-direction properties and other paper structural parameters.

In-Plane Curl, Bending Factor, and Curvature If one examines fiber F in Fig. 4, it is possible to measure its in-plane curl by the calculation mentioned on p. 313 or according to several other criteria. With reference to Fig. 32, if the straight-line end-to-end distance AB is referred to as L_1, the longest straight-line dimension AC is referred to as L_2, and the total fiber length (contour length) is designated by L, then curl can be variously and arbitrarily defined by relations such as the following:

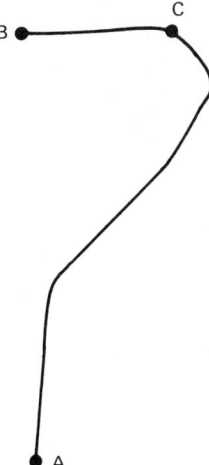

Fig. 32 In-plane length measurements made on a bent fiber and used for calculation of fiber curl.

$$\text{Curl I} = \frac{L_1}{L}$$

$$\text{Curl II} = \frac{L_2}{L}$$

$$\text{Curl III} = \frac{L}{L_2} - 1 \tag{34}$$

$$\text{Curl IV} = 1 - \frac{L_1}{L}$$

It can be noted that numerical curl factors that employ definitions I, II, and IV will always be ≤ 1. Isenberg [56], Kibblewhite [74], Kallmes and Corte [68], and Perez and Kallmes [106] used the reciprocal of Curl I, similar to the usage employed in polymer chain terminology (ratio of total length to linear end-to-end distance), and thus their curl factors are always ≥ 1. Isenberg refers to this reciprocal as the *bending factor*. Curl III can be any number from zero up. It is a measure of the fractional increase in the straight-line end-to-end length of a fiber that would result if the fiber were completely uncurled but not stretched. This definition of curl is widely employed in the textile field and is known as the *crimp ratio*.

As fibers become straighter, curl factors I and II increase and factors III and IV decrease. It is evident that in evaluating the literature, one must be careful to know clearly how the investigator has defined curl. When experiments are designed, thorough consideration should be given to the feasibility of measurements and the information that needs to be extracted from the data. For example, if a manual or digitizer method is to be employed, any of the four curl factors can be ascertained directly on the image, but the operation will proceed faster if curl I or curl IV is selected because the fiber ends are so much easier for the operator to identify than the longest dimension (L_2).

On the other hand, fiber end recognition to obtain L_1 is difficult and more error prone in the case of automatic image analysis, whereas the longest dimension L_2 (Λ in Fig. 22, distance AC in Fig. 32) is a relatively easy measurement, executed in the digital analysis (or ADC with memory) function.* There are ways to approximate curl I or curl IV using fast logic circuitry, however. Jordan and Page used the relation

$$\text{Curl I} = 2 \times \frac{\text{convex perimeter}}{\text{simple perimeter}} - 1 \qquad (35)$$

where the (simple) perimeter and convex perimeters are as shown in Fig. 22, as such an approximation. Determination of curl I on the image analyzer was used by them as an aid in detecting crossed fibers [61,61a]. They used curl III as an evaluator of different refining and moisture regimes in pulp, and felt it might well be related to wet web extensibility.

Fiber curvature, like curl, can be measured (to some extent) by the manual and image analysis techniques that have been described, although its definition is arbitrary as well. Few fibers assume the shape of a smooth circular arc, for which the radius of curvature is defined as the perimeter length divided by the subtending angle. The normally irregular shape of a fiber can be defined by the inverse of the "radius," where the subtended arc length can be the end-to-end length, an identifiable segment of length, or the contour length. Curvature thus defined is usually measured in units of radians per millimeter; the measurement tends to give too much weight to waviness in the fibers. Perez and Kallmes [106] have discussed the relationship between their *curl factor* (inverse of curl I) and fiber curvature.

Out-of-Plane Curl The measurement of out-of-plane curl is restricted to examination of fibers within sheets that have been embedded and sectioned for microscopic examination or photomicrography or electronmicrography. While any of the manual or semiautomatic techniques that are available could be used, the most powerful tools are the computer-coupled graphic digitizers that have been described in this chapter. Further information on digitizing methods is given in Sec. IV.

In a cross section of a sheet such as Fig. 33, one can see what Page et al. [98] referred to as the undulatory structure of fibers in paper—out-of-plane deformations that may be considered in terms of wavelengths. These wavelengths can range in size from several fiber thicknesses down to molecular dimensions. The fiber labeled L in Fig. 33A and B exhibits a deformation of relatively long wavelength, whereas the fiber labeled S has a deformation whose wavelength is shorter than the fiber is wide. Short-wavelength (crimped) out-of-plane deformations are considered especially

---

*It should be borne in mind that automatic image analysis is not generally suitable for quantitative measurement of fibers in sheets. The comments pertaining to fiber curl measurements in the image analyzer refer to measurements on pulp fibers.

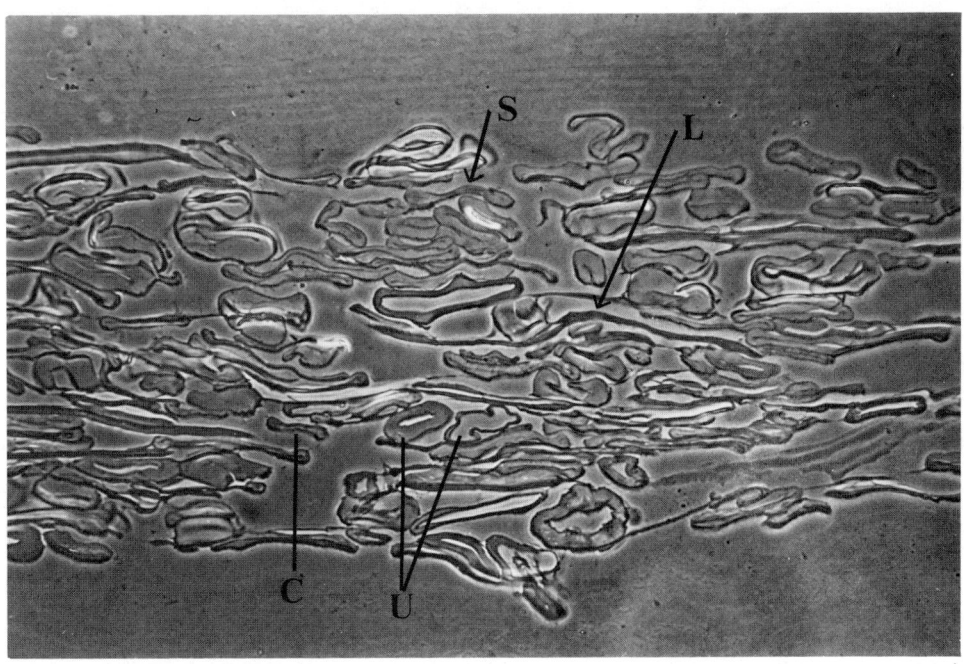

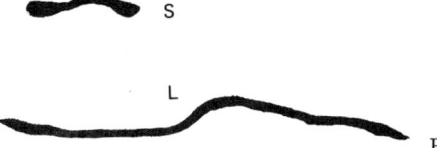

Fig. 33 (A) Cross section of 80 g/m² bleached softwood kraft embedded in epoxy resin and sectioned at 1.5 μm. C - collapsed fiber; U - uncollapsed fibers. (Photo by A. Eusufzai.) (B) Ink drawing of designated fibers of short (S) and long (L) wavelength undulations visible.

significant from the standpoint of paper shrinkage [99] and Page et al. offer polarized light microscopic evidence that the fibrillar orientation in the fiber walls is disarranged around these locations [98].

Long-wavelength undulations may significantly affect the flexural properties of paper and board, especially at or near failure [16]. A single undulation may be approximated as a segment of a circle; the ratio R of its circumferential length to the distance between its ends is

$$R = \frac{r\theta}{2z \, (\tan(\theta/2)} \tag{36}$$

where

r = radius of curvature
θ = subtended angle
z = maximum out-of-plane local displacement of the fiber

One should take parenthetical note of the fact that, while the flexibilities of two fibers of the same width and curvature might be quite similar in the plane of the sheet, those same two fibers might have very dissimilar flexibilities in the out-of-plane or Z direction. As noted in Chaps. 7 and 10 particularly, bending stiffness is a function of the cube of the thickness. Since out-of-plane curl is often of interest from the standpoint of bending deformations of the sheet, it is important that fiber thicknesses as well as curvatures be ascertained in a manner that does not result in statistical bias toward the more readily visible, easy to measure thick fibers.

IV. OTHER DIMENSIONAL PARAMETERS USED TO CHARACTERIZE FIBER RAW MATERIALS AND FIBER NETWORKS

The relationship of fiber coarseness, fiber shape, and fiber dimensions to tensile and other strength indices, folding endurance, porosity, density (and its reciprocal, bulk), opacity, and other sheet properties has been the subject of many studies and reports [4,30,52,54,98c]. There have also been many studies and reports related to the influence of the fiber fragment and nonfiber components in paper [7,8,13-15,78,103,125,134]. While the most useful data concerning fiber parameters normally come from measurements made on sheets, the expenditure of technician time in the dehydration, embedding, sectioning, mounting, photography or projection, identification of features, and manual recording of fiber measurements is generally very substantial for these operations. The advent of the graphic digitizer and other semiautomatic measurements by projection techniques has made it possible to reduce the time requirements for the recording of measurements. Although these instruments often provide the capability for huge savings in time and increased reliability of data, the proper preparation of the specimens still represents a substantial effort with good equipment by skilled personnel. Much of the actual data, therefore, on fiber dimensional parameters have been obtained (1) from fibers in wood or other source material, (2) from fibers isolated from pulp, or (3) from fibers that have been removed from sheets by hydropulping or some other mechanical action. In the case of the fiber fragment and nonfiber components of paper, our ability to gather quantitative dimensional data has been greatly enhanced by all of the recent developments cited on pp. 318-334, but especially by scanning electron microscopy [104,159], SEM combined with image analysis, and the particle counter combined with a special drainage apparatus [153].

For a discussion of fiber coarseness, the reader is referred to pp. 444-445 of Chap. 10 and the paper by Britt [12]. Other dimension-related parameters, especially those related to fiber bonding and fiber surface area, have been extensively treated in other chapters, such as Chaps. 16, 17, and 25.

For those parameters not covered elsewhere, I feel that this book requires some explanation of the most significant terms encountered in the dimensional properties testing field, combined with some brief exposition, either by discussion or by reference, that will help the reader locate or develop the specific technique or information needed. I call this "Notes on Procedures for Various Structural Parameter Determinations," which I hope will serve its intended purpose.

NOTES ON PROCEDURES FOR VARIOUS STRUCTURAL PARAMETER DETERMINATIONS

Area, Contact (See Contact area, Crossing area, and Bond area)

Area, Gross Cross-Sectional
Area, Lumen
Area, Net Cross-Sectional

These three terms refer to the space occupied by a fiber that is viewed in a transversely cut section. The area of fiber cell wall material is the net area after the area occupied by its hollow center (lumen), if any, is deducted from the gross area. The ratio of the net to the gross cross-sectional area is often termed the Mühlsteph ratio (which see) in some references, especially those dealing with wood. In wood, these areas are often measured on slide-mounted sections by photometric densitometry [45] or by a grid intersection or dot-counting method [81,87]. Automatic image analysis will probably assume the task of much of this type of determination in the future [90].

When individual fibers are measured, a planimeter or a cut-out weighing method is often used [87]. Hardacker determined net cross-sectional area of pulp fibers by compaction [48]. The compacted shape is that of a flat ribbon with rounded edges and cross-sectional area equal to $ab - [(4 - \pi)/4]^2$, where a is width and b is thickness.

When either wood or individual fibers are embedded and sectioned for mounting for purposes of cell wall area determination, it is important that the sections be thin (never over 5 µm and preferably < 1 µm) and even more important that embedding media such as methacrylates not be used, since they cause swelling of the specimen. A comparative study of 35 embedding/mounting combinations by Crosby and Mark [27] concluded that the epoxies are superior for embedding because expansion artifacts are virtually undetectable with these materials. Studies by Berlyn and Miksche [9] and Isenberg [56] support this conclusion.

The embedding technique developed by Quackenbush [113] is advantageous in that it eliminates the necessity for solvent dehydration of the specimen. Thus, possible chemical or physical alteration of the fibers is avoided. Also, the hardness of the resin medium may be varied to suit the particular material to be sectioned. Diamond or glass knives should be used for the sectioning. Clear epoxy resins are recommended for both embedding and mounting if the slide is to be permanent; if not, the choice of mounting medium is principally related to the desirability of matching the refractive indices of the embedding and mounting media as closely as possible. There are a number of commercially available mounts suitable for this purpose that can be subsequently dissolved to permit reuse of the slide.

Area, Projected This is the total area that is occupied by all the fibers contained in a known mass of pulp if they are laid out singly in a field of view. The projected area is obtained by adding the individual values of area for each fiber sampled, that is, length times width. It is not correct to calculate this area by multiplying total length by mean width or total width by mean length. The objective of ascertaining projected area is to find the maximum surface coverage that would be achieved with a known quantity of a given fiber stock. It will, of course, be different for different conditons of moisture, temperature, and pressure.

Table 2 Comparison of Thermomechanical Pulps Modified with Oxidizing Agents

Pulp	Bonded surface (contact) area of fibers (%)	Aspect ratio	Moments of inertia (mm⁴) (I_x)	(I_y)	Bonding state probability	Apparent density (kg/m³)	Tensile index (Nm/g)	Tear index (MNm²/g)
TMP no. 1 modified with								
(Untreated)	16.2 ± 11.1	3.0	NA[a]	NA	2.33	393	41.5	4.54
3.0% O_3	27.6 ± 12.9	3.1	NA	NA	2.27	452	48.9	7.37
5.8% O_3	29.4 ± 13.8	3.1	NA	NA	2.61	526	62.4	6.52
7.5% O_3	31.1 ± 10.9	3.7	NA	NA	3.01	531	62.9	6.38
TMP no. 3 modified with								
(Untreated)	21.2 ± 14.8	2.6	8.3×10^{-9}	3.1×10^{-8}	1.41	317	32.3	11.73
3.0% O_3	18.7 ± 12.9	2.5	7.1×10^{-9}	2.2×10^{-8}	1.51	337	38.5	12.24
4.2% O_3	29.2 ± 19.9	2.8	5.0×10^{-9}	2.4×10^{-8}	1.81	395	45.9	9.94
5.8% O_3	29.5 ± 17.2	2.8	6.2×10^{-9}	2.7×10^{-8}	2.11	402	48.7	9.48
CH_3COO_2H								
0.44 mol/100 g OD pulp	18.0 ± 13.9	2.7	5.4×10^{-9}	2.5×10^{-8}	1.57	386	43.6	10.14
0.50 mol/100 g OD pulp	23.7 ± 14.0	3.1	3.9×10^{-9}	2.3×10^{-8}	1.90	475	54.8	8.14
0.56 mol/100 g OD pulp	29.0 ± 17.7	3.4	3.8×10^{-9}	2.7×10^{-8}	2.24	480	65.0	8.35
1.8% ClO_2	21.6 ± 15.1	2.4	5.4×10^{-9}	2.1×10^{-8}	1.48	342	37.5	11.82
3.0% ClO_2	22.0 ± 15.5	2.8	8.7×10^{-9}	3.2×10^{-8}	1.59	367	39.8	9.81
7.0% ClO_2	27.8 ± 17.2	2.9	4.2×10^{-9}	2.3×10^{-8}	1.94	405	43.6	8.93

[a]Not available.
Source: From Ref. 117.

Aspect Ratio This is the ratio of the major to the minor axis of a fiber or other feature of interest (shive, particle, etc.) in a transversely sectioned sheet. Aspect ratios are most easily determined by digitizing photomicrographs of serial or sequential sections of paper [117,167]. Aspect ratio determination provides the investigator with information on fiber flexibility and collapse (which see) and orientation; thus, aspect ratio values will, in general, be different for sections cut normal to the machine (MD) and cross machine (CD) directions. Table 2 illustrates the way Rothenberg and Fernandez [117] utilized the aspect ratio and other sheet geometry parameters to evaluate the flexibilizing (and therefore enhanced bonding) action of various oxidizing agents on TMP handsheets as part of a larger study on the effects of bleaching, refining, and so forth, on different furnishes. It can be seen that aspect ratio correlates positively with tensile index and negatively with tear index. Aspect ratio in sheets can usually be determined more quickly and reproducibly than degree of collapse (which see) because it is less judgmental and therefore operator-dependent. It has been found that photomicrographs obtained with phase contrast illumination give much sharper boundary definition than those obtained with other methods such as polarized light. Good illumination and section thickness (not over 5 µm) are critical to this determination. See also Roundness.

Bond Area When two fibers are in optical contact (see Contact area), it is assumed that the forces of adhesion between the fibers are distributed within the optical contact area. Optical contact areas are illustrated in Fig. 7, Chap. 25. The forces, principally those of hydrogen bonds in the case of wood pulp fibers, may or may not be uniformly distributed in that area; however, the problem of hydrogen bond distribution within a fiber bond area has to this date, been intractable. For further understanding of this problem, the reader is referred to Refs. 6 and 99–102 and Chap. 25.

Bond Distance
Bond Region

These terms are with reference to the Z direction in the sheet, or normal to the fiber axis. When assemblages of fibers bonded to shives, fibers bonded to fibers, or fibers bonded to other materials are tested so as to pull the fiber from the substrate in sliding shear, the resultant degree of elongation at the contact area is often substantial. For example, Thorpe et al. [150] found an elongation of about 1 µm, equivalent to a bond strain of about 0.5%. Such deformations are far above the magnitude of hydrogen bond distances. For this reason, the "bond region" between fibers may be more appropriately thought of as some greater distance. The distances between S1-S2 interfaces of wood pulp fibers in contact with each other may be thought of as a bond region, or adhesive layer, whose properties will differ from those of the bulk of the fiber wall. Nordman et al. [94] have pointed out that increases in bonding strength related to beating time may reflect transitions from a state of bonding involving only S1 surfaces to those involving more contacts between S1 and S2 or S2 and S2 as the beating fibrillates and disrupts the fiber S1 layer (see Fig. 13 of Chap. 28). The length or depth of the bond region and the properties of that bond region are therefore variable with pulp treatment. Measurements of the bond region are best done under very high resolution light microscopy, either directly on the microscope stage, or by digitizing with high-quality micrographs. The detection of these layer thicknesses is best done by alternating the observation of the specimen between bright field and plane polarized light. A compensator (red I type) may assist in detection. Further information on the layered structure of plant fibers is given in Chap. 10.

Bonded Area, Percent
Bonded Area, Relative
Bonded Area, Total

As defined in Chap. 25, the relative bonded area (RBA) is the ratio of the total bonded area to one-half of the total external surface area of component fibers. The concept of RBA is thus tied to the idea that sheets are assemblages of two-dimensional fiber planes. As shown in Fig. 6 of Chap. 25, the possible states of bondedness of any given fiber at a point along its length are (1) that it is bonded to another fiber on one side; (2) that it is bonded to other fibers on both sides, or (3) that it is unbonded. The total bonded area is of course, the sum of the bonded areas on the fiber surface. Bonded area may also be expressed as a percent of the total surface area of the fiber.

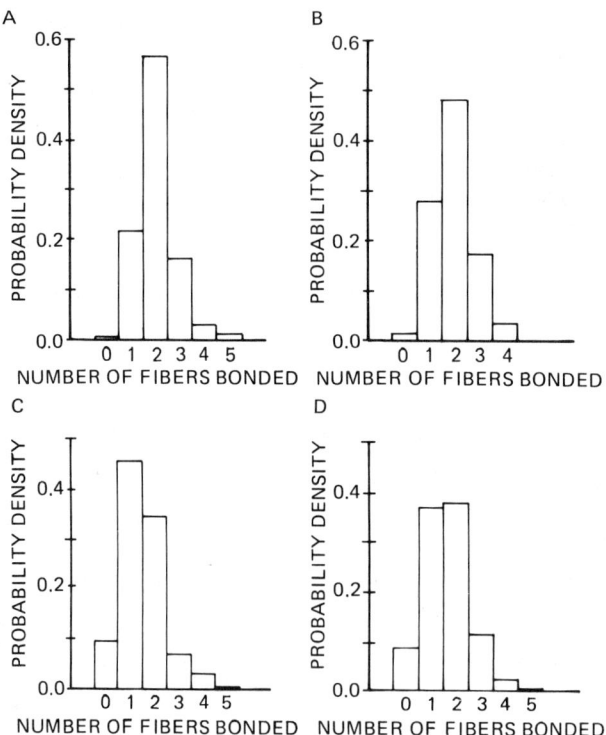

Fig. 34 Bonding state probability (number of fibers bonded to a crossing fiber within a 4 µm section of the crossing fiber). (A) Kraft CD. (B) Kraft MD. (C) Bond paper. (D) Newsprint.

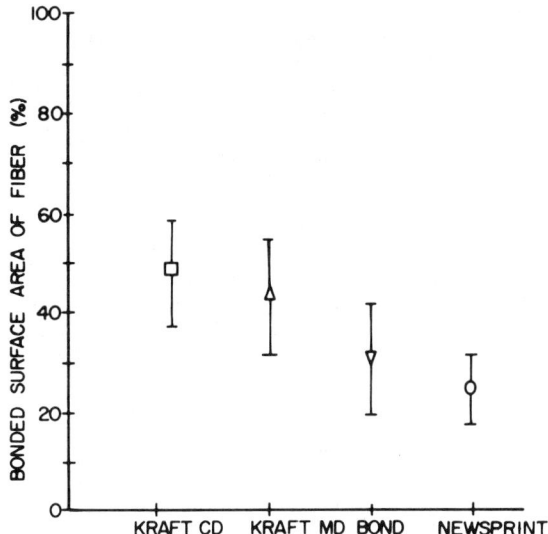

Fig. 35 Bonded surface areas of fibers in 4 types of paper (mean values ± one standard deviation).

According to the RBA concept, developed by Kallmes and coworkers [65,69], RBA has a maximum potential (RBA_{max}) that is constrained by fiber width, fiber coarseness, and sheet basis weight. Uesaka has discussed the limitations of this concept in Chap. 25. Figure 5 of Chap. 25 demonstrates that bonds vary in nature and kind, and that they have a component in the third dimension. Yang et al. [167,168] have recently shown by a digitizer technique that fibers in paper sheets are actually bonded to more than two fibers at a point in a large number of cases and this is shown in Fig. 34A to D. By their technique, they also determined the bonded surface area of fibers directly on the same sheet cross sections, as shown in Fig. 35.

The determination of percent bonded area of fibers by Yang et al. was done by calculating the ratio of the sum of that portion of the fiber perimeter in contact with other fibers in a section to the entire perimeter of the fiber, then summing up those ratios for all the sections for each fiber measured. It is assumed that the bonded portion of the fiber perimeter is constant through the section thickness, which should as thin as possible—preferably less than 1 μm. The calculation is expressed by

$$B = \frac{1}{N_f} \sum_{i=1}^{N_f} \frac{\sum_{j=1}^{N_s} S_b}{\sum_{j=1}^{N_s} S_t} \tag{37}$$

where

B = percent average bonded surface area of fiber
N_f = number of fibers whose trajectories are followed through successive sections

N_s = number of sections used for measurements
S_b = perimeter length of fiber within area of contact with other fibers
S_t = total perimeter of fiber

If one compares Figs. 34 and 35 one can see that when the bonded surface area of kraft fibers reaches 50%, there is essentially no free fiber length (which see) left; at virtually every point along the fiber length, at least one other fiber is in contact (bond area is assumed equal to contact area for such measurements). In contrast, an RBA of 50% in Fig. 6 of Chap. 25 results in a theoretical free fiber length of over 20%. The digitizer results of Yang et al. merely confirm what was reported much earlier by Page et al. [101]—that when pulp is well beaten and fibers conform well, then at virtually any point along the fiber length, the fiber is bonded at one or more points on its surface to other fibers. It is important for the investigator to bear in mind that the results of many tests such as those described in Chap. 25 are often interpreted on the basis of the multiplanar Kallmes model, despite the contradiction of that model with structural measurements such as those that have been mentioned.

Bonded Fiber Surface Area This term is best expressed as a percentage of the total surface area of all fibers measured. In Table 2 the results of some digitizer measurements by Rothenberg and Fernandez [117] given in the first column of data include the mean values plus or minus one standard deviation. The deviation indicates that a larger sample size might be advantageous in this direct method. The direct method of microscopic examination and indirect methods—optical, gas adsorption, and so on—are described in Chap. 25.

Bonding State This is the term used to describe the bondedness of a fiber at any point along its length. Conceptually, one must visualize a fiber with a series of circumferentially drawn lines on it and infer from data whether one or more crossing fibers are bonded at that imaginary line. From digitized data, Yang et al. [167] prepared the histogram of probability distribution for different types of sheets shown in Fig. 34. Similar histograms were developed by Rothenberg and Fernandez [117]; mean values for several types of thermomechanical pulp sheets are given in Table 2. Figure 34 may be considered an experimental bonding state diagram; Fig. 6 of Chap. 25 is a type of theoretical bonding state diagram, as discussed under Bonded area. For experimental work, it is critical that sections of sheets be made as thin as practicable, and never over 5 μm; otherwise, depth of focus limitations will prevent good resolution of the contact surfaces between fibers and introduce spurious data.

Brushing out (See Fibrillation)

Centroid
Centroidal Axis

The location of the center of gravity of a line, a plane area, or a solid is called the centroid. An axis through the centroid parallel to a coordinate axis is called a centroidal axis. Algorithms are available for automatic image analysis that locate the centroid of a feature (such as the fiber shown in Fig. 35, Chap. 10) with respect to either coordinate axis. The object can be of any shape that will fit in the scanned image.

Circularity Used primarily to evaluate cross-sectional shapes of textile fibers (see for example Ref. 109a), the circularity or fullness is determined by the ratio $4\pi A:P^2$, where A is the cross-sectional area measured with a planimeter and P is the fiber perimeter distance as determined by a map measuring wheel on enlarged photomicrographs. See also Roundness.

Collapse
Conformability

Collapse represents change of cross-sectional shape in a fiber from its "original" state, for example, in the stock chest, to its generally more flattened appearance in the sheet. It results from a combination of surface tension, which causes the opposing fiber walls to draw together as water leaves the fiber, and external forces of compression on the sheet, principally those developed in the press and drying sections [112]. Page et al. [98] examined fibers in sheet cross sections and arbitrarily classified as collapsed any fiber whose inner walls touched along more than half the length of the lumen.

Kallmes and Eckert [70] used a 50% criterion in an entirely different way. They viewed individual fibers in extremely thin sheets of the type shown in Fig. 36 under dark field illumination microscopy. Those fibers that were generally dark across more than half of a scan line traversing the fiber were considered collapsed; the brighter fibers—less than 50%

Fig. 36 Very low grammage "two-dimensional" sheet showing many collapsed (flattened) fibers and significant free fiber segment lengths. (Courtesy Empire State Paper Research Institute, Syracuse, New York.)

dark—were considered uncollapsed. Higher percentages of dark, collapsed fibers were found in sheets pressed at high pressure. Page [96] investigated the collapse behavior of individual pulp fibers by covering them with a nonpenetrating medium (cellophane tape adhesive) that had a refractive index similar to the fiber surface. In this way, those fibers (or portions thereof) that were collapsed appeared invisible in transmitted light; open lumina in the uncollapsed areas were revealed by reflection and refraction at the cellulose/air interface. With such a procedure, fibers can be classified as to their susceptibility to collapse upon drying.

Higgins et al. [52] adopted a criterion for collapse based on the relationship of lateral force F (the combined force of surface tension and press application) to conformability, which has been defined [2] as the ability of the fiber to conform to the shape of a crossing fiber, for example, as in Fig. 5 of Chap. 25. The conformability of a fiber by the Higgins et al. definition is the reciprocal of the load C required for collapse; they define the degree of collapse as the proportion of fibers in which $1/C > 1/F$. Collapse in this view is an all-or-nothing phenomenon, and the value of C is correlated directly to the Luce shape factor [83] for the fibers in the original wood. It is the opinion of the writer that except for the pulp fiber experiment of Page [96] (which did not involve fibers in sheets), these definitions impose arbitrary structural appearance or shape criteria that are unnecessary in the light of today's data-gathering equipment. The definition of Page et al. [98] is too dependent on operator judgment and requires a great expenditure of time for sufficient data acquisition. The Kallmes and Eckert procedure, which is dubious on theoretical grounds, is restricted to extremely thin sheets or sheets that have been peeled aprat, and peeling certainly causes some internal disruption. It is also very time consuming if a sufficient sample is to be obtained.

The Luce shape factor (which see) does not adequately predict fiber collapse, since it is based on circular shapes for both fiber wall and lumen—features seldom possessed by real fibers. Whether a fiber collapses or not may be influenced by its location within the sheet, for example, by its proximity to a thick-walled fiber. The two uncollapsed fibers marked U in Fig. 33 have markedly different wall thicknesses. The one on the right would certainly have a much lower original Luce shape factor than the collapsed fiber marked C close to it. Accordingly, for studies on sheets it is felt that the aspect ratio (which see) is a more suitable paramter than collapse or conformability. It can be obtained rapidly via digitizing or other semiautomatic means, it is much less prone to operator judgment, and it appear to be sensitive in relating structure to properties (Table 2).

Contact Area Although the concept of two fibers making contact with one another seems simple enough, the contact area is very difficult to measure satisfactorily. It is possible to observe the formation of optical contacts in paper under controlled moisture conditions by vertical (incident) polarization microscopy with the use of a porous plate apparatus (Fig. 37); Page and Tydeman [100] carried this experiment out very elegantly. They considered that bonding between fibers occurred when dark areas (the optical contact areas) appeared in the fiber crossing when the fibers were viewed normal to

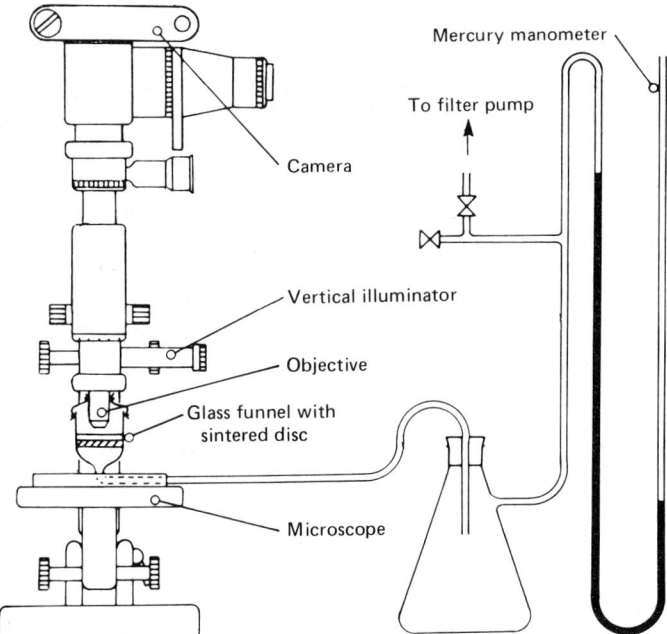

Fig. 37 A porous plate apparatus modified to follow bond formation in a drying web. (From Ref. 100.)

the sheet plane (see Bond area). For most purposes, bonding in sheets is more easily measured by the surface contact that appears when sheets are embedded and sectioned transversely, as in Fig. 33A. A semiautomatic method such as digitizing will produce information on bonded (contact) surface area and bonding state probability quite rapidly and efficiently (Table 2). It is essential to make thin sections to avoid artifacts. Figure 33A was cut at 1 µm with a glass knife. Phase contrast illumination works best in most cases. The relation between the area of optical contact by light and SEM microscopy and the actual bond area is discussed by Algar [2].

Count This is the number of fibers present in a unit mass, usually of pulp. A standard procedure for determination is to disperse 0.5 g of dyed pulp in 1 liter of water, then further subdivide and dilute until the suspension contains 10 µg (if softwood) or 5 µg (if hardwood); known amounts of the suspension are pipetted onto slides, where the dyed fibers are counted under the microscope. No fibers or fiber segments less than 0.1 mm are counted. By proportion, the total count is usually given as number of fibers per gram. Typical counts might be 5,000,000 for a softwood such as pine, and 25,000,000 for a hardwood such as oak. Dependent on the degree of processing, the count may be increased very slightly or more than doubled, for example, for highly beaten, bleached, low yield, or degraded fibers. With improved procedures, particle counters or image analyzers should be able to perform these measurements.

Crill (See Fines)

Crimp Ratio This is the fractional increase of the linear extent of a fiber as it undergoes a change from a curled to a fully straightened configuration without stretching. It is expressed mathematically as curl III in Eq. (34).

Crossing Area If one considers Figs. 1, 2, and 7 of Chap. 25, it is evident that the crossing area S in Fig. 2 is not equivalent to the contact area (which see), since fiber contact is often partly obstructed and the projected overhang of the curved fiber edge on the neighbor fiber in plan view results in a larger overlap than the actual contact area. In fact, fibers that cross may not be in contact at all. Thus, the ease of making this measurement (on fibers lying near the sheet surface) bears little relation to the usefulness of the data.

Crushing A state of fiber compaction in which the fiber is pressed beyond collapse (which see), with resultant mechanical disruption of the fiber wall layers and the microfibrillar structure, and plastic deformation [112]. Occurrences of crushing may be observed and counted by microscopy.

Diameter (See Width)

Dislocations These are areas of the fiber wall that show evidence of abrupt changes in direction of the cellulosic microfibrils in the S2 layer (see Fig. 38). Dislocations, also called *slip planes*, may not be evident unless the fiber is viewed under polarized light. They tend to be formed at a characteristic angle [72]. They are the result of excessive axial (compressive) forces and are formed because of shear instability (sliding delamination) between the sheets of microfibrils. Chipping, refining, sectioning, and other mechanical action on fibers increases their frequency of occurrence [40,72, 74]; however, sometimes they are present in the standing tree [32,72]. Dislocations may be counted on (a) a per fiber, (b) per unit length of fiber, or (c) per unit mass of fiber basis, under the polarizing microscope. Dislocations often occur concentrated together. Kibblewhite [74] calls these concentrations "zones of dislocation" and believes that they are added to or extended through the development of local compression failures at sites of bond formation as the fibers shrink and mutually restrain each other during drying. Such zones he considers identical with the microcompressions (which see) described and photographed by Page and Tydeman [99].

Felting Coefficient The ratio of fiber length to diameter or width (which see).

Feret Diameter A statistical diameter (which see) defined as the mean length of the distance between two tangents on opposite sides of the image of a particle (see Figs. 24 and 28). This parameter is readily obtainable by automatic image analysis.

Fiber Composition The different types of fibers present in a sheet of paper or fiber stock can be expressed in terms of fractional proportion. A count (which see) is made of each fiber type present, determined typically by the anatomical characteristics of the fibers and their staining reactions; each type is then expressed as a percent of the total.

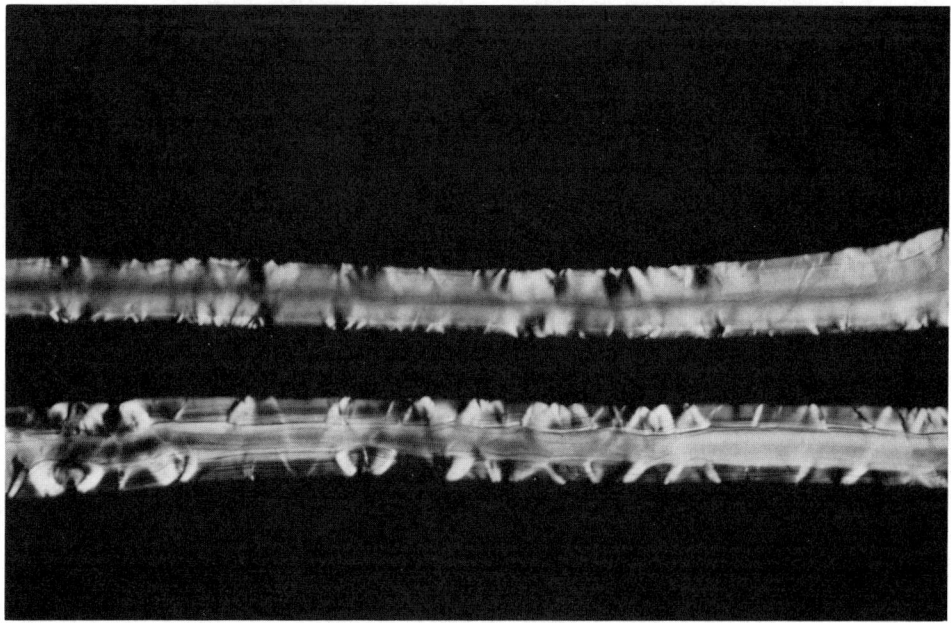

Fig. 38 Intense concentrations of dislocations in the fiber walls of a softwood, *Pinus radiata*. These fibers have undergone sufficient refining to create dislocation zones, but not enough to cause visible fibrillation. The upper fiber is a thin-walled springwood tracheid; the lower summerwood tracheid exhibits some evidence of possible delamination in the microfibrillar structure. (Courtesy of R. P. Kibblewhite, Forestry Research Institute, Rotorua, New Zealand.)

Fiber Segment This item can have more than one meaning and, therefore, more than one measure. As seen in Fig. 5B and C, the fiber marked F in Fig. 4 can be subdivided into five segments of markedly different orientation in the sheet. Measurements of length can easily be made on such fiber segments via a projected image or photograph on a digitizer tablet. Another operator, however, might distinguish additional short segments and thus divide the fiber further. From the standpoint of network mechanics (Chap. 2), it is the ability of the fiber to transmit axial load that determines segment length. Thus, one or more dislocations (which see) in an unbonded length might increase the tabulation of segments; conversely, fiber curvature through a densely bonded region may be treated satisfactorily as a single structural element, thus decreasing segment tabulation. Fiber segment length has also been used as a synonym for free fiber length (which see).

Fibril Angle The predominant helical angular orientation of the microfibrils in the fiber wall. The term is usually used with reference to the S2 layer. For further details on fibril angle and its measurement see Chap. 28 and pp. 424–431 of Chap. 10.

Fibrillation The mechanical actions to which papermaking fibers are subjected, particularly those in the refining stage, result in the loosening of microfibrillar elements from the fiber wall and provision for greater fiber-to-fiber bonding surface as a result of the fibrillation. Fibrillation of fibers can be observed most clearly in the scanning electron microscope.

Filament Winding Angle (See Fibril angle and Micellar spiral angle)

Fines The *Dictionary of Paper* [3] describes fines as "very short pulp fibers or fiber fragments and ray cells. They are sometimes referred to as flour or wood flour." A closely related term is *crill* [130], which is ordinarily used to designate the extremely fine particles (microfibrils and wall fragments) that are abraded from the surface of cellulosic fibers during refining. Thus, the term *fines* is somewhat more inclusive. Fines or crill can be quantitatively determined by fractionation through some type of filtering system or plug flow separation combined with weighing or microscopy. Further definition of fines fractions and a relatively new method for their determination is given in the next section. The retention of fines in the sheet has been shown to affect many of its properties [13,54c].

Flexibility Frequent attempts have been made to relate the shape of the fiber, for example, in the original wood, to the amount of collapse and conformability (which see) that the fiber undergoes in web formation [4,30,31]. The ratio of lumen diameter to exterior fiber diameter has been termed the *flexibility ratio* and also the *coefficient of flexibility*. However, Horn [54] felt that the ratio of fiber length to cell wall thickness could be used as a quantitative index to fiber flexibility.

Robertson et al. [116] devéloped a device that evaluated the flexibility of fibers on the basis of their rotational motion in fluids such as corn syrup and silicone oil; 100 fiber samples were subjected to laminar shear and observed by stereo microscope. A *flexibility index* was calculated for each fiber type on the basis of weighting the observed orbital motions by number and class. Thus, three different approaches to flexibility, two anatomical and one an evaluation by physical test—none of which involve measurements on fibers in sheets—have been used. The first of these, the flexibility ratio (coefficient of flexibility) has received the widest acceptance and usage and is one of the eight basic wood anatomical characteristics for papermaking suitability criteria used by the UN Food and Agricultural Organization (FAO) [38]. For further information on fiber flexibility, see pp. 463—471 of Chap. 10.

Free Fiber Length
Free Fiber Length Aspect Ratio
Mean Free Fiber Length

Considerable attention has been given in the paper physics literature to the concept that fibers in a sheet have segments of their length that are bonded to other fibers and segments that are not bonded. In extremely thin sheets (e.g., Fig. 36) or low-density materials such as tissue and nonwovens, this is a reasonable distinction. For typical sheets of moderate and high density, it was first shown microscopically by Page et al. [101] and subsequently by image digitizing by others [117,167,168], that virtually the whole of the fiber length is bonded at some point on its periphery. Thus the concept of "free" or unbonded fiber segment length does not have much relevance to most grades of paper and board. A distinction should be made between the actual unbonded distance (the free length) and the distance between midpoints or centroids of adjacent bonds, especially as the frequency of contacts (see Contact area) increases. Algar [2] and others term the midpoint distance the *fiber segment length*, but the free fiber length has also been described by that term. Kallmes and Bernier [66] discuss several definitions of free fiber length. Fiber segment (which see) also has several other meanings. Finally, it should be noted that, for different purposes, free fiber length could refer to the maximum, minimum, or average length between fiber contacts (Fig. 39A). Kallmes and Bernier [66] define the mean free fiber length as the mean total free fiber length per fiber divided by the mean number of free fiber lengths per fiber. Thus, the mean is affected by the defined choice of free fiber length. In Fig. 39B, the distance D bears a statistical relation to the free fiber length distribution. In another paper, Kallmes and Bernier [67] use the term *free fiber length aspect ratio*, which they mean to be the ratio of free fiber length to fiber width in their simulation study of idealized fiber networks. In this glossary, the term *aspect ratio* (which see) refers to actual measurement of major and minor axes of fiber cross sections.

Free Fiber Surface Area The percentage of the total surface area of a group of measured fibers that is not bonded to any other fiber; 100% less the bonded fiber surface area (which see).

Fullness (See Circularity)

Inertia (See Moment of inertia)

Kink An abrupt change (sharp bend) in the direction of a fiber, usually caused by beating or other mechanical action on pulp fibers. In Figs. 5B and C, four kinks are recognized, dividing the fiber into five segments.

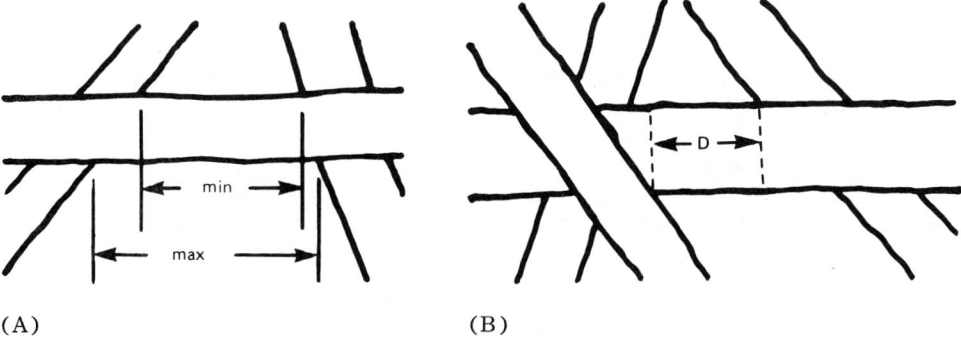

(A)   (B)

Fig. 39 Schematic of fibers crossing. (A) The minimum and maximum free fiber lengths are shown. (B) A free minimum distance D; the mean of such distances equals one-half the mean free fiber length by derivation in Ref. 66.

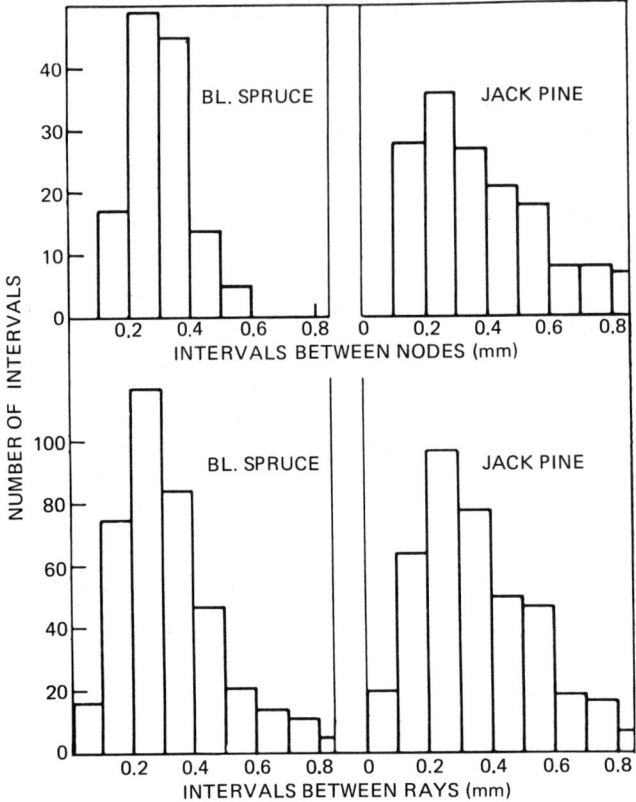

Fig. 40 Distribution of intervals between adjacent nodes in tracheids of 48% yield spruce and 48% yield jack pine sulfite pulps compared with the distribution of intervals between ray crossings in the corresponding woods. (From Ref. 40; used with permission. Copies available from TAPPI, Technology Park/Atlanta, PO Box 105113, Atlanta, GA 30348.)

Forgacs [40], who used the term *nodes*, did an extensive microscopic study on their cause and occurrence; he concluded that their frequency of occurrence was related to the spacing of rays in the pulpwood, and their cause was the relative weakness of the S2 layer in the vicinity of the rays (see Fig. 40). In his description of the nodes, Forgacs included zones of dislocation (which see) that could serve as sites for kink formation. See also Microcompression.

L/D Ratio
L/W Ratio

The ratio of fiber length to width. It is one of the eight basic wood fiber characteristics considered as significant for papermaking by the FAO [38]. The determination of length and of width is discussed on pp. 313—334 and pp. 357-358, respectively.

L/T Ratio This is the ratio of fiber length to cell wall thickness, which Horn [54] gives as "a quantitative index to fiber flexibility" (see Flexibility).

Luce Shape Factor Luce [83] made a series of tests on tubes of various shapes (illustrated in Fig. 32 of Chap. 10) and found that the lateral force required to collapse the tubes was proportional to $(d_o^2 - d_i^2)/(d_o^2 + d_i^2)$, where d_o and d_i are the outside and inside diameters of the model. This shape factor has been used by others, such as Higgins et al. [52] to evaluate hardwood fibers for collapse (which see) in papermaking.

Major Axis
Minor Axis

As defined by Yang et al. [167], a fiber viewed in a sheet cross section has as its major axis the maximum distance between any digitized coordinates lying on the perimeter of the fiber; its minor axis is taken as the distance across the fiber normal to the major axis. The information is used to calculate the aspect ratio (which see) for the fiber in the section.

Martin Diameter A statistical diameter (which see) defined as the mean length of a line that intercepts the boundary or perimeter (Fig. 22) of the image of a fiber, particle, or other object and divides the image into two portions of equal area. It is a dimension that can be obtained with an appropriate algorithm via automatic image analysis. The bisecting line is parallel to or coincident with the scan lines, irrespective of the orientation of the subject in the image.

Mean Bond Area The determination of the bond area (which see) for various pulps yields characteristic means for such measurements. For example, Page et al. [101] found microscopically that mean bond area of a bleached spruce sulfite increased by over 60% when beaten to 310 mℓ Csf.

Micellar Spiral Angle This is an archaic and erroneous term for the predominant helical orientation of the microfibrils in the S2 layer of the plant fiber wall. More acceptable terms are fibril angle, S2 angle, and filament winding angle. The determination of this parameter is discussed in Chap. 28.

Microcompression A localized compressive deformation or wrinkling of a fiber such as that shown diagrammatically in Fig. 41. Microcompressions can be introduced into fiber structure by the processes of refining, bleaching, and other mechanical actions, especially those that induce curlation. The phenomenon has also been ascribed to longitudinal contraction of one fiber caused by transverse shrinkage of a crossing fiber by Page and coworkers [99,126]. Dumbleton [33] introduced such deformations into fibers artificially (see pp. 460-461 of Chap. 10). It has been observed that the load-elongation properties of sheets having a high proportion of such fibers are

Fig. 41 Microcompressed fiber.

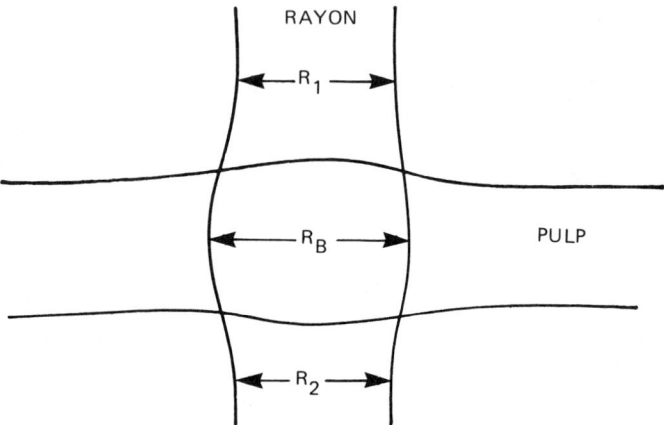

Fig. 42 Diagram to show measurements made to determine the degree of necking at a bonded crossing between a pulp and a rayon fiber. (From Ref. 100.)

similar to those exhibited by the deformed fibers [33,98b,98c,126]. Microcompressions have been studied and measured thus far by light and electron microscopy in surface views. The morphological difference between a microcompression and a slip plane or dislocation (which see) is that the dislocation is localized within the microfibrillar structure (as in Fig. 38), whereas the microcompression takes the form of three closely spaced kinks of the fiber (see Kink); both microcompressions and dislocations are phenomena associated with or induced by axial compression of the fibers.

Moment of Inertia A discussion of fiber moment of inertia and the computation of its magnitude is given in Chap. 10, pp. 463–466. A determination of the probability density or relative probability of finding fibers with a given moment of inertia in sheet cross sections was made, using data obtained by digitizer, by Yang et al. [167]. Rothenberg and Fernandez [117] have found this calculation useful in the interpretation of mechanical behavior of sheets, as shown, for example, in their data in Table 2. Its significance was also recognized by Duncker et al. [34] and Schniewind et al. [124] in their studies on stiffness properties of fibers.

Mühlsteph Ratio This is the ratio of net to gross cross-sectional area (which see) of the fiber. It is principally used with respect to the dimensions of fibers as they exist (unmodified) in wood, as a means of evaluating the suitability of the wood for technical uses such as papermaking. It can be calculated by any of the techniques mentioned under Area, cross-sectional. In the future it will probably be determined principally by automatic image analysis [90].

Necking When certain dissimilar fibers bond together, for example, wood pulp and rayon fibers, a characteristic enlarged neck is created (Fig. 42). Page and Tydeman [100] assessed this phenomenon microscopically and evaluated the degree of necking N by the relation

$$N = \frac{2R_B}{R_1 + R_2} - 1 \tag{38}$$

where R_1, R_2, and R_B are as shown in Fig. 42.

Node (See Kink)

Perimeter The perimeter of a fiber viewed laterally is often difficult to measure because of the very high ratio of length to width in many fibers. If the fiber is viewed at a magnification small enough to contain the entire fiber, the width (which see) is small and hard to measure. However, the boundary or surface perimeter and the convex perimeter (Figs. 22 and 24) can both be determined by planimetry, digitizing, and automatic image analysis under appropriate conditions—specifically, that boundary resolution is preserved as a low-magnification image is projected and enlarged for measurements to be performed on it. The convex perimeter corresponds to the distance required to wrap a string around the fiber (see Fig. 22). As such, it can be used to derive the Feret diameter (which see) shown in Figs. 24 and 28. The boundary perimeter can be used to describe shape (for example, perimeter squared divided by gross area) of fiber surfaces in cross sections and also is useful in obtaining certain other derived size measurements as indicated in Fig. 28.

Roundness The *roundness factor* of pulp fibers in cross section is defined by Isenberg [56] as the ratio of the gross area of the fibers to the area of their circumscribed circles. Thus, the roundness factor of the uncollapsed fiber on the left in Fig. 43 is greater than the collapsed fiber on the right. The factor has been used as a measure of relative collapse (which see) of the fibers of different pulps. Three other measures of fiber roundness are given by Isenberg: (a) the ratio of the gross fiber area to the area of a rectangle in which it is inscribed with the greatest width and thickness of a fiber parallel to the sides of the rectangle; (b) the ratio of the area of the cross section of the fiber to the area of a circle having a circumference equal to the perimeter (which see) of the fiber; (c) the ratio of the width to thickness of the fiber cross section. Definition (b) is identical to that for circularity (which see). Definition (c) closely approximates the definition of aspect ratio (which see), which the author believes to be the most useful measure of

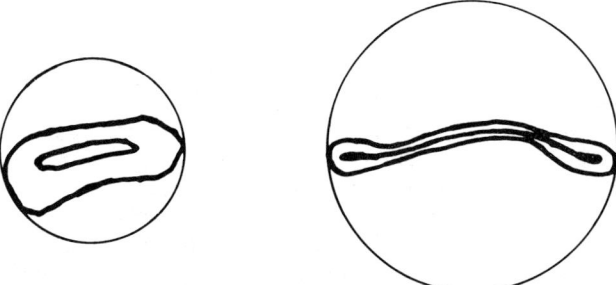

Fig. 43 Two fibers, one similar to one of the fibers marked U in Fig. 33A and the other similar to the fiber labeled S, in circumscribed circles.

fiber collapse in sheets and which requires less human or machine computation. Except for (b) and (c), the roundness factors described above do not lend themselves well to rapid techniques such as digitizing.

Runkel Ratio This is the ratio of twice the fiber cell wall thickness to the lumen diameter. It is principally used with respect to the dimensions of fibers in wood for evaluating the suitability of a species for pulp wood. It is one of the eight basic wood characteristics reported by the FAO [38] in regard to the properties of plantation wood for papermaking. In general, values less than 1 are sought, although it has been pointed out [30] that there is no material improvement in using this ratio instead of the density of the wood as a predictor; the Runkel ratio and the basic density should have a roughly constant proportionality. Densitometry [45] or fluorescence [59] measurement techniques have been used, but the determination seems amenable to automatic image analysis [90].

Segment (See Fiber segment)

Shive
Sliver

The *Dictionary of Paper* [3] defines *shive* as "a bundle of incompletely separated fibers which may appear in the finished sheet as an imperfection," and a *sliver* as "a small splinter of wood, longer than a shive and of smaller diameter in proportion to its length." Mechanical pulp usually contains shives and tends to have slivers much more often than does chemical pulp. Methods for their determination are given in the next section. They are readily observed in sheet cross sections, so data concerning them can be digitized efficiently.

Slip Planes (See Dislocations)

Specific Surface This is defined as the exposed area per unit mass of moisture-free fiber and thus provides a measure of the degree of fibrillation. Since its measurement is covered in Chap. 17, it will not be discussed here. It is important to distinguish between this term, which refers to the external surface, and *specific internal surface*, which relates to the fine capillary system within the fiber wall.

Statistical Diameter This is the statistical average of a specified parameter such as Feret or Martin diameter (which see) from microscopic or image analysis.

Thickness The thickness of the fiber cell wall in wood can be measured by a stage micrometer on a microscope [9], by densitometry [45], or by fluorescence [59]. Isolated fiber wall thicknesses may be determined by pressing part of the fiber between transparent anvils as discussed on p. 449 of Chap. 10, and making the measurement by variable permeance transducer [48] or interference microscopy [71,97]. In paper sheets, the measurement is best made by digitizing the fibers that lie along scan lines drawn on micrographs of sheet cross sections in a statistically acceptable sequence, as was done by Kimura and Mark [77].

Twist A freely dried pulp fiber undergoes extensive twisting. However, bond formation in sheets prevents extensive twist. Therefore twist as meas-

Table 3 Effect of Recycling on Fiber Dimensions: Scotch Pine Kraft Pulp

	Fiber diameter (μm)	Lumen diameter (μm)	Single wall thickness (μm)	
	Simple recycling			
Never dried vs. Dried	41.8 32.4	20.6 12.0	10.6 10.2	N.S.[a]
Never dried vs. Rewetted	41.8 34.0	20.6 N.S. 12.6	10.6 10.7	N.S.
Dried vs. Rewetted	32.4 N.S. 34.0	12.0 N.S. 12.6	10.6 10.7	N.S.
	Deinking recycling			
Never dried vs. Dried	36.1 29.4	15.1 9.5	10.5 10.0	N.S.
Never dried vs. Rewetted	36.1 32.6	15.1 12.4	10.5 10.0	N.S.
Dried vs. Rewetted	29.4 32.6	9.5 12.4	10.0 10.0	N.S.

[a]No significant difference at 95% confidence interval is indicated by N. S.
Each number represents the average of measurements on 40 fibers.
Source: From Ref. 89a.

ured in fibers (as described in Chap. 10, pp. 471-478) is not appropriate as a measure of this parameter in sheets. In Fig. 36 several fibers are seen to be twisted in a low-density, thin sheet. These twists can be counted under the microscope or on the image, but the (rarer) twists that occur in denser, thicker sheets are hard to quantify. In a large number of serial sections, one can count twists of fibers where they occur by noting the orientation of the major and minor axes (which see) of a given fiber in cross section, which rotate as the serial sections pass through the twist.

Unbonded Fiber Segment Length (See Free fiber length)

Width A width of a plant fiber generally changes substantially from its roughly polygonal shape in wood or other plant tissue to its generally more compacted shape in the paper sheet. The width and other dimensions change again upon recycling (see Table 3). Pulp fibers shrink laterally upon drying, as was demonstrated in an elegant microradiographic technique by Tydeman et al. [152]. If one is measuring original fiber diameter, examination of the wood or plant tissue in cross section is appropriate. For embedding media, see Area, cross-sectional. The techniques of stage micrometry [9], fluorescence [59], densitometry [45], and projection micrometry [56] are all available; however, the future may lie with determination by automatic image analysis [90]. Individual wet fiber widths have been determined by projection micrometry of undried, stained fibers; the procedure is given by Iseberg [56]. Essentially the same procedure has been employed for cotton fibers [50]. Individual dry fibers may be measured as freely dried or compacted (if it is desired to ascertain the lumen-free cross-sectional area) by employing an image-splitting eyepiece [48]. Such an eyepiece was used to measure the fiber diameters in Table 3. The width of fibers in sheets can be done by stage or projection micrometry, planimetry or semiautomatic image analysis (digitizing), the speed of data acquisition generally being in ascending order among those methods. Jordan and Page [61] have made determinations of pulp fiber width by automatic image analysis, using two separate approaches designed to overcome the problem that fibers are generally too long to fit in the field of view if an adequate magnification is maintained for accurate width determination. The problem of width measurement of very fine fibers (asbestos fibers) has been dealt with by Taylor and Dixon [137] by taking brightness profiles at many points along the length of the fiber, but always in a direction perpendicular to the local tangent to the fiber. The reason for using brightness information rather than measurement of width directly is that the latter procedure yields inaccurate results owing to the fact that the image of an asbestos fiber is rarely more than a few pixels wide. With brightness information it is possible to interpolate between pixels and obtain a more continuous measurement of width. Uncharacteristically large values (evidence of crossed fibers) are automatically deleted from the data base. A further discussion of width determination in the automatic image analyzer is given on p. 330.

V. DIMENSIONAL CHARACTERIZATION OF SHIVES, SLIVERS, AND FINES

This section deals with particles of fibrous nature that are present to some extent in most wood pulps and mixed pulps. Shives and slivers, which are defined in the preceding section, consist of fiber bundles and their associated cellular elements (axial and ray parenchyma, vessels, etc.) that were present in the original wood or plant vascular tissue. They are accordingly always larger than fibers, whereas the fines fractions are smaller. At the present time, these fractions are principally described by size classes, numbers of particles present, and weight proportion.

A. Shives and Slivers

Several useful methods have been published by TAPPI that contain procedures for determining shives in pulp by screening or fractionation.

UM 240 [146] calls for a dispersion of about 2.5 lb ($\sim$ 1.13 kg) of dry pulp in 66 gal (300 ℓ) of water to be slurried onto a screen, which retains any shives and slivers present. The procedure takes over 30 min. The retained particles are oven dried and weighed to obtain the result as a percent of the oven dry weight of the whole pulp.

UM 241 [147] relates to a device that fractionates by varying the width of a slot or orifice through which stock at 0.3 to 1.0% consistency passes. A particle analyzer (digital counter) with a recorder monitors the stock flow. Although this device does not provide an exact measure of the shive and sliver content either by weight or number, it is widely used, since it is relatively fast and gives the operator a control guideline for pulp quality.

UM 242 [148] prescribes the dilution of about 100 g of pulp to 0.5% consistency. Samples that comprise about 25 OD g pulp are added to water that is passed under pressure through a porous screening plate, which retains the shives and slivers. The fibers and fines can be collected from the overflow. The procedure requires about 20 min. The shives are reported as a percent of OD weight of the whole pulp.

The shives content of a flash dried pulp bale or other dry lap can be determined by counting the number of fiber bundles observed (by stereomicroscope) in two 60 g/m^2 handsheets prepared from a reslushed sample taken from the bale. The prodecure is described in UM 239 [145]. The number of shives per 100 g of moisture-free pulp is reported.

Since shives are readily observable in sheet surfaces (and to some extent beneath the surface) because of their size and (frequently darker) color, there is certainly potential for rapid determination of their size and occurrence in sheet surfaces by automatic image analysis in the future if such procedures have not already been done. Dirt specks, halftone dots, and other dark particles can be observed in light-colored sheets easily this way.

In sheet cross sections, shives and other large particles are distinguishable without difficulty to a trained technician. The size, perimeter length, aspect ratio, moment of inertia, percent bonded area, and bonding state of such fiber bundles and aggregations can all be determined with the use of a graphic digitizer, as was done by Rothenberg and Fernandez [117] for various high-yield pulps. The more classic methods of particle size and number determination by microscopy, microprojection, and hemacytometer particle counting have been described by Isenberg [56].

B. Fines

The Nature of Fines The dimensional characterization of the fines fraction in pulp or in paper sheets is a rather new endeavor. Until about 1960 there was little appreciation of the fact that the fines that arise from beating or refining action differ dramatically in shape, size, and contribution to sheet properties from the fines that result from pulping and defiberization. What are today known as primary fines were fairly well understood. These are mainly the axial and ray parenchyma cells, broken or cut ends of fibers; and miscellaneous storage, meristematic, protective, and so on, cells from the stems of trees and other plants. They are mostly separated from the fibers in pulping. Figure 44 shows the primary fines collected after passing a 200-mesh screen (opening width: 76 µm) from a Southern pine (*Pinus spp.*) pulp that had not been refined (715 mℓ Csf). Fibers do not generally pass through a 200-mesh spacing.

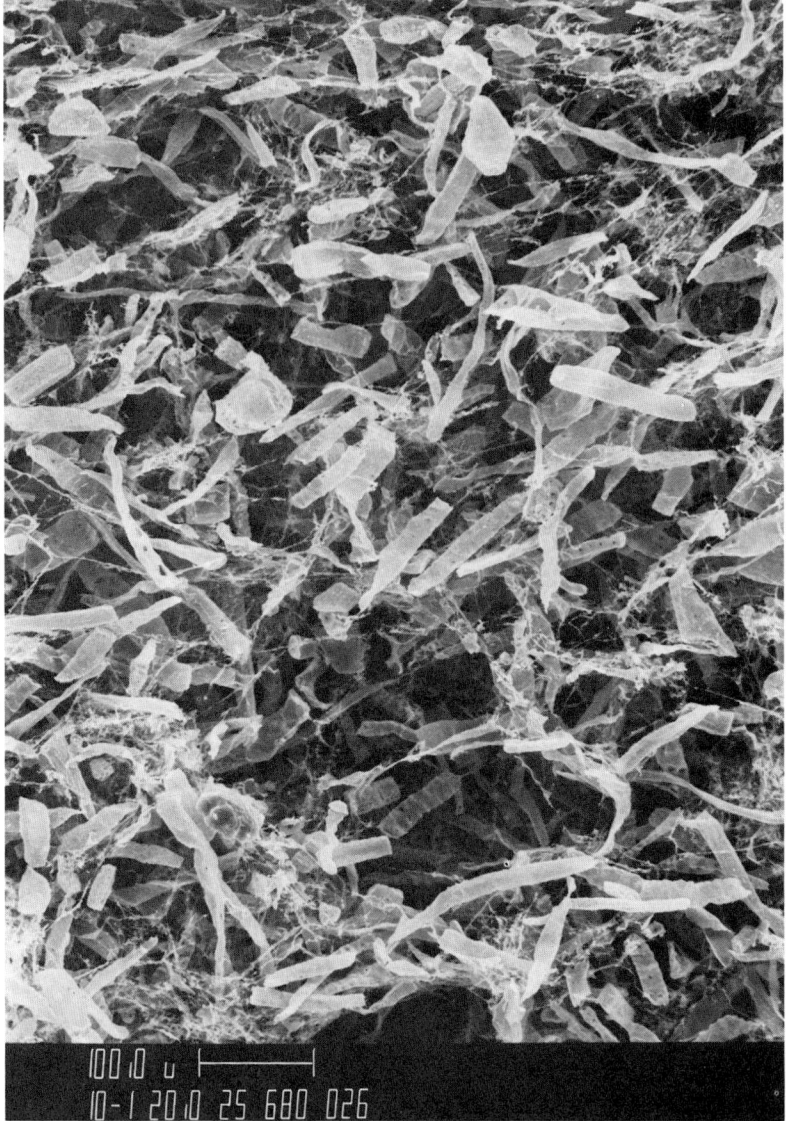

Fig. 44 Primary fines of unrefined, unbleached Southern pine pulp. (From Ref. 54a.)

Although most of the particles in Fig. 44 are cellular in nature, some finer particles of a more threadlike or reticulate form are visible, especially near the upper right of the picture. These are more typical of the fines created by beating and other mechanical action, which Steenberg et al. [130] described and photographed, giving the fraction the name *crill*. Crill, now more frequently referred to as secondary fines, is essentially material that has been removed from the fiber walls by the rubbing and crushing action of

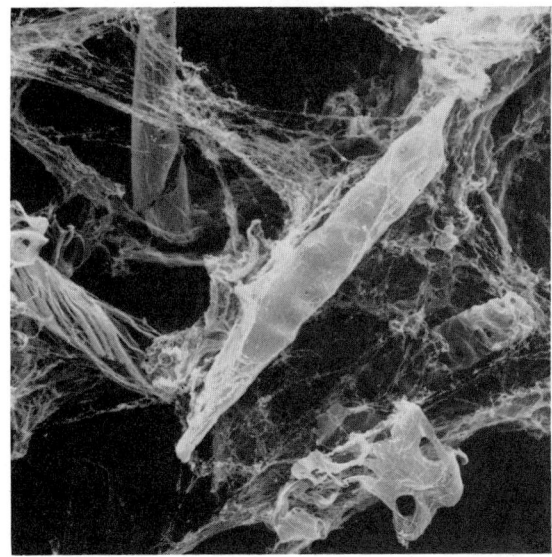

⊢⊣ 10 μm

Fig. 45 Secondary fines fraction 2° IA [8]. Bar = 10 μm. (From Ref. 54a.)

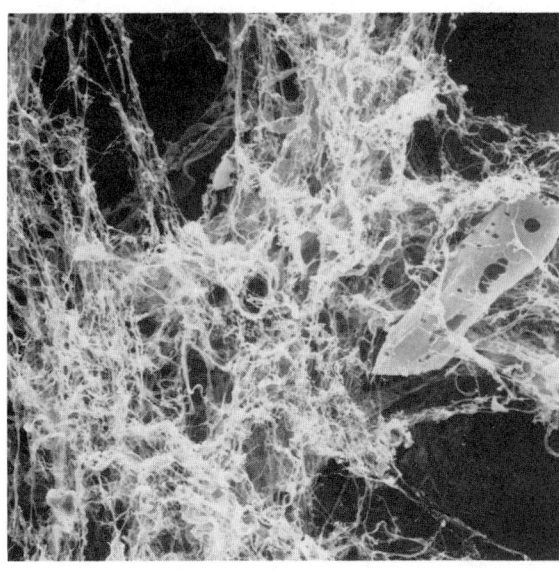

⊢⊣ 10 μm

Fig. 46 Secondary fines fraction 2° IB [8]. Bar = 10 μm. (From Ref. 54a.)

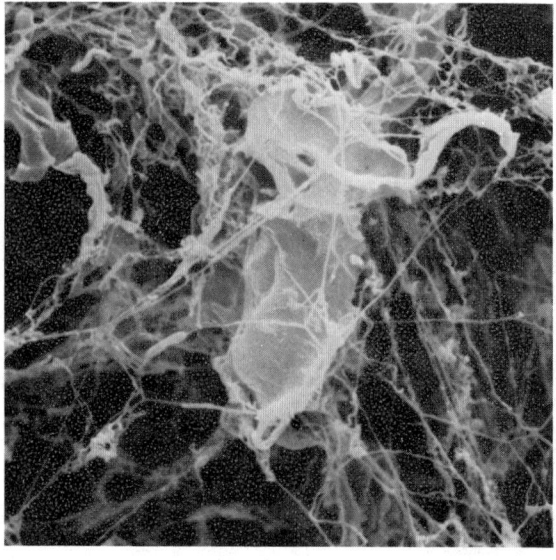

Fig. 47 Secondary fines fraction 2° IC [8]. Bar = 1 μm. (From Ref. 54a.)

beater bars and refiner plates and the rolling and rubbing of one fiber upon another. It is composed almost exclusively of slender, fibrillar particles derived from the splitting and tearing off of microfibrillar strands, principally from the S1 layer. Steenberg et al. [130] observed such particles under the phase contrast microscope and noted that earlier characterizations of such material as flourlike or slimelike were erroneous. The early descriptions were probably based on the tendencies of secondary fines to retard drainage.

In fact, secondary fines or crill materials seldom, if ever, assume a compact shape such as exists in flour or meal, even with prolonged beating. It has been shown that very long beating times result in longer, not shorter, secondary fines as the S1 layer is stripped away and the beater begins to strip microfibrillar material from the S2 layer [8,73].

Figures 45 to 48 show secondary fines from pulp that was refined for 10 s. The primary fines and the fibers in the pulp had previously been removed. Figure 45 shows particles that pass 76 μm but are retained at 38 μm spacing. Figure 46 shows particles that pass 38 μm but are retained at 19 μm spacing. Figure 47 shows particles that pass a 19 μm spacing screen and which settle after 48 h. Figure 48 shows particles that pass a 19 μm screen but which are still suspended after 48 h. Figures 49 to 52 show, respectively, secondary fines that were refined for 45 s and subsequently were fractionated in the same manner and order as the fines shown in Figs. 45 to 48.

What is evident in these eight micrographs is that the fibrillar shape is retained generally; the size is reduced as finer mesh openings are passed. Most of the cellular particles do not go through openings smaller than 38 μm, however.

Structure and Structural Anisotropy 363

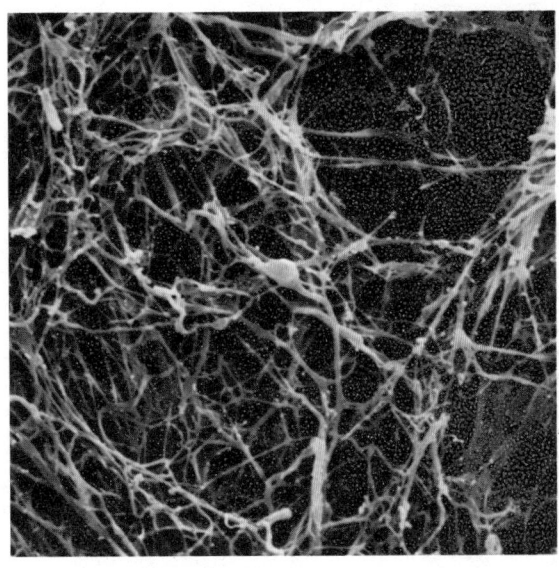

⊢⊣ 1 μm

Fig. 48 Secondary fines fraction 2° ID [8]. Bar = 1 μm. (From Ref. 54a.)

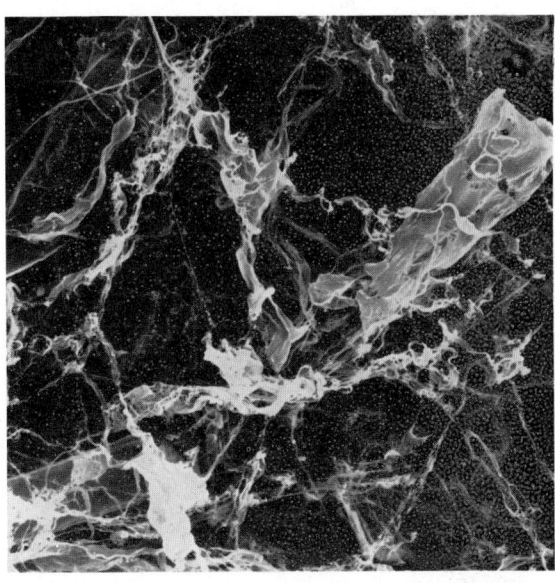

⊢⊣ 10 μm

Fig. 49 Secondary fines fraction 2° IIA [8]. Bar = 10 μm. (From Ref. 54a.)

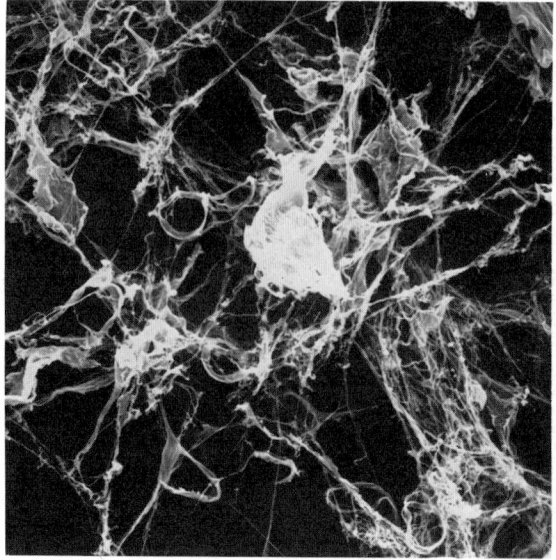

⊢⊣ 10 μm

Fig. 50 Secondary fines fraction 2° IIB [8]. Bar = 10 μm. (From Ref. 54a.)

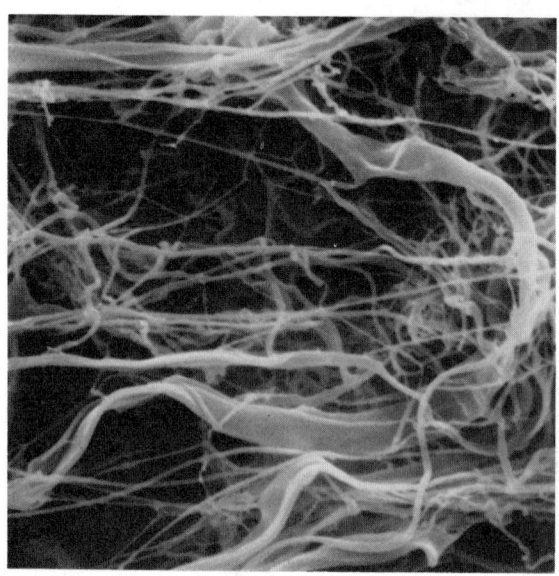

⊢⊣ 1 μm

Fig. 51 Secondary fines fraction 2° IIC [8]. Bar = 1 μm. (From Ref. 54a.)

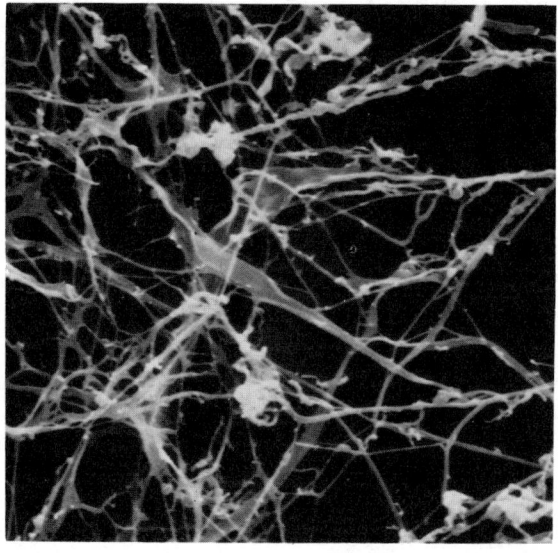

⊢⊣ 1 μm

Fig. 52 Secondary fines fraction 2° IID [8]. Bar = 1 μm. (From Ref. 54a.)

Measurement of Fines In general, the way in which fines are reported is to express them as a percent by weight (mass) according to fraction. The following relation is used:

$$F_f = \left(1 - \frac{L_m}{S}\right) \times 100 \tag{39}$$

where

F_f = fines fraction (%)
L_m = mass of the long fiber in the pulp
S = mass of all solids in the pulp

However, Kibblewhite et al. [73a,76] have developed a weighted *fines index* (F.I.), based on their examination of the effect (number of occurrences) of selected structural details of mechanical treatment on fiber surfaces that result from mechanical treatment, which may be useful for fundamental research purposes:

$$F.I. = S_1 + 2S_{1-70} + 3S_{70-30} + 4S_2 \tag{40}$$

where

S_1 = percentage of fibers in the long fiber fraction showing primary wall largely removed to reveal S1 layer with microfibrils mainly perpendicular to fiber axes.
S_{1-70} = percentage of fibers showing S1 layer partly removed to reveal

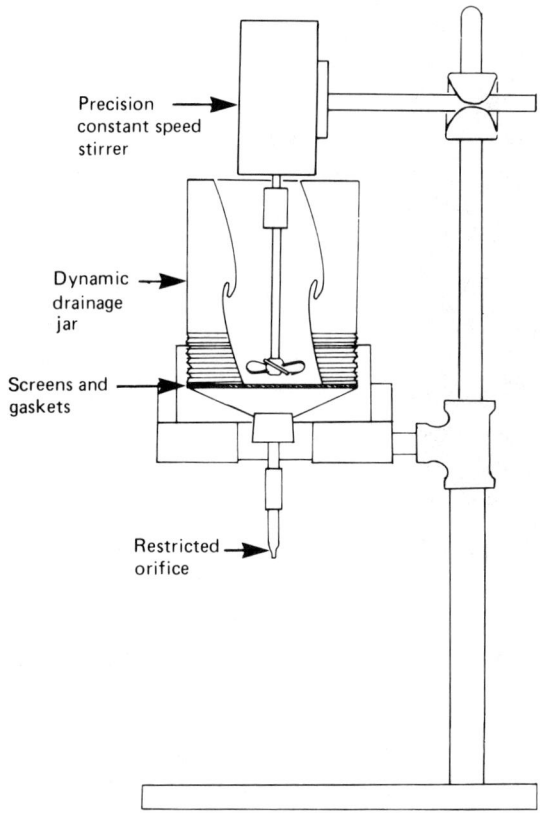

Fig. 53 The dynamic retention/drainage jar.

                     microfibrils at angles of 90° to 70° to the fiber axes.
S_{70-30} = percentage of fibers showing S1 layer partly removed to reveal microfibrils at angles of 70° to 30° to the fiber axes.
    S_2 = percentage of fibers with some S1 layer removed to reveal S2 layer

Access to scanning electron microscope facilities is needed to obtain the data for this calculation. Electron micrographs that demonstrate beating action on fiber surfaces are seen in Figs. 13 and 14 of Chap. 28.

The procedures used to fractionate fines for use in Eq. (39) have only recently been developed. Kibblewhite et al. [76] have used a Clark classifier [141].

Britt and Unbehend [14] and Unbehend [153] have reported the development of a *dynamic retention/drainage jar* (DDJ) which is designed for wet screening of fibers, fines, and additives. Shown in Fig. 53, it consists of a barrel that screws into a base having a recessed bottom. A perforated plate, screen, and gaskets are positioned between the two. The screen holes are conical in both retention and fractionation plates, which prevents binding of particles in the openings.

The procedure for determining the fines content in a sample of headbox or other stock is carried out using one 76 μm screen for operations such as quality control or several screen sizes for research purposes. Fibers and fines show good separation. However, the accuracy of the fines fractionation is critically dependent on knowing the exact consistency of the stock suspensions, since small errors in consistency can lead to large errors in the calculation of the fines fractions. Consistency should be determined by filtering a weighed quantity of stock through a weighed filter and drying in an oven. If there are fillers or other materials present, they must be accounted for precisely. Inorganic materials can be determined by ashing [139], and extractives content can be determined by following a standard such as TAPPI T 204 [138].

After fractionation, a size determination can be carried out with the use of a Coulter counter, described earlier in this chapter. Neither the wet screening in the DDJ nor the size determinations in the particle counter are without uncertainties due to the irregularity of the shape of fines particles. Some elongated particles will find their way through screen openings that are smaller than their long dimension. The Coulter counter yields results in terms of equivalent spherical diameters that hardly fit the extremely slender shape of secondary fines material. However, the DDJ-particle counter procedure is decidedly an advance over previous procedures, which ranged from simple comparison of headbox and tray consistencies to tedious screening or sedimentation procedures. The new DDJ procedure currently has the status of a TAPPI provisional method [143].

The amount of fines in machine-made paper or handsheets can be determined by soaking the sheets in hot water for 30 min at approximately 0.5% solids. The suspension is carefully defibered at reduced speed for 5 min under mild agitation in a blender. Microscopic examination will tell the investigator if defibration and fines separation are complete. When this stage is reached, the fines determination is carried out by the same procedure as used for pulp [153]. A distribution of fines through the thickness can be determined by splitting the sheet into layers (Chap. 27) and making the determination on each layer separately.

REFERENCES

1. Aaltio, E. A., Prins, W. and Hermans, J. J. (1959). X-ray investigation into the orientation of cellulose fibers in paper with respect to the plane of the sheet. *Tappi 42*(2):162A−163A.
2. Algar, W. H. (1966). Effect of structure on the mechanical properties of paper. In *Consolidation of the Paper Web* (F. Bolam, ed.), British Paper and Board Makers Assn., London, pp. 814−851.
3. American Paper and Pulp Assn. (1965). *Dictionary of Paper*, 3rd ed., New York.
4. Amidon, T. E. (1981). Effect of the wood properties of hardwoods on kraft paper properties. *Tappi 64*(3):123−126.
5. Anon. (1979). Image analysis at the Institute. *Trend 29*:8-9.
6. Asunmaa, S., and Steenberg, B. (1958). Beaten pulps and fiber-to-fiber bond in paper. *Svensk Papperstidn. 61*:686−695.

7. Balodis, V., McKenzie, A. W., Harrington, K. J., and Higgins, H. G. (1966). Effects of hydrophilic colloids and other non-fibrous materials on fibre flocculation and network consolidation. In *Consolidation of the Paper Web* (F. Bolam, ed.), British Paper and Board Makers Assn., London, pp. 639–691.
8. Bambacht, J. P., Hsu, T. H., and Unbehend, J. E. (1981). Analysis of fines fractions and their influence on sheet properties. (Submitted to *Appita*.)
9. Berlyn, G. P., and Miksche, J. P. (1976). *Botanical Microtechnique and Cytochemistry*. Iowa State Univ. Press, Ames.
10. Borch, J., and Marchessault, R. H. (1969). Light scattering by cellulose. 1. Native cellulose films. *J. Polymer Sci., C* 28:153–167.
11. Brecht, W., and Volk, W. (1958). Instrumental procedures of the fiber length measuring technique. *Das Papier* 12:196–200 (in German).
12. Britt, K. W. (1965). Determination of fiber coarseness in wood samples. *Tappi* 48(1):7–11.
13. Britt, K. W. (1980). Physical and chemical relationships in paper sheet formation. *Tappi* 63(5):105–108.
14. Britt, K. W., and Unbehend, J. E. (1976). New methods for monitoring retention. *Tappi* 59(2):67–70.
15. Bublitz, W. J., and Knutsen, D. P. (1980). Effects of deshive refining on high-yield kraft linerboard pulp. *Tappi* 63(5):109–113.
16. Carlsson, L. (1980). A study of the bending properties of paper and their relation to the layered structure. Ph.D. thesis, Chalmers Univ. of Technology, Göteborg, Sweden.
17. Carnahan, B., Luther, H. A., and Wilkes, J. O. (1969). *Applied Numerical Methods*. John Wiley and Sons, New York.
18. Carroll, C. W. (1962). Joint probability function relating fibre segmental length and orientation. In *Formation and Structure of Paper* (F. Bolam, ed.), British Paper and Board Makers Assn., London, pp. 243–245.
19. Charrier, J. M., and Marchessault, R. H. (1972). Light scattering by random and oriented anisotropic rods. *Fibre Sci. Technol.* 5:263–284.
20. Clark, J. d'A. (1942). The measurement and influence of fiber length. *Paper Trade J.* 115(26):36–42.
21. Clark, J. d'A. (1962). Effects of fiber coarseness and length. *Tappi* 45(8):628–634.
22. Clark, J. d'A. (1962). Weight average fiber length: A quick, visual method. *Pulp Paper Mag. Can.* 63(2):T53–T60.
23. Corson, S. R., and Uprichard, J. M. (1972). Fiber length of Clark screen fractions. *Tappi* 55(11):1620.
24. Corte, H., and Kallmes, O. J. (1962). Statistical geometry of a fibrous network. In *Formation and Structure of Paper,* vol. I (F. Bolam, ed.), British Paper and Board Makers Assn., London, pp. 13–52.
25. Cowan, W. F. (1975). Short span tensile analysis. Pulmac Instruments Ltd., Montreal, Canada.

26. Cox, H. L. (1952). The elasticity and strength of paper and other fibrous materials. *Brit. J. Appl. Phys. 3*:72–79.
27. Crosby, C. M., Eusufzai, A. R. K., Mark, R. E., Perkins, R. W., Chang, J. S., and Uplekar, N. V. (1981). A digitizing system for quantitative measurement of structural parameters in paper. *Tappi 64*(3):103–106.
28. Crosby, C. M., and Mark, R. E. (1975). The effects of commom embedding and mounting media on fiber cross-sectional area. Unpublished report.
29. Danielsen, R., and Steenberg, B. (1947). Quantitative determination of fibre orientation in paper. *Svensk Papperstidn. 50*:301–305.
30. Dinwoodie, J. M. (1965). The relationship between fiber morphology and paper properties: A review of literature. *Tappi 48*(8):440–447.
31. Dinwoodie, J. M. (1966). The influence of anatomical and chemical characteristics of softwood fibers on the properties of sulfate pulp. *Tappi 49*(2):57–67.
32. Dinwoodie, J. M. (1974). Failure in timber. 2. The angle of shear through the cell wall during longitudinal compression stressing. *Wood Sci. Technol. 8*:56–67.
33. Dumbleton, D. P. (1972). Longitudinal compression of individual pulp fibers. *Tappi 55*(1):127–135.
34. Duncker, B., Hartler, N., and Samuelsson, L. G. (1966). Effect of drying on the mechanical properties of pulp fibers. In *Consolidation of the Paper Web* (F. Bolam, ed.), British Paper and Board Makers Assn., London, pp. 529–537.
35. Elderton, W. P., and Johnson, N. L. (1969). *Systems of Frequency Curves*. Cambridge Univ. Press, London.
36. Ericsson, T., and Rudgård, A. (1981). Experiments towards an on-line measuring system for paper-pulp fibres. IEEE paper in press.
37. Evans, R., Franczek, W., and Luner, P. (1977). Monitoring of flocculation by means of the Coulter counter. Unpublished Report.
38. Food and Agriculture Organization, United Nations (1975). Pulping and papermaking properties of fastgrowing plantation wood species. FO:MISC/75/31, Rome, Italy.
39. Forest Biology Subcommittee No. 2 (1968). New methods of measuring wood and fiber properties in small samples. *Tappi 51*(1):75A–80A.
40. Forgacs, O. L. (1961). Structural weaknesses in softwood pulp tracheids. *Tappi 44*(2):112–119.
41. Forgacs, O. L., and Strelis, I. (1963). The measurement of the quantity and orientation of chemical pulp fibres in the surfaces of newsprint. *Pulp Paper Mag. Can. 64*(1):T3–T13.
42. Glynn, P., Jones, H. W. H., and Gallay, W. (1959). The fundamentals of curl in paper. *Pulp Paper Mag. Can. 60*(10):T316–T323.
43. Graff, J. H., and Feavel, J. R. (1944). Projection arrangement for determination of fiber dimensions. *Paper Trade J. 118*(7):140–145.
44. Graminski, E. L., and Kirsch, R. A. (1977). Image analysis in paper manufacturing. Proceedings IEEE Computer Society Conference on Pattern Recognition and Image Processing, New York, pp. 137–143.

45. Green, H. V., and Worrall, J. (1964). Wood quality studies. 1. A scanning microphotometer for automatically measuring and recording certain wood characteristics. *Tappi* 47(7):419–427.
46. Hagemeyer, R. W. (1960). The effect of pigment combination and solids concentration on particle packing and coated paper characteristics. 1. Relationship of particle shape to particle packing. *Tappi* 43(3):277–288.
47. Hamming, R. W. (1962). *Numerical Methods for Scientists and Engineers*, McGraw-Hill, New York.
48. Hardacker, K. W. (1969). Cross-sectional area measurement of individual wood pulp fibers by lateral compaction. *Tappi* 52(9):1742–1746.
49. Hardacker, K. W. (1970). Effects of loading rate, span, and beating on individual wood fiber tensile properties. In *The Physics and Chemistry of Wood Pulp Fibers* (D. H. Page, ed.), TAPPI STAP 8, pp. 201–216.
50. Harpham, J. A., and Hock, C. W. (1958). The fine paper properties of cotton linters. 1. The relationship of fiber composition to fine paper properties. *Tappi* 41(11):625–629.
51. Hastings, N. A. J., and Peacock, J. B. (1975). *Statistical Distributions*, John Wiley and Sons, New York.
52. Higgins, H. G., de Yong, J., Balodis, V., Phillips, F. H., and Colley, J. (1973). The density and structure of hardwoods in relation to paper surface characteristics and other properties. *Tappi* 56(8):127–131.
53. Hogg, R. V., and Craig, A. T. (1970). *Introduction to Mathematical Statistics*, 3rd ed. Macmillan, New York.
54. Horn, R. A. (1974). Morphology of wood pulp fiber from softwoods and influence on paper strength. U. S. Forest Products Laboratory, Madison, Wis.
54a. Hsu, T. (1981). Classification and characterization of fines produced from Southern pine kraft pulp. M. S. Thesis, SUNY College of Environmental Science and Forestry, Syracuse, N.Y.
54b. Htun, M. (1980). The influence of drying strategies on the mechanical properties of paper. Ph.D. Thesis, Royal Institute of Technology, Stockholm.
54c. Htun, M., and de Ruvo, A. (1978). The implication of the fines fration for the properties of bleached kraft sheet. *Svensk Papperstidn.* 81(16):507–510.
55. Ilvessalo-Pfaffli, M.-S., and Alfthan, G. V. (1957). The measurement of fiber length with a semi-automatic recorder. *Paperi ja Puu* 39(11):509–516.
56. Isenberg, I. H. (1967). *Pulp and Paper Microscopy*, 3rd. ed., Institute of Paper Chemistry, Appleton, Wis.
57. Jagels, R., Gardner, D. J., and Brann, T. B. (1981). Improved techniques for handling and staining wood fibers for digitizer assisted measurement. Paper submitted to *Wood Science*.
58. Japan Industry Standards Committee JIS P-8207. Method of screening test of paper pulp.

59. Jayme, G., and Bauer, G. (1957). Differentiation of early- and late-wood fibers by secondary fluorescence. *Holzforsch.* 2:16−18 (in German).
60. Jones, A. R. (1967). An experimental investigation of the in-plane elastic moduli of paper. Ph.D. Thesis, Institute of Paper Chemistry, Appleton, Wis.
61. Jordan, B. D., and Page, D. H. (1980). Application of image analysis to pulp fibre characterization, 1. Proceedings of the Fiber Society Symposium, Baltimore, Md. (Sept.).
61a. Jordan, B. D., and Page, D. H. (1981). Application of image analysis to pulp fibre characterization, 1. Proceedings 7th Fund. Res. Symposium, Technical Division British Paper and Board Makers Assn. (in press).
62. Judt, M. (1958). Fiber alignment in paper. *Das Papier* 12(21/22): 568−578 (in German).
63. Judt, M. (1959). The effect of the shake of paper machines on sheet formation and fiber orientation. *Das Papier* 13(3/4):46−54 (in German).
64. Kallmes, O. J. (1969). Technique for determining the fiber orientation distribution throughout the thickness of a sheet. *Tappi* 52(3):482−485.
65. Kallmes, O., and Bernier, G. (1962). The structure of paper. 3. The absolute, relative and maximum bonded areas of fiber networks. *Tappi* 45(11):867−872.
66. Kallmes, O., and Bernier, G. (1963). The structure of paper. 4. The free fiber length of a multiplanar sheet. *Tappi* 46(2):108−114.
67. Kallmes, O., and Bernier, G. (1964). The structure of paper. 8. Structure of idealized nonrandom networks. *Tappi* 47(11):694−703.
68. Kallmes, O., and Corte, H. (1960). The structure of paper. 1. The statistical geomtry of an ideal two-dimensional fiber network. *Tappi* 43(9):737−752.
69. Kallmes, O., Corte, H., and Bernier, G. (1963). The structure of paper. 5. The bonding states of fibers in randomly formed paper. *Tappi* 46(8):493−502.
70. Kallmes, O., and Eckert, C. (1964). The structure of paper. 7. The application of the relative bonded area concept of paper evaluation. *Tappi* 47(9):540−548.
71. Kallmes, O. J., and Perez, M. (1966). Load/elongation properties of fibres. In *Consolidation of the Paper Web* (F. Bolam, ed.), British Paper and Board Makers Assn., London, pp. 507−537.
72. Keith, C. T., and Côté, W. A., Jr. (1968). Microscopic characterization of slip lines and compression failures in wood cell walls. *Forest Products J.* 18(3):67−74.
73. Kibblewhite, R. P. (1972). Effect of beating on fibre morphology and fibre surface structure. *Appita* 26(3):196−202.
73a. Kibblewhite, R. P. (1975). Interrelations between pulp refining treatments, fibre and fines quality, and pulp freeness. *Paperi ja Puu* 57(8):519−526.
74. Kibblewhite, R. P. (1977). Structural modifications to pulp fibers: Definitions and role in papermaking. *Tappi* 60(10):141−143.

75. Kibblewhite, R. P., and Brookes, D. (1975). Factors which influence the wet web strength of commercial pulps. *Appita* 28(4):227–231.
76. Kibblewhite, R. P., Brookes, D., and Allison, R. W. (1980). Effect of ozone on the fiber characteristics of thermomechanical pulps. *Tappi* 63(4):133–136.
77. Kimura, M., and Mark, R. E. (1981). Mechanical properties in relation to network structure for press-dried paper. Submitted to *J. Japan Wood Res. Soc. (Mokuzai Gakkaishi)*.
78. Kobar, L., Hajduczki, I., and Reinicz, E. (1975). The improvement of flat crush strength by the use of modified starch and lignin. *Zellstoff Papier* 24(9):269-270 (in German).
79. Komori, T., Ujihara, Y., Matsunaga, Y., and Makishima, K. (1979). Crossings of curled fibers in two-dimensional assemblies. *Tappi* 62(3):93–95.
80. Koning, J. W., Jr., and Haskell, J. H. (1978). Papermaking factors that influence the strength of linerboard weight handsheets. U. S. Forest Products Laboratory, Madison, Wis.
81. Ladell, J. L. (1959). A method of measuring the amount and distribution of cell wall material in transverse microscope sections of wood. *J. Inst. Wood Sci.* 3:43–46.
82. Lim, Y. W., Sarko, A., and Marchessault, R. H. (1970). Light scattering by cellulose. 2. Oriented condenser paper. *Tappi* 53(12):2314–2319.
83. Luce, J. E. (1970). Tranverse collapse of wood pulp fibers: Fiber models. In *The Physics and Chemistry of Wood Pulp Fibers* (D. H. Page, ed.), TAPPI STAP 8, pp. 278–281.
84. Lynch, L. J., and Thomas, N. (1971). Optical diffraction profiles of single fibers. *Textile Research J.* 41:568–572.
85. Marchessault, R. H. (1973). Light scattering by oriented native cellulose systems. In *Structure and Properties of Polymer Films* (R. W. Lenz and R. S. Stein, eds.), Plenum, New York, pp. 25–37.
86. Mardia, K. V. (1972). *Statistics of Directional Data*. Academic Press, London and New York.
87. Mark, R. E. (1967). *Cell Wall Mechanics of Tracheids*. Yale Univ. Press, New Haven, Conn.
88. Marton, J. (1974). Fines and wet end chemistry. *Tappi* 57(12):90–93.
89. Marton, R., Alexander, S. D., Brown, A. F., and Sherman, C. W. (1965). Morphological limitations to the quality of groundwood from hardwoods. *Tappi* 48(7):395–398.
89a. Marton, R., Brown, A., Granzow, S., Koeppicus, R., and Tomlinson, S. (1974). Recycling and fiber structure. Unpublished report.
90. McMillin, C. W. (1981). Application of automatic image analysis to wood science. Paper submitted to *Wood Science*.
91. Morton, R. (n.d.) An introduction to automatic image analysis. Bausch & Lomb Analytical Systems Div., Rochester, N.Y.
92. Morton, R. (n.d.) Practical considerations in automatic image analysis. Bausch & Lomb Analytical Systems Div., Rochester, N.Y.

93. Muggli, R., Marton, R., and Sarko, A. (1971). Light scattering by cellulose. 5. Anisotropic scattering by wood fibers. *J. Polymer Sci., C* 36:121–139.
94. Nordman, L., Aaltonen, P., and Makkonen, T. (1966). Relationship between mechanical and optical properties of paper affected by web consolidation. In *Consolidation of the Paper Web* (F. Bolam, ed.), British Paper and Board Makers Assn., London, pp. 909–927.
95. Öhrn, O. E. (1969). Fiber length measuring gauge with an "easy to use" measuring probe. *Svensk Papperstidn.* 72(20):667-668.
96. Page, D. H. (1967). The collapse behavior of pulp fibers. *Tappi* 50(9):449–455.
97. Page, D. H., El-Hosseiny, F., Winkler, K., and Lancaster, A. P. S. (1977). Elastic modulus of single wood pulp fibers. *Tappi* 60(4):114–117.
98. Page, D. H., Sargent, J. W., and Nelson, R. (1966). Structure of paper in cross-section. In *Consolidation of the Paper Web* (F. Bolam, ed.), British Paper and Board Makers Assn., London, pp. 313–352.
98a. Page, D. H., and Seth, R. S. (1980). The elastic modulus of paper. 2. The importance of fiber modulus, bonding and fiber length. *Tappi* 63(6):113–116.
98b. Page, D. H., and Seth, R. S. (1980). The elastic modulus of paper. 3. The effects of dislocations, microcompressions, curl, crimps and kinks. *Tappi* 63(10):99–102.
98c. Page, D. H., Seth, R. S., and De Grace, J. H. (1979). The ealstic modulus of paper. 1. The controlling mechanisms. *Tappi* 62(9):99–102.
99. Page, D. H., and Tydeman, P. A. (1962). A new theory of the shrinkage, structure and properties of paper. In *Formation and Structure of Paper* (F. Bolam, ed.), British Paper and Board Makers Assn., London, pp. 397–421. (See also comments by J. G. Buchanan and O. V. Washburn, pp. 422-423.)
100. Page, D. H., and Tydeman, P. A. (1966). Physical processes occurring during the drying phase. In *Consolidation of the Paper Web* (F. Bolam, ed.), British Paper and Board Makers Assn., London, pp. 371–396. (See also discussion pp. 950-951.)
101. Page, D. H., Tydeman, P. A., and Hunt, M. (1962). A study of fibre-to-fibre bonding by direct observation. In *Formation and Structure of Paper* (F. Bolam, ed.), British Paper and Board Makers Assn., London, pp. 171–193.
102. Page, D. H., Tydeman, P. A., and Hunt, M. (1962). The behavior of fiber-to-fiber bonds in sheets under dynamic conditions. In *Formation and Structure of Paper* (F. Bolam, ed.), British Paper and Board Makers Assn., pp. 249–263.
103. Papermaking Additives Committee, TAPPI (1975). Commercially available chemical agents for paper and board manufacture, Committee assignment report no. 60.
104. Parham, R. A., and Kaustinen, H. M. (1974). *Papermaking Materials. An Atlas of Electron Micrographs.* Institute of Paper Chemistry, Appleton, Wis.

105. Parker, J., and Mih, W. C. (1964). A new method for sectioning and analyzing paper in the transverse direction. *Tappi* 47(5):254–263.
106. Perez, M., and Kallmes, O. J. (1965). The role of fiber curl in paper properties. *Tappi* 48(10):601–606.
107. Perkins, R. W., Jr. (1978). Prediction of the elastic properties of paper from a knowledge of network geometric parameters. In *General Constitutive Relations for Wood and Wood-based Materials* (R. W. Perkins, B. A. Jayne, and J. A. Johnson, eds.). Report of National Science Foundation Workshop by Syracuse Univ., pp. 1–16.
108. Perkins, R. W., Jr. (1980). Mechanical behavior of paper in relation to its structure. In *The Cutting Edge*, Institute of Paper Chemistry, Appleton, Wis., pp. 89–111.
109. Perkins, R. W., and Mark, R. E. (1981). Some new concepts of the relation between fibre orientation, fibre geometry and mechanical properties. To be presented at 7th Fundamental Research Symposium organized by Technical Division, British Paper and Board Industry Federation, Cambridge, England.
109a. Petkar, B. M., Oka, P. G., and Sundaram, V. (1980). The cross-sectional shapes of a cotton fiber along its length. *Textile Res. J.* 50:541–543.
110. Prud'homme, R. E., Hien, N. V., Noah, J., and Marchessault, R. H. (1975). Determination of fiber orientation of cellulosic samples by X-ray diffraction. *J. Appl. Polymer Sci.* 19:2606–2620.
111. Prusas, Z. C. (1963). Laboratory study of the effects of fiber orientation on sheet anisotropy. *Tappi* 46(5):325–330.
112. Pye, I. T., Washburn, O. V., and Buchanan, J. G. (1966). Structural changes in paper on pressing and drying. In *Consolidation of the Paper Web* (F. Bolam, ed.), British Paper and Board Makers Assn., London, pp. 353–370.
113. Quackenbush, D. W. (1971). Faults in paper coatings and their relationship to base sheet structure. *Tappi* 54(1):47–52.
114. Ranger, A. E., and Hopkins, L. F. (1962). A new theory of the tensile behaviour of paper. In *Formation and Structure of Paper*, vol. 1 (F. Bolam, ed.), British Paper and Board Makers Assn., London, pp. 311–318.
115. Reed, A. E., and Clark, J. d'A. (1950). An instrument for rapid fractionation of pulp. *Tappi* 33(6):294–298.
116. Robertson, A. A., Meindersma, E., and Mason, S. G. (1961). The measurement of fibre flexibility. *Pulp Paper Mag. Can.* 62(1):T3–T10.
117. Rothenberg, S., and Fernandez, J. M. (1981). Geometric parameters of TMP fibers treated chemically between refining stages. Unpublished report.
118. Ruck, H. and Krässig, H. (1958). The determination of fiber orientation in paper. *Pulp Paper Mag. Can.* 59(6):183–190.
119. Rudström, L., and Sjölin, U. (1970). A method for determining fibre orientation in paper using laser light. *Svensk Papperstidn.* 73(5):117–121.

120. Sadowski, J. (1976). Measurement of fibre orientation in paper by optical Fourier transform. Thesis, Helsinki Tech. Univ.
121. Saka, S., Thomas, R. J., Gratzl, J. S. (1978). Lignin distribution: Determination by energy-dispersive analysis of X-rays. *Tappi* 61(1):73–76.
122. Sauret, G. (1963). Anisotropy in the paper sheet. 1. Influence of fiber orientation. *Tech. Rech. Papetieres Bull.* 2:3–14 (in French).
123. Scandanavian Pulp, Paper and Board Testing Committee. Standard SCAN-M6:69. Fiber fractionation of mechanical pulp in the McNett apparatus.
123a. Schafer, E. R., and Santaholma, M. (1933). Effect of different-sized fibers on the physical properties of groundwood pulp. *Paper Trade J.* (TAPPI Sect.) 97:224–229.
124. Schniewind, A. P., Ifju, G., and Brink, D. L. (1966). Effect of drying of the flexural rigidity of single fibers. In *Consolidation of the Paper Web* (F. Bolam, ed.), British Paper and Board Makers Assn., London, pp. 538–543.
125. Schwalbe, H. C. (1966). Effects of sizing, adhesives and fillers on the formation and consolidation of paper webs. In *Consolidation of the Paper Web* (F. Bolam, ed.), British Paper and Board Makers Assn., London, pp. 692–740.
126. Seth, R. S., and Page, D. H. (1981). The stress-strain curve of paper. To be presented at 7th Fundamental Research Symposium, organized by Technical Division, British Paper and Board Industry Federation, Cambridge, England.
127. Setterholm, V., and Kuenzi, E. W. (1970). Fiber orientation and degree of restraint during drying. Effect on tensile anisotropy of paper handsheets. *Tappi* 53(10):1915–1920.
128. Silvy, J. (1971). Effects of drying on web characteristics. 1. Web structure and the physical characteristics of the sheet. *Paper Technol.* 12(5):T181–T191, T194–T196.
129. Silvy, J. (1980). Structural study of fiber networks: The cellulosic fiber case, D. Sci. Thesis, Univ. Grenoble (in French).
130. Steenberg, B., Sandgren, B., and Wahren, D. (1960). Studies on pulp crill. 1. Suspended fibrils in paper pulp fines. *Svensk Papperstidn.* 63(12):395–397.
131. Stein, R. S. (1973). Optical studies of the morphology of polymer films. In *Structure and Properties of Polymer Films* (R. W. Lenz and R. S. Stein, ed.), Plenum, New York, pp. 1–24.
132. Sugden, E. A. N. (1968). A semi-automated method for the determination of the average fiber length by projection. *Pulp Paper Mag. Can.* 69:T406–T411.
133. Sundararajan, P. R. (1981). X-ray studies on the orientation of inorganic materials in paper. *Tappi* 64(10):111–114.
134. Swanson, J. W. (1966). Effects of soluble non-fibrous materials on formation and consolidation of paper webs. In *Consolidation of the Paper Web* (F. Bolam, ed.), British Paper and Board Makers Assn., London, pp. 741–776.

135. Takahashi, H., Suzuki, H., and Endoh, K. (1979). The effect of fiber shape on the mechanical strength of paper and board. *Tappi* 62(7):85–88.
136. Tasman, J. E. (1972). The fiber length of Bauer-McNett screen fractions. *Tappi* 55(1):136–138.
137. Taylor, C. J., and Dixon, R. N. (1981). Image analysis applied to fibre images. Paper prepared for 7th Fundamental Research Symposium, Cambridge, England.
138. TAPPI T 204. Alcohol and dichloromethane solubles in wood and pulp.
139. TAPPI T 211. Ash in pulp.
140. TAPPI T 232. Fiber length of pulp by projection.
141. TAPPI T 233. Fiber length of pulp by classification.
142. TAPPI T 234. Coarseness of pulp fibers.
143. TAPPI Provisional Method T 261 pm-79. Fines fraction of paper stock by wet screening.
144. TAPPI former suggested method T 481 (withdrawn August 1972). Fiber orientation and squareness of paper (zero-span tensile strength).
145. TAPPI Useful Method UM 239. Fiber bundles in baled flash dried pulp.
146. TAPPI Useful Method UM 240. Shive content of mechanical pulp (laboratory flat screen).
147. TAPPI Useful Method UM 241. Shive content of mechanical pulp (von Alfthan shive analyzer).
148. TAPPI Useful Method UM 242. Shive content of mechanical pulp (Somerville Fractionator).
149. TAPPI Useful Method UM 808. Plybond peeling strength of linerboard and corrugated board.
150. Thorpe, J. L., Mark, R. E., Eusufzai, A. R. K., and Perkins, R. W. (1976). Mechanical properties of fiber bonds. *Tappi* 59(5):96–100.
151. Toroi, M. (1959). The preparation of fibre oriented sheets on a laboratory scale. *Paperi ja Puu* 41(5):271–279.
152. Tydeman, P. A., Wembridge, D. R., and Page, D. H. (1966). Transverse shrinkage of individual fibres by microradiography. In *Consolidation of the Paper Web* (F. Bolam, ed.), British Paper and Board Makers Assn., London, pp. 119–144.
153. Unbehend, J. E. (1977). The "dynamic retention/drainage jar." Increasing the credibility of retention measurements. *Tappi* 60(7):110–112.
154. Unger, E., and Unger, E. W. (1963). A contribution to the rapid determination of mean fiber length and fiber length distribution in paper stocks. *Zellstoff Papier* 12(1):4–10 and (2):40–45.
155. Unger, E. W., and Freund, F. (1975). New developments in fiber length analysis via projected images. *Zellstoff Papier* 24(5):143–146, 160 (in German).
156. Valley, R. B., and Morse, T. H. (1965). Measurement of fiber length using a modified Coulter particle counter. *Tappi* 48(6):372–376.

157. Vecchi, E. (1969). Quality control of poplar groundwood: Factors related to the structural composition of the pulp. *Tappi* 52(12): 2390-2399.
158. Visconti, S., Hien, N. V., Borch, J., and Marchessault, R. H. (1976). Light scattering be helical fiber structures: Experimental models. *J. Polymer Sci., Polymer Phys.* 14:631—641.
159. Walbaum, H. H., and Zak, H. (1976). Internal structure of paper and coatings in SEM cross sections. *Tappi* 59(3):102—105.
160. Wang, P. H., and McKimmy, M. D. (1977). The effect of pulping mixed species on paper properties. *Tappi* 60(7):140-143.
161. Ward, K., Jr., Voelker, M. H., and Maclaurin, D. J. (1965). Cotton linters as papermaking fibers: Comparative studies on rag, linters and cotton lint pulps. *Tappi* 48(11):657—650.
162. Watson, A. A., and Dadswell, H. E. (1961). Influence of fibre morphology on paper properties. 1. Fibre length. *Appita* 14(5): 168—178.
163. Wilson, J. W. (1954). Fiber length mensuration: a comprehensive history and new method. *Pulp Paper Mag. Can.* 55:84—91.
164. Yan, J. F. (1975). A method for the interpretation of fiber length classification data. *Tappi* 58(8):191-192.
165. Yang, C. F. (1975). Plane modeling and analysis of fiber systems. Ph.D. Thesis, Univ. Washington, Seattle.
166. Yang, C. F., Crosby, C. M., Eusufzai, A. R. K., and Mark, R. E., (1977). Laser determination of fiber orientation distribution in sheets. Unpublished report.
167. Yang, C. F., Eusufzai, A. R. K., Sankar, R., Mark, R. E., and Perkins, R. W., Jr. (1978). Measurements of geometrical parameters of fiber networks. 1. Bonded surfaces, aspect ratios, fiber moments of inertia, bonding state probabilities. *Svensk Papperstidn.* 81(13): 426—433.
168. Yang, C. F., Mark, R. E., Eusufzai, A. R. K., and Perkins, R. W., Jr. (1981). Measurements of geometrical parameters of fiber networks. 2. Mechanical properties in relation to network geometry for press-dried paper. *Svensk Papperstidn.* 84(9):R55—R60.

25
DETERMINATION OF FIBER-FIBER BOND PROPERTIES

TETSU UESAKA*

Empire State Paper Research Institute
State University of New York
College of Environmental Science and Forestry
Syracuse, New York

I.	Introduction	379
II.	Bonding States of Fibers	381
	A. Description of Bonding States in the Fiber Network	381
	B. Experimental Determination of Bonding States	385
III.	Mechanical Properties of Fiber-Fiber Bonds	390
	A. Bond Strength Measurement (Indirect Methods)	390
	B. Bond Strength Measurement (Direct Methods)	393
	References	398

I. INTRODUCTION

Because of the unique fiber network structure in paper, the formation and properties of fiber-fiber bonds have been the main subject of interest for many papermakers and researchers. Microscopic studies show that a considerable area of the fiber surface is covered by other fibers, and that the length of a free fiber segment, which is to say a completely unbonded segment, is of the same order of size as the fiber width in the case of a normally consolidated paper. For such a strongly interacting fiber network sys-

---

*Current affiliation: Oji Paper Co. Ltd., Central Research Laboratory, Tokyo, Japan.

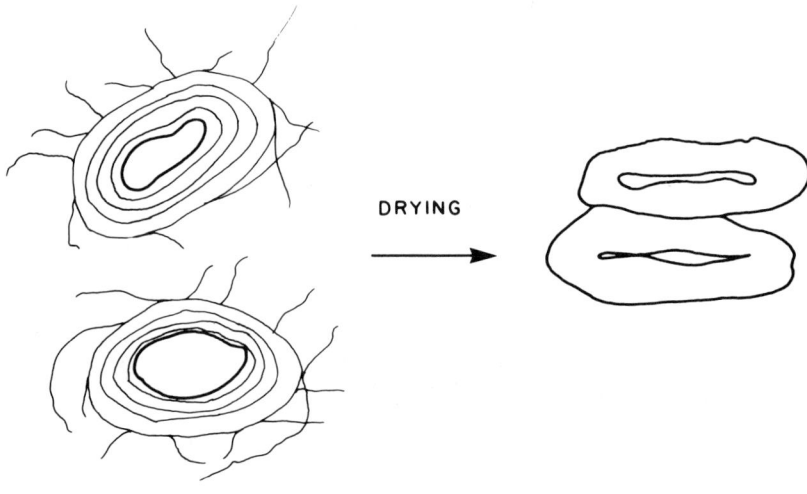

Wet beaten fibers

Fig. 1 Fiber-fiber bond formation during drying.

tem, the structure and properties of the bonded region are essential factors sure to affect almost all of the mechanical, optical, thermal, and electrical properties of the paper sheet.

The nature of the forces holding fibers together in the paper sheet has attracted considerable interest. After early mechanical force theories such as mechanical entanglement and frictional forces [10], the hydrogen bond force and other van der Waals forces theories [6,40,54] were proposed. One of the generally accepted models of fiber-to-fiber bond formation during the consolidation process is the one suggested by An den Akker [63]. It is illustrated in Fig. 1. In a fiber-water suspension, fiber walls are swollen and more or less delaminated. Surface macro- and microfibrils are partly torn away during various mechanical actions (refining, screening, etc.) [8,63]. During water removal in papermaking, surface tension of the water is associated with three important events:

- It brings the fibers and the fibrils into close contact [5].
- • It causes the fibrils to be brought back upon the surface of the parent fiber.
- • • The combined effect results in the shrinkage of the fiber cell wall.

The final structure of the fiber-fiber bond region (see pp. 341-342 of Chap. 24) is the combined S1-S1, S1-S2, or S2-S2 layers; there may be some entanglements of fibrils between those layers. Hydrogen and other molecular bonds can exist between the entangled fibrils and between the fiber surfaces.

Although considerable efforts have been devoted to the study of the elementary process of fiber-fiber bond formation, the concepts of structure and function of the fiber-fiber bond in paper have not been well recognized or developed.

The purposes of this chapter are to clarify the concept of the fiber-fiber bond as a structural entity and to review existing methods for determining structure and function of the fiber-fiber bond. For detailed descriptions of various experimental techniques, readers are referred to Chaps. 16, 17, 21, 22, and 24 of this text.

II. BONDING STATES OF FIBERS

A. Description of Bonding States in the Fiber Network

The most basic process of papermaking may be the one in which fibers dispersed in water are deposited on a wire screen. This unique process yields variation in spatial distribution of fibers (formation variation) and variation of fiber-fiber bonding states in the paper sheet because of the random nature of fiber deposition. The fact of randomness indicates the necessity of some statistical methods for describing bonding states.

A simple example of a fiber crossing is illustrated in Fig. 2. Since the crossing angle can vary according to the statistical orientation distribution (Chap. 24), the overlapped area, which may be larger than the actual contact area between the fibers, can only be described in a statistical manner. For the case of random orientation (the case wherein the value of the distribution function of the crossing angle is constant over the range), the mean overlapped area $<S>$ is given by

$$<S> = \frac{\pi \omega^2}{2} \quad \text{when } \lambda \gg \omega \tag{1}$$

where λ and ω are fiber length and fiber width, respectively. The effective length of the overlapped area can be calculated as $<S>/\omega = \pi\omega/2 \ (= 1.57\omega)$. This length can be considered as an average length of the possible bonding area. In this area various chemical, physical and mechanical interactions between fibers take place, such as hydrogen or other bond formation and the setting up of residual stresses caused by different shrinkages in the cross-

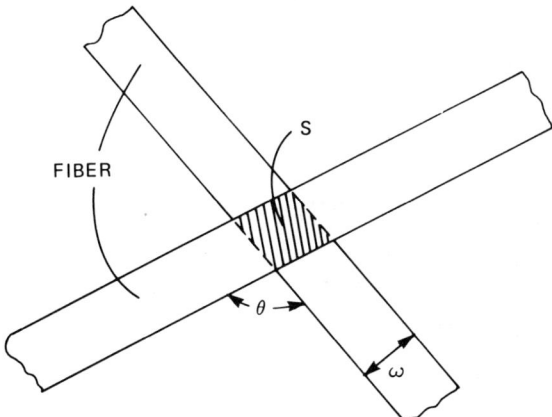

Fig. 2 Simple fiber crossing. (θ) crossing angle; (ω) fiber width; (S) area of overlap.

ing fibers. This simple example indicates that the possible bonding length is statistically about 60% larger than the fiber width, and that the width of a fiber is a critical parameter in determining the bonding area.

The statistical description of the bonding states in paper began with a series of papers on the structure of paper by Kallmes and coworkers [21–26]. Assuming the process of fiber deposition on a wire screen as a Poisson process, they developed a two-dimensional fiber network model (2-D sheets), and further extended their model to the three-dimensional case, calling it the multiplanar model (MP sheet). Although it is difficult to directly compare the prediction from the Kallmes et al. model with the actual system because of difficulties of determining parameters involved and assumptions employed, this model presents some important statistical features of paper structure.

The Poisson distribution [20] most successfully describes the following process. When N_f fibers are deposited randomly into an area A that is sufficiently large compared with fiber length, the probability that k fibers pile up on a point in the area A is given by

$$P(k) = \frac{e^{-H} H^k}{k!} \quad (2)$$

where H is the mean number of overlapping fibers given by

$$H = \frac{N_f \lambda \omega}{A} \quad (3)$$

The probability P(k) can also be regarded as the fraction of the sheet area that is covered by k fibers in the plane. One of the most important parameters of bonding states, relative bonded area (RBA), which is defined as the ratio of the total bonded area to one-half the total external surface area of fibers, can be derived on the basis of the above distribution function.

The maximum relative bonded area RBA_{max}, the upper bound of RBA, can be calculated by assuming that k fibers piled upon a point are completely bonded to each other [20].

$$RBA_{max} = 1 - \frac{1}{H}(1 - e^{-H}) \quad (4)$$

The parameter H can be expressed in terms of the basis weight (B.W.) and the fiber weight w_f by the formula $H = (B.W./w_f)\lambda\omega$ [see Eq. (3)]. As basis weight increases, RBA_{max} increases and approaches unity as B.W. $\to \infty$ or H $\to \infty$. Results of sample calculations of RBA_{max} versus basis weight are shown in Fig. 3 [22].* The potential of fiber-fiber bonding (i.e., RBA_{max}) is basically determined by three parameters: basis weight B.W., fiber weight per unit length w_f/λ (coarseness), and fiber width ω. Since for actual fiber networks there is always a strong geometric interaction between fibers, the relative bonded area is lower than the RBA_{max}.

---

*In this model the effect of fiber thickness is not taken into account explicitly (for two-dimensional fibers); the predicted RBA_{max} and RBA given here may be greater than actual values.

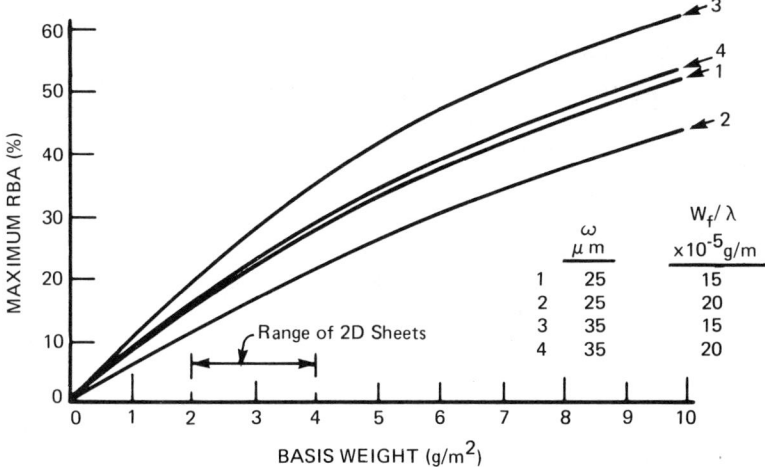

Fig. 3 RBA$_{max}$ versus basis weight for multiplanar sheet model. (From Ref. 21.)

Kallmes and Bernier [21] obtained an expression of RBA for a multi-planar fiber network consisting of N_L layers of two-dimensional sheets ($N_L \gg 3$).

$$\text{RBA} = \frac{1}{H_{2D}} \left(\beta_1 [H_{2D} - 1 + P(0)] + [1 - P(0)]^2 \times \right.$$

$$\left. \left\{ \left(1 - \frac{1}{N_L}\right) \left[\beta_2 + \frac{DP(0)^2}{1 - P(0)^2} \right] - \frac{DP(0)^2 [1 - P(0)^{2(N_L-1)}]}{N_L [1 - P(0)^2]^2} \right\} \right) \quad (5)$$

In this equation, H_{2D} is the mean number of overlapping fibers for 2-D sheets and $P(0) = \exp\{-H_{2D}\}$. The parameters β_1, β_2, and D represent the completeness of fiber-fiber bonds within a 2-D sheet, between two contiguous layers, and between two layers separated by other layers, respectively. These parameters are, therefore, dependent on fiber flexibility and shape of fiber cross section. Kallmes and Bernier [21] determined β_1, β_2, and D experimentally by examining the crossings of a three-layer sheet, each of which was dyed differently. Figure 4 shows the relative bonded area as a function of H calculated for different β_1, β_2, and D values (mildly beaten kraft and unbeaten sulfite pulps). As expected, the values of RBA are highly influenced by fiber flexibility through the parameters β_1, β_2, and D. The effect of the mean number of overlapping fibers H (or basis weight) on RBA is more significant when H is less than 5 (i.e., the mean number of overlapping fibers is less than 5). Beyond this value, RBA approaches its limiting value very slowly as H (or basis weight) increases. The value of 100% for RBA is only attainable for sheets with infinitesimally small fiber thickness and an infinite number of fibers per unit area (infinite basis weight). These results suggest that

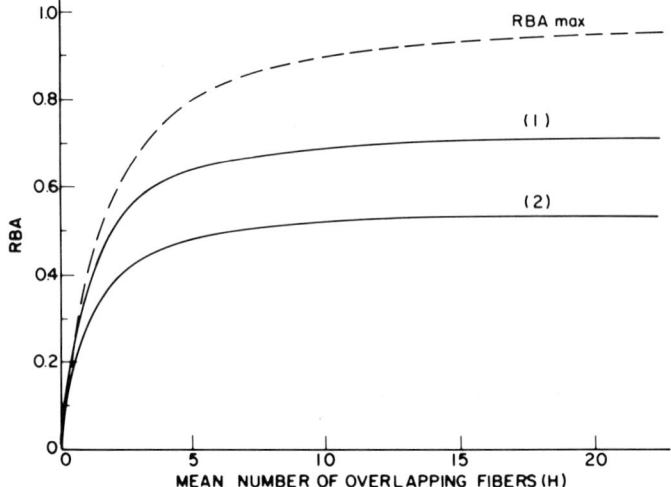

Fig. 4 RBA as a function of mean number of overlapping fibers (H) for a multiplanar sheet model. (1) Mildly beaten kraft ($\beta_1 = 0.98$, $\beta_2 = 0.82$, $D = 1.54$); (2) unbeaten sulfite ($\beta_1 = 0.84$, $\beta_2 = 0.60$, $D = 1.08$). (Data for β_1, β_2, and D from Ref. 21.)

(1) The basis weight for paper is not simply a scaling parameter such as thickness for a homogeneous continuum, but indeed is one of the important structural parameters.
(2) The degree of fiber-fiber bonding in a paper sheet generally varies with basis weight.

The concept of RBA only refers to the overall condition of the bonding states in the sheet. If one looks at a typical fiber in the paper sheet (Fig. 5), one can see three different bonding states along the fiber, i.e., free on both surfaces of the fiber (0-state), bonded on one side (1-state), and bonded on two sides (2-state). The fiber segments that correspond to the above three bonding states in Fig. 5, may have different mechanical, optical, and electrical properties. An analytical relation between RBA and the three bonding states was developed by Stone et al. on the basis of the multiplanar

Fig. 5 Idealized bonding states of a typical fiber. (From Ref. 24.)

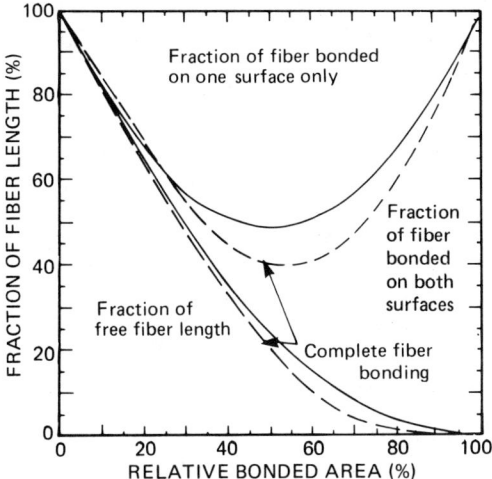

Fig. 6 Bonding state diagram for multiplanar sheet model. (From Ref. 26.)

fiber network model (see Appendix 1 in Ref. 26). The relationships between the bonding states and RBA are illustrated in Fig. 6. As Kallmes indicated [20], the basic feature of the relations does not change very much, even for the case of complete interfiber bonding. It should be noted that the value of each fraction shown in Fig. 6 is the mean value for the entire sheet. The bonding states of fibers, however, generally vary through the sheet thickness, that is, the layers near the sheet surface have different bonding states than the inner layers [22,26]. This inhomogeneity of bonding states through the thickness suggests the possibility of basis weight dependence of some properties.

Although the statistical model mentioned above describes the basic structure of randomly deposited fiber networks, there are some limitations inherent in the basic assumptions of the model. For example, it is assumed that the fiber is a flattened ribbon and the bonding is possible only on the upper and lower surfaces of the fiber, which is not the situation observed experimentally [68]. Development of a general three-dimensional statistical model would be highly helpful in analyzing experimental results obtained by an advanced technique such as image analysis or digitizing.

B. Experimental Determination of Bonding States

Work to date on the experimental determination of bonding state has focused on the size and the shape of the bonded area and the total relative bonded area. Although a number of techniques have been employed to determine bonding state, those techniques still have some problems to be overcome in various aspects. In this section, we present some basic principles and underlying assumptions of those techniques briefly, and then discuss some of the results and the problems involved. For more detailed description of the principles and the instrumentation, readers are referred to the appropriate

386 Uesaka

chapters in this text, such as Chaps. 17 (nitrogen adsorption), 16 (optical properties), and 21 and 22 (electrical properties).

Microscopic Observation The most direct way to examine the bonding state in a paper sample is to make observations by light and electron microscopy [2,47–49]. The technique used by Page et al. [48] consists of an examination of the paper specimen at medium power in polarized vertical illumination. In order to enhance the contrast in the image, 70% of the fibers in the sheet were dyed. Figure 7 shows line drawings traced from photographs of typical crossings in softwood pulps, where the black area represents an "optical" contact area between the fibers, which is assumed to be bonded. The photographs show that the fiber-to-fiber contact area is obviously influenced by the environment in the vicinity of the crossing (obstructed fiber-to-fiber bond) as well as by the shapes of individual fibers. With this technique, Page et al. examined effects of beating and drying tension on the degree of bonding [48], and also changes in the bonding states related to tensile loading and drying [47,49].

Although these studies provide some important information on the nature of fiber-fiber bonds, some difficulties in the measurement and uncertainties in the explanation of the results were also reported [48,49]. Since a microscopic image of paper is rarely perfectly clear and examination of the

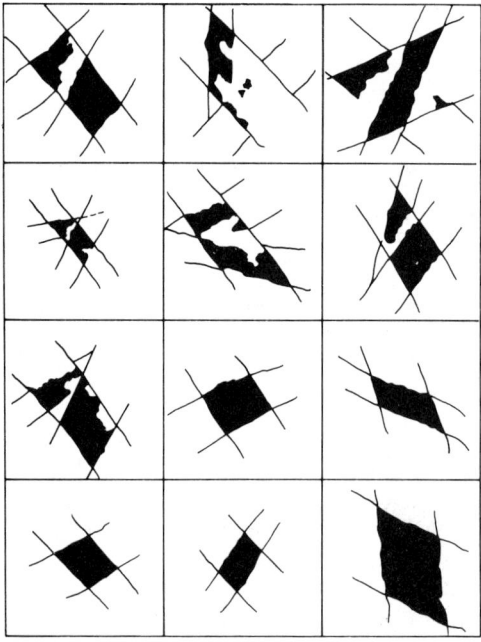

Fig. 7 Observation of optical contact areas between fibers. The first seven drawings show the bonds obstructed by neighboring fibers. (From Ref. 48.)

image is tedious, there is a tendency to be biased toward selecting the field that contains more identifiable bonds, and further toward selecting those fiber-fiber bonds having easy visibility. It should also be noted that microscopic examination of the sheet in its natural plane (by convention, the X,Y plane) is usually made only on one side of the fibers and also tends to be selective toward those fibers located near the sheet surface (surface fibers). Selection of fibers located near the surface for examination always yields a less dense image of fiber networks than exists throughout the sheet generally. This fact should be taken into account whenever free fiber segment length and frequency of fiber-fiber bonds along the fiber are discussed.

Gas Adsorption and Optical Scattering Methods Indirect methods for the determination of bonding states, such as the optical and gas adsorption methods, are based on the measurement of the surface area of the constituent fibers. Considerable work has been done both on development of the methods and examination of the validity of the basic assumptions. The early literature on these techniques has been reviewed by Swanson [59].

The nitrogen adsorption technique was employed by Haselton [16,17] to determine the bonded area with the use of the Brunauer, Emmett and Teller (BET) equation (Chap. 17). The bonded area was defined as the difference between the specific surface area (internal and external) area of the sheet and the specific surface of the "unbonded" sheet, in which there is no fiber-fiber bond. In order to obtain the unbonded sheet, two drying techniques were compared: (1) solvent exchange drying from water to benzene (or butanol) and (2) spray drying, in which a 0.20% consistency pulp slurry was sprayed onto a horizontal sheet of polyethylene and dried so that the drying condition was close to that of the usual water-dried sheet [17]. Because of the highly porous structure of benzene-(or butanol-) dried pulp fibers [17, 18], the solvent-exchange drying technique consistenly gave higher values of BET area than those for the spray-dried fibers. Also, electron micrographs of slightly beaten sulfite pulp fibers [17] showed that the surface of butanol-dried fibers still have pronounced surface fibrillation, whereas the surface of the water-dried fibers was almost free of projecting fibrils. Such evidence is consistent with and supportive of the results of BET area measurements, which suggests that the application of spray-drying techniques to the determination of the specific surface area of unbonded sheets has further potential.

The most commonly used method is the optical method [18,29,31,32,50, 52,60], which utilizes the specific scattering coefficient calculated from an adaptation of the Kubelka-Munk theory (see Chap. 16). The basic assumptions of this method can be summarized as follows [60]:

1. The specific scattering coefficent s is a linear function of the "external" surface area A.

$$s = kA + i \qquad (6)$$

 where k and i are constants.
2. The specific scattering coefficient s_t for a completely unbonded sheet is determined by the extrapolation of an s versus tensile strength (or elastic modulus) curve to zero tensile strength (or elastic modulus).

Therefore the total dry surface area A_t is calculated from the corresponding specific scattering coefficient s_t by using Eq. (6).

On the basis of these assumptions, the bonded surface area A_b and the relative bonded area RBA are given by

$$A_b = A_t - A = \frac{s_t - s}{k} \tag{7}$$

$$RBA = \frac{A_b}{A_t} = \frac{s_t - s}{s_t - i} \tag{8}$$

The relation given in Eq. (6) has not been generally proved from the physics of light-paper sheet interaction. Haselton [16,17] and later Swanson [60], however, showed that there was an excellent linear relationship between the scattering coefficient s and surface area A_{BET} measured by the nitrogen adsorption technique. They also showed that k and i in Eq. (6) are essentially constants for sheets with varying degrees of refining and wet pressure but are dependent on the type of pulp and the wavelength of the light employed. Therefore, the use of the assumption 1 is limited to the same pulp type and the same wavelength.

With respect to assumption 2, several criticisms have been presented concerning the validity of the extrapolation of a single s versus tensile strength curve for sheets with different degrees of refining [32,60]. The underlying assumption of the extrapolation technique is that the functional relationship between tensile strength (or elastic modulus) and bonded area does not change during refining. In other words, the variables, other than the bonded area, that can affect the tensile strength are to remain constant during the treatment.

Experimental results, however, showed that pulps with various degrees of beating gave different extrapolated values [32]. Therefore, assumption 2 can be used only for pulps prepared by varying wet pressure at each freeness level.

Rennel [53a] conducted a systematic study on bonded area determination by optical methods and compared light scattering values for solvent exchange-dried fibers, freeze-dried fibers and spray-dried fibers. The light scattering coefficient was greater for solvent exchange-dried fibers than for freeze-dried fibers, and the latter was greater than for the spray-dried fibers. Hasuike [17a] found a remarkable agreement between the light scattering values of spray-dried unbonded sheets and those obtained from the extrapolation of sonic modulus to its zero value by varying wet pressure.

Digitizing Method A more straightforward way of analyzing internal fiber network geometry was recently developed by Yang and coworkers [68] and Eusufzai [9]. This method essentially consists of the analysis of photomicrographs of sheet cross sections on a digitizer interfaced with a computerized data processing system [6a,68].

In a point mode digitizer, particular points of interest in the fiber cross sections, such as shown in Fig. 8, are digitized. In a continuous mode digitizer, the stylus or cursor obtains data points at very short intervals for integration. Numerical position information (e.g. the coordinates of features

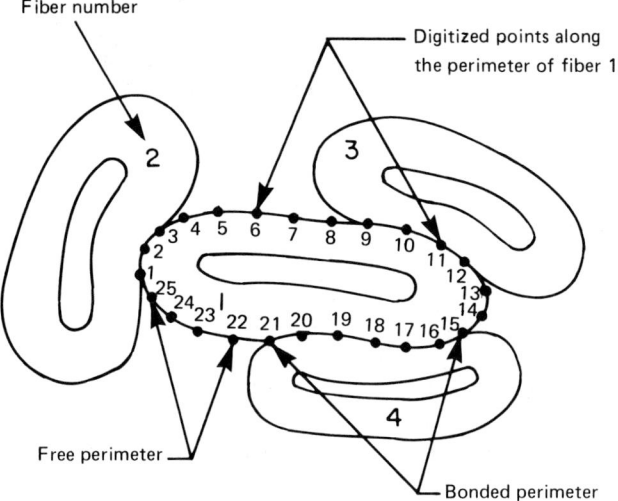

Fig. 8 Point mode loci of fiber cross sections. (From Ref. 68.)

that may represent bonded or unbonded surface) obtained from the digitizer is processed by the computer system. Various geometric parameters may be determined, such as fiber aspect ratio, fiber moments of inertia, relative bonded area, and bonding state probability [9,68]. (See also pp. 339-358.)

The main advantage of this method are the following:

- There is little or no ambiguity in the measurements as long as the cross sections are thin (< 1 μm).
- • Various data processing operations can be easily conducted through the computer system. This feature is especially useful for the statistical description of bonding states.

Other Methods The DC electrical conductivity method was first employed by Smith [57,58] to determine relative bonded area. The basic idea is that electrical conductivity is a measure of the intensity of hydrogen bonding between external (or internal) surfaces. The relative bonded area RBA was defined as the ratio of the measured conductivity σ to the extrapolated conductivity σ_∞ for infinite beating time, that is, RBA = σ/σ_∞ [57]. The underlying assumption of this definition is that the conductivity σ is "proportional" to the external surface area of the sheet in different beating stages. The applicability of this assumption, however, has not been fully explored.

Byrd [4a] introduced a new quantity called web shrinkage energy as an index of fiber-fiber bonding. This energy was defined as the area under the curve of stress measured during uniaxially restrained drying versus shrinkage strain measured during free drying. Since the external force does not work on paper during completely restrained drying, this quantity does not have the usual meaning of energy, such as the work done by external force in a tensile test, but may be a complex function of moisture-expansion coefficients and relaxation moduli, both of which are also functionals of drying history. In order to establish a sound relationship between the web shrinkage energy and fiber-to-fiber hydrogen bonding, more work would be needed on the constitutive stress-strain-moisture relation from structural and phenomenological (thermodynamic) standpoints.

III. MECHANICAL PROPERTIES OF FIBER-FIBER BONDS

Determination of mechanical properties of fiber-fiber bonds, in particular bond strength, has certainly been of considerable interest for many years. The concept of bond strength, however, has not been well defined or clarified in many practical instances. This term appears to be widely used in the paper technology field as the interface strength of fiber-fiber bonds, a counterpart of fiber strength. Quite often, work on "bond strength" is reported with the intention of explaining experimental results for various effects of manufacturing variables on sheet strength properties. As will be seen in the following sections, the "bond strength" is, however, essentially the structural strength of a fiber-fiber bond system rather than the strength of the interface between bonded fibers. Therefore, the definition and physical meaning of bond strength should be fully clarified in each experimental determination.

A. Bond Strength Measurement (Indirect Methods)

One of the outstanding contributions in the paper physics field is the fundamental study on optical changes in sheet structure during tensile straining cycles that was conducted by Nordman et al. [41–44]. They found a linear relationship between the net work W done by external force during a tensile straining cycle (dissipated energy) and the change in the scattering coefficient Δs (Fig. 9). The observed change in the scattering coefficient was attributed to the creation of new surface as a result of fiber-fiber bond breakage (partial or total); the "bonding strength," which is often called Nordman bonding strength, was defined as the gradient of the dissipated energy W versus the scattering coefficient change Δs curve (Fig. 9).

Since this remarkable finding, considerable discussion has ensued [27, 30,41,64] on the physical meaning of Nordman bonding strength. Van den Akker indicated that there is a great discrepancy between the theoretically estimated rupture energy of the fiber-fiber bond (on the oder of 10^{-2} to 10^{-1} J/m^2) and the Nordman energy (23 J/m^2), which was corrected on the basis of nitrogen adsorption data [17] to convert the scattering coefficient to surface area.

This discrepancy is not surprising in light of fracture mechanics studies. For ductile materials having defects or cracks, most of the energy input is usually consumed in viscoelastic (linear or nonlinear) and plastic deformation of the surrounding material rather than in the creation of new surface. The dissipated energy also depends on the deformation mode around defects. Therefore, Nordman bonding strength may be considered as a quantity representing an energy-dissipative characteristic of the sheet related to structural changes in it during uniaxial tensile straining, rather than the interface strength of the fiber-fiber bond.

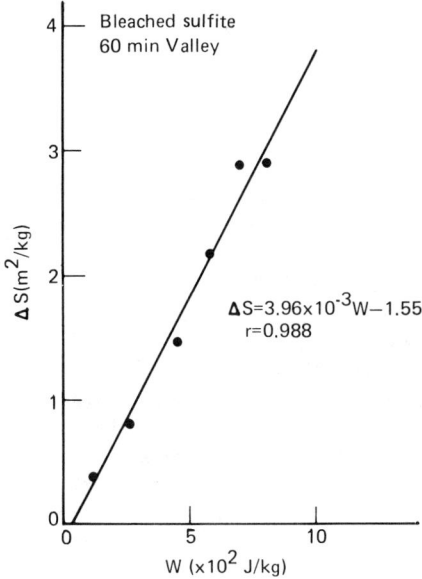

Fig. 9 Relationship between change in scattering coefficient Δs and dissipated energy W during a tensile straining cycle. (From Ref. 43.)

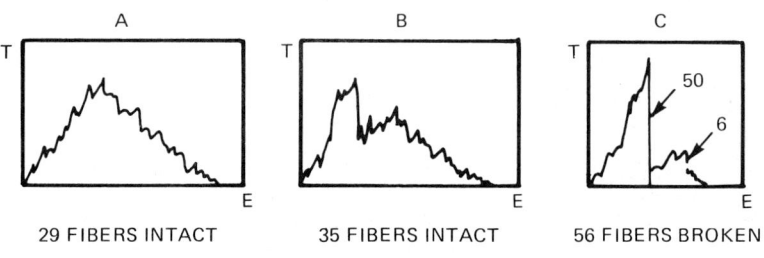

Fig. 10 Typical load-elongation curves from fiber extraction tests. (From Ref. 7.)

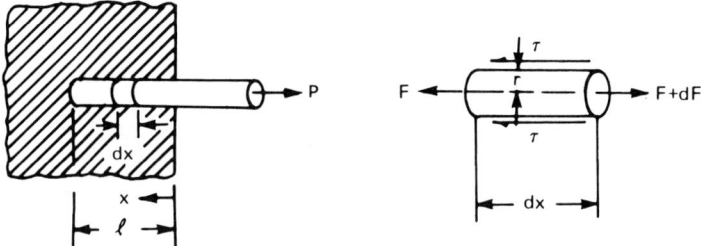

Fig. 11 Pull-out test of fiber-matrix composite. (From Ref. 14.)

Single-fiber extraction tests were performed by Davison [7] and Katsura et al. [28] in order to evaluate the respective roles of fiber and fiber-fiber bond strength in development of the entire sheet strength. A handsheet was gently torn and the fibers at the carefully teased edge were cut, except for only one protruding fiber, which was subsequently extracted from the sheet. Typical extraction load-displacement curves are shown in Fig. 10. Various determinations, such as the number of fibers extracted, peak resistance load, and work-to-break were made from these tests.

Similar testing principles can be seen in analogous tests of fiber-matrix composites [14] in which a filament embedded in a matrix is pulled out, as shown in Fig. 11.

Assuming that the shear stress in the fiber-matrix interface is uniformly distributed along the interface, we determine the shear strength τ of the interface by the relation

$$\tau = \frac{p_{max}}{2\pi r \ell} \tag{9}$$

where

p_{max} = maximum applied load
r = radius of the filament
ℓ = embedded length of the filament

Solution of Eq. (9) for p_{max}, that is, $p_{max} = 2\pi r \tau \ell$ shows that the maximum load is influenced by the embedded length ℓ of the filament as well as the shear strength τ of the interface; thus the number of fibers pulled out or broken during loading is determined by both τ and ℓ. An approximate analysis of Greszczuk [14], however, showed that the shear stress distribution along the interface is not uniform but has a maximum at the edge ($x = 0$, Fig. 11). A mechanical analysis similar to Greszczuk's could be applied in a single fiber extraction test to obtain a measure of the interface strength of a single fiber-network system.

A semiempirical equation for tensile strength (breaking length) was proposed by Page [46] and later applied in the evaluation of unbleached kraft pulps by Jones [19]. This equation is expressed in terms of the breaking length T as

$$\frac{1}{T} = \frac{9}{8Z} + \frac{12A_f \rho g}{bPL(RBA)} \qquad (10)$$

where

  Z = zero-span tensile strength
  A_f = average fiber cross section
  ρ = density of the fibrous material
  g = gravitational acceleration
  b = shear bond strength per unit bonded area
  P = perimeter of the fiber cross section
  L = fiber length
  RBA = relative bonded area

By substituting known or assumed values for T, Z, A_f, ρ, P, L, and RBA, it is possible to obtain the shear bond strength b that appears in Eq. (10). Sample calculations given by Page [46] showed that the order of magnitude of the derived value of b was comparable to those measured by direct methods, which will be described later. Although Eq. (10) appears to describe the general trends of experimental data quite well, the validity of this equation should also be examined from a mechanics point of view.

Z-direction strength tests are widely used in paper testing laboratories to determine a practical measure of bond strength [5a,18a,18b,61,61a,61b,61c, 61d,61e,65,67]. The testing modes being used are either tensile [18a,61, 61c,65,67] or delamination type [5a,18b,61a,61b,61d,61e]. For the Z-direction tensile test, Wink and Van Eperen [67] reported a complex dependence of the strength value on both basis weight and thickness. The internal bond test (delamination type) [3,51,53], which is being used extensively in quality control and for other practical purposes, measures an impact delamination strength of the fiber layers in a sheet.

Effects of specimen size and testing conditions on these tests, however, have not yet been fully examined. Because of the complexity of the micro-deformation mode of a fiber network, the relationship between Z-direction strength and fiber-fiber bond strength has not been established. Experimental details of the methods are covered in Chap. 5 in this text.

B. Bond Strength Measurement (Direct Methods)

A number of studies have been conducted to obtain the magnitude of bond strength quantitatively. A fiber-shive system (Fig. 12) was employed by

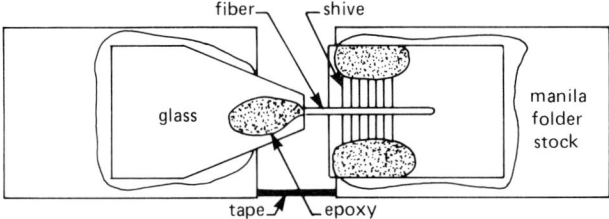

Fig. 12 Bond test (fiber-shive system). (From Ref. 62.)

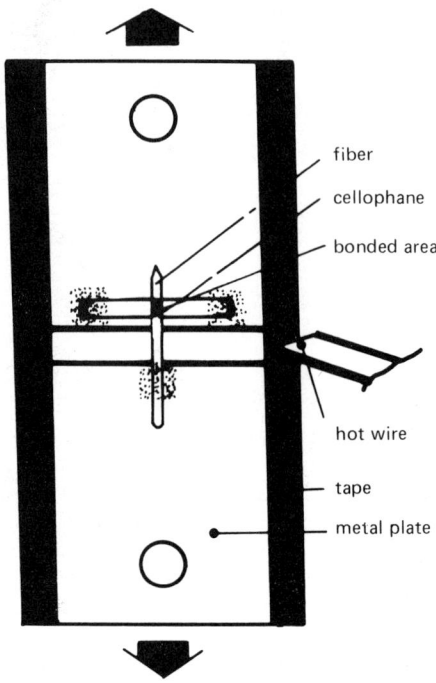

Fig. 13 Bond test (fiber-cellophane system). (From Ref. 36.)

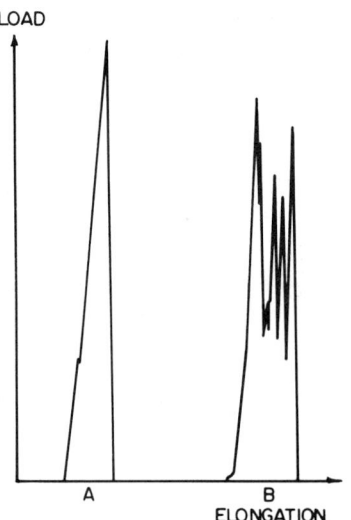

Fig. 14 Typical load-elongation curves of fiber-cellophane bond system. (A) Curve observed for relatively stiff fibers. (B) Curve more frequently observed. (From Ref. 36.)

McIntosh and Leopold [35], McIntosh [34], and later Thorpe and coworkers [62]. A single fiber was bonded to a shive consisting of 5 to 15 fibers in order to get a large contact area. The bonded area was measured under polarized transmitted light [34,35]. The bond (shear) strength was defined as the maximum load to separate the single fiber bonded to the shive divided by the bonded area. In addition to the bond shear strength, Thorpe et al. [62] determined the strains both in the free fiber segment and the bonded region of the fiber-shive system by measuring the displacements of reference points consisting of xerography toner particles that had been deposited on the fiber surface.

The bond shear strength of a single fiber crossing was determined by Mayhood et al. [33] and Schniewind and Nemeth [56]. Test specimens were prepared in one of two ways:

1. From fiber crossings deposited on a wire from an extremely dilute suspension [33]
2. By crossing two wet fibers at an angle of 90° [56]

Mohlin [36] employed a fiber-cellophane bond system instead of the fiber-shive bond (Fig. 13). Two distinct load-elongation curves were obtained (Fig. 14): one is the curve that has only one peak, observed for relatively stiff fibers (Fig. 14A), and the other consists of a series of abrupt reductions in load due to partial breakage of the bond, which was more frequently observed (Fig. 14B). In the latter case Mohlin defined bond shear strength as the sum of cumulative load increments of the curve divided by the bonded area [36].

Because of difficulties in specimen preparation, handling, and testing, a fair number of specimens are rejected in actual tests. The reported strength values fall within a range of 0.3 to 8.1 $\times$ 10^6 Pa and generally have a large variation within the same sample [33–36,56,62]. Effects of various manufacturing variables on bond shear strength have been reported: pulp and fiber type [15,33–36,56,62], pulp yield [15,34], drying [37,56], beating [33,37], chemical modification [38], wet strength resin [55], and corona treatment [13]. One of the interesting results is that bond shear strength of summerwood fibers has been reported as larger than that of springwood fibers [34,35,56], whereas Mohlin showed that if the cumulative bond strength defined above was used for comparison, there was no significant difference between them [36].

A basic assumption of the determination procedure for bond shear strength, that is, dividing failure load by bonded area, is that the shear stress distribution is uniform over the bonded area. However, recent studies of adhesive bond systems show that stress distribution in the bonded area is highly nonuniform and, furthermore, significantly affected by various factors [1]. Goland and Reissner [11] made an approximate analysis of a single lap joint (Fig. 15) by using the finite deflection theory of a cylindrically bent plate. Figure 16 shows the distribution of the shear stress τ_0 and the tearing stress σ_0 (where σ_0 is the normal stress acting perpendicular to the joint plane) in the adhesive layer along the joint plane for a relatively inflexible adhesive [11]. In this particular case the shear stress τ_0 and the tearing stress σ_0 are highly localized in small regions near the end of the joint. For the case of relatively flexible layers, these stresses are distributed more uniformly but still have a maximum value at the joint edges [11].

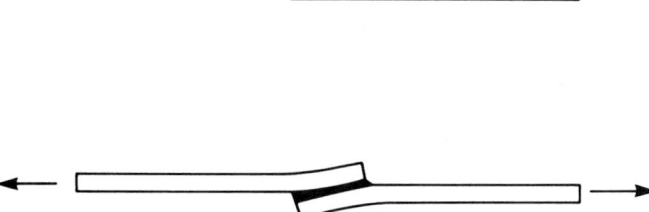

Fig. 15 Single lap joint. The upper figure shows the joint under no load. Loading causes rotation of the joint plane (lower figure).

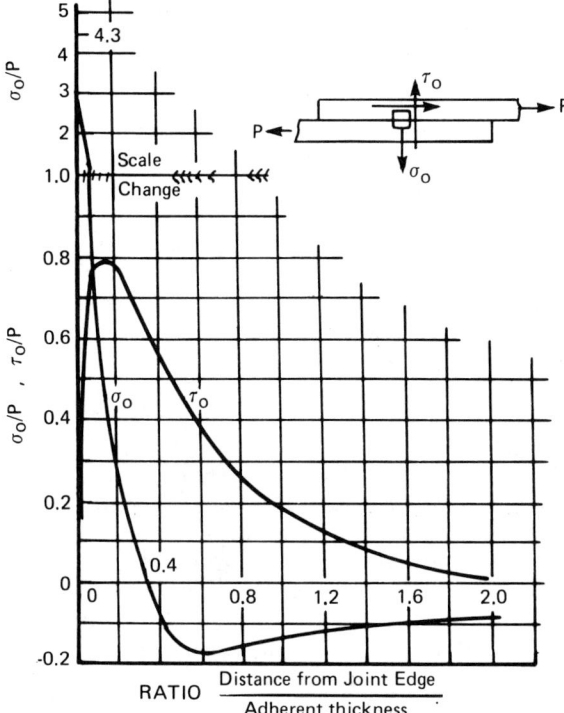

Fig. 16 Stress distribution along joint plane with relatively inflexible adhesive. (τ_0) shear stress; (σ_0) tearing stress; (P) mean tensile stress of the adherend away from the joint. (From Ref. 11.)

The stress distribution near the edge is also influenced by lap length, adherend and adhesive thicknesses, and geometry of the adherend near the end (for example, a tapered end) as well as adherend and adhesive properties [11,66]. The problem of load transfer of a strip that is lap-jointed to a plate, such as in the fiber-shive or the fiber-cellophane systems, was investigated by Goodier and Hsu [12] and later solved for the more general case by Muki and Sternberg [39]. In either case, the region near the edge of the bonded area plays an important role in the load transfer. For example, if the Poisson ratios of two bonded materials are the same, the load is transmitted by the periphery of the bonded region [12].

For the fiber-shive system, Thorpe and coworkers [62] were the first to model these mechanical aspects and make an approximate stress analysis of a two-dimensional rectangular bonded region. Button [4] conducted a systematic study of a single lap joint of cellophane fibers and developed a linear elastic model of the joint system by using a finite element analysis and fracture mechanics. Effects of fiber width, fiber thickness, fiber axial modulus, bond length, and bond asymmetries (dimensional and fiber modulus asymmetry) were examined in detail. It was found that among the above-mentioned factors, the bond length has a remarkable influence on bond shear strength (Fig. 17).

In the short bond length region (short overlapped length, < 300 µm), the bond shear strength rapidly decreases with increasing bond length, as shown in Fig. 17. This strong dependence on bond length was predicted from a linear elastic fracture mechanics model on the basis of a fracture

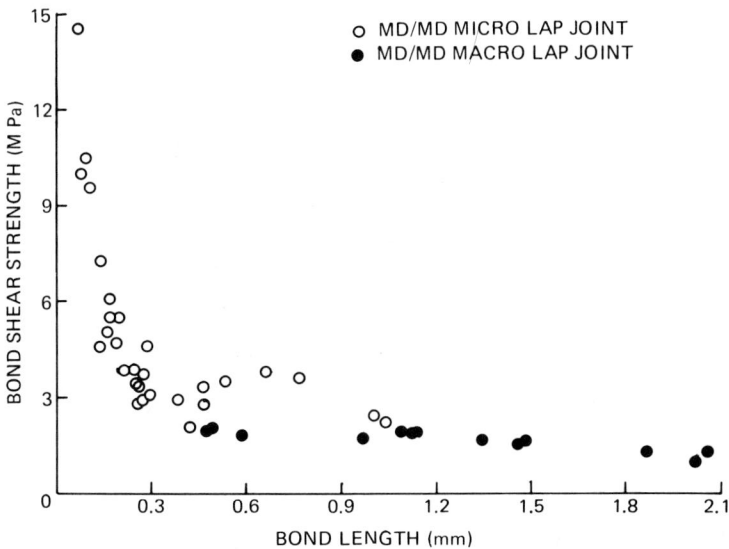

Fig. 17 Relationship of bond shear strength to bond length (overlap length) for cellophane lap joints. Macro and micro lap joints were prepared from 12 mm wide and 0.1 to 0.5 mm wide cellophane film strips, respectively. (From Ref. 4.)

toughness value determined from the cracked micro-lap joint data. The modulus asymmetry, such as MD/CD configuration of the single lap joint, and the dimensional asymmetry, such as thickness difference between overlapped fibers, were found to result in a reduction in the bond strength through the formation of more intense stress fields than in the symmetric case [4].

The above mechanical analyses for fiber-fiber bond models show that bond shear strength is the strength of the bond structure rather than the interface bond strength (or inherent bond shear strength). Even in the case of simple structures such as discussed in this section, various factors can affect the strength value just as much as the (unknown) interface strength. This feature of the bond shear strength may partly explain and reconcile the more or less conflicting results that have been obtained so far in model studies of the fiber-fiber bond.

REFERENCES

1. Anderson, G. P., Bennet, S. J., and DeVries, K. L. (1977). *Analysis and Testing of Adhesive Bonds*, Academic Press, New York.
2. Asunmaa, S., and Steenberg, B. (1958). Beaten pulps and fiber-to-fiber bond in paper. *Svensk Papperstidn.* 61(186):686–695.
3. Blockman, A. F., and Wikstrand, W. C. (1958). Interfiber bond strength of paper. *Tappi* 41(3):191A–194A.
4. Button, A. F. (1979). Fiber-fiber bond strength: A study of a linear elastic model structure. Ph.D. thesis, Institute of Paper Chemistry, Appleton, Wis.
4a. Byrd, V. L. (1974). Web shrinkage energy: An index of network fiber bonding. *Tappi* 57(6):87–91.
5. Campbell, W. B. (1933). Forest Service Bulletin 84, Department of Interior, Canada.
5a. Canadian Pulp and Paper Association Useful Method D.8U (1977). Plybond strength of paperboard (using Jumbo Muller tester).
6. Corte, H., and Schaschek, H. (1955). The physical nature of paper strength. *Das Papier* 9(11):519–530 (in German).
6a. Crosby, C. M., Eusufzai, A. R. K., Mark, R. E., Perkins, R. W., Chang, J. S., and Uplekar, N. V. (1981). A digitizing system for quantitative measurement of structural parameters in paper. *Tappi* 64(3):103–106.
7. Davison, R. W. (1972). The weak link in paper dry strength. *Tappi* 55(4):567–573.
8. Emerton, H. W. (1957). *Fundamentals of the Beating Process*. British Paper and Board Industry Research Association, Kenley, England.
9. Eusufzai, A. R. K. (1982). Sheet structure in relation to internal network geometry and fiber orientation distribution. M.S. thesis, State University of New York, College of Environmental Science and Forestry, Syracuse, New York.
10. Gallay, W. (1958). Some aspects of the theory of the beating process. In *Fundamentals of Papermaking Fibers* (F. Bolam, ed.), British Paper and Board Makers Assn., pp. 377–387.

11. Goland, M., and Reissner, E. (1944). The stresses in cemented joints. *J. Appl. Mech.* 11:A17–A27.
12. Goodier, J. N., and Hsu, C. S. (1954). Transmission of tension from a bar to a plate. *J. Appl. Mech.* 21(6):147–150.
13. Goring, D. A. I. (1967). Surface modification of cellulose in a corona discharge. *Pulp Paper Mag. Can.* 68(8):T372–T376.
14. Greszczuk, L. B. (1969). Theoretical studies of the mechanics of the fiber-matrix interface in composites. In *Interfaces in Composites*, ASTM STP 452, American Society for Testing and Materials, Philadelphia, pp. 42–58.
15. Hartler, N., and Mohlin, U-B. (1975). Cellulose fiber bonding. 2. Influence of pulping on interfiber bond strength. *Svensk Papperstidn.* 78(8):295–299.
16. Haselton, W. R. (1954). Gas adsorption by wood, pulp, and paper. 1. The low-temperature adsorption of nitrogen, butane, and carbon dioxide by sprucewood and its components. *Tappi* 37(9):404–412.
17. Haselton, W. R. (1955). Gas adsorption by wood, pulp and paper. 2. The application of gas adsorption techniques to the study of the area and structure of pulps and the unbonded and bonded area of paper. *Tappi* 38(12):716–723.
17a. Hasuike, M. (1973). On the physical properties and macrostructure of earlywood and latewood sheets. *J. Japan Wood Res. Soc. (Mokuzai Gakkaishi)* 19(11):547–553 (in Japanese).
18. Ingmanson, W. L., and Thode, E. F. (1959). Factors contributing to the strength of a sheet of paper. 2. Relative bonded area. *Tappi* 42(1):83–93.
18a. Japan TAPPI Standard Method No. 18–77 (1981). Determination of internal bond strength of paper and board (in Japanese).
18b. Japan Tappi Standard Method No. 19–77 (1981). Determination of plybond strength of paperboard (in Japanese).
19. Jones, A. R. (1972). Strength evaluation of unbleached kraft pulps. *Tappi* 55(10):1522–1527.
20. Kallmes, O. (1972). A comprehensive view of the structure of paper. In *Theory and Design of Wood and Fiber Composite Materials*, (B. A. Jayne, ed.), Syracuse Univ. Press, Syracuse, N.Y., pp. 157–175.
21. Kallmes, O., and Bernier, G. (1962). The structure of paper. 3. The absolute, relative, and maximum bonded areas of random fiber networks. *Tappi* 45(11):867–872.
22. Kallmes, O., and Bernier, G. (1963). The structure of paper. 4. The free fiber length of a multiplanar sheet. *Tappi* 46(2):108–114.
23. Kallmes, O., and Corte, H. (1960). The structure of paper. 1. The statistical geometry of an ideal two dimensional fiber network. *Tappi* 43(9):737–752.
24. Kallmes, O., and Eckert, C. (1964). The structure of paper. 7. The application of the relative bonded area concept to paper evaluation. *Tappi* 47(9):540–548.
25. Kallmes, O., Corte, H., and Bernier, G. (1961). The structure of paper. 2. The statistical geometry of a multiplanar fiber network. *Tappi* 44(7):519–528.

26. Kallmes, O., Corte, H., and Bernier, G. (1963). The structure of paper. 5. The bonding states of fibers in randomly formed paper. *Tappi* 46(8):493–502.
27. Kärnä, A. (1961). Studies on the relationships between the optical and stress-strain properties of paper, 1. *Paperi ja Puu* 43(8):465–472.
28. Katsura, T., Murakami, K., and Imamura, R. (1978). Single fiber extraction test using long fiber sheets. *J. Japan Wood Res. Soc. (Mokuzai Gakkaishi)* 24(8):552–557 (in Japanese).
29. Kenney, F. C. (1952). Physical properties of slash pine semichemical kraft pulp and of its fully chlorited component. *Tappi* 35(12):555–563.
30. Lathrop, A. L., and Hardacker, K. W. (1962). Light scattering by individual fibers under mechanical strain. *Tappi* 45(2):169–172.
31. Leech, H. J. (1954). An investigation of the reasons for increase in paper strength when locust bean gum is used as a beater adhesive. *Tappi* 37(8):343-349.
32. Luner, P., Kärnä, A. E. U., and Donofrio, C. P. (1961). Studies in interfiber bonding of paper: The use of optically bonded areas with high yield pulps. *Tappi* 44(6):409–414.
33. Mayhood, C. H., Kallmes, O. J., and Cauley, M. M. (1962). The mechanical properties of paper. 2. Measured shear strength of individual fiber to fiber contacts. *Tappi* 45(1):69–73.
34. McIntosh, D. C. (1963). Tensile and bonding strengths of loblolly pine kraft fibers cooked to different yields. *Tappi* 46(5):273–277.
35. McIntosh, D. C., and Leopold, B. (1962). Bonding strength of individual fibers. In *Formation and Structure of Paper* (F. Bolam, ed.), vol. 1, British Paper and Board Makers Assn., London, pp. 265–270.
36. Mohlin, U-B. (1974). Cellulose fiber bonding. Determination of interfiber bond strength. *Svensk Papperstidn.* 77(4):131–137.
37. Mohlin, U-B. (1975). Cellulose fiber bonding. 3. The effect of beating and drying on interfiber bonding. *Svensk Papperstidn.* 78(9):338–341.
38. Mohlin, U-B. (1975). Cellulose fiber bonding. 4. Effect of chemical modification on rayon fiber bonding ability. *Svensk Papperstidn.* 78(10):373–375.
39. Muki, R., and Sternberg, E. (1968). On the stress analysis of overlapping bonded elastic sheets. *Intl. J. Solids Structures* 4(1):75–94.
40. Nissan, A. H. (1959). Fundamentals of adhesion from molecular forces in cellulose. *Tappi* 42(12):928–933.
41. Nordman, L. (1958). Bonding in paper sheets. In *Fundamentals of Papermaking Fibers* (F. Bolam, ed.), British Paper and Board Makers Assn., London, pp. 333–347.
42. Nordman, L., Gustafsson, C., and Olofsson, G. (1952). On the strength of bonding in paper. *Paperi ja Puu* 34(3):47–52.
43. Nordman, L., Gustafsson, C., and Olofsson, G. (1954). The strength of bonding in paper, 2. *Paperi ja puu* 36(8):315–320.

44. Nordman, L., Gustafsson, C., and Olofsson, G. (1955). Optical measurement of bond breaking during a tensile test. *Tappi* 38(12):724–727.
45. Page, D. H. (1960). Fiber-to-fiber bonds. 1. A method for their direct observation. *Paper Technol.* 1(4):407–411.
46. Page, D. H. (1969). A theory for the tensile strengh of paper. *Tappi* 52(4):674–681.
47. Page, D. H., and Tydeman, P. A. (1962). A new theory of the shrinkage, structure nad properties of paper. In *Formation and Structure of Paper* (F. Bolam, ed.), vol. 2, British Paper and Board Makers Assn., London, pp. 397–413.
48. Page, D. H., Tydeman, P. A., and Hunt, M. (1962). A study of fiber-to-fiber bonding by direct observation. In *Formation and Structure of Paper* (F. Bolam, ed.), vol. 1, British Paper and Board Makers Assn., London, pp. 171–193.
49. Page, D. H., Tydeman, P. A., and Hunt, M. (1962). The behavior of fiber-to-fiber bonds in sheets under dynamic conditions. In *Formation and Structure of Paper* (F. Bolam, ed.), vol. 1, British Paper and Board Makers Assn., London, pp. 249–263.
50. Parsons, S. R. (1942). Optical characteristics of paper as a function of fiber classification. *Paper Trade J.* 115(25):34–42.
51. Parsons, S. R. (1969). Effect of interfiber bonding on tearing strength. *Tappi* 52(7):1262–1266.
52. Ratliff, F. T. (1949). The possible correlation between hemicelluloses and the physical properties of bleached kraft pulps. *Tappi* 32(8):357–367.
53. Reynolds, W. F. (1974). New aspects of internal bonding strength of paper. *Tappi* 57(3):116–120.
53a. Rennel, J. (1969). Opacity in relation to strength properties of pulps. 1. Method for producing unbonded fibers and determining their light-scattering coefficient and surface area. *Svensk Papperstidn.* 72(1):1–8.
54. Robertson, A. A. (1970). Interactions of liquids with cellulose. *Tappi* 53(7):1331–1339.
55. Russell, J., Kallmes, O. J., and Mayhood, C. H. (1964). The influence of two wet-strength resins on fibers and fiber-fiber contacts. *Tappi* 47(1):22–25.
56. Schniewind, A. P., Nemeth, L. J., and Brink, D. L. (1964). Fiber and pulp properties. 1. Shear strength of single fiber crossings. *Tappi* 47(4):244–248.
57. Smith, W. E. (1965). Determination of the relative bonded area of handsheets by direct-current electrical conductivity. *Tappi* 48(8):476–480.
58. Smith, W. E. (1969). Investigation of a method for measuring interfiber bonding in pulp handsheets based on sheet and fiber D.C. electrical conductivities. Ph.D. thesis, North Carolina State University, Raleigh.
59. Swanson, J. W. (1956). Beater adhesives and fiber bonding—the need for further research: A review of the literature on beater or wet-end adhesives. *Tappi* 39(5):257–270.

60. Swanson, J. W., and Steber, A. J. (1959). Fiber surface area and bonded area. *Tappi* 42(12):986–994.
61. TAPPI Standard T 506 su-68 (1978). Internal bond strength of paper and paperboard as measured by a Z-directional tensile test.
61a. TAPPI Useful Method UM 403 (1978). Test for interfiber bond using the internal bond tester.
61b. TAPPI Useful Method UM 522 (1978). Plybond strength of paperboard.
61c. TAPPI Useful Method UM 528 (1978). Internal bond strength of paperboard (Z-directional tensile).
61d. TAPPI Useful Method UM 569 (1978). Internal bond strength of paper strip delamination.
61e. TAPPI Useful Method UM 808 (1978). Plybond peeling strength of linerboard and corrugated board.
62. Thorpe, J. L., Mark, R. E., Eusufzai, A. R. K., and Perkins, R. W. (1976). Mechanical properties of fiber bonds. *Tappi* 59(5):96–100.
63. Van den Akker, J. A. (1959). Structural aspects of bonding. *Tappi* 42(12):940–947.
64. Van den Akker, J. A. (1969). An analysis of the Nordman bonding strength. *Tappi* 52(12):2386–2389.
65. Van Liew, G. P. (1974). The Z-direction deformation of paper. *Tappi* 57(11):121–124.
66. Westmann, R. A. (1975). Geometrical effects in adhesive joints. *Intl. J. Eng. Sci.* 13(4):369–391.
67. Wink, W. A., and Van Eperen, R. H. (1967). Evaluation of z-direction tensile strength. *Tappi* 50(8):393–400.
68. Yang, C. F., Eusufzai, A. R. K., Sankar, R., Mark, R. E., and Perkins, R. W. (1978). Measurements of geometrical parameters of fiber networks. 1. Bonded surfaces, aspect ratios, fiber moments of inertia, bonding state probabilities. *Svensk. Papperstidn.* 81(13):426–433.

26
DIMENSIONAL PROPERTY MEASUREMENTS

VANCE C. SETTERHOLM

U.S. Forest Products Laboratory*
Forest Service
U.S. Department of Agriculture
Madison, Wisconsin

I.	Introduction	404
	A. Basic Properties	404
	B. Breaking Length	405
	C. Force per Cross-Sectional Area	405
II.	Methods to Be Used for Physical Tests in a Constant Environment	407
	A. Effective Thickness Concept and Supporting Theory	407
	B. Methods for Obtaining Effective Thickness t_e	408
	C. Mercury Pycnometer	409
	D. Microscope	409
	E. Modified Micrometers	409
	F. Comparison of FPL Thickness Micrometer with Standard Caliper Method	411
	G. Measurement of Length and Width	412
III.	Techniques for Quantifying Changes in Specimen Dimensions Due to Changing Environments	413
	A. Thickness	413
	B. Length and Width	413
	References	414

This article was written and prepared by U.S. Government employees on official time, and it is therefore in the public domain.
*Maintained at Madison, Wisconsin, in cooperation with the University of Wisconsin.

I. INTRODUCTION

Measurement is the key to process change and product improvement. Unless we can measure the functional properties of paper accurately and precisely, we are limited in what we can do to improve its performance and utility. For example, how do we change opacity or stiffness unless we can measure these properties? While visual inspection and the feel of materials give us a sensory estimate of utility, if we are to be objective we know we must quantify or measure these properties. The more basic and precise our measurments of dimensional properties, the more universal will be their application.

Accuracy and precision are well defined (see Chap. 1), and failure to achieve these objectives is perhaps one of choice or expediency. From a casual examination of the existing body of technical information relevant to the paper industry, it is apparent that we have not placed great value on dealing with basic properties when it comes to the evaluation of the physical properties of paper. Hence, the paper industry has grown up with properties and units of measurement that are unique to this industry. As a result, those of us who work with these concepts are restricted in the extent to which we can relate to technology from other industries, and thus we are limited in our ability to make process and product changes. While research has been a useful adjunct to the paper industry, it has not been the vital companion of production that research is found to be in the chemical and electronics industries.

If research has not played its part, if research has not received its share of the budget, it has to be because it lacks utility. But this lack of utility stems directly from the researchers' inability to make the right measurements.

A. Basic Properties

In testing the strength or elastic properties of any material, we conventionally push it, pull on it, or twist it; these tractions are reflected by deformations in three basic modes—tension, compression, and shear (see Chaps. 2, 4, and 5).

The expression of these properties in a materials evaluation sense depends on how one views the material as well as the intended use of the resultant test data. Tensile strength, for example, can be stated as a force to failure, newton (N); force per unit width, newton per meter (N/m); stress or load per cross-sectional area under stress, newton per meter square (N/m^2); pascals (Pa); or breaking length in kilometers (km) (Chap. 5).

The choice of units in which to express the tensile strength of paper is an unresolved controversy. However, there is probably some general agreement that each of these units can be appropriate for a particular application. The unit of tensile load or force to failure says nothing about the dimensions of the material tested and, therefore, is perhaps least useful. However, such a unit may be sufficient for an in-plant evaluation of samples having nearly identical form and composition. Similarly, force per unit width is most applicable to those materials having a common thickness and, as a consequence, is more useful in quality control investigations.

The remaining controversy is essentially between those who prefer specific tensile properties, or breaking length, and those who prefer force per unit of cross-sectional area. The advantages of breaking length (meters) are that (1) it presumes a direct comparison of the intrinsic tensile strength of materials and (2) it does not require careful measurement of or concern for thickness.

B. Breaking Length

If we picture a long strip of material (uniform in width and thickness) suspended from one end, it is clear that as the length of the strip is increased, the weight of the strip will increase to the point where the supported end will eventually fail in tension. The length of such a strip is the breaking length for that specific material. In terms of a simple equation, the weight per unit area of the sheet (W/A) times its length (L) and width (w) would equal the total force required to break the strip (P):

$$\frac{W}{A} L w = P \tag{1}$$

Then

$$L = \frac{P/w}{W/A} \tag{2}$$

and breaking length is seen as the ratio of breaking force per unit of width to basis weight (sheet weight per unit of area).

Breaking length is frequently referred to as a specific property because it may also be interpreted as being derived by dividing strength (in terms of force per unit area) by density. But it must not be inferred that the density variable has thus been removed from consideration or that breaking length is constant as a function of density. The linear relationship implied by this calculation is the exception rather than the rule for any structural materials made from wood fiber.

For a material such as paper where properties generally do not vary linearly as a function of density, the use of specific parameters may, in fact, be more misleading than helpful.

C. Force per Cross-Sectional Area

The expression of properties in terms of force per unit area (pascals) has two main advantages:

1. It allows for the use of data in common structural engineering design calculations.
2. It allows for the comparison of paper with other structural materials.

Those who disagree with the use of force per cross-sectional area usually do so on the assumption that we do not know how to define paper thickness in a way that adequately handles surface roughness. If we limit our definition of thickness to those standard tests (e.g., Appita, Japan TAPPI, SCAN, TAPPI, etc.) that are promulgated by the major paper industry societies, those people are correct. However, it will be seen that it is possible to arrive at a more basic definition of thickness that overcomes the surface roughness problem.

It is frequently suggested that, because the wood substance or number of fibers varies along the length of the uniform cross section, a specific property such as breaking length is a more appropriate choice of unit. But even if nonlinear density versus strength and stiffness relationships could be circumvented, there remains the problem of dealing with the engineering design of structures. The utility of using engineering units greatly enhances our ability to treat paper as a structural material. On the other hand, even for nonstructural applications such as studies of the internal structure of paper or fibers, stress units can be useful.

The chief difficulty in expressing strength or stiffness in units that involve thickness, that is, pascals, comes when we attempt to evaluate the pulp strength potential in terms of paper properties. This becomes a complicated process because we know that any densification or thickness reduction of the dry sheet of paper will yield an apparent increase in strength. The load-carrying ability or strength in force per unit width, however, may or may not be changed. Thus, when we test two pieces of paper for purposes of inferring something about the quality of pulp from which the paper was made, we always run the risk of an invalid comparison because one may have been pressed to a different thickness during manufacture. Accordingly, it is often wise to make such comparisons on the basis of equal density.

A proper correction for density is most important. One of the most common mistakes made by both paper scientists and technicians is to assume a simple linear relationship when, in fact, we know that many properties vary with density in a nonlinear fashion. There is no assurance, moreover, that comparisons on the basis of load or nondimensional terms will lead to a valid comparison because such variables as restraint during drying, fiber orientation, moisture at the time of pressing, moisture profile differences during pressing, and drying stresses all influence such a comparison. As a consequence, each of these factors needs to be understood or controlled.

There are objections to the quantification of paper properties in units of sheet thickness, width, and length because paper is not a solid. It contains voids and, therefore, presumably should not be studied in these terms (as a continuum). The rebuttal to the nondimensional approach is based on three main arguments:

- No material is a true solid, and all contain voids to a varying degree; practical results have shown that paper can be treated as a continuum.
- The paper industry has for too long used test units, that is, breaking length, tear, fold, and so on, that preclude or prevent direct comparison with other materials of commerce. [The paper industry should use units of stress (thickness based) in order to enter the arena of structural engineering.]
- Effective thickness methods are available.

As a final comment, it seems apparent that there is a necessity for both approaches to paper and pulp evaluation and that the wise and careful technician will not close the door to either approach but will select that which best serves the purpose at hand.

II. METHODS TO BE USED FOR PHYSICAL TESTS IN A CONSTANT ENVIRONMENT

For all practical purposes, there is but one problem in determining the gross size of paper and that is in estimating a thickness value. The determination of the length and width dimensions of paper under conditions of constant humidity is easily and accurately made with a common steel scale graduated in millimeters or hundredths of inches. With careful work, one can usually obtain an accuracy of ±2%. The determination of paper thickness within ±2% is another matter. A problem arises in attaining thickness measurement precision comparable with that obtained in measuring the length. Measurement of thickness within ±2% requires the ability to discriminate to within 0.006 mm (approximately). When viewed on an appropriate scale for such precision, paper is no longer a flat, smooth sheet, but rather a material whose surface topography is one of enormous hills and valleys. The error in thickness as determined with a standard micrometer may run as high as 80% [3]. Calculated values for attendant properties such as density must then also be in error by 80%. Errors of this kind will not be compensatory because the anvils of a standard micrometer caliper rest on the tops of peaks in the paper surface and, therefore, measure extremities rather than the average thickness.

Trade and industry associations throughout the world have long standardized the platen size and pressure applied in the caliper measurement of paper, pulp handsheet, or board thickness—for example Appita Standard P426, CPPA D.4, ISO R534, JIS P-8118, SCAN-P7, and TAPPI T 411. These standards have value in that they are repeatable and reproducible from laboratory to laboratory. As such, they can serve as useful referee purpose. Their deficiency stems from the fact that they are arbitrary and fail to account for the surface irregularity and void volume.

A. Effective Thickness Concept and Supporting Theory

A most useful approach is to define thickness in terms of engineering mechanics or, specifically, that value obtained from the simultaneous solution of the equations for bending and extensional stiffness.

$$S_b = EI = \frac{Ewt^3}{12} \tag{3}$$

$$S_e = EA = Ewt \tag{4}$$

where

A = cross-sectional area
E = elastic modulus
I = moment of inertia
S_b = bending stiffness
S_e = extensional stiffness
t = thickness of sample
w = width of sample

Therefore,

$$t = \sqrt{\frac{12(EI)}{(EA)}} \tag{5}$$

It has been shown that a thickness value so obtained (that is, by making bending tests for EI and tension tests for EA) will agree well with thickness values obtained by volume displacement of mercury or by examining thickness profiles with a microscope [1]. The thickness value of Eq. (5) is termed *effective thickness*, since it represents the quantity of paper that is effective in the performance of paper under mechanical load. The concept of effective thickness presumes that not all the fibers in a sheet of paper will be positioned so they can contribute equally well to the load-carrying capacity of the sheet. Those fibers resting on the hills of the paper surface may not be stressed to the same extent that occurs in fibers lying within the body of the sheet or even those lying within the valleys of the surface roughness.

Conceptually, effective thickness is the distance between two imaginary planes cutting through the surface roughness at a distance halfway below the peak surfaces and halfway above the valleys of the actual contours of the paper's surface.

An effective thickness concept, therefore, obviates the previous necessity for dealing with paper strength in such thickness-independent units as breaking length. It also allows easy access to traditional engineering approaches for paper and paperboard evaluation and makes it possible to compare properties of paper with other structural materials in classical engineering units.

B. Methods for Obtaining Effective Thickness t_e

While the effective thickness concept is rooted in the simultaneous solution of equations for bending and extensional stiffness, this is not the preferred way of measuring t_e. Not only are these mechanical tests slow and tedious, but, to conform with theory, they require homogeneous material properties throughout the thickness of the sheet. An experimental determination for Eq. (5) will be influenced by variation in elastic modulus from top to bottom of the sheet. A good example of the experimental problem would be the determination of the thickness of multi-ply cylinderboard where material properties normally change from ply to ply. A mathematical analysis of this problem can be found in Rosenthal's report [2]. This report examines the effect of density gradients within the sheet on effective thickness. Rosenthal concludes that "the effect of physically reasonable density gradients within a sheet has no significant effect on its effective thickness. The unfinished multi-ply sheet with a definite elastic modulus gradient may minimally increase the effective thickness, but not to the extent that it becomes physically unacceptable [limits shown in report]. In spite of any sheet inhomogeneities in structure or material, the effective thickness concept produces a mechanistically equivalent section with a good approximation for purposes of applying the classical equations of mechanics to paper." Experimental evidence provided by Uesaka et al. [4] reinforces this conclusion to the extent that the differences in elastic modulus per unit mass may be much less significant than uneven thickness profile (local variations in thickness, basis weight, or density) from the standpoint of effect on bending stiffness EI and consequently on effective thickness t_e and other calculations based on it.

C. Mercury Pycnometer

A second, direct method to determine effective thickness is mercury displacement. A good analysis of this method is provided by Kimura et al. [1]. Mercury pycnometers are standardized vessels providing accurate determination of liquid volume. Since mercury will not enter the pore structure of paper (except under pressure), it is possible to use a pycnometer to measure paper volume. By carefully noting the overflow volume, mercury pycnometers can also be made to measure paper thickness.

The mercury pycnometer, which depends on volume displacement of the mercury, can be expected to give quite accurate results provided caution is taken to see that the vessel seals are tight, no air is entrapped and the mercury is cleaned regularly to remove dirt and debris that gradually accumulate. (Special care must be taken to avoid spillage that might result in a buildup of mercury vapor within the laboratory environment.)

Displacement of mercury is a technique that should be considered mainly as a device for calibration. It will work well for a few specimens, but it is sufficiently complicated and tedious that it is unlikely to be used for routine laboratory analysis. The Forest Products Laboratory mercury pycnometer is shown schematically in Fig. 1.

D. Microscope

Effective thickness of paper can also be determined by direct examination of its cross section. Effective thickness is the distance between two imaginary planes that pass midway between the peaks and valleys of the surface roughness profile. Location of these planes requires that accurate cross sections be available for examination photographically or by microscope. These cross sections can be obtained by embedding the paper in a nonswelling plastic matrix and then slicing sections through the supported sheet. While there is reason to believe this, too, could be an accurate procedure with sufficient sampling, it is more time consuming than mercury volume displacement and thus should be considered only as an alternative calibration procedure.

E. Modified Micrometers

In 1974 the Forest Products Laboratory (FPL) introduced a new concept to paper thickness measurement and reported on a modified micrometer developed for routine laboratory use. Since then, a similar device was constructed by and for use at the Swedish Forest Products Research Laboratory. This apparatus is constructed with a roller feed mechanism (Fig. 3). The State University of New York College of Environmental Science and Forestry has also initiated design and construction of an effective thickness apparatus. Pictures of the FPL micrometer and Swedish Forest Products Research Laboratory thickness micrometer are shown in Figs. 2 and 3, respectively.

The FPL micrometer utilizes a standard dial-type micrometer adjusted so that the combination of platen radius and load yields the same thickness value as that obtained with the mercury pycnometer or from experimental determination of Eq. (5). The specimen is passed slowly beneath the platen, and the micrometer records the variation in surface profile. It is the operator's task to strike the mean through the peaks and valleys. This task is

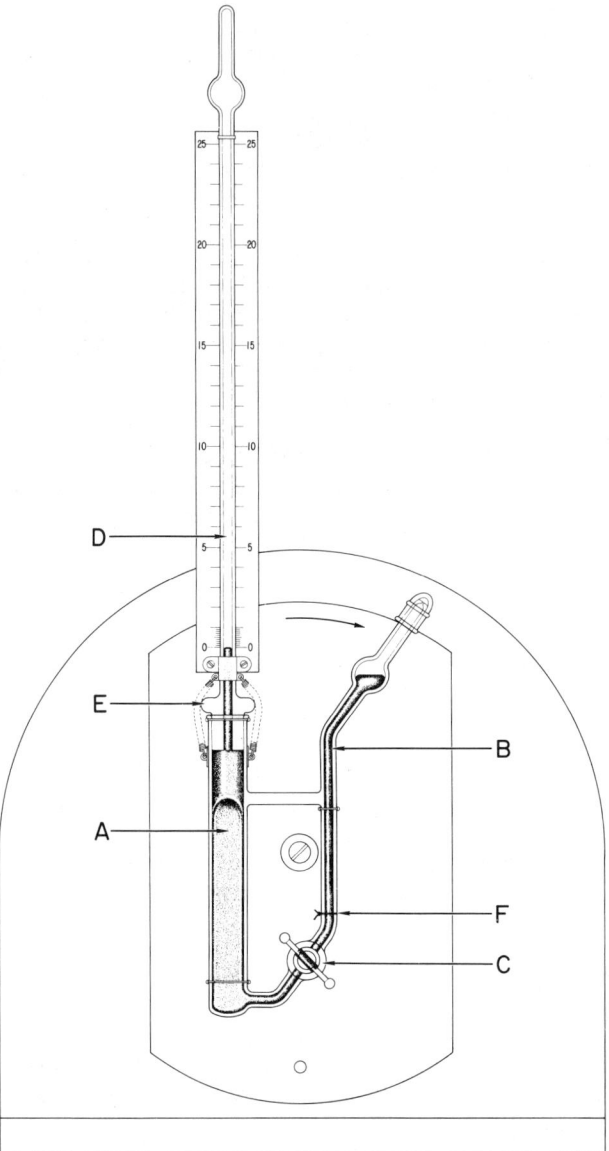

Fig. 1 Drawing of mercury pycnometer showing: (A) displacement tube to accept test sample; (B) reservoir; (C) valve; (D) overflow tube calibrated for thickness; (E) glass stopper; (F) indicator to insure proper volume is added from reservoir to displacement tube.

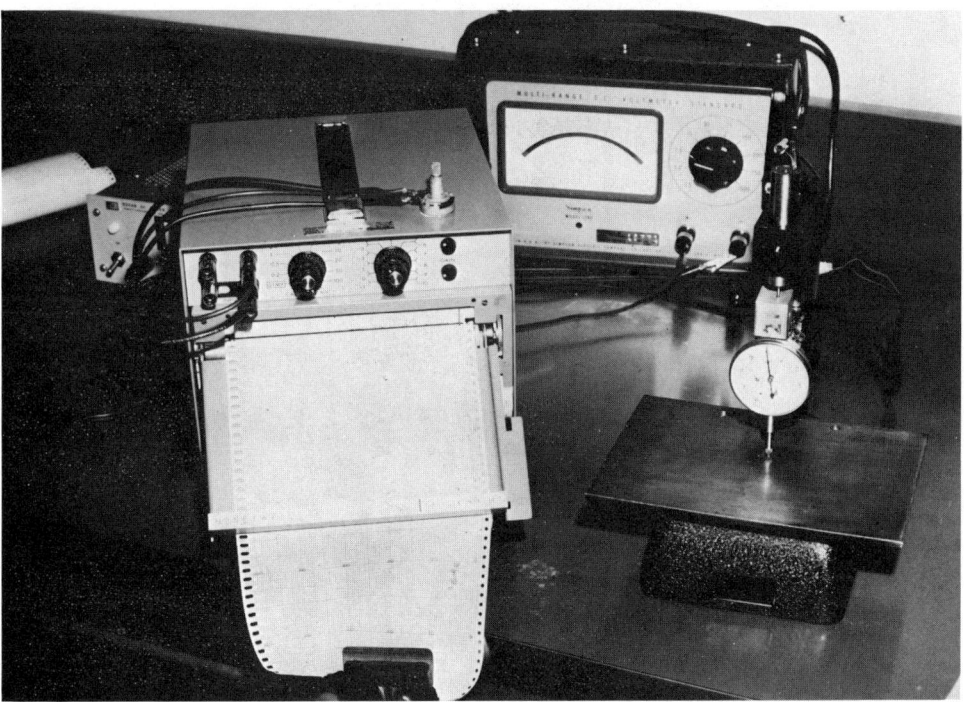

Fig. 2 U.S. Forest Products Laboratory effective thickness micrometer.

greatly simplified if the micrometer is augmented with a position transducer and recorder system so that the operator need merely look at the recorded thickness variations and note the mean. Platens of the FPL and Swedish Forest Products Research Laboratory's devices are both spherical and each has a radius of about 2.3 mm. The stem load required for the FPL apparatus to achieve thickness values comparable to that obtained with the mercury pycnometer is about 60 g. This value was arrived at experimentally. Similar stylus-type gages, designed with linear variable differential transformers (LVDT), Moiré light scaling, or magnetic scaling devices are becoming commercially available. However, some laboratories may opt to design and build a special apparatus. This is a relatively simple procedure, but one requiring painstaking care on the part of the builder.

F. Comparison of FPL Thickness Micrometer with Standard Caliper Method

Studies at FPL conducted over several years indicate that the surface roughness error in the standard TAPPI method T 411 using one sheet is generally less than 25%. However, with very coarse and rough papers such as corrugating medium, errors in thickness measurement using the standard methods and devices can be as large as 80% [1]. We find that for a typical linerboard having a basis weight of 200 g/m^2, the effective thickness will usually be

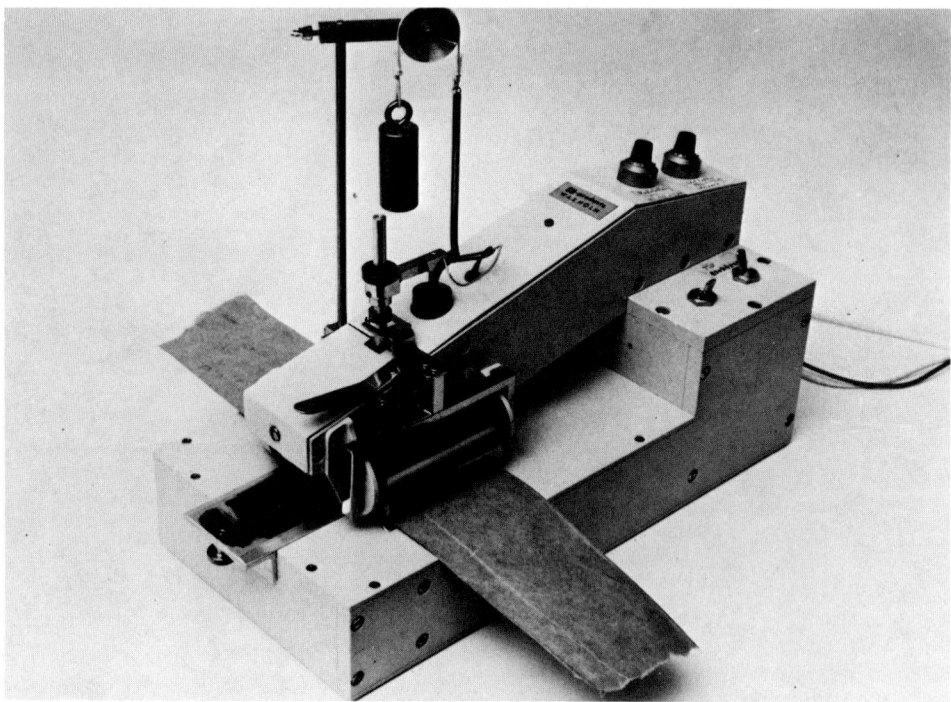

Fig. 3 Swedish Forest Products Research Laboratory effective thickness micrometer.

from 5 to 15% less than the caliper thickness. This corresponds to a surface roughness of from 0.01 to 0.05 mm.

An alternative standard method for handsheet thickness determination recognizes but does not fully account for surface roughness. It prescribes taking an average from a nested stack of five or more sheets. Procedures are given in standards such as Appita P427, CPPA D.4, ISO R438, and SCAN-P7.

G. Measurement of Length and Width

For the determination of basis weight of paper samples, it is well to measure length and width as accurately as possible. Usually this means measuring using a standard sheet size with a steel scale graduated in 0.01 in. or to the nearest 0.25 mm. For tensile, tear, or fold specimens, standard cutting devices that produce the same width of specimen each time are often used. Once a sample from the cutting device is measured to the nearest 0.001 in. or 0.025 mm with a traveling microscope or machinist's micrometer, these measurements need not be repeated but can be accepted as the standard performance of that particular specimen cutter.

For the laboratory measurement of random samples of paper for various purposes, standard steel scales have proven satisfactory and there has been little incentive to develop anything better for standard laboratory practice.

III. TECHNIQUES FOR QUANTIFYING CHANGES IN SPECIMEN DIMENSIONS DUE TO CHANGING ENVIRONMENTS

A. Thickness

Measurement of paper thickness under varying moisture conditions becomes a problem of how to make measurements on a soft, uneven surface. In recent years, the electronics industry has provided a number of transducers that are more than sensitive enough for our purposes. The most sensitive of these will discriminate to the nearest one-millionth of an inch or 0.000025 mm. Dial-type micrometers are commonly available that measure to the nearest 0.0025 mm.

Many standard micrometers apply far too much pressure; therefore, the technician must either be very judicious in selecting or modifying a standard device or else devise a thickness gage and standards for proper performance. The problem is one of getting a good sample that is representative of the material being evaluated—that is to say the paper must not be densified by the thickness measuring apparatus. A good experimental approach might utilize a pair of rollers of suitable diameter coupled with a displacement transducer and a drive to feed the sample through. The proper pressure would then be determined when the thickness profile obtained is repeatable and undiminished.

B. Length and Width

The problems in the measurement of changing length and width are similar to those for changing thickness. Generally, the sample is suspended with a light tensile load applied to it. It is essential that this sample be able to tolerate light load so that the specimen will hang straight, but that the load be matched to the sample strength so that deformation due to load will be negligible. Under these conditions, a transducer attached to the end of the specimen with a simple clamp will allow as close a measurement of changing length as could be desired.

One may also consider the use of printed grids. With printed grids, there are a number of approaches that can be taken to measuring deformation on a paper sample. The finer the grid, the more precisely deformations can be measured on the sample. This technique is used more frequently where it is desired to learn something about the strain distribution across the surface of the sample. It is possible then to measure point-to-point changes in dimension using a traveling microscope.

For run-of-the-mill measurement of dimensional changes due to moisture or temperature, a steel scale measuring to 0.01 in. on a 10 in. sample will give a precision of 0.1%. This is more than adequate for most cases; however, such a method has a disadvantage of not easily relating dimensional changes to a time base. It also requires costly manpower for measurement. Even the older type Neenah expansimeters should be considered for labor-intensive means for measuring dimensional stability of paper. There are a number of electrical transducers that can be modified to provide a continuous readout of change in length, width, or thickness of paper subjected to varying humidity for various times. These have the advantage that they can be coupled with a recorder to provide continuous measurement, the principle limitation being one of the cost of instrumentation. The Instron Corporation manufactures a clip-on tension-compression strain gage that is quite suitable for many dimensional change measurements. Further attention to problems of measurement of dimensional change is given in Chap. 27. In the past 10 years, we have witnessed a rather pronounced change from manual measurement of dimensional movement in paper to one of increasing use of automated instrumentation. Most of these devices, however, are limited only by the imagination and budget of the technician conducting the research.

Wholesale improvement in our measurement techniques are in order, and it is well within our ability to make these improvements. For example, work has been done to develop computer programs with the objective of obtaining the periodic patterns of thickness variation that occur in some machine-made papers, using Fourier analysis of thickness versus distance. Aggressive research in the pulp and paper area using instrumentation that allows us to measure basic properties can lead to significant advances for the paper industry.

REFERENCES

1. Kimura, M., Oda, M., Iwasaki, Y., and Kadoya, T. (1979). Study on determination of paper thickness by mercury buoyancy method. *J. Japan Wood Res. Soc. (Mokuzai Gakkaishi)* 25(2):139-144. (In Japanese.)
2. Rosenthal, M. R. (1977). Effective thickness of paper: Appraisal and further development. USDA Forest Service Research Paper no. 287.
3. Setterholm, V. C. (1974). A new concept in paper thickness measurement. *Tappi* 57(3):164.
4. Uesaka, T., Murakami, K., and Imamura, R. (1979). Elastic properties of paper and its inhomogeneity with special reference to the basis weight dependence of elastic moduli. *J. Soc. Mat. Sci. Japan* 28(310): 629-634. (In Japanese.)

27

CURL, EXPANSIVITY, AND DIMENSIONAL STABILITY

CHARLES J. GREEN, JR.
Paper Technology Center
Xerox Corporation
Webster, New York

I.	Introduction	415
II.	Dimensional Stability	416
	A. Concepts	416
	B. Evaluation of Dimensional Stability	420
III.	Sheet Curl	426
	A. Concepts	426
	B. Curl Evaluation	429
IV.	Future Needs	440
	Equipment Suppliers	441
	References	441

I. INTRODUCTION

Dimensional stability of paper can be thought of as encompassing the effects of hygroexpansivity and thermal expansion and contraction; the effects of mechanical loading; or any combination of these [37]. In this chapter all aspects will be considered; the last-named effect will be included when mechanical loading interacts with hygroexpansivity properties.

The measurement of dimensional stability appears deceptively simple. For example, one may suspend a strip of paper in a small vessel and pass air, conditioned to a specific relative humidity (RH), through the apparatus [15]. In this apparatus, the change in length with relative humidity may be measured by various means, for example, precision micrometer [30] or light projection on a screen [7].

There are several precautions that have to be observed to obtain precise results. For example, tension placed on the sample during humidity cycling can cause significant creep [15,39]. Besides this, dimensional changes are a function of paper moisture content, which in turn exhibits a hysteresis with humidity cycling. In addition, dimensional movements of paper are affected by the specific humidity cycle used [1,5]. Up to 80% RH, the usual reversible dimensional changes are observed [26]. However, above 80% RH the additional moisture taken up by the paper acts as a plasticizer and releases *internal stresses* within the sheet structure [1,5,33]. In this case, permanent shrinkage takes place within the plane of the paper which can be measured at a *reference moisture content* by comparing the final dimensions to the original dimensions (usually at a RH below 50%).

Because of this dual characteristic of *reversible* and *nonreversible* dimensional changes, the measurements by which dimensional stability is evaluated must be selected according to the end-use treatments applied to the paper. For instance, if the paper moisture content rises above 10% (corresponding to about 80% RH) or if the paper is heated [21], measurements that can evaluate permanent shrinkage tendency will be needed.

Another topic of interest is the possibility that mechanical measurements can be used independently to evaluate hygroexpansivity. Correlation of elastic modulus [12] and breaking strain [32] to hygroexpansivity have already been demonstrated. This topic will be discussed further on pp. 425-426.

A subject closely allied to dimensional stability is curl, the second topic of this chapter. *Curl* is a manifestation of differences in dimensional stability within the sheet thickness [16]. The factors that are important in the measurement of dimensional stability can be equally important in evaluating curl. Therefore, dimensional stability measurements will be treated initially, followed by methods for evaluating paper curl.

II. DIMENSIONAL STABILITY

A. Concepts

The dimensional behavior of paper can be divided into *reversible* and *recoverable* (or nonreversible) components [3,4,10,28,38]. *Reversible* dimensional changes are those in which there is a one-to-one correspondence between moisture content and paper dimensions; at any given moisture content, whether arrived at by increasing or decreasing the moisture, a sample will exhibit only one set of characteristic dimensions. This behavior is illustrated for length in Fig. 1. The paper sample at moisture contents A, B, and C will exhibit corresponding lengths a, b, and c. This type of dimensional movement is found in paper that has been dried without constraint to normal drying shrinkage (i.e., freely dried) or in paper that has been rewetted and dried without constraint a number of times [2]. It is also observed below the plasticization relative humidity (usually 65 to 80%) [26]. It should also be noted that, like most mechanical properties, comparison of dimensions with moisture content do not exhibit the hysteresis that is usually found

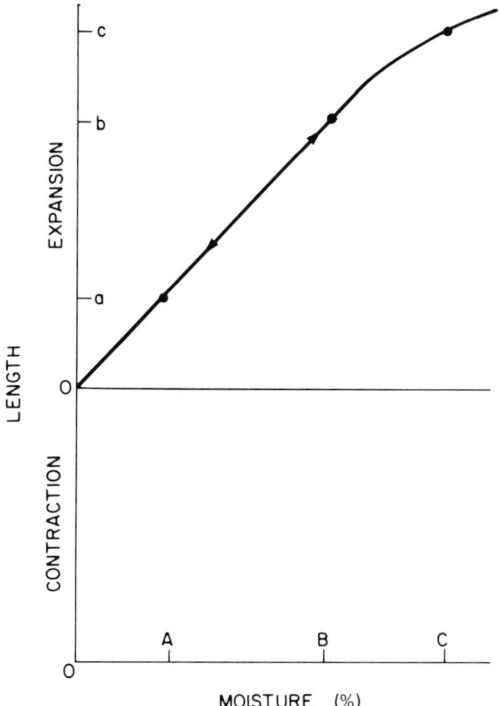

Fig. 1 Reversible expansivity.

when such comparisons are made with relative humidity as the independent variable.

Above 80% RH, sheet dimensions at the same moisture content do not correspond between the humidification and dehumidification cycles because of nonreversible shrinkage (illustrated in Fig. 2 by lengths a and a_2 at moisture content A [1,5,33]). If the specimen is rehumidified to an equal or higher relative humidity than it was subjected to the first time, an additional shrinkage can take place, as shown in Fig. 3 by the second dehumidification (II) compared to the first (I). Therefore, at one moisture content A the sample has exhibited three different lengths, a_1, a_2, and a_3. Any number of lengths can be exhibited at a given moisture content, dependent on sample history and according to the restraint applied during drying and the relative humidity to which the sample has been subjected. The total available shrinkage can be determined by repeated wetting and drying without constraint to shrinkage (see p. 424).

Another principle illustrated by Fig. 3 is that as irreversible shrinkage increases, the rate of reversible dimensional chage per unit change in moisture also increases (slope II > slope I > initial slope). The dimensional in-plane behavior discussed in the preceding paragraphs can be explained by the use of Eq. (1) (Fig. 4):

$$\delta_W = \delta_E + \delta_R \tag{1}$$

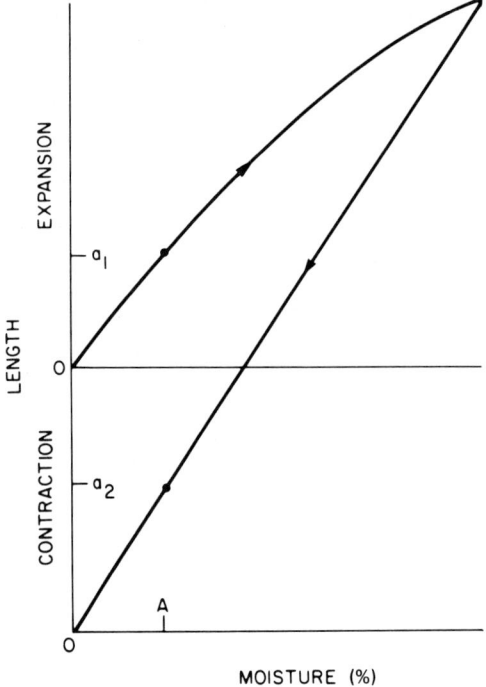

Fig. 2 Nonreversible shrinkage.

where

δ_W = total expansion from dry to wet of a paper structure rewet and dried without constraint $(d_w - d_0)$

δ_E = reversible expansion from dry to wet $(d_w - d_r)$

δ_R = available or recoverable shrinkage $(d_r - d_0)$

Equation (1) can be applied using engineering strains or percent strain using the freely dried dimensions as the unit length, or even simply as true lengths (for sheets of the same size).

The physical meaning of Eq. (1) is illustrated in Fig. 4. Reversible expansion δ_E is given by the wet length minus the dry length $(d_w - d_r)$, and the total available shrinkage δ_R by the dry length minus the stress-free dry length $(d_r - d_0)$. (To obtain engineering strains these values are divided by the dry length d_0.) From the basic relationship in Eq. (1) we can see that as available shrinkage is increased, reversible expansion decreases if total expansion remains constant.

Equation (1) can be applied to any direction in the plane of the paper, although normally it is used in the machine or cross directions. Total expansion δ_W is a function of type of fiber, refining, materials added to the furnish, and fiber orientation [3,4,7]. Typical variations in total expansion because of fiber type and refining are illustrated by the data in Table 1 (freely dried samples) [7]. However, it should be noted that when samples

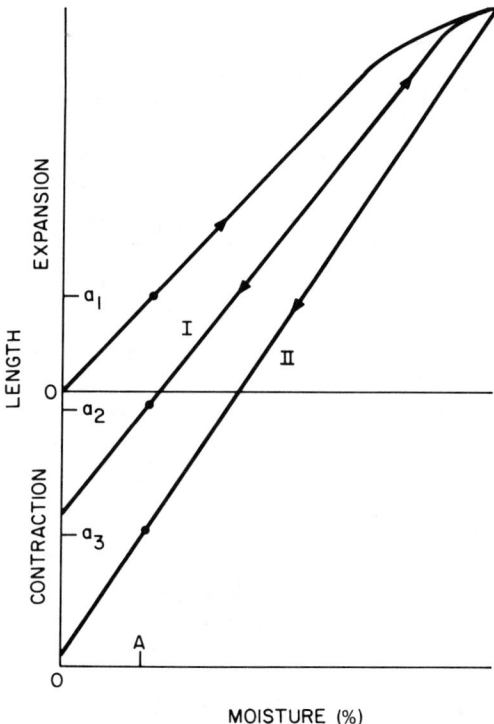

Fig. 3 Multiple humidity cycling.

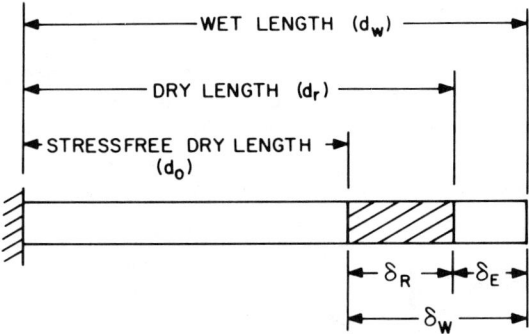

Fig. 4 Model of dimensional properties.

Table 1 Effect of Pulp, Refining, and Drying Restraint on Wet Expansion

Bleached pulp (Kraft)	Canadian standard freeness (ml)	Expansion, 30%RH to wet (%)	
		Freely Dried	Full Restraint
Douglas fir	355	4.08	1.00
	700	3.23	1.19
Southern pine	300	6.25	1.08
	700	3.92	1.08
Sweetgum	300	5.60	1.17
	700	2.36	0.97

Source: Adapted from Ref. 14.

are dried with full restraint (such as those obtained by making handsheets that are dried in intimate contact with chrome plates), there is little, if any, difference in wet expansion properties [2]. Available shrinkage δ_R increases as drying stresses are increased [29], but the extent to which available shrinkage develops also depends on factors that influence the Young moduli and other elastic constants. Such factors include the fiber and furnish variables mentioned above, as well as moisture content and temperature. Paper with a higher Young modulus requires a proportionately higher stress to obtain an equivalent amount of available shrinkage.

The distribution of total expansion between available shrinkage and reversible expansion can also be modified by mechanical treatments, such as creep-inducing sustained loads, and by high humidity, wetting, or heat treatments. Creep increases available shrinkage [5], while humidity, wetting, and heat decrease it [21]. Because mechanical treatments can affect the balance between shrinkage and reversible expansion, we can also observe creep in humidification experiments, if sufficient stress is applied during humidification. A net increase in length can be observed at the end of the treatment, as illustrated by Fig. 5.

This type of dimensional movement can also be observed in sheet thickness if the sheet has been calendered [21]. The calendering mechanically produces a recoverable expansion instead of a shrinkage. The physical relationships between the dimensional variables in thickness are depicted in Fig. 6, with the permanent changes in dimensions being observed as increases in thickness.

B. Evaluation of Dimensional Stability

Equipment Three important features to consider in equipment used to measure dimensional stability are: precision of dimension measurement, control and measurement of sample moisture content at the time of measurement, and tension applied during sample conditioning and measurement. Expansimeters such as the NBS Expansimeter [7] and Neenah Expansimeter [35] incorporate

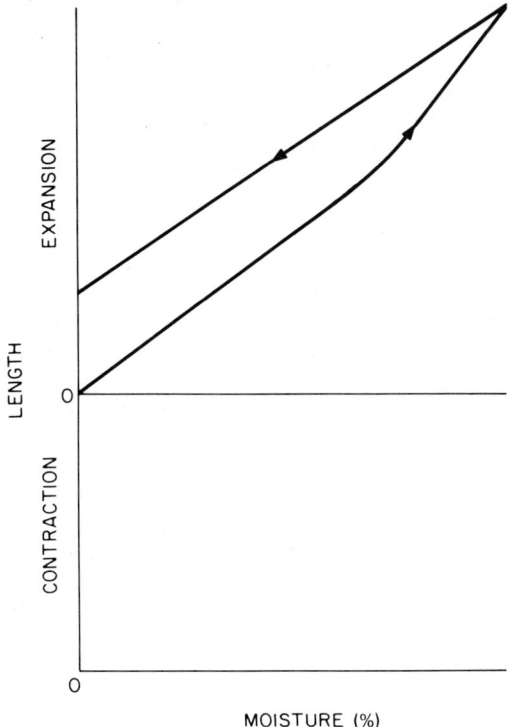

Fig. 5 Nonreversible expansion.

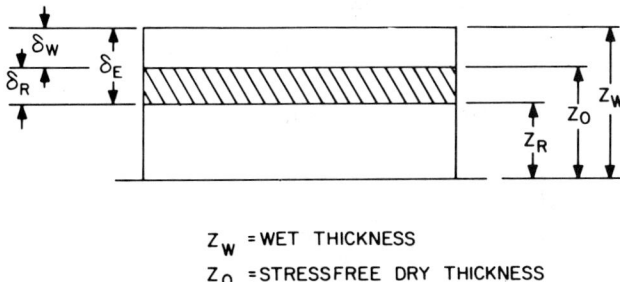

Z_W = WET THICKNESS
Z_O = STRESSFREE DRY THICKNESS
Z_R = RESTRAINED DRY THICKNESS
δ_R = RECOVERABLE OR AVAILABLE EXPANSION

Fig. 6 Model of dimensional properties (thickness).

both precision dimension measurement and a means of controlling the RH of the atmosphere surrounding the samples to effect moisture control. Samples are cut in strips 13 mm wide or less in the direction of the desired measurement. The NBS Expansimeter uses weights to keep samples under tension while the Neenah Expansimeter relies on the weight of leveling bulbs to supply the tension. The desired relative humidity is obtained by using saturated salt solutions [41]. Dry air can be obtained by using commercial drying compounds such as Drierite [7].

In measuring dimensional stability with the above equipment, it is important to be able to repeat any given moisture content so that data from several tests can be compared. This can be done by the use of a control sample. One or two strips of the control sample are mounted in the expansimeter and cycled a number of times through the highest relative humidity that will be used, to properly condition them. Control sample length can be used to determine when a desired condition has been reached, since the measurements reflect control sample moisture content [30].

There are several types of sheet-size dimension gages, including the Quick Skan. The Quick Skan can be used to measure dimensional stability if two or more humidity-controlled rooms are available. The use of the gage is simple in that sheets are placed on a backlighted surface with one edge against a reference edge. The crossing of a gage line and the opposite edge of the paper is a measure of the sheet dimension compared to nominal length. For example, in Fig. 7, the nominal length is 11 in. (279 mm), and the sample edge crosses the calibrated scale at −40, which is −0.040 in. (1 mm) from nominal. The sample length, therefore, is 10.960 in. (278 mm).

For certain types of measurements the Quick Skan offers some advantages. Whole sheet dimensions can be measured quite rapidly and they can be made without tensile stresses. The moisture content of the samples can be measured rapidly with a microwave attenuation detection instrument, such as a Moistrex. Rapid conditioning of samples is also possible.

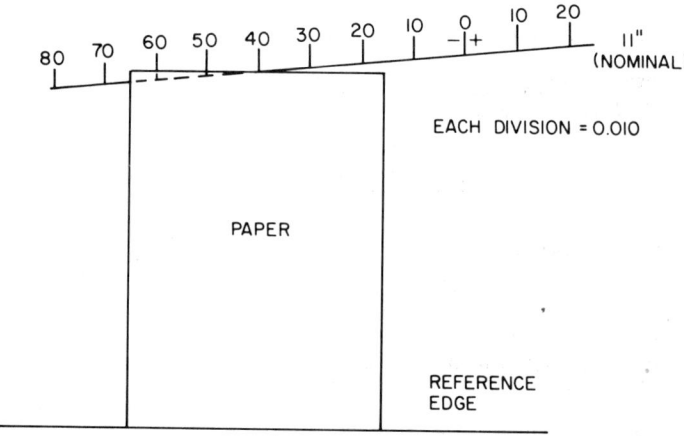

Fig. 7 Quick Skan schematic scale (exaggerated nonparallelism).

Test Methods It is important to know the relative humidity conditions that paper is exposed to in end use, so that appropriate test procedures can be defined. Usually a low relative humidity test procedure is of little value when a paper is subject to high humidity or is wetted during use. On the other hand, a procedure that uses high relative humidity or measures permanent changes may not accurately reflect reversible dimensional changes.

Low Relative Humidity End-Use Procedures For low relative humidity end-use problems, reversible expansivity measurements should be made at relative humidities below 60%. A transition range of 60 to 80% is also largely in the reversible range, except for papers made under high tensile stresses [26]. For greater precision, a range of humidity as wide as possible should be selected for testing. Sample lengths are measured at two relative humidities using an expansimeter device. The difference in length is calculated and can be expressed as a coefficient of moisture expansion by

$$\lambda_m = \frac{\Delta d}{d_0 \Delta m} \qquad (2)$$

or as a coefficient of relative humidity expansion using

$$\lambda_{10H} = \frac{10 \Delta d}{d_0 \Delta H} \qquad (3)$$

where

Δd = change in dimension
Δm = change in moisture content
ΔH = change in relative humidity

For relative humidities up to 60%, λ_m and λ_{10H} should be approximately the same for most uncoated papers. Measurements on uncoated papers will typically have values of λ_m = 0.00027 to 0.00044 for MD (machine direction) samples and λ_m = 0.0008 to 0.0017 for CD (cross direction) samples [7].

Measurements of reversible expansivity can be made using a portable sheet-size dimension gage if two humidity-controlled rooms are available. The unique advantage of this procedure is that the moisture content of the sample can be measured using a microwave meter.

High Relative Humidity End-Use Procedures If paper is exposed to relative humidity levels above 80%, significant permanent shrinkage takes place when it is redried using stress-free conditions [21]. It is possible that significant irreversible shrinkages can occur as low as 70% RH if a paper has high available shrinkage [26]. The extent of shrinkage will be reduced if tension is applied during humidification. If the stress applied is high enough, extensions will be observed instead of shrinkage, as discussed in Sec. II.A.

Available shrinkage can be evaluated by soaking cut-to-size sheets in water (minimum 15 min, after which the excess water is blotted off and samples are dried without constraint between dry blotters). Sample dimensions of dry paper are measured before and after the wetting procedure using a sheet-size dimension gage. Moisture content of the dry samples is checked with a microwave moisture meter to determine that they agree within 0.2%. After air drying it may be necessary to dry the treated samples at 105°C for 10 to 15 min to remove the humidity hysteresis effects to obtain comparable moisture contents. An alternative method is to make a correction for moisture difference using a known coefficient of moisture expansion λ_m for that paper. When making measurements of the redried paper, care must be taken to flatten the sample along the measured dimension to remove the effect of wrinkles.

Flattening of the sample can be accomplished by laying a clear heavy plastic rule on the paper along the line of measurement. The effect of wrinkles can be minimized by making comparative measurements at 50% RH rather than at lower values. The same points on the paper should be used for both measurements, which can be facilitated by drawing lines in the direction of measurement to mark the points [31]. For surface-sized machine-made paper, the variation in shrinkage for one wetting cycle is on the order of 0.64 to 1.1% in the machine direction and 0.29 to 1.23% in the cross direction [21]. (Strictly speaking, a sample must be rewet and dried a number of times to obtain complete shrinkage of a sheet [24], but for the practical evaluation of paper properties, one set of wetting and drying should be adequate for most mill process control and end-use evaluations. If it is deemed necessary, multiple cycles of wetting and drying for measurements discussed on p. 417 can be made until shrinkage ceases to increase.) Available shrinkage can also be measured with an expansimeter, using the following procedure [27]: Specimens are mounted in the expansimeter and the relative length recorded at a reference moisture content. They are then carefully removed with the clamps attached and placed in water for 15 min. The samples are carefully blotted to remove excess water and allowed to dry without constraint to ambient conditions. They are then replaced in the expansimeter and conditioned to the reference moisture content. Shrinkage is calculated as the difference between the initial and final measurements.

In an end-use process, dimensional changes can occur that cannot be attributed to moisture dimensional stability alone. For example, offset printing can cause dimensional changes that are independent of moisture expansion [14]. Often, inadequate information is available to directly duplicate the effect of tensile stresses on dimensional stability in a laboratory experiment. By interpolating data obtained in expansimeter measurements using varying tensions, the process-related irreversible changes can be duplicated. Samples of a particular paper are run through the process and tested in the expansimeter. In the expansimeter, various weights are applied to different strips of the same paper. The expansimeter is then cycled through an RH regime as nearly equivalent to process conditions as possible. The dimensional changes of the process paper are measured with a sheet-size dimensional gage. By comparison of the process data and expansimeter results, an equivalent stress for expansimeter experiments can be determined. Several other papers that exhibit differences in end-use results should also be tested in this manner to determine how well laboratory data reproduce process data. Care must be exercised that process conditions are not changed in the ex-

Table 2 Effect of Stress on Permanent Change in Length with High-Humidity Treatments

Paper	Change in Length (%) with Applied stress	
	14 kPa[a]	160 kPa
A	−0.25	0.28
B	−0.20	0.12
C	−0.47	−0.22

[a]Calculated using caliper thickness.

periments. An example of the effect (of two levels of tension) on permanent dimensional changes applied in an expansimeter experiment using xerographic bond papers is shown in Table 2. Note that the effect of increased tension varies for the three samples as well as the absolute level of permanent dimensional change. One sample exhibits permanent contractions at both tensions, while two samples change from permanent contraction to permanent extension when tension is increased.

The effect of tension on sheet extension depends on factors such as Young modulus, coefficient of expansion, relative humidity, and available shrinkage. For example, if the same tension is used in both machine and cross directions, the effect of reducing irreversible shrinkage will be greater on the cross-direction sample if (as is usual) the CD Young modulus is lower.

Total Expansion Total expansion is a measure of the combined effect of fiber, furnish, refining, additives, fillers, and fiber orientation [2-4,7,13]. Total expansion can be evaluated in the following manner. Sheets are soaked in water for 15 min and the dimensions measured with a sheet dimension gage. After stress-free drying, the dimensions are measured again (see pp. 423-424) and the difference is calculated. If the measurements are done in conjunction with initial sample dimensions, both total expansion and available shrinkage can be evaluated on the same samples. For surface-sized machine-made MD paper, total MD expansion measured in this way can vary between 0.6 and 1.05%. The CD expansion can vary between 2 and 3.7%.

Approaches That Utilize Mechanical Testing Several mechanical properties have potential for predicting dimensional stability. Creep and stress relaxation rates increase as reversible expansivity increases (and as available shrinkage decreases). It has been found that elastic moduli increase as reversible expansivity decreases [12]. Internal stress as measured by the Johanson-Kubát technique [23,24] increases as available shrinkage increases. (This method has the disadvantage that a relationship between internal stress and shrinkage needs to be determined first.)

Perhaps the most practical method of using a mechanical property to predict dimensional stability is based on a study reported by Silvy [32], in which breaking strain and elastic expansivity were correlated to drying

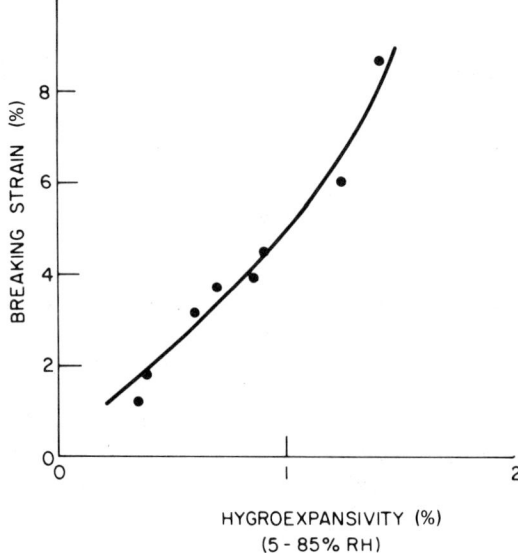

Fig. 8 Breaking strain versus hygroexpansivity.

shrinkage. In Fig. 8 a nearly linear correlation between expansivity and breaking strain can be observed over a substantial range. Although the correlation in Fig. 8 has not been established as universal, it appears that correlations can be developed for papers of interest. Conceivably, routine predictions of reversible expansivity will be made using breaking strain in the future.

III. SHEET CURL

A. Concepts

Paper curl can be described as an out-of-plane deformation so that a sheet is curved instead of flat. Curl is intimately related to the dimensional properties within the layered structure of a sheet of paper. If the dimensional difference changes between two or more layers are reversible, then the resultant curl will be reversible. Likewise, to the extent dimensional changes are permanent, curl will also be permanent.

Relatively small changes in the dimensions of one layer with respect to another cause rather severe curls. (The relationship between radius of curvature and differential strain is illustrated in Fig. 9 for 0.1 mm thickness paper. A radius of curvature of 0.1 m is produced by a differential strain of 0.067%.) The extent of curvature, K in m^{-1}, can be calculated (for a two-layer structure) by the application of Eq. (4) [9]:

$$K = \frac{24(\Delta d_1 - \Delta d_2)/d_0}{Z(E_2/E_1 + E_1/E_2 + 14)} \tag{4}$$

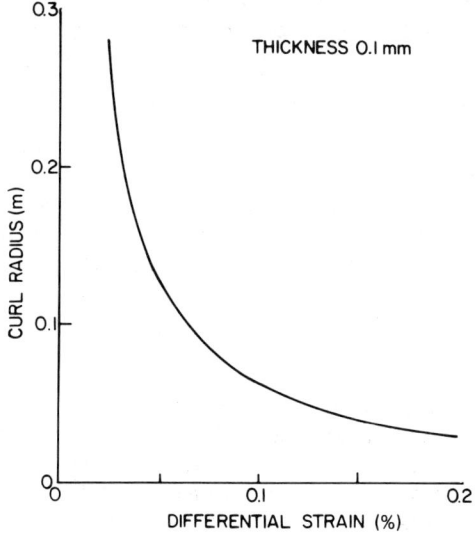

Fig. 9 Relationship between curl radius and differential strain.

where

Δd = dimensional change of each layer
E = Young modulus
Z = sheet thickness
1,2 = layers

A method of calculating curl that takes into account a multilayer structure has recently been published [8].

The dimensional changes described by Δd in Eq. (4) can be divided into reversible and permanent changes for each layer by Eq. (5):

$$\frac{\Delta d}{d_0} = \lambda_m \Delta m + \frac{\Delta \delta_R}{d_0} \qquad (5)$$

where

λ_m = moisture coefficient of expansion
Δm = change in moisture content
$\Delta \delta_R$ = change in available shrinkage

(A term $\lambda_T \Delta T$, where λ_T is the coefficient of thermal expansion and ΔT is change in temperature could also be added to the right-hand side of Eq. (5); However, rapid heat dissipation quickly reduces the curl effect of this term before any practical measurements can be made.)

The term $\lambda_m \Delta m$ represents reversible changes that will result in the reversion of curvature to the original configuration observed at the original moisture content. The term $\Delta \delta_R$ represents either changes caused by bending stresses applied to the sheet or by shrinkage of the particular layer in question.

Equation (4) can be simplified considerably to Eq. (6) if $0.5 \leq E_1/E_2 \leq 2.0$, because the error in calculating K is less than 2.5% for that range of values:

$$K = \frac{1.5(\Delta d_1 - \Delta d_2)/d_0}{Z} \qquad (6)$$

Each layer of a sheet of paper can have a unique combination of reversible expansivity and available shrinkage, the sum of which depends on the free wet expansion $(d_w - d_0)$ of that layer. Therefore, all three values can vary because layers within a sheet structure may vary in composition and fiber orientation. For further discussion of these properties, the reader is referred to Chap. 24.

The degree of observed curl will depend on which factors in Eq. (5) have been affected. Reversible curl is caused only by differential coefficients of moisture expansion and a change in moisture content. On the other hand, permanent curl is produced when a *change* in available shrinkage in one layer differs from that which takes place in one or more of the other layers. Available shrinkage can increase as a result of creep or stress relaxation or decrease as a result of strain recovery caused by exposure to heat or high moisture [5,21].

The degree of both permanent curl and reversible curl will increase as the difference in total expansivity between the (two) layers increases. This has been illustrated by an experiment in which two-ply handsheets were prepared from two pulps with various degrees of refining [19]. Samples were then heated in an oven. Increasing degrees of curl were observed as the differential expansivity of the samples increased (see Fig. 10). Both permanent and reversible curl (difference between total and permanent curl) increased, illustrating the principle that both factors are closely related to differential expansivity.

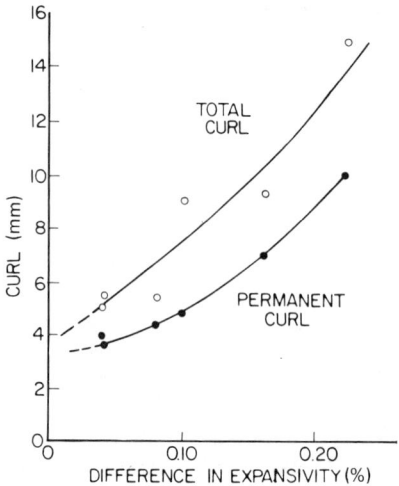

Fig. 10 Effect of expansivity of curl of two-ply sheets.

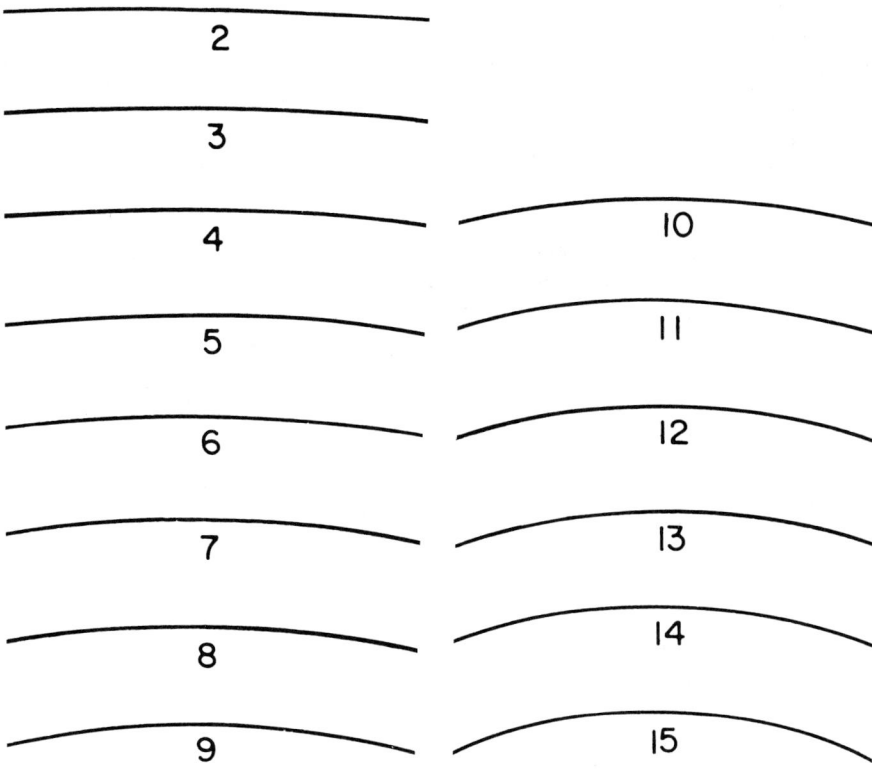

Fig. 11 Curvature template for curl measurement (curvature in m^{-1}).

B. Curl Evaluation

Measurement of Curvature Whenever possible, paper sheet curvature should be measured with the sheet in a hanging position to eliminate effects of gravity. Samples are supported at a center point along the top edge and curvature is matched to a template. Template curvatures are typically from 2 to 15 m^{-1} in unit increments. (A full-size example is given in Fig. 11.) Curvature obtained in this manner can be used to calculate the differential strain producing this curl by using Eq. (4).

Several sample configurations can be used. Which is best depends on which curl properties are being studied. Spitz and Blickensdorfer [34] used the sample configuration and mounting shown in Fig. 12A. A small hole is punched at the top of a 75 × 75 mm sample that is supported on a horizontal rod or hook passed through the hole. On the other hand, Hendry and Newman [22] recommended cutting narrow strips 3 to 6 mm wide and about 80 mm long. (If samples are exposed to high humidity, samples only 60 mm long should be used.) Samples can be mounted conveniently on a vertical circular rod with a rubber band stretched around the length of the rod, as shown

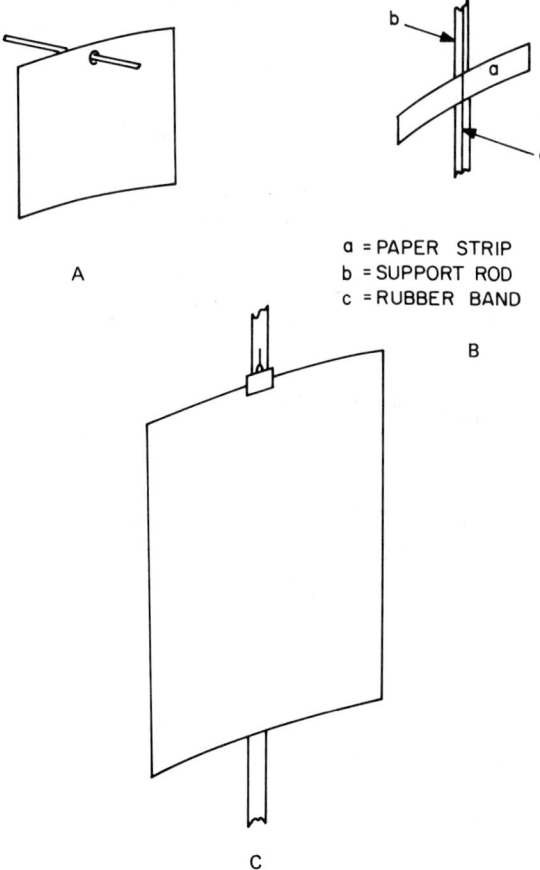

Fig. 12 Sample mountings (schematic).

in Fig. 12B. Sample preparation technique can be important, especially for those samples with small dimensions. Vaughan [36] recommended the use of sharp dies, a wooden cutting block, and a hammer to ensure minimum and equal edge stresses.

It is sometimes desirable to measure curl of a cut-size sheet (usually no larger than 216 × 279 mm). In this case samples are mounted as shown in Fig. 12C. The sheet is clamped at the top with a narrow clamp or clip. The vertical center line of the sheet rests against the vertical backing support.

Systematic Analysis of Curl Various procedures can be used to interpret the mechanisms by which paper has curled as a result of a given set of process variables. An orderly analysis includes measurement of

- Sheet properties
- Layer to layer properties
- Properties related to composition variables

Sheet Properties Measurements of sheet properties provide basis information on sheet behavior and evaluation of end-use suitability. Measurements can be divided into three general types: low-humidity, high-humidity, and heating procedures.

Low Humidity Low-humidity procedures are used to measure reversible curl.

Procedures. Samples are mounted as shown in Fig. 12A or B. They are then placed in conditioned atmospheres of 15% and 65% RH (or other convenient relative humidities between 0 and 65%) for a sufficient time in each to attain equilibrium. A measurement of curvature is taken at each condition. The difference in curvature is a measure of reversible curling tendency.

Sets of samples should be prepared so that horizontal orientations for both MD and CD are tested. At least two replications should be used for each paper tested.

Interpretation. If narrow strips are used, additional information can be obtained from the data. If the MD and CD horizontal mounted strips curl to opposite surfaces (e.g., wire and felt sides concave) the most likely cause of curl is a differential fiber orientation between wire and felt sides. The side toward which the CD horizontal strip curls proceeding from high to low RH has the greater proportion of fibers aligned with the machine direction.

If only the CD horizontal strip develops curl, differences in wire to felt side composition can also be a contributing factor. Likewise, if both MD and CD horizontal strips curl in the same direction, composition factors apparently dominate.

High Humidity High-humidity procedures are used to measure permanent curl caused by the release of available shrinkage in the sheet structure. Results of this type of test have been reported for several size press treated sheets [19] (see Table 3).

Table 3 Radius of Curvature of 99% RH Humidified Paper Strips

Paper	Concave side	Long axis of strip	Radius of curvature (in.)	(m)
B	FS	CD	1.0	0.0254
C	FS	CD	2.5	0.0635
D	FS	MD	4.0	0.1016
E,F,G,H,I	—	MD and CD	8.0	0.2032
J	FS	CD	2.0	0.0508
K	WS	CD	2.2	0.0559
L	WS	CD	2.8	0.0711
L	FS	MD	2.8	0.0711
M	FS	CD	3.4	0.0864
N	WS	MD	3.4	0.0864

Source: From Ref. 20. © 1981, American Chemical Society. Reprinted with permission.

Procedures. Samples may be mounted by the methods illustrated in Fig. 12A or B; however, the use of narrow strips will provide more meaningful data in most cases. When strips are subject to high humidity, their lengths should be limited to 60 mm or less to prevent creep-induced sagging or drooping. Initial curl should be measured at some low reference RH of betwen 0 and 50%. Specimens are then conditioned at 99% RH and their curvature measured. Samples are then reconditioned at the reference relative humidity and the curvature measured again. The change in curvature at the reference relative humidity is a permanent change in curl.

If a sample has significant curl at 99% RH then a second sample should be prepared by conditioning a sample at 99% RH with the curl side down on a flat surface. Then it should be mounted in the usual way and reconditioned at the reference relative humidity. The change in curvature by this procedure should also be measured. The reason for this modification is to overcome the stiffened beam effect caused by the partial curl of the specimen. The U shape of the specimen inhibits the formation of the true curvature.

As a confirmation procedure or an alternate procedure, 75 × 75 mm samples may be immersed in water for 5 min to thoroughly wet them. They are then repeatedly blotted with fresh dry blotters (usually four to five times is sufficient) until the samples are dry and form a curl. One sample is blotted wire side up, and the second felt side up. The axis and extent of the curvature are recorded.

Interpretation. The extent of curvature in m^{-1} of permanent curl can usually be grouped into the following categories:

$5\ m^{-1}$ usually insignificant
5 to $10\ m^{-1}$ significant, possible problems
$10\ m^{-1}$ usually very problematic
$40\ m^{-1}$ about maximum observed [20]

As with reversible curl, curl to opposite surfaces for MD and CD horizontal samples indicates differential fiber orientation as the cause. The results should qualitatively match the results obtained for reversible curl.

For measurements of curl of wetted samples, the axis of the cylinder of curvature should be at right angles for wire side and felt side up samples. The more MD oriented fibers will be in the layer with the axis of the cylinder in the MD. If only MD or CD horizontal samples exhibit permanent curl, wire to felt side composition can also be contributing to curl.

Heating Methods Methods that use heat to measure curl have been reported [20,36]. (Typical results of this type of test are given in Table 4 for strips of paper heated by means of a radiant heater.) Test results obtained using radiant heat are a combination of the reversible curl of drying and the permanent curl of strain recovery through plasticization. The degree of permanent curl obtained by the heating method is not as great as that with a high-humidity procedure. However, a heat curl method is generally faster and easier to perform.

Procedures. Samples may be prepared in the same manner as for high-humidity testing (Fig. 12A and B). They should be mounted with MD horizontal and CD horizontal on alternate samples. The most desirable heating method is to place a heater on each side of the sample an equal distance away. The heat source can be nichrome wire element or a quartz heater

Table 4 Radiant Curl of CD Paper Strips

Paper	Curl height (mm)	
	WS to heater	FS to heater
B	45 FS	45 FS
C	18 FS	30 FS
E	0	0
I	11 WS	2 FS
J	23 FS	34 FS
L	40 WS	18 WS
N	9 WS	6 WS
O	13 WS	6 WS

Source: From Ref. 20. © 1981, American Chemical Society. Reprinted with permission.

placed at such a distance so that maximum curl develops in 2 to 4 s. The distance will depend on the power of the heat source. If only one heater is used, tests should be run on two specimens with wire side and felt side facing the source on alternate specimens.

After curl is formed the heaters are turned off and the curvature measured immediately (immediate curl). After 7 min (or perhaps longer for samples thicker than 0.15 mm) recovery time to rehumidify, the samples are measured again (conditioned curl). All tests should be run at the same relative humidity conditions, generally at 50% RH if possible.

Interpretation. The reaction of paper to heating is generally similar to humidity treatments except that the degree of permanent change is usually smaller. The permanent curl is taken as the difference between conditioned and untreated curl. Reversible curl is the difference between immediate curl and conditioned curl.

Layer-to-Layer Properties The measurement of layered properties, the most elementary being two equal parts, requires equipment that can split the paper in the dry state or grind away half of the sheet thickness. Both methods are feasible. Dry splitting requires a less costly apparatus, but it will disrupt the interfiber bonding of about the center third of the sheet. Since the outer layers have more influence on curl properties than the center portion of the sheet thickness, this is not a serious problem and is probably advantageous. On the other hand, precision grinding can provide samples whose structure is essentially undisturbed.

Sheet Splitting and Grinding Sheet splitting is performed by passing the paper sample strips between steel rollers after affixing double-backed tape to both surfaces. A pair of rollers such as those found in a spring loaded Beloit sheet splitter is satisfactory. Samples can be cut to any convenient length and width; however, when strips more than 25 mm wide are split, the results may be unsatisfactory. Strips must be cut in widths that match the double-backed tape being used.

For expansivity measurements, strips should be cut no wider than 12.7 mm. Scotch 400 double-coated tape (or equivalent) is applied to each side of the sample. When the sample is passed between the steel rollers, the strip splits approximately in half. Samples are then immersed in toluene for a few seconds, which softens the adhesive so it can easily be removed from the strips. They are allowed to air dry before use.

Grinding can be performed with a surface grinder of a kind commonly used in a machine shop, employing a technique used at the Institute of Paper Chemistry [40]. The paper is held flat with a vacuum plate and the grinding performed with a Norton Carburundum coarse grinding wheel No. 32A46-I8VG (or equivalent). A grinder such as the fully hydraulic Covel H10 manufactured in Benton Harbor, Mich., can be used. The vacuum plate contains two narrow grooves approximately 2.54 mm wide, 5.08 mm deep, and 160 mm long, that are spaced about 100 mm apart. The plate surface is finish-ground after installation to assure parallelism between the plate and the grinding wheel.

Expansimeter Measurements Expansimeter measurements combine both low- and high-humidity procedures (e.g., measurement of reversible and permanent changes) into one test series. Split or ground samples are mounted in the expansimeter. One complete set consists of four strips; wire and felt side machine direction and wire and felt side cross direction.

Procedures. Initial conditioning and length measurements are made at low relative humidity (0 to 15%). Samples are then conditioned at high relative humidity (96 to 99%) and sample lengths remeasured and recorded. It is important to remove as much tension as possible from the sample strips during the humidification cycle to prevent creep from taking place, especially for samples whose basis weight is 75 g/m^2 or less. After humidification samples are reconditioned to the low relative humidity reference.

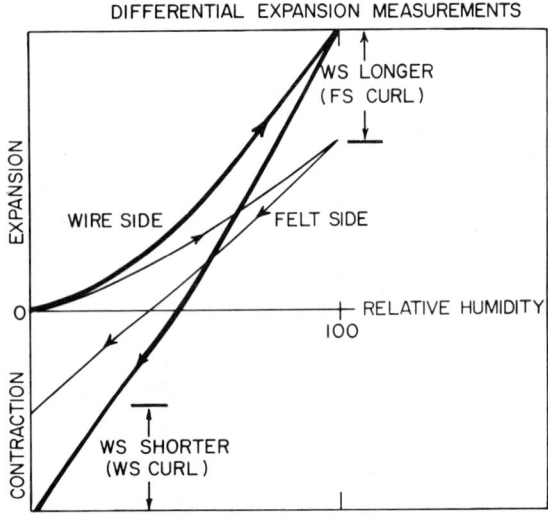

Fig. 13 Illustration of wire side to felt side dimensional changes.

Interpretation. The normal response of wire and felt side strips is for them to change dimensions at different rates. A typical response of a sample is illustrated in Fig. 13, where the wire side strip increases in length faster than the felt side as the humidity is increased (corresponding to felt side curl at high humidity). When the samples are returned to the low relative humidity reference the wire side strip typically contracts more than the felt side strip, corresponding to wire side curl at low humidity.

When there is a significant difference in wire versus felt side fiber orientation, the MD and CD strips generally develop opposite responses [20]. For wire side-oriented paper the wire side becomes longer in the MD as relative humidity is increased and the felt side of the CD samples becomes longer, as a rule. The opposite set of responses will occur for felt side-oriented papers. It should be noted that although many papers exhibit greater wire side orientation, there are some papers for which the degree of felt side orientation is decidedly greater.

A convenient way to assign a value to relative wire to felt side orientation is to perform the following calculation:

$$\theta = \frac{(\Delta y / \Delta x)_{ws}}{(\Delta y / \Delta x)_{fs}} \qquad (7)$$

where

θ = wire to felt side expansivity ratio
Δx = MD contraction (high to low RH)
Δy = CD contraction (high to low RH)
ws = wire side
fs = felt side

A change in length is taken as the difference between the high relative humidity dimension and the final low relative humidity dimension. A value of θ greater than one indicates that the wire side has a higher proportion of fibers aligned in the MD, while a θ less than one is obtained if the felt side has the greater proportion of MD-oriented fibers.

Mechanical Measurements Two measurement techniques that can be used to evaluate differential fiber orientation are (a) short span tensile strength and (b) sonic velocity. Kallmes [25] described the use of the zero-span tensile strength to measure differences in fiber orientation within the sheet thickness. By using sonic velocity as a measure of the Young modulus [11], measurement on split sheets can be used to differentiate between papers which have more MD-oriented fibers on the wire side layer and those which have this condition in the felt side layer. The data in Table 5 compare the results of tests made on split sheets of two size press-treated commercial papers using short-span tensile strength, sonic velocity, and moisture expansivity to calculate orientation ratio. For further information on zero-span and short-span tensile testing, refer to Chap. 5.

Each paper has very similar wire to felt side ratios by all three methods, though the papers have different curl properties (see papers A and B, Table 7). (Note that the ratio for moisture expansivity is a CD/MD ratio, compared to an MD/CD ratio for mechanical measurements, because of the inverse rela-

Table 5 Wire- to Felt-Side Property Ratios of Two Uncoated Papers

Property ratio	Paper a	Paper b
$(Z_x/Z_y)_W/(Z_x/Z_y)_F$	1.21	0.73
$(v_x/v_y)_W/(v_x/v_y)_F$	1.23	0.78
θ [Eq. (7)]	1.30	0.84

where

Z = short-span tensile strength
v = sonic velocity
x = machine direction
y = cross direction

tionship between expansivity and mechanical properties such as tensile strength and Young modulus.)

Composition Variables Although differences in wire to felt side fiber orientation has been cited as the main cause of curl [3,17], another source of differences in expansivity is composition. Size press-treated sheets, such as those used in office duplicators and coating raw stocks, can have variations in the distribution of surface size, filler, and fines. The variation in starch and filler content can be determined by analyzing split sheets (see pp. 433-434 for methods of splitting sheets). Starch can be determined colormetrically [6] by the method described in ASTM standard D 591 or TAPPI standard T 419. Filler content can be determined by methods routinely used to measure ash content (TAPPI T 411).

For coated grades the quantity of moisture-sensitive binder on each surface is important in reversible curl reactions [5]. Curl at low humidities will generally develop toward the surface with high binder content because binders usually have higher coefficients of expansion. The quantity of binder on each surface is most easily determined by measuring the coating weight applied to each surface (by the usual production techniques, e.g., by difference in basis weight of raw stock, one side coated and two side coated sheets) and calculating the binder weight from the fraction of binder used in the coating formulation for each coating.

Evaluation of Curl Derived from End Use This chapter has previously dealt with the evaluation of curl properties that result directly from the structure of the paper. Equally important is the interaction of the processes of end use with the structural variables. The usual imaging (e.g., reproduction) process inputs include heat, moisture, and mechanical bending. As is often the case, the inputs of heat and moisture between the two surfaces also differ. If paper is mechanically restrained from bending during this differen-

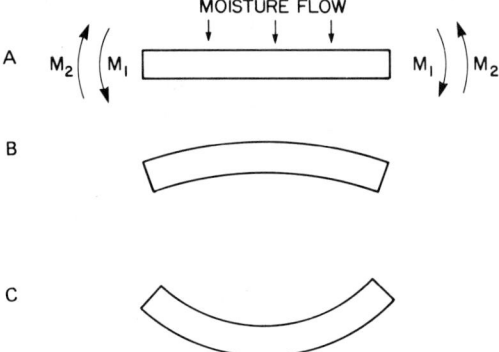

Fig. 14 Restraint to bending during treatment.

tial input of heat or moisture, within the sheet a differential stress is created that produces an effect that can be treated analytically as a bending moment. For example, consider the sequence of events shown in Fig. 14. A moisture uptake by one surface of a sheet of paper causes a bending moment M_1 which tends to curl the sheet downward as in Fig. 14B. But if the sheet is constrained in a flat position, an equal and opposite bending moment M_2 acts to counterbalance bending moment M_1. In this configuration stress relaxation can take place such that the sheet becomes nearly flat even after the bending moment M_2 is removed. However, with excess moisture still in the top surface, much of that moisture now migrates into the rest of the sheet, causing expansion of the bottom surface and an upward curl, as shown in Fig. 14C.

Sheet bending may also occur when heat is applied to one surface. Initially, a thermal expansion of the heated surface takes place. However, if the paper contains sufficient moisture, a second reaction that takes place is a loss of moisture from the heated surface, causing it to contract. Therefore, sequential bending moments may occur in opposite directions by the application of heat to a restrained sheet. Moisture content is often an important factor, since higher moisture contents tend to increase curl away from the heated surface. The curl reactions of a xerographic fuser, described on p. 439, are effects that suggest a dual bending relaxation dependent upon an initial moisture content.

Three processes—offset printing, xerography, and paper drying—have been chosen to describe paper process versus curl interactions and how they can be evaluated. These are contained in the pages that follow. The effects described are summarized in Table 6.

Offset Printing Curl In offset printing, one side of the paper becomes moistened by the offset printing blanket. The initial curl away from the printed side is caused by the expansion of that side. The extent of curling will depend on the amount of moisture added (a process variable) and the moisture expansivity of the paper (a paper variable). The final curl after drying is caused by shrinkage of the printed side, causing curl to that surface. The

Table 6 Effect of Uneven Process Treatment of Paper

Process	Type of process effect	Result
Offset printing	Wetting	Initial curl away from wetted surface
		Final curl toward wetted surface
Xerography (roll fusers) (75 g/m² paper)	Heating	Low moisture paper (3 to 4%): curl toward heated surface
		High moisture paper (6 to 7%): curl away from heated surface
Cylinder drying (papermaking) (75 g/m² paper)	Heating	Lower dryer temperatures: curl toward dryer surface
		Higher dryer temperatures: curl away from dryer surface

extent of curl is governed by the amount of moisture added and the available shrinkage of the paper.

A test method that can be used to compare the effect of wetting on two or more papers consists of a four-step procedure [31]:

1. Apply a thin uniform film of water to a cleaned glass or zinc plate.
2. Contact the sheet of paper (about 200 × 250 mm in size) by firmly rubbing it against the plate.
3. Remove the paper from the plate, measuring the maximum curl observed away from the wet side.
4. Hang the paper to dry and measure the final curl toward the moistened side.

A similar test can be performed using strips of paper 100 mm long by 12.7 mm wide [20]. One surface is moistened by pressing a wet blotter against it for 5 s. The strip is allowed to dry and the final curvature measured. The advantage of this method is that machine and cross direction strips can be tested for relative wire and felt side curl. What is usually observed is a preferential curl toward the wire side in one direction (MD or CD) and more curl toward the felt side in the other direction (Table 7) [20].

Xerographic Fuser Curl In the typical medium- or high-speed xerographic process a roll fuser is used to fix the toner image to the paper. At relatively low moisture contents (3 to 4%), curl is toward the printed side, which contacts the hot fuser roll. On the other hand, at higher moisture contents (6 to 7%) curl is away from the printed side. Test methods using heat such as

Table 7 Effect of Differential Wetting on Paper Strips

	Curl toward moistened side after drying (mm)			
	MD strip		CD strip	
Paper	Wire[a]	Felt[a]	Wire[a]	Felt[a]
A	12	2	34	27
B	32	0	3	38
C	19	2	5	21
D	0	30	11	22
E	0	60	10	6

[a]Side moistened
Source: From Ref. 20. © 1981, American Chemical Society. Reprinted with permission.

described on pp. 432-433 [20] have been used to estimate fuser curl at lower moisture content levels [19]. But because fuser variables such as temperature, speed, and contact arc have such a strong influence on the performance of any particular fuser system, it has been difficult to develop systematic test methods to predict curl performance of paper for all fusers. It has been suggested that factors influencing the coefficient of expansion and available shrinkage be measured in the study of xerographic curl [19].

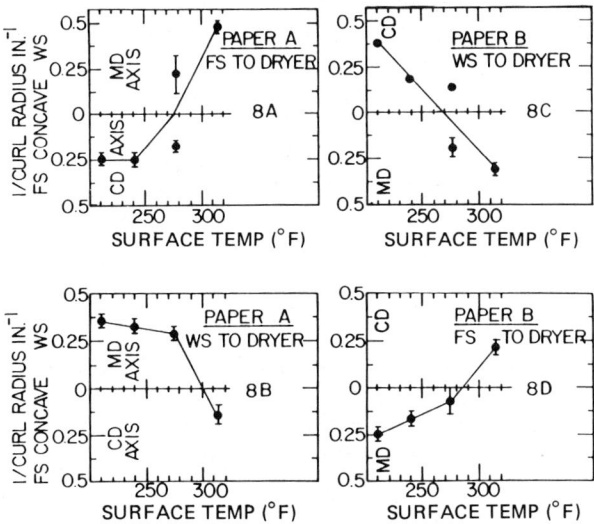

Fig. 15 Effect of drying temperature on paper curl (Reprinted with permission from Ref. 20. © 1981 American Chemical Society.)

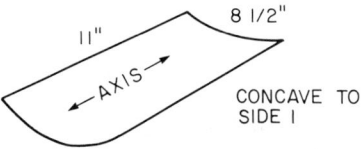

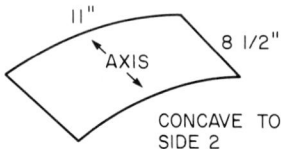

Fig. 16 Dual curl characteristic of paper. (Reprinted with permission from Ref. 20. © 1981, American Chemical Society.)

Drying of Paper The drying of paper with cylindrical driers is a process interaction that can be used to illustrate the effect of temperature on curl behavior (Fig. 15) [20,24]. When the drying temperature is relatively low, curl is toward the dryer surface. At sufficiently higher temperatures, curl is away from the dryer surface.

On the other hand, the axis of curvature of curl is determined by fiber orientation, regardless of whether curl is toward or away from the dryer. If the expansivity ratio θ_E is greater than one, the axis of curvature of the curl cylinder is in the machine direction for wire side curl and in the cross direction for felt side curl. For an expansivity ratio less than one, the axes are at 90° to those just described (Fig. 16).

IV. FUTURE NEEDS

A discussion of future needs for the study of curl and dimensional stability can be divided into three categories:

- Standardization of test procedures
- • Commercial test equipment
- • • New technology

A prerequisite to standardized commercial test equipment is the adoption of formal standard procedures by technical associations. There are few standard procedures available today for measuring dimensional stability and curl. The problems of standardizing these types of tests in the past included a lack of understanding of the nature of dimensional stability and thereby an oversimplification of the test procedures. The information that has been published in the last 10 to 15 years has contributed greatly to an increased understanding of the subject. Therefore, it is more practical today than formerly to standardize test procedures. This standardization should include measurements of both reversible and permanent dimensional stability and curl.

When standardized procedures become available, equipment manufacturers will have an incentive to develop and manufacture testers conforming to these standards. Equipment for measuring dimensional properties is already available to meet the needs of standardized procedures. However, to measure curl instruments are needed that can combine humidity control with a gage.

Since dimensional and mechanical properties have been shown to be intimately related, the use of mechanical properties to assess dimensional stability is a distinct possibility.

One other aspect of curl evaluation deserves mention. This is the effect of moisture diffusivity within the sheet structure (Chap. 17). It is potentially an important property in the interaction of paper with end use. If moisture can diffuse from one surface faster than another, there is a potential for curl in that curl will tend to form toward the surface from which moisture is removed last [18]. What needs to be developed are methods of measuring the moisture diffusion rates of each surface. It may be possible to use the methods applied to packaging materials; however, the rate of diffusion for many papers is very high and these methods may not work well for such papers.

EQUIPMENT SUPPLIERS

Expansimeters
Testing Machines, Inc., Amityville, N. Y. 11701
    Neenah Paper Expansimeter
    P. A. T. R. A. Paper Expansion Measuring Apparatus
    Brown Expansion Meter

Dimension Gages
Quick Skan, Berwyn, Ill. 60402
    Quick Skan Gages

Short Span Tensile Strength
Pulmac Instruments International, Montpelier, Vt. 05602

Sonic Velocity
H. M. Morgan Co., Inc., Cambridge, Mass. 02140
    Dynamic Modulus Tester PPM-5R

Microwave Moisture Gage
Neotec, Silver Spring, Md. 20910

REFERENCES

1. Back, E. L., and Klinga, L. O. (1963). The effect of heat treatment on internal stresses and permanent dimensional changes of paper. *Tappi* 46(5):284-288.

2. Brecht, W. (1958). Beating and hygrostability of paper. Transactions of Symposium on Fundamentals of Papermaking Fibres, pp. 241-262.
3. Brecht, W. (1961). Effect of structure on major aspects of paper behaviour with fluids. Transactions British Paper and Board Makers Assn. Symposium on Formation and Structure of Paper (Oxford), pp. 427-460.
4. Brecht, W., Gerspach, A., and Hildenbrand, W. (1956). The effect of drying tensions on some paper properties. *Das Papier* 10(19/20):454-458.
5. Brezinski, J. P. (1956). The creep properties of paper. *Tappi* 39(2):116-128.
6. Browning, B. L., Publitz, L. O., and Baker, P. S. (1952). Determination of starch in paper. *Tappi* 35(9):418-420.
7. Callinan, T. D., Crimi, J. S., Swartz, P. M., and Wirtz, L. H. (1961). The hygroexpansivity of tabulating cards. *Tappi* 44(6): 163A-171A.
8. Carlsson, L. (1979). Theory of out-of-plane hygroinstability of multi-ply sheets. 1979 International Paper Physics Conference preprint, pp. 31-40.
9. Carlsson, L., Fellers, C., and Htun, M. (1980). Curl and two sidedness of paper. *Svensk Paperstidn.* 83(7):194-197.
10. Coles, D. R., and Hudson, F. L. (1961). The dimensional stability of paper: *Paper Technol.* 2(5):473-479; discussion, p. 480.
11. Craven, J. L., and Taylor, D. L. (1965). Nondestructive sonic measurement of paper elasticity. *Tappi* 48(3):142-147.
12. DeRuvo, A., Lundberg, R., Martin-Löf, S., and Söremark, C. (1973). Influence of temperature and humidity on the elastic and expansional properties of paper and the constituent fibre. Transactions of Symposium on the Fundamental Properties of Paper Related to Its Uses (Cambridge), pp. 785-806.
13. Fahey, D. J., and Chilson, W. A. (1963). Mechanical treatments for improving dimensional stability of paper. *Tappi* 46(7):393-399.
14. Fink, P. (1965). Dimensional stability of offset paper. Proceedings of 8th International Conference of Printing Research Institutes. Pergamon, New York, pp. 261-275.
15. Gallay, W. (1973). Stability of dimensions and form of paper, 1. *Tappi* 56(11):54-63.
16. Gallay, W. (1973). Stability of dimensions and form of paper, 2. *Tappi* 56(12):90-95.
17. Glynn, P., Jones, H., and Gallay, W. (1959). The fundamentals of curl in paper. *Pulp Paper Mag. Can.* 60(10):T316-T328.
18. Glynn, P., Jones, H., and Gallay, W. (1961). Drying stresses and curl in paper. *Pulp Paper Mag. Can.* 62(1):T39-T48.
19. Green, C. J. (1979). Characteristics of paper which contribute to curl. 1979 International Paper Physics Conference Preprint, pp. 41-45.
20. Green, C. J. (1981). Curl properties of paper structures. *Ind. Eng. Chem. Prod. Res. Dev.* 20:147-150.
21. Green, C. J. (1981). Dimensional properties of paper structures. *Ind. Eng. Chem. Prod. Res. Dev.* 20:147-150.

22. Hendry, I. F., and Newman, J. A. S. (1963). Effects of machine variables on the curl of paper. *Paper Technol.* 4(4):381-388.
23. Htun, M., and deRuvo, A. (1978). Correlation between drying stress and the internal stress of paper. *Tappi* 61(6):75-77.
24. Johanson, F., and Kubát, J. (1964). Measurements of stress relaxation in paper. *Svensk Paperstidn.* 67(20):822-832.
25. Kallmes, O. J. (1969). Technique for determining fiber orientation distribution throughout the thickness of a sheet. *Tappi* 52(2):482-485.
26. Larocque, C. (1936). The extension of paper by adsorbed water vapour. *Pulp Paper Mag. Can.* 37(4):199-209.
27. Maynard, C. R. G., and Newman, J. A. S. (1957). The prevention of machine shrinkage and its effect on moisture expansion. *Tappi* 40(7):177A-180A.
28. Nordman, L. S. (1958). Laboratory investigations into the dimensional stability of paper. *Tappi* 41(1):23-30.
29. Nuttall, G. H. (1972). Importance of paper shrinkage and its control. *Drying of Paper and Paperboard* (G. Gavelin, ed.), Lockwood, New York, 266-276.
30. Rance, H. F. (1954). Effect of water removal on sheet properties: The Water evaporation phase. *Tappi* 37(12):640-654; discussion; pp. 681-683; *Pulp Paper Mag. Can.* 55(13):210-223.
31. Reed, R. F. (1970). *What the Printer Should Know About Paper.* Graphic Arts Foundation, Pittsburgh, pp. 126-174.
32. Silvy, J. (1971). Effects of drying on web characteristics. *Paper Technol.* 12(6):445-451.
33. Smith, S. F. (1950). Dried-in strains in paper sheets and their relation to curling, cockling and other phenomena. *Paper-Maker* 119(3):185-188, 190-192.
34. Spitz, D. A., and Blickensdorfer, P. S. (1963). The cause and cure of paper curl. *Tappi* 46(11):676-680.
35. Van den Akker, J. A., Root, C. and Wink, W. A. (1942). Multiple-specimen Neenah expansimeter. *Paper Trade J.* 115(24):33-36.
36. Vaughan, J. M. (1980). Practical quality considerations, communications base paper. Proceedings of 1980 TAPPI Printing and Reprography Conference, pp. 113-119.
37. Weiner, J., and Byrne, J. (1965). Hygroexpansivity and hygroscopicity of natural and synthetic fibers. Bibliographic series 216, Institute of Paper Chemistry, Appleton, Wis.
38. Wink, W. A. (1961). The effect of relative humidity and temperature on paper properties. *Tappi* 44(6):171A-178A.
39. Wink, W. A. (1968). What is dimensional stability and how it is measured and controlled? *Tappi* 51(6):170A-171A.
40. Wink, W. A. (1971). Private communication.
41. Young, J. F. (1967). Humidity control in the laboratory using salt solutions: A review. *J. Appl. Chem.* 17(11):241-245.

28
FIBER STRUCTURE

RICHARD E. MARK

Empire State Paper Research Institute
State University of New York
College of Environmental Science and Forestry
Syracuse, New York

I.	Introduction	445
II.	Wood Structure	447
III.	The Chemical Nature of Plant Fibers	451
	A. The Physical and Chemical Nature of Fiber Molecular Components	452
	B. Relative Abundance	458
IV.	Fiber Alteration in Processing	460
V.	Fibril Angle Determination	464
	A. Methods Used	466
	B. Comparisons Between Methods	475
	References	480

I. INTRODUCTION

It seems appropriate, in a reference book devoted to paper testing, to include a short chapter that will provide those who enter this field with backgrounds in engineering, chemistry, and other areas not associated particularly to wood and plant fibers with some fundamentals of the fiber structure of these materials. A section on the determination of the S2 fibril angle is included because the reference sources for this important parameter are scattered in the literatures of several disciplines.

Table 1 Classification of Papermaking Fibers

Natural fibers	Plant (cellulosic)	Xylem (wood)	Softwood (coniferous)
			Hardwood (dicotyledonous)
		Phloem (stem bark or bast)	
		Seed or fruit hairs	
		Leaf	
		Monocotyledonous (grass, bamboo, etc.)	
	Animal (proteinaceous)	Wood and hair	
		Silk and tussah	
		Leather	
	Mineral	Asbestos	
Manufactured organic fibers	Cellulosic	Regenerated	Viscose rayon
			Cuprammonium rayon
			Cellulose nitrate
		Etherified	
		Esterified	Cellulose acetate
	Noncellulosic	Acrylic (polyacrylic)	
		Carbon	
		Elastomeric (rubber)	
		Modadrylic	
		Polyamide (nylon)	
		Polyester	
		Polyethylene	
		Polypropylene	
		Polyurethane	
		Proteinaceous animal and plant derivatives	
		Synthetic rubber	
		Vinyl derivaties	
Manufactured inorganic fibers	Metallic	Fine wire and tinsel	
	Nonmetallic (mineral)	Fibrous glass	
		Rock and slag wools	
		Whiskers and other inorganic fibers	

All the major sources of fibers currently used for the manufacture of paper come from the plant kingdom. Although the material in this chapter is principally devoted to fibers of wood origin, it should be emphasized that there is a great similarity of structure among all plant fibers (see pp. 424-431 of Chap. 10). Most papermaking fibers are derived from wood. However, even those fibers classed as "nonwoody," including bagasse, cereal straws, bamboo, esparto, sabai, kenaf, jute, sunn, hemp, abaca, henequen, sisal, and so on, contain some lignin; among the plant fibers in general, only those fibers derived from certain seed hairs, such as cotton, and the cellulose derived from algae are devoid of association with the chemical constituent that makes a cellulosic cell woody, that is, lignin.

The attributes that primarily distinguish the plant fibers that fall into the topmost category, xylem (wood) in Table 1 from the other categories of plant fibers are:

1. The range of other plant cell types in the tissues associated with the fibers
2. The chemical and physical makeup of those associated tissues
3. The amount of lignin normally present in the fiber walls

Seed hairs such as cotton and kapok are essentially pure fiber; at the other extreme, some of the leafy raw materials from which paper has been made may yield no more than 10 to 20% of usable cellulosic fiber.

II. WOOD STRUCTURE

In Table 1 it is to be noted that there are two major types of wood that serve as fiber sources. The forest products industries refer to these two major categories as hardwood and softwood. But the names *hardwood* and *softwood* really refer to the botanical origin of the particular wood and not to the hardness of a particular wood on either an absolute or a relative scale. It is true that the wood of a softwood is usually lighter and softer than a typical hardwood. However, there is great variability in both categories, especially the latter. Thus, some of the world's softest woods are hardwoods.

Hardwoods come from the dicotyledonous branch of the botanical class Angiospermae, and include such woods as the aspens and poplars (*Populus*), Northern hemisphere beech (*Fagus*), Southern hemisphere beech (*Nothofagus*), maples (*Acer*), birches (*Betula*), eucalypts (*Eucalyptus*), and oaks (*Quercus*). A three-dimensional scanning electron micrograph (SEM) of a small block of the wood of a rather typical hardwood, birch (*Betula*), is shown in Fig. 10, Chap. 10.

Softwoods derive from the order Coniferales (cone-bearing trees) of the botanical class Gymnospermae, and include such woods as fir (*Abies*), hemlock (*Tsuga*), larch (*Larix*), Douglas fir (*Pseudotsuga*), spruce (*Picea*), pine (*Pinus*), and hoop and Paraná pines (*Araucaria*). Three-dimensional SEM views of the wood of *Pinus strobus* L. are shown in Figs. 1 to 3 [50].

The woods of hardwoods and softwoods contain various cell types, as shown in Table 2, although usually a given wood in either group contains only some of the cell elements found in the group as a whole. By way of example, *Pinus strobus* possesses tracheids, longitudinal epithelial parenchyma, ray tracheids, ray parenchyma, and horizontal epithelial parenchyma. As is typical for a conifer, some 90% of the volumetric composition and approximately 95% of the dry mass of this wood can be attributed to its tracheids.

The three recognized planes of wood, which are evident for the hardwood in Fig. 10 of Chap. 10 and for the softwood shown in Fig. 1 are (1) the transverse plane, revealed when wood is cut across the gain, (2) the tangential-longitudinal plane, which is parallel to a plane tangent to the surface of the tree, and (3) the radial-longitudinal plane, which runs from the center of the tree radially to its surface. The wood rays are aligned radially and thus appear as rows of short cells with orientation normal to the tracheids in Fig. 1. *Pinus strobus* possesses two kinds of cells in its narrow rays. These are

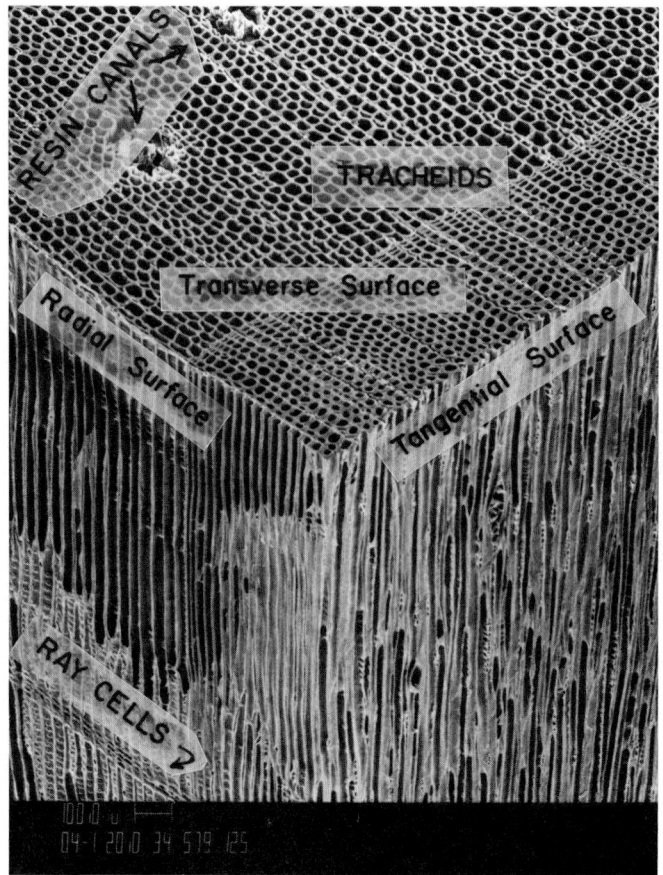

Fig. 1 A three-dimensional scanning electron micrograph of a conifer wood, eastern white pine. Parts of two annual rings are included. Most of the cellular elements in this wood are hollow, tubular fibers known botanically as tracheids. Tracheids serve two functions in the living tree—mechanical support and water conduction. The grain of wood (vertical in this picture) arises from cellular alignment of the tracheids. The bar scale at the lower left represents 100 μm. (Courtesy of W. A. Côté, Jr., State College of Environmental Science and Forestry, Syracuse, New York.)

ray tracheids, which are nonliving, short fibrous cells, and ray parenchyma, which serve as food storage and conduction tissue in the outer (sapwood) part of the wood in the living tree stem. The other type of parenchyma in *P. strobus* is epithelial, meaning that is surrounds the resin ducts that occur both within occasional wide rays and at scattered locations longitudinally among the tracheids of this species. Longitudinal resin ducts or canals are evident both in Fig. 1 and in Fig. 2. The surrounding epithelial cells produce the resin (pitch) that flows in the canals. Many softwood and some hardwood species possess resin or gum ducts, which provide physiological

Table 2 Cell Types of Hardwoods and Softwoods

Hardwoods	Softwoods
Longitudinal fiber and vessel elements	
Vessel elements[a]	Tracheids[a]
Fiber-tracheids[a]	Resinous tracheids
and/or	Strand tracheids
Libriform fibers[a]	
Vascular tracheids	
Vasicentric tracheids	
Longitudinal parenchyma elements	
Strand parenchyma	Strand parenchmya
Fusiform parenchyma	Epithelial parenchyma
Epithelial parenchyma	
Horizontal fiber elements	
None	Ray tracheids
Horizontal parenchyma elements	
Ray parenchyma[a]	Ray parenchyma[a]
Upright cells	Epithelial parenchyma
Procumbent cells	
Epithelial parenchyma	

[a]These elements are almost always present.
Source: From Ref. 17.

advantages to the tree but create chemical and engineering problems in pulping. Resin canals are often large enough to be seen by the unaided eye, especially when they are plugged with pitch or gum that has a color different from the wood.

In Fig. 2 and especially in Fig. 3, another feature of wood fiber and other cell walls is evident—the intercellular pits that provide passage for air, water and dissolved minerals, and other materials in tree sap to move from one cell to another.

The pits that form between adjacent cells vary markedly in size and shape according to (a) the nature of the adjoining cells that share a pit opening and (b) the species of the tree or other plant. Intertracheid pits are identified in Fig. 3 for the wood of *Pinus strobus*. Pits with overhanging margins such as those marked in Fig. 3 are termed bordered pits. In this same wood a very different pit shape is readily visible in the radial surface of Fig. 2, where a ray adjoining the rows of tracheids has been cut to reveal large windowlike pit openings between tracheids and ray parenchyma. The

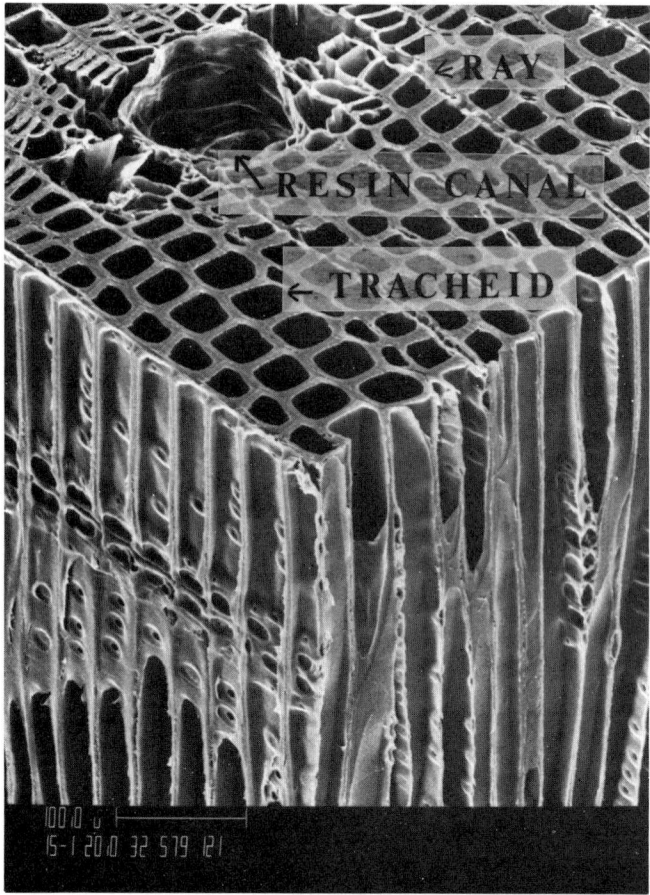

Fig. 2 Three-dimensional scanning electron micrograph of eastern white pine wood at higher magnification than Fig. 1. Additional details of the tracheids, resin canals and other cellular elements are evident. Bar scale represents 100 μm. (Courtesy of W. A. Côté, Jr., State College of Environmental Science and Forestry, Syracuse, New York.)

pits between tracheids and ray tracheids are again different, and so on. Each wood, hardwood or softwood, has characteristic pit sizes, shapes, and orientations that aid greatly in identification of the fiber source, and these characteristics are incorporated into such industry standards as TAPPI T 8 [67] and T 259 [68]. The reader is also referred to Refs. 7, 17, 18, 25, 41, 58, 58a, and 65. For other means of fiber identification, one may refer to CPPA Standard B.7 [14] and Refs. 6, 12, 16, 34, 39, 40, and 53.

The foregoing description of wood and its fibrous elements will help, it is hoped, in relating the structure of woods to the structure of their fiber constituents. As discussed in Chap. 10, pp. 424-431, Tables 2 and 3, and Figs. 1, 2, 6−17, and 22 of that chapter, the basic cell wall architecture is

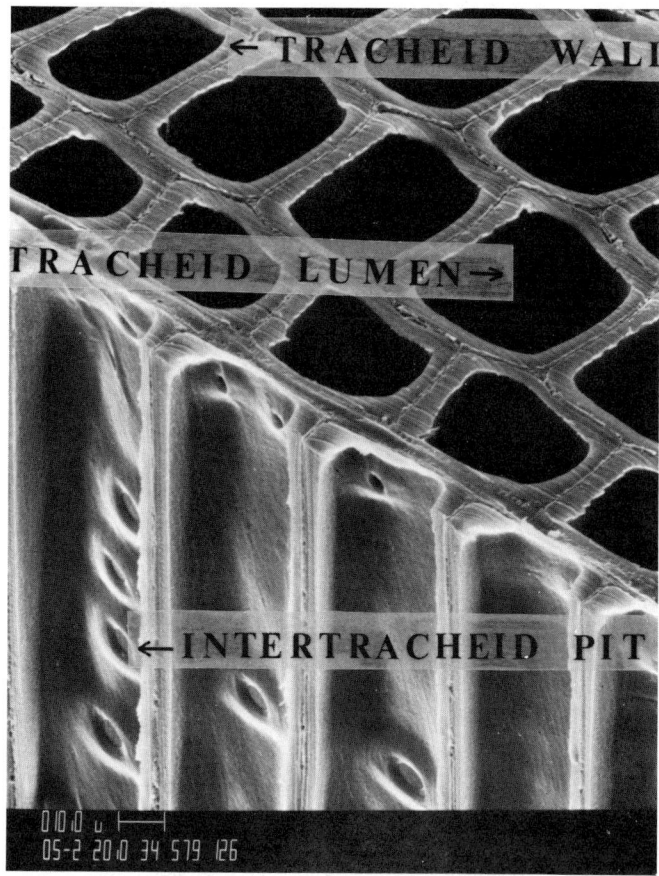

Fig. 3 High magnification scanning electron micrograph of eastern white pine wood. The fiber cell walls surround the hollow center (lumen) of each fiber. Large bordered intertracheid pits and smaller pits of the type that connect a tracheid and a ray tracheid are visible on the radially cut surface. Bar scale represents 10 μm. (Courtesy of W. A. Côté, Jr., State College of Environmental Science and Forestry, Syracuse, New York.)

strikingly similar for fibers derived from any and all species of the plant kingdom. Some major nonwood plant fiber raw materials sources are enumerated in Table 3 of this chapter.

III. THE CHEMICAL NATURE OF PLANT FIBERS

In this section the molecular components of plant fibers are discussed in relation to the architecture of the fiber wall. Here some differences as well as similarities are brought out. The suitability of a given type of *fiber* for pulping and papermaking is related to, but not identical with, the suitability of a *species* for these purposes, since associated tissues play a large role in determining questions of technical and economic feasibility.

Table 3 Examples of Some Common Plant Fibers

Type of fiber	Genera from which derived
Phloem (bark, bast)	Flax (*Linum*), hemp (*Cannabis*), jute (*Corchorus*), ramie (*Boehmeria*), kenaf (*Hibiscus*), mitsumata (*Edgeworthia*), kozo (*Broussonetia*), sunn (*Crotalaria*)
Seed or fruit hairs	Cotton (*Gossypium*), kapok (*Ceiba*), coir (*Cocos*)
Leaf	Abaca (*Musa*), henequen and sisal (*Agave*), New Zealand flax (*Phormium*), pineapple (*Ananas*), caroa (*Neoglazovia*).
Monocot	Rice (*Oryza*), sugarcane bagasse (*Saccharum*), corn (*Zea*), esparto (*Stipa, Lygeum*), sabai (*Eulaliopsis*), bamboo (*Bambusa, Dendrocalamus, Pseudostachyum*, etc.), cereal straws (*Triticum, Avena, Hordeum, Secale*)

A. The Physical and Chemical Nature of Fiber Molecular Components

One may think of the cell wall of a plant fiber as a composite material, for indeed it is. It has a filamentous reinforcement material (principally cellulose), a matrix of polymers that holds these filaments in place, and a few fillers in the matrix that modify the matrix properties somewhat. Structurally, the simplest way to envision how the filamentous reinforcement relates to the cell wall of the fiber is to imagine that you have a skein of yarn looped in a loose coil hanging from your fingertips. Now take the bottom of the loop with your other hand and give the bundle of yarn a twist as you pull on it. You are left with something that resembles a rope except for the end loops that curve between your fingertips (Fig. 4). If you were to twist this skein a few more times and then encase the resultant structure in some rigid matrix material, you would have a rudimentary model of the way a plant fiber is constructed. Your filaments of yarn are analogs of microfibrils, the actual structural filaments in the fibers (refer to Fig. 5). We will consider first the molecular structure of the microfibrillar and matrix constituents, then examine how they are assembled in greater detail. One must differentiate between the terms *microfibril* and *fiber*. The microfibril represents just one strand of the "yarn" that makes up the structural framework (the twisted skein) of the fiber cell wall.

Physical Nature of the Polymeric and Other Constituents The most important of the chemical constituents of all plant fiber cell walls is cellulose, a long-chain, unbranched condensation polymer of β-D glucose units. The structure of a glucose unit is illustrated in Fig. 6A and B. It is a ring structure formed by the bonds between five carbon atoms and one oxygen atom. In Fig. 6A, a planar schematic drawing of the glucose monomer illustrates the α and β isomers of D-glucose. Of the two possible D-glucose structures, one

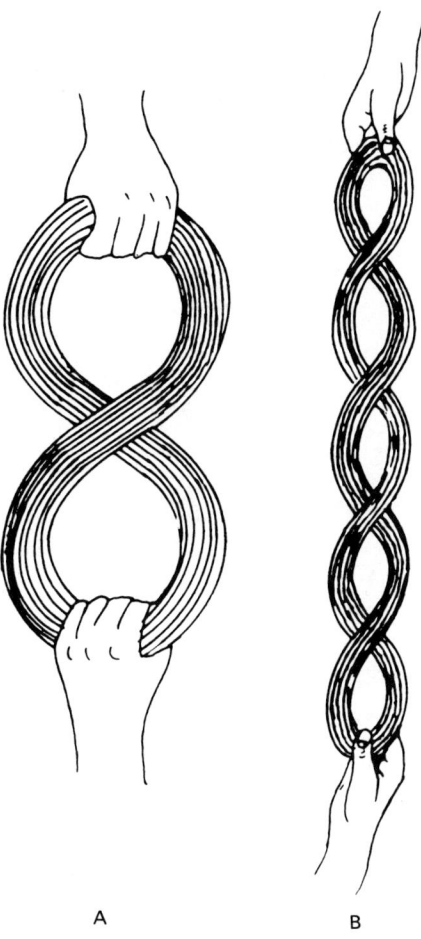

Fig. 4 Twisted skein models. (From Ref. 50.)

has the hydroxyl (OH) groups on the top two carbon atoms on the same side of the ring (the α form) and the other (β) on opposite sides. The D (for dextro-) notation comes from the convention of also describing sugar molecule isomers by the position of the OH group on the next-to-last carbon atom in the planar projection. The reference OH group in Fig. 6A is to the right (dexter) side of the carbon chain; hence the monomer is D-glucose.

Under certain conditions, solutions of D-glucose will undergo changes from α or β form to some equilibrium mixture of the two via a transient intermediary form in which the ring is broken open, as also shown in Fig. 6A. But the stable form of glucose is the ring with side groups as shown. A ring formed from five carbon and one oxygen atoms is said to have a *pyranose* form. Figure 6B provides a schematic of the α and β ring structures in which the rings are illustrated as planar (they are actually bent). The car-

Fig. 5 Cellulosic microfibrils. (A) Microfibrils of the cell wall of the marine alga *Valonia*. Plant fibers possess homologous microfibrillar textures. (From Ref. 31, courtesy of K. Mühlethaler, Swiss Federal Technical Institute (E. T. H.), Zurich.) (B) Aggregations of microfibrils at the end of a crushed tracheid from which the matrix has been removed. (From Ref. 13.)

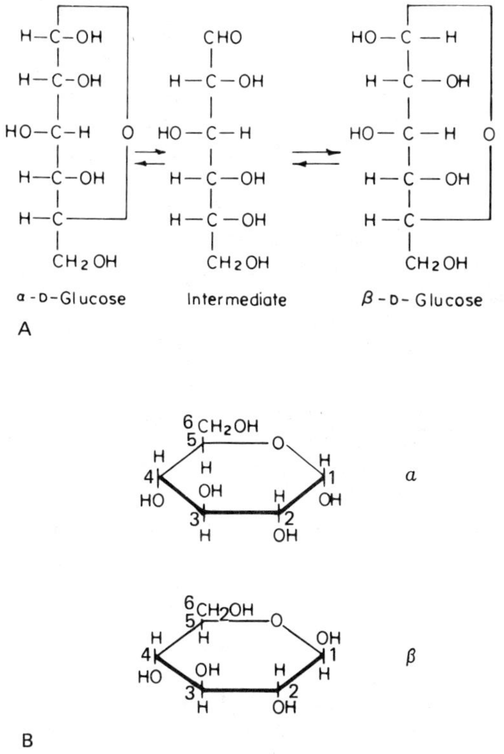

Fig. 6 Structure of glucose. (A) Planar schematic for D-glucose, showing the two isomeric forms (left and right) and an open-ring intermediary form (center). (B) Planar schematic of ring structures for α and β D-glucose, showing conventional numbering of carbon atoms.

bon atoms in Fig. 6B are numbered according to convention. The α and β structures form polymers that are distinctly different in many chemical and physical properties.

In a native cellulosic fiber, the number of cellulose molecules that pass through a given cross section is on the order of one or two billion. The cellulose that is present in a plant fiber differs substantially from the "cellulose" that is found in such industrial products as rayon tire cord or cellophane. The differences are in some cases chemical but more usually physical in nature. For this reason, the cellulose found in plant tissues is designated *native* cellulose to distinguish it from the modified industrial celluloses. The supermolecular arrangement of native cellulose (as it is normally found) is referred to as cellulose I.

A very significant physical aspect of cellulose I is that it is extensively aggregated into monoclinic crystalline arrays. These crystalline arrays (*crystallites*) are the principal, if not exclusive, constituent of microfibrils, which are the observable (by microscopy) filamentary structures that form the structural framework of the wood cell wall.

Until quite recently, it was widely believed that the chain molecules in cellulose I were antiparallel, that is, that chains packed together in a crystallite of native cellulose were arranged in an alternating pattern as to chain directionality. Conclusive proof that all chains within the microfibril are oriented in the same direction has been obtained by the development of computer-assisted crystallographic and conformational analysis, and by the development of methods to determine the minimum energy packing arrangement of the molecules [8,9,32,63,73].

The typical length of a chain molecule of native cellulose is unknown but probably well exceeds the value of 10^4 nm (100,000 Å) that may be determined experimentally for the substance. The reason for this statement is that the physical removal of one chain from its associates in the cell wall (by whatever means) in order to measure it almost certainly results in scission (breakage) of primary valence bonds linking the glucose units together; thus, we actually measure chain fragments. As chemical techniques for removing cellulose from fibers have become progressively less harsh, measured values for the degree of polymerization (DP) have risen from numbers in the hundred to numbers sometimes exceeding 30,000.

While nearly one-half of the dry mass of wood is cellulose, there is a family of related polymers present whose structures are based on various combinations of sugar residues of xylose, glucose, mannose, galactose, and arabinose. These other polysaccharides are collectively known as hemicelluloses. Typically the hemicelluloses are branched polymers having DP values in the low hundreds. They vary from relatively unbranched, alkali-resistant species such as glucomannan to the relatively nonlinear, soluble types such as the arabinogalactans, which can form modified sugar polymers containing pectic and uronic acid residues. The hemicellulose content of some plant fibers, for example cotton, is quite small. Cotton fibers are composed over 90% of cellulose, whereas wood fibers are normally less than 50% cellulose. Other plant fibers generally have cellulose contents lying between these values. The nature and proportions of the hemicelluloses found in different woods varies, although there are broadly consistent patterns. For example, the predominant hemicellulose in hardwoods is a glucuronoxylan whose structure is remarkably consistent from wood to wood. In softwoods the main hemicelluloses are the galactoglucomannans, a family of closely related polysaccharides that differ principally in the proportions of their component

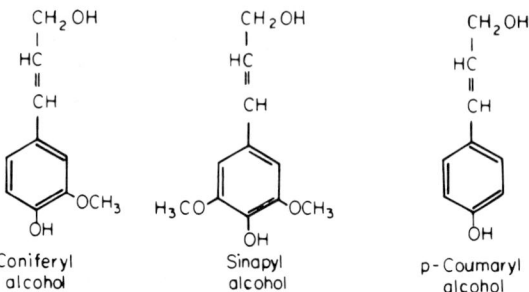

Fig. 7 Precursor monomers of lignin.

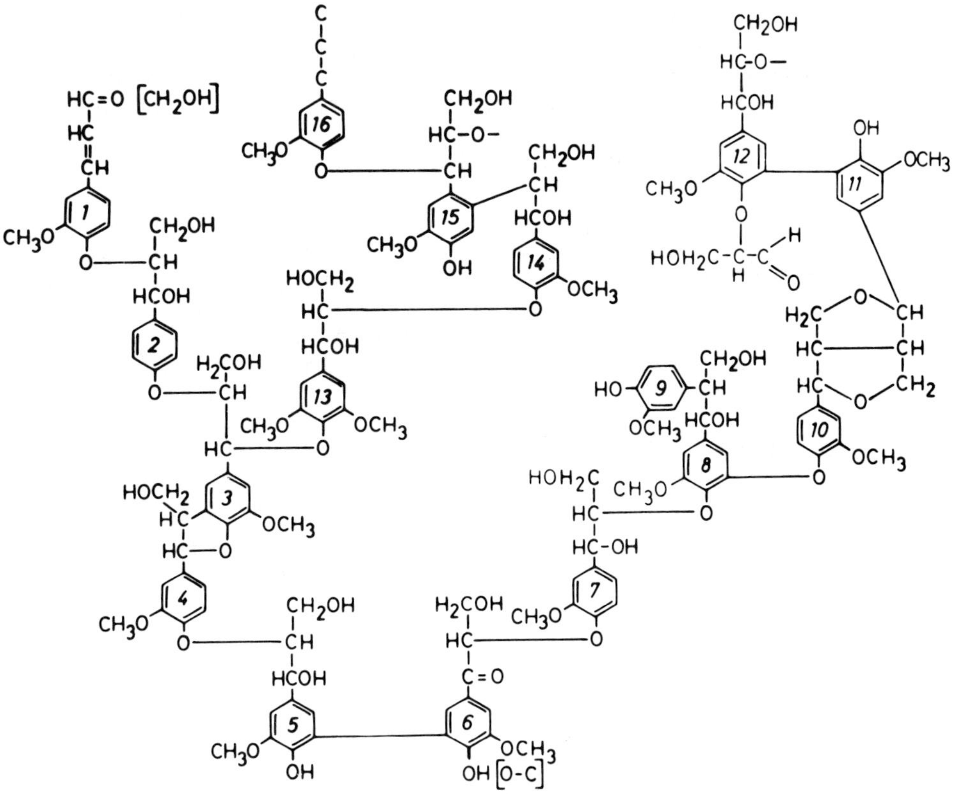

Fig. 8 Proposed model for a typical softwood lignin. (From Ref. 1.)

sugar residues. There are two types normally found—a relatively soluble polymer containing mannose, glucose, and galactose units in the ratio 3:1:1 and a much more resistant type found in close association with cellulose, in which the ratio approximates 30:10:1. Other galacto-glucomannans with somewhat different polymer compositions may also be present. Hemicelluloses may exhibit some degree of orientation and crystallinity, particulary when they are in close association with cellulose.

An amorphous polymeric material called lignin comprises the third major molecular component found in all woods and most other plant fibers, excluding seed hairs such as cotton and kapok. Algal cellulose is also devoid of association with lignin. As with the hemicellulose group, the actual composition of "lignin" varies between woods even though there are fairly consistent general differences between the lignins in hardwoods as compared with softwoods.

Lignins are polymerized enzymatically in plant cell walls from three primary precursor monomers: with reference to Fig. 7, these precursors are (1) trans-coniferyl, (2) trans-sinapyl, and (3) trans-p-coumaryl alcohols. These alcohols form the most common polymer structures found in various lignins. The polymerization is dehydrogenative in nature. A typical conifer lignin might be composed of units such as the 16-monomer unit shown in Fig. 8. The majority of these units are of the guaiacyl propane type, which derives from the trans-coniferyl monomer. Unit no. 13 exhibits two methoxy (OCH_3) side groups; it is a syringyl propane unit, derived from trans-sinapyl alcohol. The derivative of the trans-p-coumaryl unit, the third precursor in Fig. 7, is coumaryl propane (unit no. 2 in Fig. 8).

In spruce (*Picea* sp.), which has a rather typical conifer lignin structure, the lignin originates from coniferyl, sinapyl, and coumaryl alcohols in a ratio of approximately 80:6:14. The lignin of beech (*Fagus*), which is fairly typical for hardwoods, has ratios of 49:46:5. Thus, the softwood lignins are predominantly of the guaiacyl type, while guaiacyl and syringyl groups are about equally represented in the hardwoods [62].

Just as the removal of native cellulose chain molecules in an intact condition from a fiber cell wall is physically difficult, if not impossible, it has not yet been possible to remove lignin from wood without some alteration of the chemical structure during contact with the solvent. The molecular weights of soluble lignin derivatives cover a range of over three orders of magnitude ($< 10^3$ to $> 10^6$). Goring [33] notes, "Apparently, lignin can be dissolved as an entity small enough to be a pure chemical compound or as a particle large enough to show the behavior of a high polymer or a colloid. ...Such wide polydispersity leads to a certain degree of indeterminacy in ascribing a single value of the molecular weight to a given sample of lignin. A number average molecular weight, M_n, may be an order of magnitude smaller than the weight average molecular weight, M_w." Determination of both M_w and M_n and computation of the ratio of the two gives a measure of the polydispersity of the lignin fractions that are isolated. Average values for these two numbers can be determined from

$$M_n = \frac{\sum_i w_i}{\sum w_i/M_i} \quad \text{and} \quad M_w = \frac{\sum_i w_i M_i}{\sum_i w_i} \tag{1}$$

where the w_i values are the weights and M_i values are the molecular weights of the individual fractions obtained by some technique such as dialysis or precipitation. Experimental results show that M_w/M_n ratios of between 2 and 7 are found for the lignins removed from various woods. For a monodisperse system, $M_w/M_n = 1$. If a linear polymer is degraded by random chain scission, $M_w/M_n \to 2$.

Cellulose, hemicellulose, and lignin are by far the major components found in most plant fibers. As can be noted in Fig. 9, some proteins, inorganic substances, and so on, are also to be found in wood cell walls, and other materials such as starch, fats, and resins may be extracted from the cell lumens. The total of these minor constituents rarely exceeds 10% and is usually closer to 3 or 4%. Good papermaking plant fibers tend to be high in cellulose and low in protein and inorganic matter.

B. Relative Abundance

Aside from the small and often highly variable proportins of inorganic matter and other minor constituents, the pattern of relative abundance of chemical species in woods is quite consistent. The cellulose content of both hardwoods and softwoods is normally in the range of 42 ± 2%.

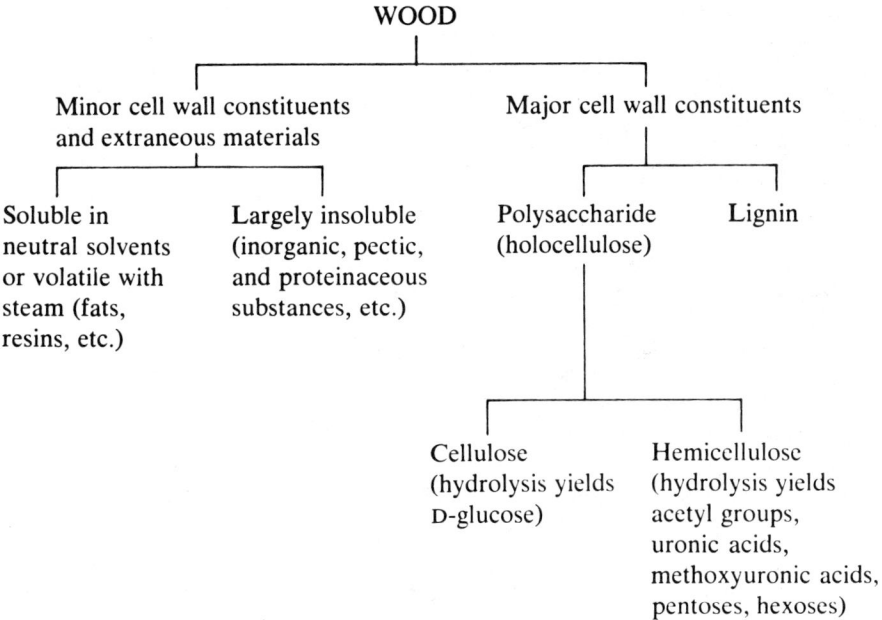

Fig. 9 Constituents of wood. (Adapted from Ref. 11.)

Hemicelluloses and lignin tend to be found in complementary proportions. For example, the lignin content of various conifer woods lies in the general range of between 24 and 33%. Since the proportion of cellulose changes but little from wood to wood, the woods that have high lignin contents have low (about 25%) hemicellulose, and vice versa. Hardwoods contain generally less lignin and more hemicellulose than the conifers. The lignin contents for temperate-zone hardwoods are usually determined to fall in the 16 to 24% range if one bases the determination on the lignin that remains as an insoluble residue after the carbohydrates have been hydrolyzed with H_2SO_4. Even if the "soluble" lignin is counted, the above hardwood lignin percentage only rises to the range of 19 to 28%. Further detail concerning the proportions of chemical constituents in other fibers has been given in Chap. 10, pp. 410-412.

Abundance of different molecular species also varies within a given species of plant or tree, because of soil fertility and other environmental factors, and within the wood of one individual tree. As an example of the latter, the lignin content of the fibers in one annual ring of a temperate-zone tree tends to be higher in the fibers formed at the start of the growing season (*springwood*) than in those formed later (*summerwood*).

Within the fiber itself, the abundance of various polymer species and minor constituents varies across the cell wall. While polysaccharides that contain glucose, mannose, and xylose residues tend to be distributed throughout the cell wall, the galactan, arabinan, and other pectopolyuronide polymers are typically concentrated in the outer part of the fiber and in the middle lamella, whch is the cementing layer that bonds adjacent cells to-

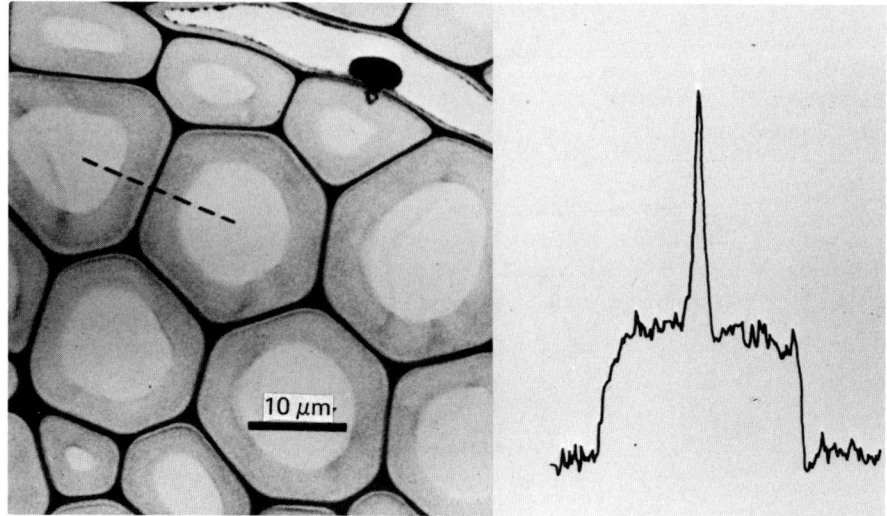

Fig. 10 Wood fibers in cross section together with a curve of the lignin distribution across the double cell wall. The distribution curve is obtained by a quantitative microspectrographic technique applied along the dashed line superposed on the photograph. (From Refs. 29, 30.)

gether. Although the concentration of lignin is also high in the middle lamella, that structure does not usually exceed 12% of the volume of normal wood. Since the middle lamella is less dense than the cell wall and the lignin content of wood is on the order of 20 to 30%, it is apparent that less than half of the total lignin is located in the middle lamella [4,49]. Actual quantitative determinations by microradiography and various microscopic and microspectrographic techniques (see Fig. 10) have shown that some 15 to 25% of the total lignin is contained therein, the balance being found in the walls of the fibers and other cells [29,30]. Lignin is found throughout the fiber wall, although the distribution varies among the layers.

IV. FIBER ALTERATION IN PROCESSING

The topochemical distributions that have been described in the preceding section have great significance relative to the effectiveness of various pulping and papermaking procedures. Obviously for a pulping chemical to be effective in fiber separation, it must penetrate to the middle lamella and remove or soften the concentration of lignin there. If the pulping process is thermomechanical, it is again the lignin-rich middle lamella area that must be plasticized sufficiently to permit the fiber separation to proceed mechanically. Figure 11 of Chap. 10 is a scanning electron micrograph of a small chip of birch whose fibers and ray cells are in the process of separating as a result of the action of the pulping liquor; in general, complete separation of cells is required for maximum quality of the furnish.

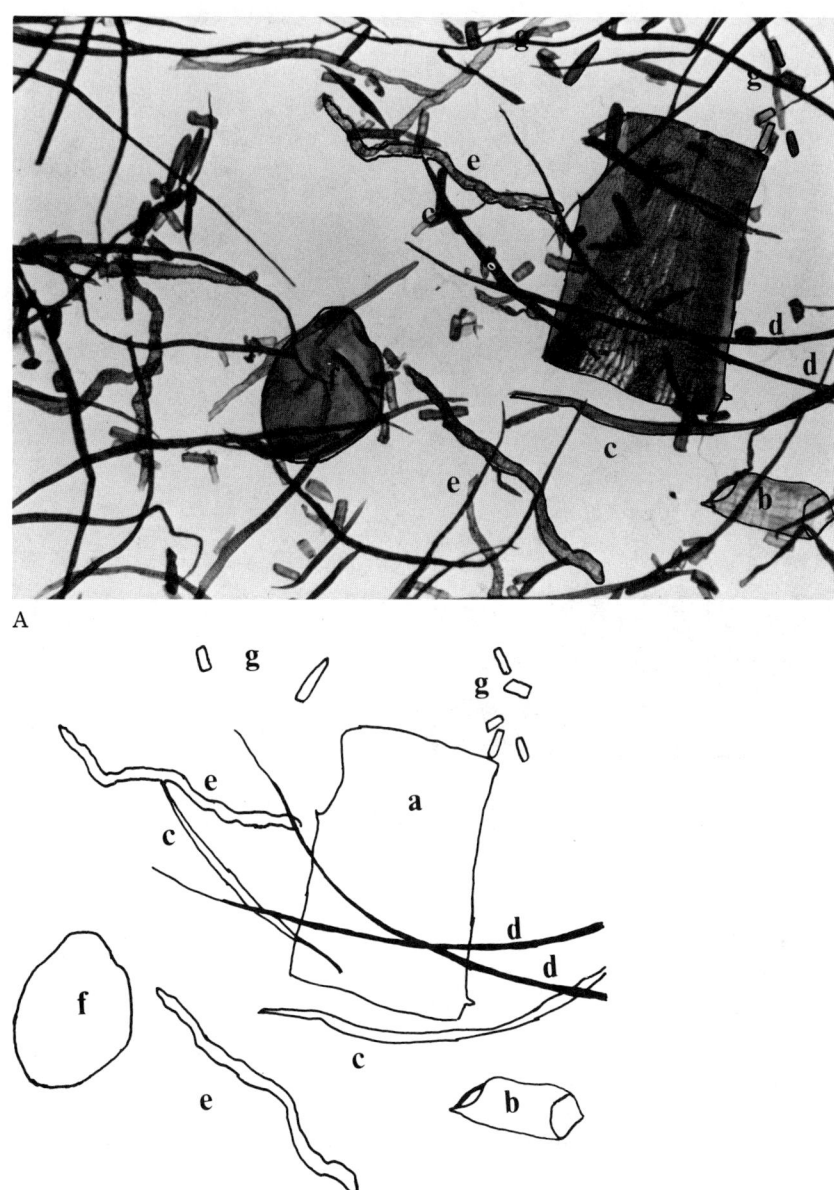

Fig. 11 Cellular elements in a pulp of oak (*Quercus* sp). A. Photomicrograph at 90 ×. (Courtesy of the Institute of Paper Chemistry, Appleton, Wisconsin.) B. Elements in photograph identified by author as to cell type: (a) springwood vessel element, heavily pitted; (b) summerwood vessel element; (c) fiber-tracheid; (d) libriform fibers; (e) vasicentric tracheids; (f) tylosis; (g) ray parenchyma cells.

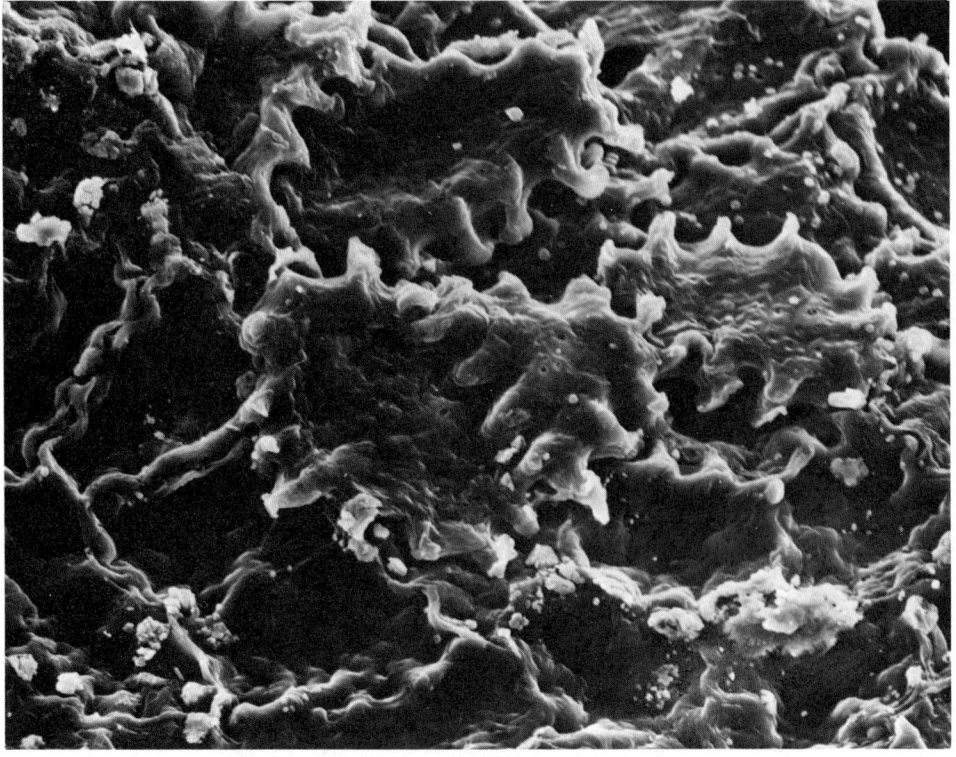

Fig. 12 Sclereids (stone cells) from oxygen-pulped pine bark at magnification 570×. Courtesy of C. M. Crosby, State College of Environmental Science and Forestry, Syracuse, New York.)

When fiber-containing raw material is pulped, the pulp contains all of the cell types that were present in the original plant. Figure 11 shows the various cell elements derived form a pulped chip of oak. A comparison of the identified cell types with Table 2 demonstrates that this particular oak possesses all the cellular elements usually present in a hardwood plus some cell types (vasicentric tracheids) that are not found in some hardwoods. It is the large springwood vessel elements that tend to create low pick resistance and reduced Z-direction strength in paper that contains them. However, vessel-containing pulps can be improved with proper refining [52]. Similarly, the advent of "whole-tree" pulping has introduced, in many cases, a large number of cell elements into furnish with unknown or deleterious papermaking characteristics. Tree bark is extraordinarily diverse in its cell composition; the barks of some species are resinous, others are corky, still others are composed largely of food storage and conduction tissue. Relatively few barks contain large quantities of useful papermaking fiber. Figure 12 shows a type of stone cell that is found in the barks of some conifer species. Cell material of such nature creates special problems in stock preparation.

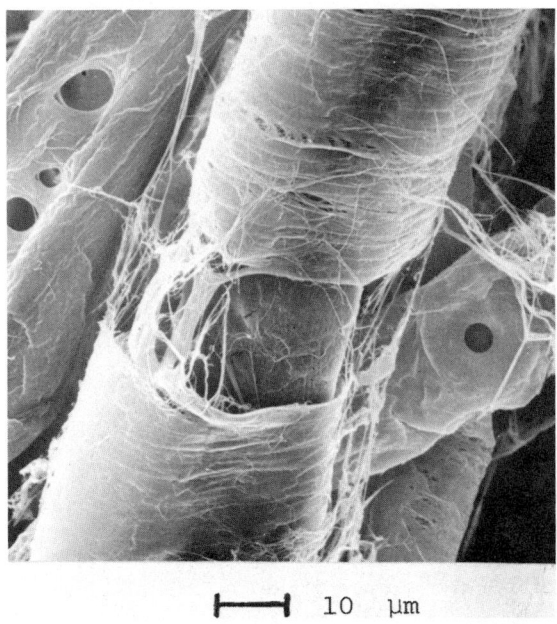

⊢━━┥ 10 μm

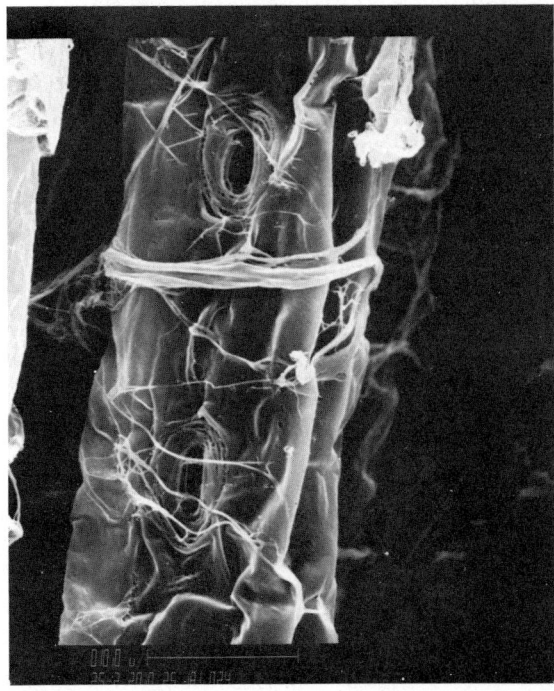

Fig. 13 Scanning electron micrographs of refined fibers [3]. A. Moderately heavily refined. Bar scale represents 10 μm. (Courtesy of T. H. Hsu, State College of Environmental Science and Forestry, Syracuse, New York.) B. Very heavily refined. Bar scale represents 10 μm.

Fig. 14 Dislocations and internal delamination in a refined fiber [46]. (Courtesy of R. P. Kibblewhite, Forestry Research Institute, Rotorua, New Zealand.)

The smaller cell elements that may be present in pulp, such as those in Fig. 12, or the parenchyma cells identified by "G" in Fig. 11B, tend to form the bulk of the primary fines fraction (see, for example, Fig. 44 of Chap. 24).

The fiber elements of pulp usually undergo extensive internal and external change as the chips or other raw material are processed. The greatest effect on fiber structure, both internal and external, occurs during the beating or refining operation. Secondary fines (see Figs. 45 to 52 of Chap. 24) are formed as material from the outer parts of the fiber is fibrillated and sometimes torn off. Figure 13A shows a moderately heavily refined fiber whose S1 layer has been partly removed and extensively loosened. A very severely refined fiber with the S1 layer mostly removed and the microfibrillar reinforcement around the pit openings loosened is shown in Fig. 13B. Other changes include cutting, kinking, the creation of dislocations or slip planes (Fig. 38 of Chap. 24), and localized wall distortions, for example, microcompressions (Fig. 14). Internal delamination also takes place, as shown in Fig. 14, which enlarges the internal surface area (see Fig. 8 of Chap. 10) and makes the fiber more susceptible to collapse or flattening with enlarged potential bonding area in the sheet as drying proceeds. The terms used in the foregoing discussion are defined and further discussed in "Notes on Procedures for Various Structural Parameter Determinations," Chap. 24, pp. 339-358.

V. FIBRIL ANGLE DETERMINATION

Earlier in this chapter (e.g., Figs. 4 and 5) and in Secs. II to IV of Chap. 10, it has been pointed out that the cellulosic microfibrillar organization is typically helical in any layer of the plant cell wall, that each layer has angular characteristics that set it apart from the other layers, and that some layers have a crossed helical structure, at least in some types of fiber. The characteristic layered structures for various fibers are shown in Figs. 9, 12-14, 16, and 22 of Chap. 10. In wood fibers and most other plant fibers used for papermaking, the central layer of the fiber—the S2 layer—makes up the largest part of the total cell wall (see Tables 2 and 3 of Chap. 10). The properties of that layer dominate but do not exclusively determine the properties of the fiber.

In recent years there has been an increasing awareness of the importance of the helical microfibrillar organization of plant cell walls and the strongly anisotropic properties they impart to the fibers; microfibril angles are now recognized to be a prime determinant of quality in wood and pulp fibers [10].

All fundamental studies of the strength of paper, as well as its other mechanical properties, depend upon a knowledge of the Young moduli of fibers. These moduli in turn are extremely sensitive to variations in S2 angle. If the S2 filament winding angle is less than 10° from the fiber axis, a variation of just 1° can change the fiber axial modulus by 10 GPa ($\sim 1.4 \times 10^6$ psi) or more (see Fig. 18 of Chap. 10 and Ref. 50a).

The reason for this profound effect of S2 angle can be demonstrated by assuming for simplicity that the fiber consists only of the S2 layer and that each microfibril in this layer has precisely the same orientation. This approximation makes more evident the relationship between the stiffness of the fiber wall in the microfibril direction and its stiffness in the fiber axis direction. The two-dimensional stiffness relation for the above assumption is

$$\frac{E_P}{E_A} = \cos^4\alpha + \frac{E_P}{E_N}\sin^4\alpha + \left(\frac{E_P}{G_{PN}} - 2\mu_{PN}\right)\sin^2\alpha\,\cos^2\alpha \qquad (2)$$

where

E_A = modulus of elasticity of the fiber parallel to its axis

E_P, E_N = moduli of elasticity of the S2 layer parallel (P) and normal (N) to its cellulosic microfibrils. Both are assumed constant for a given fiber.

G_{PN} = S2 shear modulus of rigidity associated with directions P and N. This is also a constant.

μ_{PN} = Poisson ratio for S2, a constant that represents the ratio of contraction in the direction indicated by the second subscript to extension under stress in the direction indicated by the first subscript, in the plane of the layer.

α = angle at which the microfibrils are oriented with respect to the fiber axis

It can be seen immediately that if all the microfibrils in this one-layer fiber are aligned at 0° to the fiber axis, then the last two terms on the right side of the equation become 0 and $E_P/E_A = 1$. But when the S2 filament winding angle is greater than 0°, all three right-hand terms are operative. The large fixed ratios E_P/E_N and E_P/G_{PN} come into play so that the variable E_A rapidly drops in relation to E_P (E_P/E_A increases with α). In real pulp fibers with the usual S1, S2, and S3 layers, the relationship is somewhat more complex, but the above example serves to illustrate the general relationship.

As will be brought out in this discussion, it is of critical importance that whatever method is used to make S2 angle determination (or any fibrillar orientation measurement), it should be confirmed or calibrated against a *direct* measurement method of unquestioned reliability. Although *reliability* is a vague term, it is the opinion of the author that a demonstration of the most reliable direct method will serve as a guide to selection of other methods—that is, the suitability of other methods can be evaluated against direct observation of cell wall microfibrillar striations, the first method to be discussed.

A. Methods Used

Many methods have been employed by various investigators to ascertain mean fibrillar angles in plant fiber cell walls. These are listed below in several categories:

Direct Measurement Methods

Observations of Cell Wall Striations With particular types of fibers that are prominently striated, such as compression wood tracheids, it is quite easy to observe S2 angle in either the light or electron microscope [17,18a,55,69]. In a light microscopy, special techniques such as metal shadowing, phase contrast, polarized light, or the use of ultraviolet light (UV) with or without fluorochrome stains, can enhance the appearance of the striations or render them visible where they would otherwise be too indistinct to measure (which is usually the case in normal fibers). In most of these methods, fibrillar angles may be directly measured with a protractor eyepiece or rotating stage. In the case of UV light, the angle is measured on a developed film or plate of a type particularly sensitive to UV wavelengths. For eye safety, direct observation is not possible with UV, of course.

Crosby and Mark [22] reported on a method that utilizes a combination of phase contrast with mercury burner illumination filtered to pass only wavelengths between 330 and 475 nm (BG3 filter). This is principally a near-ultraviolet source with some visible light transmitted in the blue-violet range. To obtain the greatest resolution of the striations, photographic exposures are made directly on metallographic plates. Examples of pulp fibers photographed under these illumination conditions are shown in Figs. 15A and B. The S2 angles can be measured directly on the photographs by protractor. The advantages of this method are:

- The fiber does not require any staining or other special processing, nor does the UV light affect the fiber in any way, leaving it undamaged and available for any other type of test or measurement desired.

- •Sufficient resolution is obtained on the exposed plates to make measurements that would not be possible by other methods. This improved resolution comes from the relatively narrow band of wavelengths to which the panchromatic plates are sensitive.

- ••In areas where pits occur, the circular microfibril border in the S1 layer (Fig. 13B) tends to cause scatter in X-ray beams. But striations observable in the S2 layer are little affected by the presence of pits.

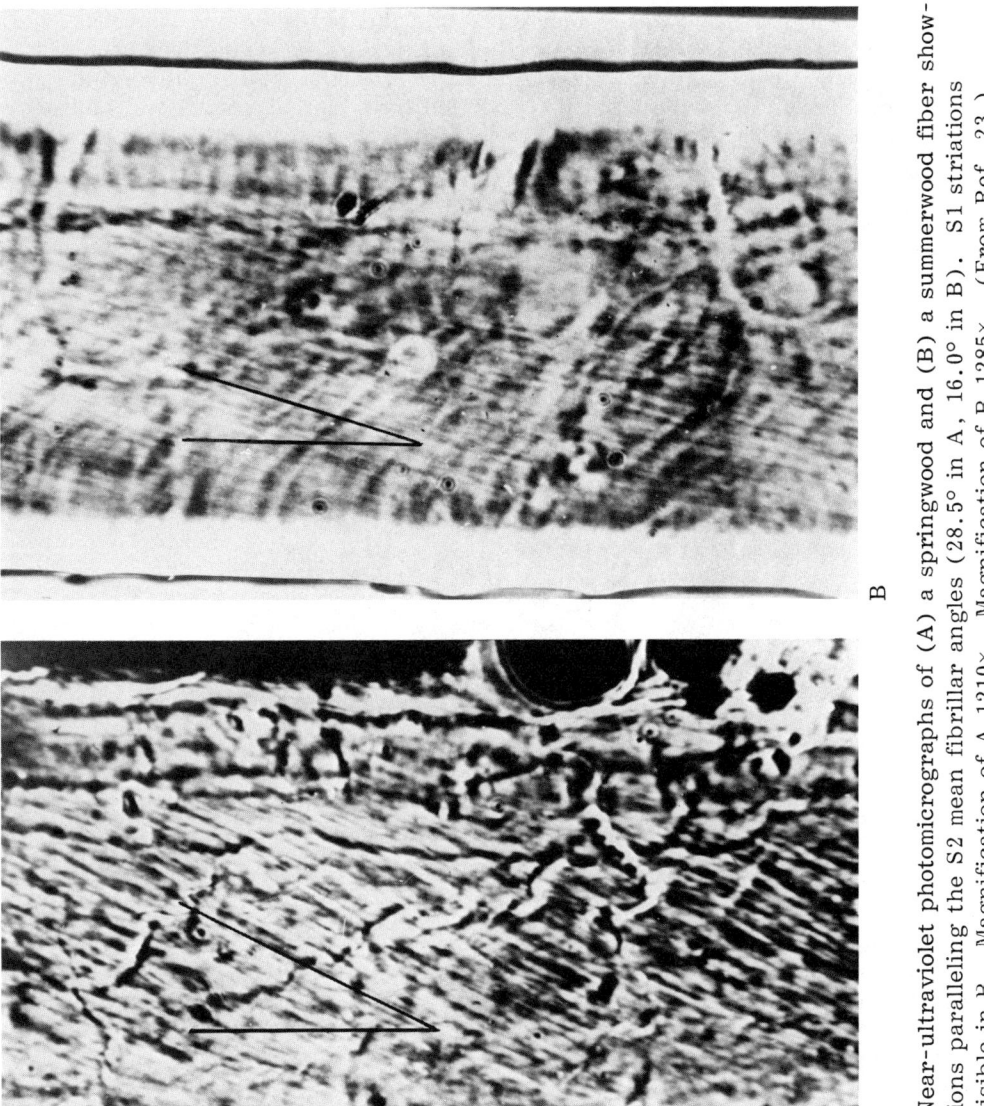

Fig. 15 Near-ultraviolet photomicrographs of (A) a springwood and (B) a summerwood fiber showing striations paralleling the S2 mean fibrillar angles (28.5° in A, 16.0° in B). S1 striations are also visible in B. Magnification of A 1210×. Magnification of B 1285×. (From Ref. 23.)

The S2 microfibrils really deviate only a small amount as they pass around the pit due to the fact that a pit *changes shape* as it passes through the wall of the fiber, as shown in Fig. 16. This fact imparts a special advantage, in terms of accuracy, to the methods involving measurement of orientation of deposited crystals, of cracks and splits, and especially to measurement of striations as in UV light.

- ●●The method is especially useful in thick-walled summerwood fibers with small S2 angles, which are the most difficult to test by other methods.

- ▬The striations of the S2 layer are distinguishable from all other layers in the UV Method, and are not influenced by them even in cases where striations of a layer such as S1 are visible.

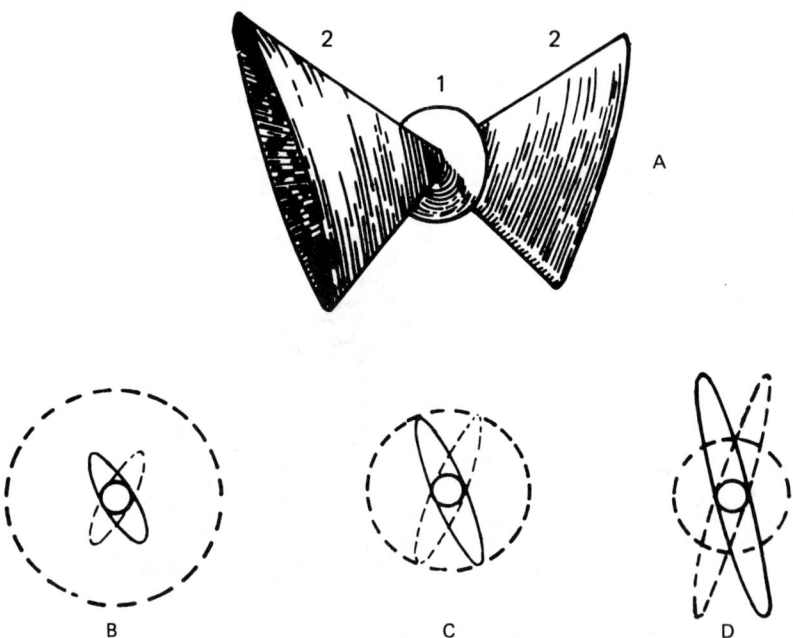

Fig. 16 The shape of a typical bordered pit pair. (A) Oblique lateral view of a pit pair: (1) pit chamber; (2) apertures. (B,C,D) Diagrams of various types of pit pairs in frontal view, showing the changes in form that typically occur in fiber walls of increasing thickness. At the inner wall of the fiber, the aperture is quite elongated and is usually canted from the vertical. The broken lines indicate portions of the pit pair on the side of the compound wall away from the observer. The outermost broken-line circle represents the periphery of the pit chamber. (Adapted from Ref. 10a.)

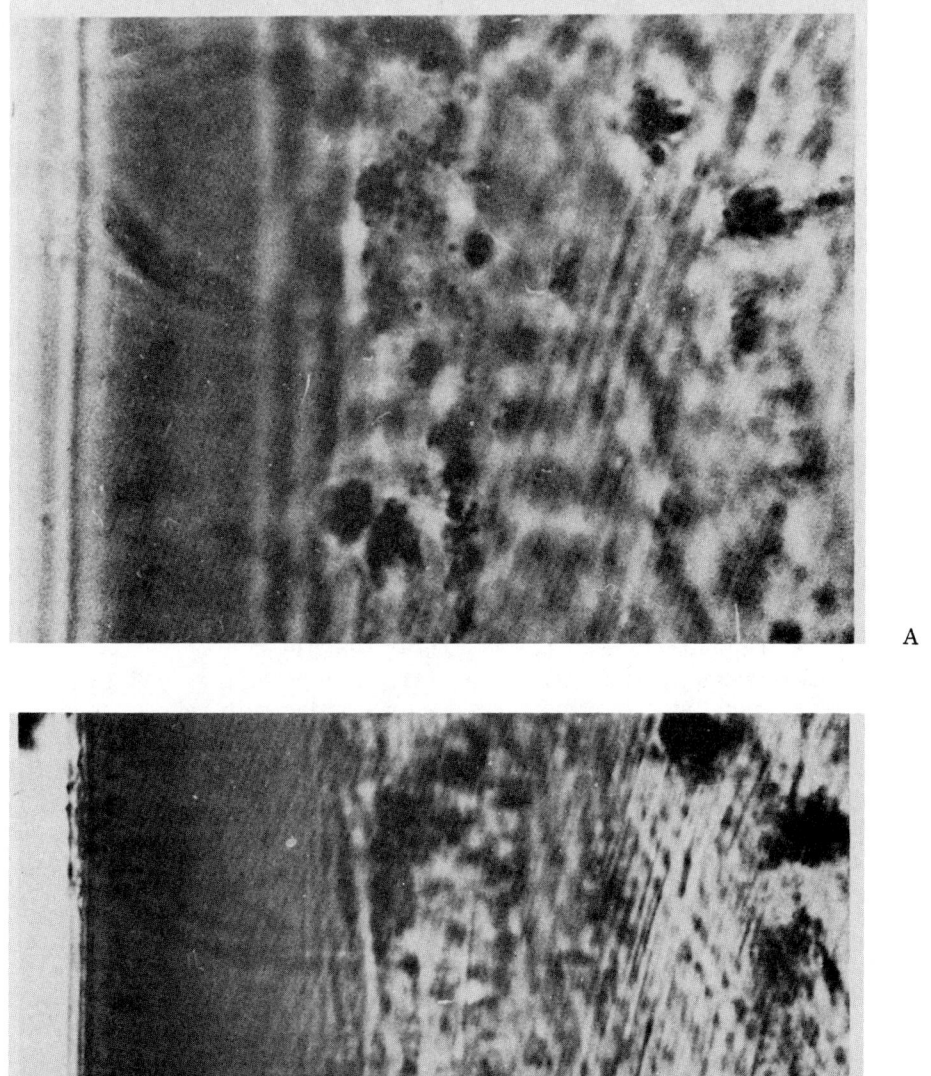

Fig. 17 Loblolly pine (*Pinus taeda*) summerwood fiber, detail photograph. (A) Conventional phase contrast illumination, 1300×. (B) The same location, bright-field ultraviolet at wavelength 270 nm, 1300×.

The disadvantages of the method are that slide mounting of the fibers is necessary and that high-resolution photographic plates must be used.

Subsequently, Crosby and Mark [23] investigated the use of very low wavelength illumination, both as a check on the accuracy of the determinations made from the phase contrast-near UV procedure and also to see whether the images of the striations could be enhanced at lower wavelength.

In any microscopic system the power to resolve detail of decreasing size is inversely proportional to the wavelength of the illumination used. Thus, it is advantageous to use as short a wavelength as possible. Conventional microscopes reach their lower limit in the range of 365 to 400 nm; below that point the amount of light transmitted by glass drops rapidly.

Although further decreases in wavelength into the ultraviolet range are advantageous in terms of increased resolution, the application is not too common because of the expensive equipment required—an ultraviolet microscope equipped with a xenon burner, monochromator, and quartz optics. Although in this case the phase-contrast capability is lost, it is more than offset by (1) the short wavelength of illumination and (2) the restriction of band width to about 3 nm by the monochromator, giving essentially monochromatic light. Under these conditions, the lower limit of useful wavelength was found to be 270 nm. For thick-walled (summerwood) fibers, the exposure times are impractically long below this point. Spectrum analysis film, a high-resolution, high-contrast material, is needed.

Figure 17A and B are photomicrographs of the same spot on a loblolly pine summerwood pulp fiber; there is a small difference in focal plane. Figure 17A was photographed under conventional phase contrast, filtered to give a peak at about 440 nm. For Fig. 17B, the UV microscope was used at the 270 nm wavelength (no phase contrast). The resolution of the finer fibrillar aggregations still showing the same orientation with respect to the fiber axis is readily apparent.

Observations of Pit Aperture Angle Microfibrils in the S2 make a smooth deviation around the intertracheid pits in a manner that gives the aperture an elongated appearance. Numerous workers have measured the inclination of the long axis of the oval or ovate aperture with respect to the fiber axis and assumed this to be equivalent to the S2 angle [38,43,45,72].

Cracks and Splits in the Fiber Wall Quite often, particularly in compression wood formed late in the growing season, helical cracks will develop in the fiber wall that appear prominently when thin longitudinal sections are made. It is also possible to induce these cracks in normal wood by fungal enzyme action (Fig. 18), chemicals, or mechanical stress. These cracks follow the general fibrillar alignment of the S2 layer.

Deposition of Crystals One of the most effective techniques for obtaining direct measurements of fibrillar angles is by the induced formation of crystalline materials between clusters of fibrils in the cell wall. These elongated crystals show the orientation of microfibrils in the S2 layer clearly and occasionally reveal the filament winding angles of S1 and S3 also. Silver staining methods have proven useful for determining similar structures in

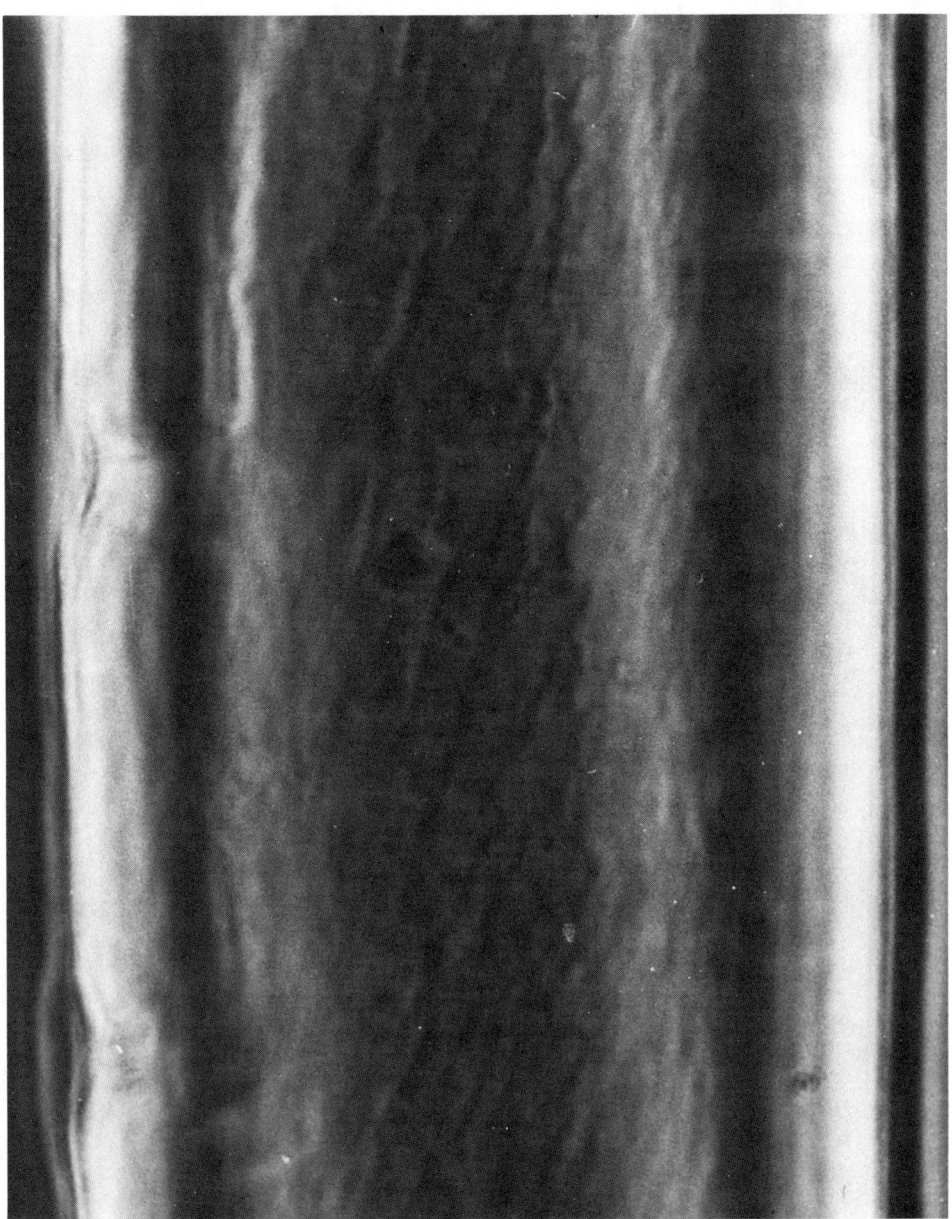

Fig. 18 Loblolly pine summerwood fiber showing enhancement of fibril pattern as a result of attack by the soft-rot fungus *Chaetomium globosum*. Phase contrast, 2000×. (Courtesy of C. M. Crosby, State College of Environmental Science and Forestry, Syracuse, New York.)

animal tissues, but iodine crystal deposition is the method that has been generally employed for woody tissue [54].

Observations of the Planes of Hydrolysis Certain soft-rot fungal enzymes create cavities within the secondary wall of cellulosic plants, both woody and nonwoody, that assume an orthorhombic dipyramidal shape. The long axis parallels the cellulosic mean microfibrillar orientation in the S2 layer. The sides of the pyramidal ends of the cavities tend to incline at 23° from the microfibrillar axis, which would coincide with the $(45\bar{4})$ plane of extended-chain crystalline cellulose I. Because of this very regular orientation of the enzyme-created cavities in relation to the fibril orientation, Cowling [21] has urged broader use of the technique for fibril angle determination, especially for the hardwood species, and progress has been made along these lines [59,71]. Figure 19 shows fibers of bamboo (*Bambusa vulgaris*) with such soft-rot cavities.

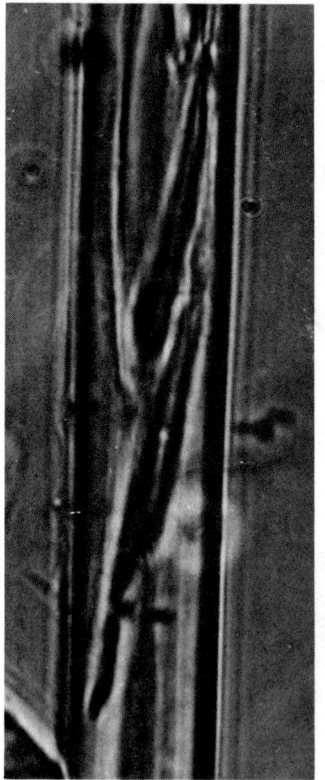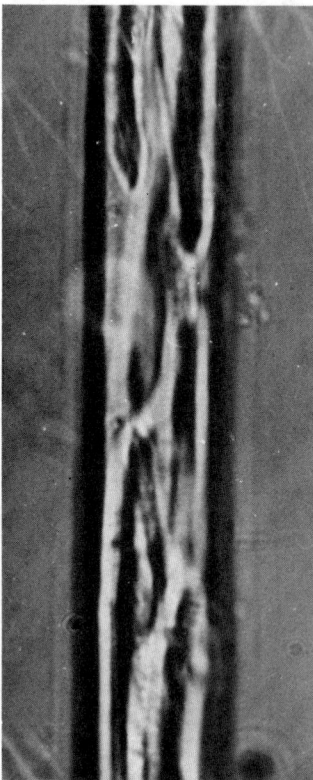

Fig. 19 Bamboo (*Bambusa vulgaris*) fibers showing cavities produced by an unidentified soft-rot fungus. Phase contrast, 950×. (Courtesy of C. M. Crosby, State College of Environmental Science and Forestry, Syracuse, New York.)

Replica Methods

Optical Replicas for Use in Light Microscopy Meylan [54] has used a technique for etching sectioned surfaces of wood with potassium hydroxide, followed by pressing on a softened sheet of cellulose acetate to obtain a replica of the surface. Meylan found, however, that the cut surface, shadowed with antimony for greater contrast, showed the microfibril structure in the S2 layer only in patches where the knife had cut into S2, and he considered it a relatively unreliable method.

Replicas for Use in Electron Microscopy Most all of the transmission electron microscopic work done with wood and fibers involves the use of carbon replicas. Attempts to observe filament winding angles directly on wood and fibers by transmission electron microscopy have proven only moderately successful in practice because of the extreme difficulty in getting ultrathin sections and lack of contrast in these sections. The replicas can be metal-shadowed for good contrast, whereas attempts to metal-shadow the fiber surfaces directly have resulted in degradation of the fiber in the intense electron beam. Among others, Harada [35] and Dunning [27] have used surface replicas to study the microfibrillar orientations in the various fiber wall layers.

Indirect Measurement Methods

Polarized Light Methods There are four general methods of determining mean fibrillar orientations with the use of the polarizing microscope. The most accurate and also the most tedious of these is the Sénarmont compensator method, which is well described by the originator, R. D. Preston, in his book [60] (see also Ref. [5]). It has been employed subsequently by Mark [48a, 49], Manwiller [47,48], Tang [66], and Crosby [24], each of whom contributed advances in technique and results. A polarizing microscope fitted with a Sénarmont compensator is used to measure birefringence of secondary cell wall layers individually by means of a series of sections cut at various angles to the long axis of the cells. A plot of birefringence versus section angle yields a curve whose maximum point corresponds with the filament winding angle of the observed layer. Unfortunately, the method is not applicable to single fibers.

The Becke line refractive index method was employed by Meredith [53a] on cotton fibers of irregular outline. It requires immersion of the fiber in a succession of liquids having different refractive indices, observing the fiber boundary under polarized light, and matching this boundary to a refractive index of one of the liquids. This value is then employed in a formula giving the relationship of the index ellipsoid of (theoretically) perfectly aligned molecules to those having an angular inclination to the fiber axis.

Several techniques have been devised to determine mean fibril angle with polarized light transmitted through a single wall. The major extinction position (M.E.P.) of the wall is then determined in relation to the fiber axis and the angular difference measured via rotating stage is taken as equal to the fibril angle. In order to obtain single fiber walls for observation, it is necessary to slice the fibers in some way. Cowdrey and Preston [20,60] have done this by macerating a wood sample, adhering the tracheids to a glass slide, and scraping the slide with a hand razor in order to cut away the uppermost cell walls and leave the lower walls on the slide, whose M.E.P. can then be measured. A similar technique has been employed by Cousins [19]. Meylan [54] accomplished the single-wall exposure by slicing a wood block at a small angle to the grain direction prior to maceration. Although this technique is more tedious than the above, it has the advantage of making it possible to pre-select the ring position to be observed and to segregate radial walls from tangential walls accurately. Sometimes it is not possible or practical to pre-slice before maceration. Alexander et al. [2] mounted undried fibers in gelatin on a glass slide. The fibers were then sliced at an angle with a hand-held razor blade to expose part of a single wall and permit the determination of M.E.P.

Leney [46a] has made a significant contribution to the M.E.P. polarized light method by a sectioning/maceration technique that produces half-sections of individual fibers. In this technique, longitudinal microtome sections approximately half the thickness of the fibers are cut. These sections are macerated by pulping or other available techniques such as the glacial acetic acid solution described by Leney. Thus, fibers with only a single wall are made available for observation after a special mounting procedure. The method has the advantage of permitting an annual ring to be sectioned serially so that separate observations can be made of the fibers in various parts of the annual ring.

A method utilizing reflected polarized light has been developed and extensively used by Page et al. [56a,57]. Mercury droplets are introduced under pressure into the pulp fiber lumen, permitting an M.E.P. measurement on the single (upper) wall of the fiber using epi-illumination in the microscope.

For fibers with thin (< 0.2 μm) S1 and S3 layers and whose S2 layers lie in the range of 1.3 to 3.4 μm thick, the errors in measurement introduced by this method are small enough that the method can be used for some purposes. However, the method fails for fibers whose S2 layers are relatively thick (> 3.5 μm), especially if there is relatively thick S1 or S3 layer present [28], as for example in the summerwood fibers of the Southern pines.

X-Ray Diffraction Analysis Because X-ray diffraction affords a measure of the mean crystallite angle of a large number of cells scanned simultaneously, it is the most rapid means of obtaining average values for fibril angle. Because of its simplicity, this method has proven very attractive and convenient; consequently, X-ray diffraction analysis has become a favorite among the available methods for fibril angle determination. Most researchers follow the procedure recommended by Hermans [36], in which it is assumed that one-half of the width of the reflection of the (002) crystallographic plane of cellulose I at a proportion (usually 40 or 50%) of peak value will be equal to the mean microfibril angle. This latter value in turn is assumed equal to the S2 angle, although some authors recognize that this is not strictly so [37,42,61].

The method just described is so universally used in textile fiber research that the term *X-ray angle* is generally used in place of *fibril angle*, the most common term in wood research. Evaluation of the (002) arc is also very often employed as the standard technique for determining S2 angle in wood and wood fibers.

Table 4 Comparative Measurements of Mean Fibrillar Angle by Jurbergs

Tree	Fibril Angle by Long Axis of Pit Aperture (degrees)	Fibril Angle by X-ray (002) Arc (degrees)
1	33.7	35.0
2	40.5	36.5
3	32.9	31.5
4	30.5	30.0
5	32.8	28.5
6	34.0	35.5
7	21.7	11.5
8	31.0	42.0
9	28.7	26.0
10	32.9	28.5

Material: Slash pine (*Pinus elliotti*) summerwood tracheids isolated from 8th and 9th growth rings by boiling in aqueous 17.5% nitric acid solution.
Source: From Ref. 43.

It has been observed by Cave [15] that X-ray reflections arising from cellulose I crystal planes of the type (040) could give the microfibril angle distribution directly, at least in theory, but that these were generally too weak or poorly resolved to be used. However, Sobue et al. [64] have developed an X-ray diffractometer technique that enables measurement of the three-dimensional crystallite orientation in wood cell walls from (040) plane reflections. The method has also been used by Kalyanaraman [44] and El-osta et al. [28a].

Light Scattering Visconti et al. [70] have used small angle light scattering to estimate the mean helical winding angles in cellulosic fibers. Test and theory agree fairly well, but the experimental light scattering results have had little comparison with tests by any other method.

B. Comparisons Between Methods

It might be expected that, with a large variety of experimental methods having been used, there would exist a substantial amount of comparative data for the various procedures. Surprisingly, these comparisons are quite limited in number.

In Table 4 is reproduced a table from Jurbergs's [43] paper in which he compares eighth and ninth ring summerwood tracheids of *Pinus elliotti*, taken from 10 different trees. The fibril angle is measured by one indirect method (X-ray) and one questionable "direct" method (pit aperture orientation). Although the overall averages for the 10 trees show a difference of less than

Table 5 Comparative Measurements of Mean Fibrillar Angle by Cowdrey and Preston

Ring number	Fibril angle by M.E.P. on longitudinally cut fiber (degrees)	Fibril Angle by X-ray (002) arc (degrees)
7	20.6	21.8
10	13.3	11.8
12	9.4	9.6
14	8.8	9.3
16	7.0	8.9
19	6.5	6.5

Material: Sitka spruce (*Picea sitchensis*) tracheids isolated by maceration in 5% chromic acid
Source: From Ref. 20.

Table 6 Comparative Measurements of Mean Fibrillar Angle by Alexander, Marton, and McGovern

	Canadian standard freeness of pulp	Fibril angle by M.E.P. on diagonally cut fiber (degrees)	Fibril angle by X-ray (002) arc (degrees)
Springwood, 60% yield	710	16.0	12.5
Springwood, 60% yield	490	15.0	10.0
Springwood, 60% yield	100	13.3	8.0
Springwood, 48% y8eld	690	18.8	18.5
Springwood, 48% yield	225	14.1	14.5
Springwood, 48% yield	120	11.7	10.5
Summerwood, 60% yield	730	17.8	12.0
Summerwood, 60% yield	260	15.1	10.5
Summerwood, 60% yield	36	12.3	9.5
Summerwood, 48% yield	730	16.0	15.0
Summerwood, 48% yield	160	13.6	12.0
Summerwood, 48% yield	65	10.6	8.5

Material: Norway Spruce (*Picea abies*) kraft pulp fibers
Source: From Ref. 2.

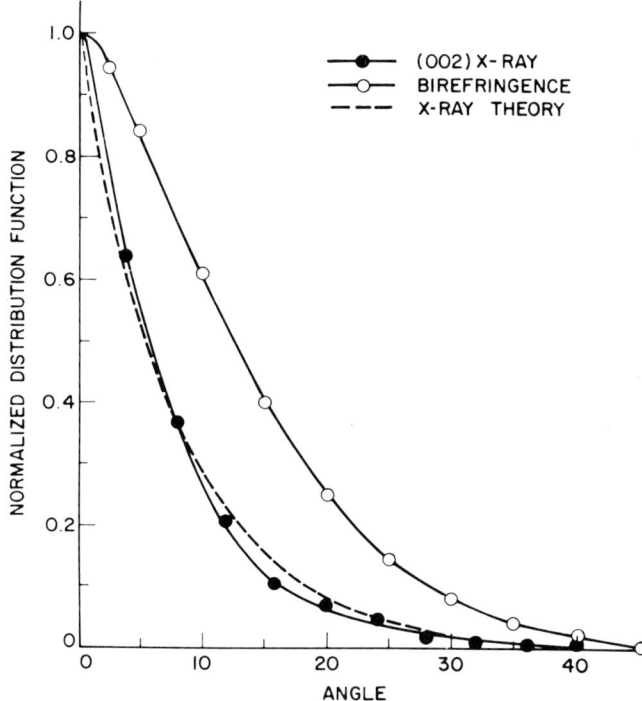

Fig. 20 Comparison between the measured fibril angle distribution function as obtained by the reflected polarized light method and the X-ray intensity data of the (002) plane for a wood sample. (From Ref. 61.)

2° between the two methods of measurement, individual trees show differences of as much as 11° (or an 88.7% discrepancy). It is not made clear in this or other works in which the X-ray method is compared to another whether the specimens are handled in the same way for the two types of measurement. As an example, Cowdrey and Preston [20] compared results for six specific rings of a stem of *Picea sitchensis* (Table 5), but they merely state that approximately 45 tracheids were used for each tabulated measurement by the M.E.P. method. For X-ray measurements, "three or four specimens from the same annual ring" were used.

It should be reiterated that neither the M.E.P. method nor the X-ray method, which were also used by Alexander et al. [2] (see Table 6), are direct methods. Individual comparative values range up to a difference of 27% in Cowdrey and Preston's results and 66% in the data of Alexander et al.

Prud'homme and Noah [61] also compared two indirect methods, X-ray diffraction and reflected polarized light with mercury intrusion. They showed that X-ray intensity measurements for the paratropic reflections do not give the fibril angle distribution function directly; it is necessary to correct the curve in accordance with Cave's measurement theory [15]. A comparison of the corrected X-ray intensity data, X-ray theory, and results using the reflectance method on the polarizing microscope is shown in Fig. 20.

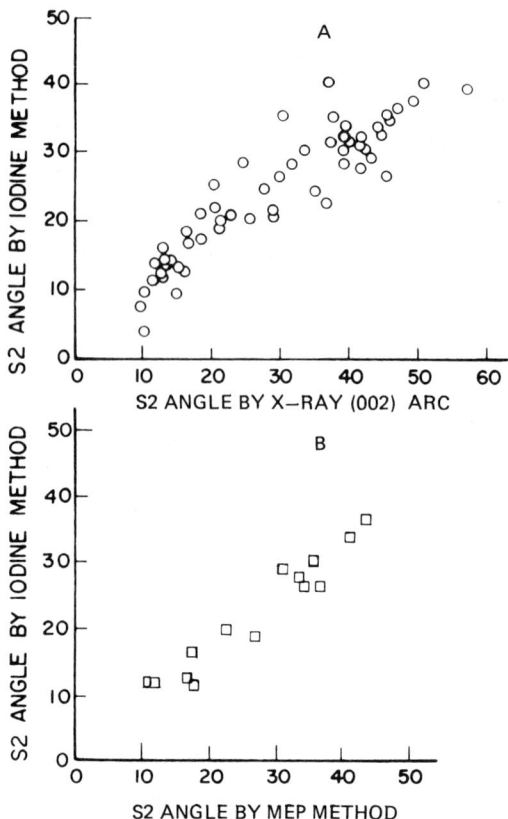

Fig. 21 Comparison of mean microfibril angle as measured by iodine staining with (A) half the angular width of the (002) X-ray arc at 40% peak height, and (B) major extinction position of polarized light transmitted through a single tracheid wall. (Adapted from Ref. 54.)

When further comparisons were made between direct and indirect methods [26,54], even more pronounced discrepancies were observed. In Fig. 21A and B, Meylan's comparisons of data obtained from iodine staining with X-ray analysis and M.E.P., respectively, show that there are highly significant differences. If the indirect methods were accurate, regression lines comparing them with the iodine method should pass through the origin and have a slope equal to unity, but these do not occur in either case. In contrast, the slope of a regression line comparing optical replicas and striations with iodine staining did *not* differ significantly from a line through the origin and having slope = 1 [54]. Meylan ascribes the contribution of the S1 and S3 layers as at least one factor in the distorted values given by the indirect methods, whereas angles determined by iodine staining, replica, and striations "all refer specifically to the S2 layer and are in close agreement." Meylan's reservations concerning the influence of the S1 and S3 layers on results from indirect methods have been confirmed by the study of El-Hosseiny and Page [28].

Table 7 Comparative Measurements of Mean Fibrillar Angle in Virginia Pine (*Pincus virginiana*) Tracheids Isolated from the 15th Ring by Glacial Acetic Acid and Hydrogen Peroxide

Location within ring	Fibril angle by direct measurements on UV plates (degrees)	Fibril angle by M.E.P. on diagonally cut fiber (degrees)	Fibril angle by reflected polarized light after mercury intrusion (degrees)
Springwood	16	24.5	21.5
Springwood	17.5	19 (24.5)	17.5
Springwood	20.5	19	18.5
Springwood	28.5	20.5 (20.5)	21
Transition zone	22	8 (13.5)	15.5
Transition zone	14.5	9.5	20
Transition zone	20	14	13
Transition zone	17.5	10 (20)	15
Summerwood	10	10 (14.5)	21
Summerwood	19	10.5 (17)	15.5
Summerwood	18.5	7.5 (12)	16.5
Summerwood	17.5	12	3

All angles are recorded as the nearest one-half degree of that actually observed. Values appearing in parentheses represent a second M. E. P. determination made in gelatin. (From Ref. 22.)

A further complication in the precise measurement of the S2 filament winding angle occurs in the case of individual fibers (as differentiated from those in fiber bundles or in wood). This complication relates to the extensive twisting that takes place when an isolated pulp fiber is allowed to dry freely [51]. The phenomenon is not restricted to wood pulp; it is general with isolated native cellulosic fibers and is a particular problem in cotton, where the presence of twists, or convolutions, affects the strength of the fibers and contributes to the discrepancies between measurements of fibril orientation made by different means.

Crosby and Mark [22] made comparative measurements of mean S2 fibrillar angle on isolated Virginia pine tracheids by the direct method of striation angles on UV plates and two indirect polarized light methods. They found that the indirect methods they tested did not yield results that followed the UV light results closely or consistently, as evidenced in Table 7. They concluded that since the agreement is poor between S2 angle measurements made by different methods, a direct method is necessary to obtain reliable data. The most accurate direct method, as shown in the preceding discussion, employs ultraviolet light, with or without phase contrast. Resolution is improved by transmitting a narrow (filtered) wavelength band to the plate or film on which measurements are made. The phase-contrast UV method permits accurate S2 angle measurements to be made on fibers whose S2 microfibrils are aligned at a small inclination to the fiber axis, whereas most other methods yield less reliable data for such fibers.

REFERENCES

1. Adler, E. (1977). Lignin chemistry: Past, present and future. *Wood Sci. Technol.* 11:169–218.
2. Alexander, S. D., Marton, R., and McGovern, S. D. (1968). Effect of beating and wet pressing on fiber and sheet properties. 1. Individual fiber properties. *Tappi* 51(6):277–283.
3. Bambacht, J. P., Hsu, T. H., and Unbehend, J. E. (1981). Analysis of fines fractions and their influence on sheet properties. (Submitted to *Appita*.)
4. Berlyn, G. P., and Mark, R. E. (1965). Lignin distribution in wood cell walls. *Forest Prod. J.* 16(3):140-141.
5. Berlyn, G. P., and Miksche, J. P. (1976). *Botanical Microtechnique and Cytochemistry*. Iowa State Univ. Press, Ames.
6. Bhattacharyya, B., Bhattacharyya, H., Banerjee, P. K., Guha, G., and Bhattacharyya, T. K. (1975). Staining of paper fibers: Use of different ink powders and naturally occurring dyestuffs for the identification of cellulose and lignified fibers. *J. Indian Acad. Forensic Sci.* 14(2):50–53. (Original not seen; cited in *ABIPC* 47(5):484(1976), abstr. no. 4667.)
7. Bhattacharyya, J., Guha, G., and Bhattacharyya, B. (1976). Quantitative microscopical analysis on cellular elements for the identification of bamboo pulp. *Indian Pulp Paper Tech. Assn.* 13(4):307–310. (Original not seen; cited in *ABIPC* 48(8):887 (1978); abstr. no. 8253.)
8. Blackwell, J., and Kolpak, F. J. (1975). The cellulose microfibril as an imperfect array of elementary fibrils. *Macromolecules* 8:322–326.
9. Blackwell, J., Kolpak, F. J., and Gardner, K. H. (1978). The structures of celluloses I and II. *Tappi* 61(1):71-72.
10. Boyd, J. D. (1974). Relating lignification to microfibril angle differences between tangential and radial faces of all wall layers in wood cells. *Drevársky Výskum* 19(2):41–54.
10a. Brown, H. P., Panshin, A. J., and Forsaith, C. C. (1949). *Textbook of Wood Technology*, vol. 1, McGraw-Hill, New York, p. 79.
11. Browning, B. L. (1963). The composition and chemical reactions of wood. In *The Chemistry of Wood* (B. L. Browning, ed.). Interscience, New York, pp. 57–101.
12. Browning, B. L. (1977). *Analysis of Paper*. Marcel Dekker, Inc., New York.
13. Bucher, H. (1958). Discontinuities in the microscopic structure of wood fibres. In *Fundamentals of Papermaking Fibres* (F. Bolam, ed.). British Paper and Board Makers Assn., London, pp. 7–26.
14. Canadian Pulp and Paper Assocaition Standard B.7. Quantitative Analysis of Fibre Mixtures.

15. Cave, I. D. (1966). Theory of X-ray measurement of microfibril angle in wood. *Forest Prod. J.* 16(10):37–42.
16. Chiaverina, J. (1964). Methods for the microscopic analysis of paper. *Rev. Forest. Franc.* 2:115–128 (in French).
17. Core, H. A., Côté, W. A., and Day, A. C. (1979). Wood structure and identification, 2nd ed., Syracuse Univ. Press.
18. Côté, W. A., ed. (1980). *Papermaking Fibers: A Photomicrographic Atlas*, Renewable Materials Institute and Syracuse Univ. Press, Syracuse, N.Y.
18a. Côté, W. A., Jr., and Day, A. C. (1965). Anatomy and ultrastructure of reaction wood. In *Cellular Ultrastructure of Woody Plants* (W. A. Côté, Jr., ed.), Syracuse Univ. Press, pp. 391–418.
19. Cousins, W. J. (1972). Measurement of mean microfibril angles of wood tracheids. *Wood Sci. Technol.* 6(1):58.
20. Cowdrey, D. R., and Preston, R. D. (1965). The mechanical properties of plant cell walls: Helical structure and Young's modulus of air-dried xylem in *Picea sitchensis*. In *Cellular Ultrastructure of Wood Plants* (W. A. Côté, Jr., ed.), Syracuse Univ. Press, pp. 473–492.
21. Cowling, E. B. (1965). Microorganisms and microbial enzyme systems as selective tools in wood anatomy. In *Cellular Ultrastructure of Wood Plants* (W. A. Côté, Jr., ed.), Syracuse Univ. Press, pp. 341–368.
22. Crosby, C. M., and Mark, R. E. (1974). Precise S2 angle determination in pulp fibers. *Svensk Papperstidn.* 77(17):636–642.
23. Crosby, C. M., and Mark, R. E. (1975). The determination of S2 angle: Further studies. Unpublished report.
24. Crosby, C. M., de Zeeuw, C., and Marton, R. (1972). Fibrillar angle variation in red pine determined by Sénarmont compensation. *Wood Sci. Technol.* 6(3):185–195.
25. Desch, H. E., and Dinwoodie, J. M. (1981). *Timber: Its Structure, Properties and Utilisation*, 6th ed., Timber Press, Forest Grove, Oreg.
26. Dumbleton, D. P. (1972). Longitudinal compression of individual pulp fibers. *Tappi* 55(1):127–135.
27. Dunning, C. E. (1969). The structure of longleaf-pine latewood. 1. Cell-wall morphology and the effect of alkaline extraction. *Tappi* 52(7):1326–1341.
28. El-Hosseiny, F., and Page, D. H. (1973). Measurement of fibril angle of wood fibers using polarized light. *Wood Fiber* 5(3):208–214.
28a. El-osta, M. L., Kellogg, R. M., Foschi, R. O., and Butters, R. G. (1973). A direct X-ray technique for measuring microfibril angle. *Wood and Fiber* 5(2):118–128.
29. Fergus, B. J. (1968). Lignin distribution and delignification of xylem tissue. Ph.D. Thesis, McGill Univ., Montreal, Canada.
30. Fergus, B. J., and Goring, D. A. I. (1970). The distribution of lignin in birch wood as revealed by ultraviolet microscopy. *Holzforsch.* 24(4):118–124.

31. Frey-Wyssling, A., and Mühlethaler, K. (1970). *Ultrastructural Plant Cytology*, Elsevier, New York.
32. Gardner, K. H., and Blackwell, J. (1974). The structure of native cellulose. *Biopolymers* 13:1975–2001.
33. Goring, D. A. I. (1971). Polymer properties of lignin and lignin derivatives. In *Lignins: Occurrence, Formation, Structure and Reactions* (K. V. Sarkanen and C. H. Ludwig, eds.) Wiley-Interscience, New York, pp. 695–768.
34. Graff, J. H. (1940). *A Color Atlas for Fiber Identification*. Institute of Paper Chemistry, Appleton, Wis.
35. Harada, H. (1965). Ultrastructure and organization of gymnosperm cell walls. In *Cellular Ultrastructure of Woody Plants* (W. A. Côté, Jr., ed.), Syracuse Univ. Press, pp. 215–233.
36. Hermans, P. H. (1949). Physics and Chemistry of Cellulose Fibres. Elsevier, New York, p. 248.
37. Hill, R. L. (1967). The creep behavior of individual pulp fibers under tensile stress. *Tappi* 50(8):432–440.
38. Hiller, C. H. (1964). Correlation of fibril angle with wall thickness in summerwood of slash and loblolly pine. *Tappi* 47(2):125–128.
39. Hubbard, L. R. (1966). Identification of textile fibers. *Amer. Dyestuff Reptr.* 55:282–286.
40. Isenberg, I. H. (1967). *Pulp and Paper Microscopy*, 3rd ed., Institute of Paper Chemistry, Appleton, Wis.
41. Jane, F. W., Wilson, K., and White, D. J. B. (1970). *The Structure of Wood*, A. & C. Black, London.
42. Jentzen, C. A. (1964). The effect of stress applied during drying on some of the properties of individual pulp fibers. *Tappi* 47(7):412–418.
43. Jurbergs, K. A. (1963). Determining fiber length, fibrillar angle and springwood-summerwood ratio in slash pine. *Forest Sci.* 9(2):181–187.
44. Kalyanaraman, A. R. (1980). Spirality of the crystallite of the natural cotton fibers. *J. Appl. Polymer Sci.* 25(11):2523–2529.
45. Kellogg, R. M., and Ifju, G. (1962). Influence of specific gravity and certain other factors on the tensile properties of wood. *Forest Products J.* 12(10):463–470.
46. Kibblewhite, R. P. (1980). Effects of pulp freezing and frozen pulp storage on fiber characteristics. *Wood Sci. Technol.* 14:143–158.
46a. Leney, L. (1981). A technique for measuring fibril angle using polarized light. *Wood and Fiber* 13(1):13–16.
47. Manwiller, F. G. (1966). Sénarmont compensation for determining fibril angles of cell wall layers. *Forest Prod. J.* 16(10):26–30.
48. Manwiller, F. G. (1967). Tension wood anatomy of silver maple. *Forest Prod. J.* 17(1):43–48.
48a. Mark, R. E. (1965). Tensile stress analysis of the cell walls of coniferous tracheids. In *Cellular Ultrastructure of Woody Plants* (W. A. Côté, Jr., ed.), Syracuse Univ. Press, pp. 493–533.
49. Mark, R. E, (1967). *Cell Wall Mechanics of Tracheids*. Yale Univ. Press., New Haven, Conn.

50. Mark, R. E. (1980). Molecular and cell wall structure of wood. *J. Educ. Modules Mater. Sci. Eng.* 2(2):251–308.
50a. Mark, R. E., and Gillis, P. P. (1973). The relationship between fiber modulus and S2 angle. *Tappi* 56(4):164–167.
51. Mark, R. E., Thorpe, J. L., Angello, A. J., Perkins, R. W., and Gillis, P. P. (1971). Twisting energy of holocellulose fibers. *J. Polymer Sci., C.* 36:117–195.
52. Marton, R., and Agarwal, A. K. (1965). Papermaking properties of hardwood vessel elements. *Tappi* 48(5):264–269.
53. McCrone, W. C., Delly, J. G. and Palenik, S. J., eds. (1979). *The Particle Atlas: An Encyclopedia of Techniques for Small Particle Identification*, vol. 5. *Light Microscopy Atlas and Techniques*, 2nd ed., Ann Arbor Scientific Publications, Ann Arbor, Mich.
53a. Meredith, R. (1953). Measurements of orientation in cotton fibres using polarized light. *Brit. J. Appl. Phys.* 4:369–373.
54. Meylan, B. A. (1967). Measurement of microfibril angle by X-ray diffraction. *Forest Prod. J.* 17(5):51–58.
55. Meylan, B. A., and Butterfield, B. G. (1972). *Three-Dimensional Structure of Wood*, Syracuse Univ. Press.
56. Okano, T. (1968). Relation between micelle angle and X-ray diffraction diagrams. *J. Japan Wood Res. Soc. (Mokuzai Gakkaishi)* 14(7):358–362 (in Japanese).
56a. Page, D. H. (1969). A method for determining the fibrillar angle in wood tracheids. *J. Micros.* 90(pt. 2):137–143.
57. Page, D. H., El-Hosseiny, F., Winkler, K., and Lancaster, A. P. S. (1977). Elastic modulus of single wood pulp fibers. *Tappi* 60(4):114–117.
58. Panshin, A. J., de Zeeuw, C., and Brown, H. P. (1964). *Textbook of Wood Technology*, vol. 1. *Structure, Identification, Uses and Properties of the Commercial Woods of the United States*, 2nd ed. McGraw-Hill, New York.
58a. Parham, R. A., and Gray, R. L. (1982). *The Practical Identification of Wood Pulp Fibers* (in press), TAPPI Press, Atlanta, Ga.
58b. Parham, R. A., and Kaustinen, H. M. (1974). *Papermaking Materials: An Atlas of Electron Micrographs*. Institute of Paper Chemistry, Appleton, Wis.
59. Pérez-Mogollón, A. (1973). Soft rot in beech caused by *Chaetomium globosum* Kunze, M. S. Thesis, SUNY College of Environmental Science and Forestry, Syracuse, N.Y.
60. Preston, R. D. (1974). *The Physical Biology of Plant Cell Walls*, Chapman and Hall, Ltd., London.
61. Prud'homme, R. E., and Noah, J. (1975). Determination of fibril angle distribution in wood fibers: A comparison between the X-ray diffraction and the polarized microscope methods. *Wood Fiber* 6(4):282–289.
62. Sarkanen, K. V., and Ludwig, C. H., eds. (1971). *Lignins: Occurrence, Formation, Structure and Reactions*. Wiley-Interscience, New York.
63. Sarko, A. (1978). What is the crystalline structure of cellulose? *Tappi* 61(2):59–61.

64. Sobue, N., Hirai, N., and Asano, I. (1971). Studies on structure of wood by X-rays. 2. Estimation of micelle orientation in cell walls. *J. Japan Wood Res. Soc. (Mokuzai Gakkaishi)* 17(2):44–50.
65. Strelis, I., and Kennedy, R. W. (1967). *Identification of North American Commercial Pulpwoods and Pulp Fibers.* Univ. Toronto Press.
66. Tang, R. C. (1973). The microfibrillar orientation in cell-wall layers of Virginia pine tracheids. *Wood Sci.* 5:181–186.
67. TAPPI T 8. Identification of wood and fibers from conifers.
68. TAPPI T 259. Species identification of nonwood plant fibers.
69. Timell, T. E. (1982). *Compression Wood in Gymnosperms* (in preparation), Springer-Verlag, Heidelberg.
70. Visconti, S., Hien, N. V., Borch, J., and Marchessault, R. H. (1976). Light scattering by helical fiber structures: Experimental models. *J. Polymer Sci. (Polymer Phys. Ed.)* 14:631–641.
71. Wardrop, A. B., and Jutte, S. M. (1968). The enzymatic degradation of cellulose from Valonia ventricosa. *Wood Sci. Technol.* 2(2):105–114.
72. Wellwood, R. W. (1962). Tensile testing of small wood samples. *Pulp Paper Mag. Can.* 63(2):T61–T67.
73. Woodcock, C., and Sarko, A. (1980). Packing analysis of carbohydrates and polysaccharides. 11. Molecular and crystal structure of native ramie cellulose. *Macromolecules* 13(5):1183–1187.

APPENDIX: INDEX OF STANDARDS, CIE PROCEDURES, AND TECHNICAL INFORMATION SHEETS

STANDARDS INDEX

American Society for Testing and Materials (ASTM)

C 177	246, 247
D 257	229
D 523	18
D 589	4
D 591	436
D 777	256
D 864	253
D 985	4
D 1117	159, 162
D 1434	83
D 1535	12
D 2244	14
D 2766	243
D 2863	258

Australia-New Zealand Pulp and Paper Industry Technical Association (Appita)

p 426	407
p 427	412

486 Appendix: Standards Index

Canadian Pulp and Paper Association (CPPA)

B.7	450
D.4	407, 412
D.8U	393
E.1	4
E.2	4
F.1	114
F.2	118
F.3	108
TS D14	85

Deutsches Institut für Normung (German Standards Institute) (DIN)

DVM 3413	85

International Standards Organization (ISO)

R 438	412
R 534	407
2469	4, 5, 30
2470	4, 5, 6, 30
2471	4, 5, 30

Japan Industry Standards Committee (JISC)

JIS C-2111	85, 236, 237
JIS P-8117	85
JIS P-8118	407
JIS P-8123	4, 6
JIS P-8138	4, 6
JIS P-8207	315
JIS Z-8902	16

Japan TAPPI Standards

No. 5	85
18	393
19	393

Scandinavian Pulp, Paper and Board Testing Committee (SCAN)

C32	165
M6	315

Appendix: Standards Index 487

P 3   4
P 7   407, 412
P 8   4
P 12   118
P 13   117, 159
P 18   108
P 19   85
P 22   89
P 26   88, 93

Svenska Industriens Standardiseringskommission (Swedish Industry Standardization Commission) (SIS)

SIS 25 12 28   158

Technical Association of the Pulp and Paper Industry (TAPPI)

RC 303   85
T 8   450
T 204   367
T 211   367
T 232   314
T 233   315, 316
T 234   315, 317
T 251   88
T 259   450
T 261   367
T 411   407, 411, 436
T 419   436
T 425   4, 6, 34
T 432   156
T 441   118
T 448   89
T 452   4, 6
T 458   108
T 460   85
T 461   256
T 462   130
T 464   89

488 Appendix: Standards Index

T 480	18, 32
T 481	286
T 482	89
T 491	117
T 492	114
T 506	393
T 508	9, 12
T 513	112
T 515	12
T 519	4
T 522	21
T 527	9
T 653	18
UM 239	359
UM 240	358
UM 241	359
UM 242	359
UM 403	393
UM 522	393
UM 524	85, 86
UM 528	393
UM 548	16
UM 569	393
UM 808	332, 393

Underwriters' Laboratory Standard (UL)

94.3	256
94.7	257

PROCEDURES OF THE COMMISSION INTERNATIONALE D'ÉCLAIRAGE (INTERNATIONAL COMMISSION ON ILLUMINATION) (CIE)

No. 15 (E-1.3.1)	3, 4, 7
CIE 45-20-195	6, 30

TAPPI TECHNICAL INFORMATION SHEETS

TIS 017-1	28
TIS 017-2	34

SUBJECT INDEX

Absorbency, liquid (*see also* Penetration):
  nonaqueous, 123-141
  rate, 130
  tissue and toweling, 143-168
  wadding, 159-160, 162
Absorptance:
  intensity of light, 3
  Kubelka-Munk Theory, 5, 20-21, 26, 32-39
Absorption, aqueous (*see also* Wetting), 104-107, 110-119, 136-137, 144-165
  Bristow apparatus, 110-111, 118
  capacity, 144, 152, 161-162
  capillary imbibition, 104
  capillary rise test methods, 154, 159-160
  coefficient, 110-111
  drop test, 114
  ESCA, 119
  factors, 162-165
  floating time test methods, 154, 158-160, 162
  heat, 119
  hydrophobic agents, 104, 112

[Absorption, aqueous]
  inclined plane test methods, 154, 160-161
  influences, 162-165
  lateral rate of, 136-137
  orifice test methods, 154-157
  porous plate test methods, 154, 158, 162
  rate, 144, 152-165
  saturation gradient, 159
  swelling, 104-105
  test methods, 152-162
  theory, 144-152
  tissue and toweling, 152-165
  wipe-dry test methods, 154-165
Absorption, light:
  coefficient 3, 33
  power 33
Absorption, non-aqueous, 136-137, 144
  lateral rate of, 136-137
Absorption current, electrical, 232
Adhesive bond stress, 395-398
  bond length dependence, 395-398
  double (symmetric) lap joint, 398
  single lap joint, 395-397

Subject Index

Aging, 164-165, 259
  accelerated, 165
Air resistance, 73, 87
Angle:
  fiber crossing, 381
  fibril, 306-308, 349, 353
  filament winding, 306-308, 350, 353
  mean orientation, 310-312
  micellar spiral, 353
Angular distribution functions:
  cosine, 291-292, 295, 297-299
  curve-fitting, 297-299
  elliptical, 291-292, 295-299
  goodness of fit test for, 297-299, 320
  mean directionality, 293-295, 310-312
  von Mises, 291-296, 299
Anisotropy:
  breaking length, 312
  elastic constants, 254, 295, 312
  electrical, 172-173, 184-185, 190-195
  equivalent pore, 308-310
  mechanical properties, 254, 286-287, 295, 312
  strength, 286-287, 312
  thermal expansion, 254
Area:
  centroid of, 344
  circularity, 345
  crossing, 348
  fiber cross section, 312, 328, 331, 339
  fiber-fiber bond, 339-344, 353, 359, 381-390
  fiber-fiber contact, 339-344, 346-347, 386-388
  fiber surface, 312, 338, 344, 382
  lumen, 339
  mean bonded, 353
  Mühlsteph ratio, 354
  overlapped, 381
  percent bonded, 342-344, 359
  projected, 339
  relative bonded, 342-344, 382-385, 388-389, 393

[Area]
  total bonded, 342-344
Ashing, 367
Aspect ratio, fiber, 313, 340-341, 350-351, 353, 359
  length, 350-351
  major axis, 353
  minor axis, 353
Atmosphere, ISO standard, 107
Autoignition temperature:
  definition, 255
  measurement, 257

B.E.T. (Brunauer, Emmett, and Teller) equation, 61-64, 387-388
Bark, 462
  composition of, 462
Basis weight, 343, 382-384
  bonding area dependency on, 382-384
Bending factor, fiber, 334-336
Bendtsen air permeability, 127
Beta-ray transmission, 22, 28
Bond:
  adhesive, 395-398
  area, 339-344, 353, 359, 381-390, 393
  completeness, 383
  density, 314
  destruction, 390-398
  distance, 341, 380
  energy of rupture, 390-391
  fiber-cellophane, 394-398
  fiber-fiber, 32-40, 314, 331, 338-344, 353, 359, 379-402
  fiber-shive, 393-398
  force theories, 380
  formation, 380, 387
  friction, 380
  hydrogen, 380, 389-390
  length effect on strength, 395-398
  maximum potential, 382
  mechanical entanglement, 380
  mechanical properties, 380, 390-398
  necking, 354-355

[Bond]
  region, 341, 380
  state, 340, 342, 344, 359, 381-390
  strength, 314, 380, 390-398
  stress distribution, 392, 395-398
  structure, 379-390
  van der Waals force theory of, 380
  Z direction test of, 393
Bond delamination test, 393
Bonding state, 340, 342, 344, 359, 381-390
  determination, 385-390
  diagram, 385
  probability, 340, 342, 344, 359
  variability, 385
Breakdown strength, electrical (see Dielectric strength)
Breaking length (see also Strength and Tensile strength), 312, 392-393, 404-406
  defined, 405
  MD/CD ratio, 312
  Page's semiempirical equation, 392-393
Bridge methods for dielectric properties tests:
  conjugate Schering bridge, 223
  Schering bridge, 221-223, 226
  transformer bridge, 223
Brightness, 4-6, 14-16, 27
  standard methods, 4-6
Bristow apparatus, 110-111, 113, 119
Britt jar (see Dynamic retention/drainage jar)
Brushing out (see Fibrillation)
Bulk, 338
Burning point, definition of, 256

Caliper (see Thickness, sheet)
Calorimetry, 243-245
Capacitance, electrical, 173-174, 225-226
  variable, 225-226
Capillarity, 65-73, 104-106, 117
  capillary imbibition, 104-106

[Capillarity]
  capillary pressure, 105-106
  capillary radius, 105, 110
  capillary rise, 104-110
  contact angle, 104-106, 108-110
  Hagen-Poiseuille equation, 70
  Laplace equation, 105
  Lucas-Washburn equation, 106, 110
Capillary:
  flow, 145-152
  fountain effect, 149-150
  imbibition, 104
  models, 146-152
  radius, 146-149, 151
  rise (see also Capillary rise), 148-150, 154, 159-160
  wetting, 146-152
Capillary rise:
  cancelling velocity, 108-109
  method, 108-110
  theory, 104-107
Capillography, 108-109, 119
Cellophane-fiber bond strength, 394-398
Cellulose:
  abundance, 458-460
  association with other polymers, 456-460
  crystallinity, 455-456
  degree of polymerization (DP), 456
  depolymerization, 259-272
  glucose units in, 452-457
  loss tangent effects of, 208-210
  microfibrils, 452-456, 463-464
  molecular weight, 456
  polymerization, 456
  swelling by interchain bond breaking, 104-105
  thermal decomposition of, 259-277
  thermal expansion of, 253-255
Centroid, 344
Chromaticity:
  coordinates, 9-16, 27-28
  diagram, 9-15
Chromatography, inverse gas, 119
Circularity, fiber (see also Roundness), 345
Classifiers, fiber length, 315-317

Coarseness:
  bonding area dependency on, 382
  constant, 317
  determination, 316-318
  fiber, 315-318, 330, 338, 343
  pulp, 316-318, 338
  wood, 317-318, 338
Coating:
  optics, 5, 34-39
  penetration, 118
Cobb test, 104, 118
Collapse, fiber, 312, 345-346, 355-356
Color (see also Colorimetry):
  charts, 12
  differences, 13-14
  hue, 9, 16
  intensity, 7
  lightness, 7
  matching, 9-16
  metamerism, 12-14
  object, 7
  saturation, 9
  space, 13
  strength, 12
  temperature, 9-12
  wavelengths of, 2, 7-8
Colorimetry, 6-16, 23-28, 42
  charts, 12
  chromaticity coordinates, 9-16, 27-28
  CIE system, 7-14
  excitation purity, 9
  instrumentation, 23-28
  metameric matching, 12-14
  Munsell system, 12
  scales, 12
  trichromatic coefficients, 9
  tristimulus values, 7-16, 42
  visual, 12
Combustion:
  burning point, 256
  energy output, 259-260
  flash point, 255, 257-258
  heat of, 258-259
  ignition temperature, 255, 257
  inhibition of, 272
  oxygen index, 256, 258
Compliance, 125-126

Composition, fiber, 348, 451-460
Compressibility, dynamic, 124-127
  creep, 125
  dwell time, 125
  printing, 127
  test apparatus, 125-126
Conduction, thermal:
  conductivity (which also see), 242, 245-251
  specific heat, 242-245
Conductivity, electrical:
  AC (alternating current), 173, 176, 203
  DC (direct current), 173, 176-184, 226-229, 389-390
  fiber, 184
  fiber bond area determined by, 389-390
  ionic, 179-182, 202, 208
  liquid penetration test, 113-115
  moisture influence, 177-179
  surface, 173, 232-235
  temperature influence, 232
  volume, 172, 176-184
Conductivity, thermal, 242, 245-251
  ASTM steady state method, 246-247
  density effects, 245-246
  determination, 246-251
  electrical insulation, 248
  heat pulse method, 249
  moisture effects, 245-251
  Terada's method, 248
  Terasaki's method, 247-248
  unsteady state method, 249
Conformability, fiber, 345-346
Consistency, pulp, 367
Contact angle, 104-106, 108-110, 118, 145-149, 152
Contact, fiber, 339-344, 346-347, 386
Cosine distribution function, 291-292, 295, 297-299
  shape parameters, 292
  single cosine term, 292
  two cosine term, 292
Coulter counter, 318-319, 367
Count, 347
Creep, 416, 420, 425, 434
Crill (see Fines)
Crimp ratio, 335, 348

Crossing area (*see also* Overlapped area), 348
Curl, fiber, 290, 313, 318, 330, 332, 334-338, 348
  bending factor, 334-336
  crimp ratio, 335, 348
  definitions, 290, 334-337
  determination, 334-338
  factors, 334-336
  in-plane, 334-336
  out of plane, 336-338
Curl, sheet, 290, 312, 416, 426-441
  analysis of, 429-441
  curvature equations, 426-428
  curvature template, 429
  differential wetting effects, 438-439
  drying effects on, 439-440
  dual, 440
  evaluation of, 429-441
  expansimeter measurements for, 434-435
  fiber orientation effects, 434-436
  fiber orientation measurements, 434-436
  layered sheet properties, 427, 433
  measurement of, 429-441
  multilayer, 427, 433
  offset printing, 437-439
  permanent, 428, 431-433
  reversible, 428, 431-433
  temperature effects on, 436-440
  xerographic, 438-439
Curliness (*see* Curl, fiber)
Curvature, fiber, 336-338
Curvature, sheet curl, 426-429
Curve-fitting methods, angular distribution of fibers, 297-299

Darcy's law, 73-74, 77-78, 84, 126-127
DC electrical conductivity, fiber bond area determined by, 389-390

Decomposition, thermal, 259-277
  bond scission, 261-265
  charring, 257-272
  DP (degree of polymerization) changes, 260-277
  high-temperature reactions, 266-277
  inhibition of, 272
  instruments for measuring, 273-277
  isothermal degradation, 259-261
  low-temperature reactions, 259-265
  molecular weight changes, 261-262
  oxidative degradation, 259-261
  pyrolysis, 260-277
  tar formation, 265-272
  weight loss, 260-261
Deinked fiber, 357
Densitometry:
  instrumentation, 23-26, 28, 32
  micro, 25-26
  thickness determination by, 356
Density, 59-61, 213-214, 217-220, 338, 340, 405-406
  determination, 59-61, 406
Depolarization currents, 192-195
Depolymerization, thermal, 259-277
Diameter:
  Feret, 322, 328, 348
  fiber (*see* Width, fiber)
  lumen, 356-357
  Martin, 353
  statistical, 356
Dielectric breakdown (*see* Dielectric strength)
Dielectric constant, 175, 177-181, 184-195, 202, 208, 211-215
  anisotropic, 190-192
  complex, 175
  density effects, 213-214
  fiber, 211-214
  measurement, 221-226, 229-235
  moisture influence, 177-181, 185, 187, 189-190
  temperature effects, 213-215
Dielectric loss tangent (*see* Loss tangent *and* Loss parameters)
Dielectric strength, 175, 180-181, 203, 215-220, 235-238
  AC (alternating current), 217-218, 235

[Dielectric strength]
  air impermeability, 216-219
  barrier effect, 217
  DC (direct current), 217, 235
  damped oscillation, 235
  density effects, 217-219
  formation effects, 217-220
  impulse, 217-219, 235
  measurement, 234-238
  moisture effects, 238
  oil impregnation effects, 216-217
  polymer composites, 220
  process effects, 219
  thickness effect, 217-218
  time effect, 235-236
  uniformity, 217-219
Differential thermal analysis, 276
Diffraction:
  light, 286, 300-303
  X-ray, 286, 305-308, 312, 325, 474-478
Diffuser, perfect reflecting, 30
Diffusion, gas:
  coefficient, 78-80, 84
  permeation, 79-81, 83-84
  solid-solution, 81
Diffusion, liquid, 106-107
  Hoyland apparatus, 107
Digitizer, graphic, 287-288, 313, 320-321, 331-336, 338, 359, 388-389
  bonding determined by, 388-389
  centroid determination by, 331
  curl (fiber) determination by, 334-336
  dyed fiber observations, 332-334
  fiber areas by, 331
  fiber length by, 332-334
  fiber perimeter by, 331
  moment of inertia determination by, 331
  operations flowchart, 320-321
  sheet cross sections, 359, 388-389
  shive identification, 359
  video interface, 331
Dimensions, fiber, 312-358
  fiber-fiber bond, 339-344, 354-355

[Dimensions, fiber]
  fiber-fiber contact, 339-344, 346-347
  swelling, 104-105
Dimensions, sheet, 403-443
  changes, 415-443
  curl, 416, 426-441
  detection of changes, 413-414
  drying restraint effect, 415-420
  environmental changes effects, 413-414
  expansion coefficient, 423-424, 427, 436, 439
  expansion, thermal, 415, 427, 439
  expansion, total, 425
  humidity effects on, 415-441
  hysteresis, 416-417, 424
  measurement of, 403-415, 420-426
  mechanical loading effects, 415-417, 420, 424-425
  moisture effects on, 413-441
  nonreversible, 415-423
  recoverable, 415-423
  reversible, 416-423
  shrinkage parameter, 417-418, 423-424
  stability of, 415-443
  temperature effects, 415, 427, 436-440
  thickness, 420-421
  variation, 415-443
Dioxane method (see Gas drive method)
Direct deflection test for DC conductivity, 227-229
Dislocations, fiber, 348-349, 352-354
Dissipation factor, electrical (see Loss tangent)
Draw, mechanical properties, effect on, 287
Drop penetration tests, 104
Drying:
  B.E.T. area, 387
  curl effects, 427-441
  cylinder, 436-440
  dimensional stability effects, 415-441
  hysteresis removal, 424
  restraint effect, 415-420

[Drying]
  sheet curl effects, 427-441
  solvent exchange, 387
  spray, 387-388
Dwell time, 125
Dynamic retention/drainage jar, 366-367

Edge effect, electrode, 236-238
EDXA (Energy dispersive X-ray analysis), 325
Elastic constants:
  anisotropy related to thermal expansion coefficient, 254
  bonding, 314
  fiber length and, 314
  fiber moduli and, 314
  fiber orientation and, 295, 312
Electrical properties, 171-240
  anisotropic, 172-173, 184-185, 190-192
  CR value (see volume resistance and insulation resistance)
  capacitance, 173-174, 225-226
  conductivity, 172-173, 176-184, 202-203, 208, 226-229, 232-235
  density effects, 213-214, 217-219
  depolarization, 192-195
  dielectric parameters, 175, 177-181, 184-195, 202-238
  dipole moment, 202, 231
  dissipation factor (see Loss tangent)
  fibers, 184, 211-214
  frequency effects, 187-190
  hardboard, 189
  insulation resistance, 173, 210-212, 220, 231-232
  ion concentration effects, 187, 202, 208-210, 220
  life endurance, 235-236
  loss parameters, 174-175, 185, 202-210, 220-235
  measurements, 221-238
  moisture influence, 177-181, 185, 187, 189-195, 233-235, 238

[Electrical properties]
  orthotropic, 172-173
  permittivity, 173-175, 184-190
  polarization, 185, 231-232
  pressboard, 236
  relaxation, 184
  reprographic papers, 232-235
  resistance, 173, 175, 210-212, 226-229
  salt effects, 189
  surface conductivity, 173, 232-235
  surface resistivity, 173, 232-235
  temperature effects, 179-182, 186-189, 192-195, 213-215, 231-232
  time effects, 183-184, 235-236
  V-t test, 235-236
  volume conductivity, 172, 176-184
  volume resistance, 173, 210-212
  volume resistivity, 173, 181-182, 210, 226
  wood, 184, 189
Electrode measurements of dielectric properties, 229-238
  edge effect, 236-238
Elliptical distribution function, 291-292, 295-299
  Fourier expansion form, 296
  shape parameter, 292
Energy:
  bond, 390
  dissipated, 390-391
  fracture, 390
  rupture, 390
  shrinkage, 389-390
  surface free (see Energy, surface free)
Energy, surface free, 145-147, 152-153, 164-165
  Gibbs adsorption isotherm, 146
  tension, 145-146
Equivalent pore, 306-311
Erlang distribution function, 332-334
  cumulative frequency, 333-334
  fiber length analysis, 332-334
ESCA (see Spectroscopy, electron)
Expansimeter, 420-425, 434-435
Expansion, thermal, 251-255
  anisotropic, 254
  density effects, 254

[Expansion, thermal]
  linear coefficient, 251-254
  measurements, 252-255
  moisture effects, 251, 254
  volume coefficient, 253-255
Expansivity, 415-426, 434-436, 439-440
  coefficient, humidity, 423
  coefficient, moisture content, 423-424
  ratio, 435, 440
Extraction tests, fiber, 391-393
Extractives, determination, 367

Felting coefficient, 348
Feret diameter, 322, 328, 348
Fiber:
  alteration in processing, 460-464
  area, contact, 339-344, 346-347, 386-388
  area, crossing, 348, 381
  area, cross-sectional, 312, 328, 331, 339
  area, overlap, 348, 381
  area, projected, 339
  area, relative bonded, 342-344, 382-385, 388-389, 393
  area, surface, 312, 338-340, 344, 351, 356, 382
  aspect ratio, 313, 340-341, 353, 359
  bending factor, 334-336
  bonding, 314, 331, 338-344, 353, 359, 379-402
  cellulose (see Cellulose)
  cell wall structure, 325
  centroid, 344
  chemical composition, 451-460
  circularity, 345
  classification, 446-447, 451-452
  coarseness, 315-318, 330, 338, 343
  collapse, 312, 345-346, 355-356
  composition of pulp, 348
  conformability, 345-346
  convex perimeter, 313, 328-329, 355

[Fiber]
  crossing angle, 381
  crushing, 348
  curl, 290, 313, 318, 330, 332, 334-338, 348
  curvature, 336-338
  deinked, 357
  delamination, 464
  density, linear (see coarseness)
  dimensions, 312-358
  dislocations, 348-349, 352-354, 464
  dyed, for measurements, 285-291, 310-311, 332-334
  end-to-end orientation, 289-290
  extraction tests, 391-393
  felting coefficient, 348
  fibril angle (see also Fibril angle), 306-308, 349-350, 353
  fibrillation, 463-464
  flattening (see collapse)
  flexibility, 314, 338, 350, 353, 383
  fragment (see segment, Fines)
  free length, 350-351
  free surface area, 351
  helical organization, 464-466
  hemicellulose, 456-459
  identification, 334, 448-451
  inorganic, 446-447
  kinking, 313, 328, 330, 334, 352, 354
  length (see also free length and segment length), 287-291, 306-334
  light scattering, 300-305
  lignin, 456-460
  Luce shape factor, 346, 353
  lumen, 339, 356-357
  mass, 316
  mean orientation angle, 310-312
  microcompression, 353-354, 464
  microfibrils in, 452-456, 463-464
  mineral, 446-447
  moduli, 314
  molecular components, 452-460
  moments of inertia, 340, 354, 359
  Mühlsteph ratio, 354
  nodes, 352
  non-woody, 446-447
  orientation distribution, 284-312, 320

[Fiber]
   perimeter, 313, 328-329, 355
   physical composition, 447-480
   pit apertures, 449-451, 470, 475
   plant, 451-452
   polymers, 452-460
   probability of bonding, 340
   processing alterations, 460-464
   recycled, 357
   refined, 463-464
   roundness, 355-356
   Runkel ratio, 356
   S2 angle, 306-308, 353
   segment length, 332-334, 349-351, 379
   segment orientation, 289-290, 306-311, 349
   shape, 338
   shape factors (see length, width, Luce shape factor, etc.)
   shrinkage, 380
   size analyzer, 318-319
   slip planes, 348, 354, 464
   springwood, 459
   strength, 314, 390
   structure, 445-484
   summerwood, 459
   surface area, 312, 338-340, 344, 351, 356
   swelling, 104-105
   synthetic, 446-447
   thickness, 338, 356-357
   topochemistry, 325
   twisting, 313, 356-357
   wall thickness, 356
   width, 312, 315, 330, 343, 357-358, 381-382
   wood, 447-450, 461-462
Fiber-cellophane bond strength, 394-398
Fiber-fiber bond, 314, 331, 338-344, 353, 359, 379-402
   factors affecting strength, 395-398
Fiber network model:
   multiplanar, 382-385
   Poisson distribution, 382
   three-dimensional, 382, 385
   two-dimensional, 382-385

Fiber-shive bond strength, 393-398
Fibril (see Microfibril and Fibril angle)
Fibril angle, 306-308, 349, 353, 464-480
   determination, 464-480
Fibrillation, 350, 380, 387, 464
Fick's laws, 78-79, 105-107
Filament winding angle (see Fibril angle)
Filler:
   determination, 367
   orientation, 312
Fines, 318, 350, 359-367
   classification, 359-365
   definition, 350
   determination, 318, 359, 365-367
   index, 365-366
   inorganic, 367
   primary, 359-360
   secondary, 360-365
Flammability, 257-258, 271-272
Flash point:
   definition, 255
   measurement, 257
Flattening, fiber (see Collapse)
Flexibility, fiber, 314, 338, 350, 353, 383
   bonding effect of, 383
   coefficient, 350
   index, 350, 353
   ratio, 350
Floating time test for absorption, 154, 158-160, 162
Floc:
   size determination, 318
   strength determination, 318
Flotation test, water penetration, 114
Flow, liquid (see also Penetration, liquid)
   boundary layer, 150-151
   capillary, 145-152
   forced, 145
   interfacial, 150
   mass rate test methods, 154-155
   no-slip, 150
   spontaneous, 145
   surface tension, 150
Flowmeter, 82, 87-88
Fluorescence, 14-16, 23, 27

Folding endurance, 314, 338
Formation, 21-26, 28-30, 42-44
  effect on opacity, 43-44
  effect on sheet reflectance, 43-44
  floc frequency, 22
  microdensitometers, 25-26
  on-line control, 28-30
  statistical geometry, 42-44
  testers, 22-23, 25-26
  wavelength spectra, 44
Fracture energy, 390
Free fiber length, 350-351, 379-380
  aspect ratio, 351
  concept, 379
  mean, 351
Free fiber surface area, 351
Frequency, electrical (see Electrical properties)
Frequency, floc, 22
Fresnel reflection, 18, 22, 41
Frictional force theory of fiber bonding, 380
Friele-MacAdams-Chickering color formula, 13
Fullness, fiber (see Circularity)

Gas adsorption:
  bonding state determination by, 387-388
  specific surface measurement by, 61-64, 387-388, 390
Gas drive method for pore size distribution, 65, 70-73
  Hagen-Poiseuille equation, 70-71, 74-75
  liquid head model, 71-73
  saturation model, 71-73
Gas liquid chromatography, 265-268, 275-276
Gas permeability, 73-92
  mass coefficient, 77-78
  mean free path, 77
  measurement, 82-88
Gloss, 16-20, 40-42
  color effects, 42
  contrast ratio, 17-20
  goniophotometers, 19, 24-25, 32

[Gloss]
  interpretation, 40-42
  luster, 17-19
  meter, 18, 23-27, 32
  sheen, 17-18
Goodness of fit test, angular distribution functions, 297-299
Grammage (see Basis weight)
Gravimetric test for water vapor permeability, 89-90
  cup test, 89-90

Hagen-Poiseuille equation, 70-71, 74-75
Heat (see also Thermal properties)
  pulse method for conductivity, 249
Hemacytometer, 359
Hemicellulose, 456-459
  abundance, 458-460
  molecular weight, 456-457
Hercules size test, aqueous liquids, 114
Hoyland's apparatus, 107
Humidity:
  curl effects, 427-441
  dimensional stability effects, 415-441
  expansion coefficient, 423, 436, 439
  hysteresis, 416-417, 424
  mechanical properties effects, 425-426
  sheet curl effects, 427-441
Hydrogen bond, 380, 389-390
Hydrophobicity, 103-104, 111-113
Hygroexpansion, 254
Hygroexpansivity (see Moisture changes and Expansivity)

Ignition temperature, 255, 257
  definition, 255
  measurement, 257
IGT print tester, 112-113
Illuminants, 9-16
Image analysis, 287-288, 313, 319-331, 338
  automatic, 319-331

[Image analysis]
  bond problem, 331
  crossed fiber problem, 329-330
  curl (fiber) determination by, 336
  digitizing, 287-288, 313, 320-321, 331-336
  illumination, 325-327
  measurements, 320-338
  pattern recognition, 319-320
  SEM-based, 321, 323-325, 338
  semi-automatic, 287, 320, 338
  spatial perception, 320
  TV-based, 321-324
Immersion test, 117-118
Impact delamination test, 393
Inclined plane test for absorption, 154, 160-161
Inertia, moments of:
  centroid, 344
  fiber, 340, 354, 359
Infrared absorption test, water penetration, 113-117
Ink transfer equation, 128
Inorganic constituents:
  determination, 367
  orientation in sheets, 312
Insulation resistance:
  composites, 220
  current effects, 231-232
  electrical, 173, 210-212, 220, 231-232
  fluoroethylene propylene, 220
  laminates, 220
  polymethyl pentene films, laminates, 220
  polypropylene film, 220
  polypropylene-laminated paper, 220
  silicone grafted polyethylene, 220
Internal bond test, 393

Kinks, fiber, 352, 354
Knudsen flow, 77-78, 81
Kozeny-Carman equation, 76, 78, 126
Kubelka-Munk theory, 5, 20-21, 26, 32-39, 62, 387

[Kubelka-Munk theory]
  equation of, 62
  limitations, 36-39

L/D ratio, 352
L/T ratio, 353
L/W ratio, 352
Lambert-Beer Law, 131
Laplace equation, 105-106, 147
Leakage current, electrical, 232
Length, fiber, 290, 306-334, 350-351, 355, 359
  aspect ratio, 350-351
  classification, 315-318, 332-334
  determination of, 287-291, 310, 313-334, 355, 359
  distribution, 290, 306-318, 328-334
  end-to-end, 332-334
  folding endurance relation, 314
  free, 350-351, 379-380
  map curvimeter (map reader), 319, 329
  number average, 315-318
  planimetry, 319
  segment, 332-334
  sheet moduli relation, 314
  sheet strength relation, 314
  weighted average, 315-318
  wet strength relation, 314
  wood, 314, 318
Length, sheet, measurement methods, 412-414
Life endurance, electrical, 235-236
Light:
  absorptance, 3, 32-39
  absorption coefficient, 33
  color wavelengths, 7-9
  diffraction (see scattering)
  dominant wavelengths, 9
  illuminant standards, 9-14
  intensity, 3, 22
  interactions with paper, 1-54
  luminance factor, 16
  luminosity, 2, 16
  reflectance, 3-20, 27-44
  scattering, 32-40, 286-287, 300-305, 387-388, 390-391, 475

[Light]
  transmittance, 3, 20-23, 28-30, 32-40
Lightness (color intensity), 7
Lignin, 456-460
  abundance, 458-460
  location in wood fiber, 459-460
  molecular weight, 458
Loss parameters, electrical, 174-175, 185, 202-210, 220-235
  angle, 174
  current, 174-175
  dielectric, 204
  factor, 175
  tangent (see Loss tangent)
Loss tangent, 174-175, 202-210, 220-235
  ash effects, 210
  carboxyl effects, 208
  cellulose effects, 208-210
  hemicellulose effect, 206-208
  lignin content effects, 204-206
  measurement, 221-235
  metallic ion effects, 210, 220, 231
  moisture effects, 233-235
  polymer composites, 220
  temperature effects, 231
Lucas-Washburn equation, 106, 110, 126, 132, 137, 149, 151
  absorption rate, 137
Luce shape factor, 346, 353
Lumen:
  area, cross-sectional, 339
  diameter, 356-357
Luminosity, 2, 16
Luster, 17-19

MD/CD/ZD properties (see Anisotropy)
Magnetic scanning, 138-139
Map curvimeter (map reader), fiber length determination by, 319, 329
Martin diameter, 353
Mass:
  fiber, 316
  pulp, 339

Mass spectroscopy, 273-275
Mechanical entanglement theory of fiber bonding, 380
Mechanical properties (see specific property of interest)
Mercury:
  compression during intrusion, 66-67
  contact angle, 67
  density determination, 59-60
  intrusion method for pore size distribution, 65-70
Microcompressions, 353-354
Microfibril, 452-456, 463-480
  angular orientation, 464-480
Micrometer, thickness-measuring, 409-413
Microprocessors:
  on-line control, 28-30
  optical measurement, 28
Microscopy:
  bonding state determined by, 386-387
  fiber dimensions determined by, 313-315, 338, 356, 359
  fluorescence, 356
  projection, 314-315, 338, 359
  scanning electron (SEM), 321, 323-325, 338
  sheet thickness determined by, 409
Mie theory, 36, 39-40
Moisture (see also Moisture changes)
  electrical properties effects, 177-181, 185, 187, 189-195, 233-235, 238
  on-line monitoring, 28
  removal, 307
Moisture changes:
  curl effects, 427-441
  differential wetting, 438-439
  dimensional stability effects, 413-441
  expansion coefficient, 423-424, 427, 436, 439
  hysteresis, 416-417, 424
  mechanical properties effects, 425-426
  sheet curl effects, 427-441
Mottle, 20
Mühlsteph ratio, 354

Multilayer sheets, curl properties of, 427, 433
Munsell color charts, 12

Necking, 354-355
Network model:
   multi-planar, 382-385
   Poisson distribution, 382
   three-dimensional, 382, 385
   two-dimensional, 382-385
Nitrogen adsorption (*see* Gas adsorption)
Nodes, fiber (*see* Kinks)
Nonfiber components (*see* Filler *and* Inorganic constituents)
Nordman bonding strength, 390

On-line control:
   beta ray, 28
   color, 30
   infrared, 28
   microwave, 28
   optical properties, 28-30
Opacity, 4-6, 20-23, 37, 43-44, 338
   factor, 37
   formation effects, 43-44
   paper backing, 5, 20
   printing, 5, 20
   standard methods, 4-6
   white backing, 5
Optical properties, 1-54, 300-307, 338, 380, 384, 386-388
   brighteners, 14-15
   character recognition, 20
   coatings, 5, 34-39
   colorimetry, 6-14
   fiber contact area, 386-388
   fluorescence, 14-16, 23
   formation, 21-23, 42-44
   gloss, 16-20
   instrumentation, 23-28
   on-line control, 28-30
   opacity (*see* Opacity)
   reflectance, 3-20, 27-44

[Optical properties]
   standardizing laboratories, 30-31
   transmittance, 3, 20-23, 28-30, 32-40
   transparency, 21
   unbonded sheets, 387
   whiteness, 14-16
Orientation, fiber:
   best fit of center points, 289-290
   crossing angle, 381
   curl effects of, 434-436
   determination, 285-312, 320
   direct methods for determining, 285-299
   distribution of fibers, 284-312, 320, 381
   distribution of reflected light, 19
   end-to-end of fiber, 289-290
   equivalent pore method, 306-311
   fiber crossing angle, 381
   fiber segment, 289-290, 306-311, 320
   filler particles, 312
   functions to represent, 291-299, 320
   indirect methods for determining, 286-287, 300-311
   inorganic constituents, 312
   mean directionality, 293-295, 310-312
   mechanical properties and, 286-287, 295, 312
   methods for determination of, 285-312, 434-436
   polar diagrams of, 286
   sampling for, 291
   sector size, choice of, 291
   sheet curl effects of, 434-436
   Silvy's method, 306-311
   strength and, 286-287, 312
   Z-directional, 312
Orifice test for absorption, 154-157
Overlapped area (*see also* Crossing area)
   effective length, 381
Oxygen index:
   definition, 256
   measurement, 258

Page's tensile strength equation, 392-393
Particle analysis, 318-319, 322, 325, 338, 359, 367
  chemistry, 325
  counting, 318-319, 338, 359, 367
  EDXA, 325
  shape, 322, 325
  size, 318-319, 325, 328, 359, 367
Pattern recognition, 319-320
Penetration, adhesive, 129
Penetration, ink, 127-134
  filtration mechanism, 128-129, 132-133
  measurement, 130-134
  oil front, 133
  phase separation, 128-129
  reflection method, 130-131
  transfer equation, 128
Penetration, liquid (see also Flow, liquid), 103-119
  absorbency rate, 130
  Bristow apparatus, 110-111, 113, 119
  capillary, 145
  Cobb test, 104, 118
  conductance test, 113-115
  diffusion, 106-107
  drop test, 104, 114
  dye test, 114
  Fick's laws, 78-79, 105-107
  forced, 124-129, 145
  gluing, 118
  Hercules size test, 114
  immersion test, 117
  infrared absorption, 113-117
  Laplace equation, 105-106, 147
  Lucas-Washburn equation, 106, 110, 126, 132, 137, 149, 151
  measurement, 130-134
  multiple internal reflection spectroscopy, 115-116
  nonaqueous, 123-141
  organic, 133-134
  polarity test, 113
  printing, 118
  rate, 113-118
  spectroscopy, 115-116
  spontaneous, 145

[Penetration, liquid]
  test methods, 110, 113-118
  theory, 124-129
  uptake, total, 117-118
  volumeter, 133-134
  wave propagation, 113, 117
  wettability, 129
Penetration, oil, 131-134
  front, 133
  measurement, 131-134
  volumeter, 133-134
Perimeter, 313, 322, 328-329, 331, 355, 359
  boundary, 355, 359
  convex, 313, 328-329, 355
Permeability:
  air, 73, 84, 88, 127, 307
  cell, 82, 83
  coefficient, 74, 83-84
  flow rate measurement, 82, 87-88
  gas, 73-92, 307
  intrinsic, 127
  measurement, 82-92
  volume coefficient, 74, 83-84
  water vapor, 88-92
Permeation:
  diffusion-type, 79-81, 83-84
  high-vacuum measurement, 83-84
  low-vacuum measurement, 83
  measurement, 83-87
  water vapor, 81, 88-92
Permittivity, electrical, 173-175, 184-190
  complex, 173-175
pH, 117
  aqueous liquid, 107
  paper, 107
Phase separation, 128-129
Pick resistance, 462
Poiseuille equation, 148-149
Poiseuille's Law, 126
Poisson distribution, 382
Polarity:
  surfactants, 105
  surface, 113
Polarized light, fibril angle measurement by, 473-474, 476-479
Pores, internal (see Pore space, Pore size distribution,

[Pores, internal], Porosity, and Voids)
Pores, surface:
　depth, 135-136
　magnetic scanning for, 138-139
　roughness related to, 135-136
　size distribution, 135-139
　width, 139
Pore size distribution, 58, 64-73, 135-139, 146-149, 151
　measurement, 64-73, 135-139
　　capillarity methods, 65-73
　　gas drive methods, 65, 70-73
　　mercury intrusion, 65-70
　　sorption, 65
　　X-ray small angle scattering method, 64
　number, 70-73
　Stone-Scallan plot, 68
Pore space:
　air permeability of, 307
　effective, 58
　equivalent, 306-311
　fiber, 58
　gas permeability of, 307
　'ink-bottle,' 68-70
　moisture removal from, 307
　optical properties related to, 307
　radius, 146-149, 151
　sheet, 57-59
　size distribution, 58, 64-73, 146-149, 151
　strength related to, 307
　volume proportion, 58, 144
Porosity, 58, 59-61, 144-145, 163, 338
　definition, 58
　influence on absorption rate, 163
　liquid transport, 144-145
　measurement, 59-61
　　gas displacement, 59-61
　　liquid displacement, 59-60
Porous plate test for absorption, 154, 158, 162
Print quality, 139
Print testers:
　IGT, 112-113
　Prüfbau, 112-113
Printing compression, 127

Probability, bonding state, 340, 342, 344, 359
Probability functions:
　fiber length distribution:
　　Erlang, 332-334
　　Rayleigh, 334
　fiber orientation distribution, 291-299, 320
　mean fraction of fibers that overlap Poisson distribution, 382
Projection:
　area of fibers in pulp, 339
　fiber dimensions determined by, 314-315, 338, 359
Prüfbau print tester, 112-113
Pulp (see also Fiber)
　coarseness, 316-318, 338
　consistency, 367
　count, 347
　deinked, 357
　extractives content, 367
　fiber composition, 348
　fines, 318, 350, 359-367
　fractionation, 315, 367
　projected area, 339
　recycled, 357
　shive, 356, 358-359
　sliver, 356, 358-359
　specific surface, 356
　wetting times, 111
Pycnometer, mercury, 409-410
Pyrolysis (see also Decomposition, thermal), 259-277
Pyrolyzers, 273

Radius, hydraulic, 75-76
　shape factor, 75-76
Recycled fiber, 357
Reflectance, 3-20, 27-44
　cell wall, 39, 44
　coating stock, 34-39
　color effects, 42
　diffuse blue factor, 6
　diffuser, perfect reflecting, 30
　directional blue factor, 6
　formation effects, 43-44
　Fresnel reflection, 18, 22

[Reflectance]
   gloss, 16-20, 27, 40
   Kubelka-Munk theory, 5, 20-21, 26, 32-39, 62, 387
   on-line control, 28-30
   reflectometry, 3-6, 20, 27, 30-32
   scattering (see also Scattering), 33-40
   spectral, 6
   specular, 16-20, 40-42
   variation with angle, 18-20
   variation with surface finish, 19, 20
Reflection method, 130-131
Refractive index, 3, 18, 34, 40
Relative bonded area, 342-344, 382-385, 388-389, 393
   experimental, 388-389, 393
   maximum potential, 382
   theoretical, 382-385
Repellency, ink, 112-113
Repellency, water, 104, 112-113
   fatty acid, 103
   hydrophobic agents, 104
   IGT tester, 112-113
   Prüfbau tester, 112-113
   resins, 103
   self sizing, 111
   sizing agents, 104, 112
Resistance, electrical:
   AC (alternating current), 175, 229-235
   DC (direct current), 226-229
   insulation, 173
   volume, 173, 210-212
Resistivity, electrical:
   surface, 173, 232-235
   volume, 173, 181-182, 210, 226
Resonance methods for dielectric properties tests, 221, 224-226
   Q meter, 224-225
   variable capacitance, 225-226
Reynolds number, 77
Rise-canceling velocity, 108-109
Roughness, sheet, 41-42, 111, 135-136, 148, 411-412
   index, 111
   optical contact measurement, 136

[Roughness, sheet]
   pore depth distribution, 135-136
   stylus method, 135-136
Roundness (see also Circularity), 355-356
Runkel ratio, 356
Rupture energy, bond, 390-391

S_2 angle (see also Fibril angle), 306-308, 349, 353, 464-480
SEM (Scanning electron microscopy), 321, 323-325, 338
Saturation, color, 9
Saturation gradient, 159
Scattering (see also Light and X-ray diffraction):
   coefficient, 33, 387-388, 390-391
   fiber contact (bond) determined by, 32-40, 387-388, 390-391
   fiber orientation determined by, 286-287, 300-308
   fibril angle determined by, 475
   heterogeneous sheets, 34-36
   homogeneous sheets, 32-34
   Kubelka-Munk theory, 5, 20-21, 26, 32-39, 62, 387
   Mie theory, 36, 39-40
   particulate, 39-40
   small angle light, 286, 301-305, 475
Schering bridge, 221-223
   conjugate, 223
Scott bond test (see Internal bond test)
Segment, fiber, 289-290, 306-311, 332-334, 349-351, 379, 384
Self-sizing, 165
Shape factors, fiber (see Length, Width, Curl, Luce shape factor, etc.)
Shear strength (see also Strength):
   bond length effect, 395-398
   fiber-cellophane bond, 394-398
   fiber-fiber bond, 391-398
   fiber-shive bond, 393-398
Shear stress distribution:
   adhesive bond, 395-398
   analysis for bonding, 393-398

[Shear stress distribution]
  fiber-fiber bond, 392, 395, 398
Sheen, 17-18
Sheet dimensions (*see* Dimensions, sheet)
Sheet splitting, 290, 332, 367
Sheet structure, 283-390
  determination, 283-367
  fiber-fiber bond in, 379-390
  models, 381-385
Shive:
  bonding, 393-398
  definition, 356
  determination, 358-359
Shive-fiber bond strength, 393-398
Shrinkage:
  energy in drying, 389-390
  fiber cell wall, 380
  sheet (*see* Dimensions, sheet)
Size analyzer, particle, 318-319, 325, 328, 359, 367
Sizing agents, 104-105, 111, 112
  self-sizing, 105, 111
Slip planes, fiber, 348, 354
Sliver, 356, 358-359
Solids content, mechanical properties, effect on, 287
Solubility coefficient, 80
Solvent exchange drying, B.E.T. area, 387-388
Spatial perception, 320
Specific heat, 242-245
  measurement methods, 243-245
Specific surface area, sheet, 58-64
  B.E.T. equation, 61-64
  definition, 58
  measurement, 61-64
    dynamic nitrogen adsorption, 62-64
    gas adsorption, 61-64
    optical method, 62
    solution adsorption, 62
Speckle, 20
Spectroscopy:
  electron, 119
  fluorescence, 15-16, 23
  multiple internal reflection, 115-116
  spectrophotometers, 23-24, 28-29

Splitting, sheet, 290, 332, 367, 433-434, 436
Spray drying, B.E.T. area, 387-388
Spreading, liquid, 103-105, 146-148, 151-152
  coefficient, 146, 148, 152
  pressure, 146
  primary, 151
Stone-Scallan plot of specific displacement volume, 68
Strain gage, 414
Strength, color, 12
Strength, mechanical (*see also* Breaking length)
  anisotropic, 286-287, 312
  basic, 404-405
  bond, 314, 380, 390-398
  delamination, 393
  factors influencing, 314, 338, 380, 395-398, 406
  fiber, 314, 390
  force per cross-sectional area, 405-406
  impact delamination, 393
  internal bond, 393
  measurement of, 404-406
  Nordman bonding, 390
  Page's tensile strength equation, 392-393
  specific, 405
  tearing, 314, 340
  tensile index, 340
  units, 404-406
  void configuration related to, 307
  wet, 144, 314
  Z direction, 334, 393
  zero span tensile, 286-287, 310-311
Stress:
  adhesive bond, 395-398
  extension, 424-425
  internal, 416, 425
Stress distribution (*see* Shear stress distribution)
Stretch, mechanical properties, effect on, 287
Subjective print quality, 139
Surface conductivity, 173, 232-235
Surface finish, 16-20, 41-42

Surface free energy, 103-107, 110, 145-147, 152-153, 164-165
  Gibbs adsorption isotherm, 146
  tension, 145-146
Surface polarity test, water penetration, 113
Surface resistivity, 173, 232-235
Surface, specific:
  determination, 61-64, 387-388, 390
  external, 387
  internal, 356, 387
Surface tension, 146, 151-152, 380
  cohesive, 146
  interfacial, 146
  spreading, 146, 151
Surfactants, 105, 109
Swelling, 104-105

Tear strength, 314, 340
Temperature, color, 9-12
Temperature, drying, 427-441
Temperature effects (*see also* Thermal properties)
  electrical, 179-182, 186-189, 192-195, 213-215, 231-232
  ignition temperature, 255
  sheet curl effects, 436-440
  thermal expansion, 415, 427, 439
Tensile index, 340
Tensile strength (*see also* Strength), 286-287, 310-311, 314, 338, 340, 392-393, 404-406
  units of measurement, 404-406
  Z direction, 334, 393
  zero-span test, 286-287, 310-311, 392-393
Tension, surface, 146, 151-152
  cohesive, 146
  interfacial, 146
  spreading, 146, 151
Thermal analysis, 270, 276
Thermal depolarization, 192-195
Thermal expansion, 415, 427, 439
Thermal properties, 241-279
  anisotropic, 254

[Thermal properties]
  burning energy, 259
  burning point, 256
  capacity, 243
  combustion, 255-259
  conduction, 242-251
  conductivity, 242, 245-251
  decomposition, 259-277
  depolymerization, 260-277
  dimensional change, 251-255
  expansion, 251-255
  fibers, 245, 254, 277
  flammability, 257-258, 271-272
  flash point, 255, 257-258
  free radical formation, 261-265
  ignition temperature, 255, 257
  linear expansion coefficient, 251-254
  measurement methods, 243-259, 273-277
  moisture effects, 245-251, 254
  oxygen index, 256, 258
  pulp, 245, 250, 254, 276-277
  pyrolysis, 259-277
  smoldering, 271-272
  specific heat, 242-245
  transition, 254
  volume expansion coefficient, 253-255
  wood, 274-275
Thermogravimetry, 261, 270-271, 276-277
Thickness, fiber, 338, 356-357, 382
  bonding area dependency on, 382
  cell wall, 356-357
  Runkel ratio, 356
Thickness, sheet, 405-414, 420-421
  caliper versus micrometer, 411-413
  dimensional variation, 420-421
  effective, 406-411
  Fourier analysis of, 414
  measurement of, 405-413
  moisture effect on, 413
  on-line control, 28-30
  profile, 408
  roughness effect on, 411-412
  stack, nested, 412
Time effects, 183-184, 235-236
  relaxation, electrical, 184

Tissue absorbency, 143-168
Toweling absorbency, 143-168
Transformer bridge, 223
Transient current, electrical, 231
Transmittance, optical, 3, 20-23, 28-30, 32-40, 44, 131-132
   cell wall, 39, 44
   coating stock, 34-39
   contrast ratio, 20
   intensity of light, 3, 22
   Kubelka-Munk theory, 5, 20-21, 26, 32-39, 62, 387
   oil-penetrated paper, 131-132
   on-line control, 28-30
   opacity, 20-23, 37
   transparency ratio, 21
Transparency, 21
Turbulence, 77
   Reynolds number, 77
Twisting, fiber, 313, 356-357

Ultrasonic propagation technique (see Wave propagation test)
Ultraviolet method for fibril angle determination, 466-472, 479-480
Uptake, water:
   electrical conductivity test, 113-115
   immersion test, 117-118
   infrared absorption test, 113-117
   swelling, 104-105

van der Waals' forces, 380
Voids, 104-105, 125-126, 135-139, 307
   air permeability of, 307
   distribution, 137-139
   gas permeability of, 307
   magnetic scanning for, 138-139
   moisture removal from, 307
   optical properties of, 307
   size, 137-139
   spectral diagram, 139

[Voids]
   strength properties related to, 307
   surface, 135-139
   swelling effects on, 104-105
Voltmeter-Ammeter method for DC conductivity, 228-229
Volume conductivity, 172, 176-184
Volume resistance, 173, 210-212
Volume resistivity, 173, 181-182, 210, 226
Von Mises distribution function, 291-296, 299
   Fourier expansion form, 295-296
   shape parameter, 293-295
V-t (Voltage-time) test (see Life endurance, electrical)

Wadding absorbency tests, 159-160, 162
Water vapor permeability, 88-92
   closed cell method, 90-91
   comparison method, 90, 92
   gravimetric method, 89-90
   sweep gas method, 90-92
   test condition, 89
Water vapor transmission, 89-90
   transfer rate, 90-92
Wavelength, 7-9, 44
   color, 7-9
   dominant, 9
   spectra, 44
Wave propagation test, water penetration, 113, 117
Web shrinkage energy, 389-390
Wettability, 129
Wetting (see also Absorption, liquid and Moisture changes)
   capillary walls, 146-147
   contact angle, 104-106, 108-110, 118, 145-149
   demand wettability, 156-157, 162
   factors (influences), 162-165
   gluing, 118
   heat, 119
   hydrophobic agents, 104
   kinetics, 148-152

[Wetting]
   Laplace equation, 105
   printing, 118
   repellency, 104, 112-113
   spreading, liquid, 103-105
   strength relation, 144
   surface free energy, 103-107, 110
   test methods, 107-113, 152-162
   theory, 104-107, 144-152
   times, pulp, 111
Whiteness:
   brighteners, 14-15
   fluorescent, 15
   formula, 16
Width, fiber, 312, 315, 330, 357-358, 381-382
   bonding area dependency on, 381-382
   determination of, 357-358
   diameter analysis, 330
   grid analysis, 330
   image analysis, 330
Width, sheet, measurement methods, 412-414
Wipe-dry test for absorption, 154, 161
Wood:
   cell types in, 447-462
   chemical composition, 451-460
   fiber types, 447-450, 461-462
   fibril angle (see Fibril angle)

[Wood]
   hardwood, 447
   parenchyma, 447-449, 461-462
   planes of, 447-451
   softwood, 447
   springwood, 459, 462
   structure, 447-451
   summerwood, 459
   vessel elements, 449-451, 461-462

X-ray diffraction:
   EDXA (energy dispersive X-ray analysis), 325
   fiber orientation determined by, 286, 305-308, 312
   fibril angle measurement by, 474-478
Xylem (see Wood)

Z direction fiber orientation, 312, 334, 336-338
   out of plane curl, 336-338
   Z direction sheet properties and, 334
Z direction strength, 334, 393, 462
Zero span tensile test, 286-287, 310-311

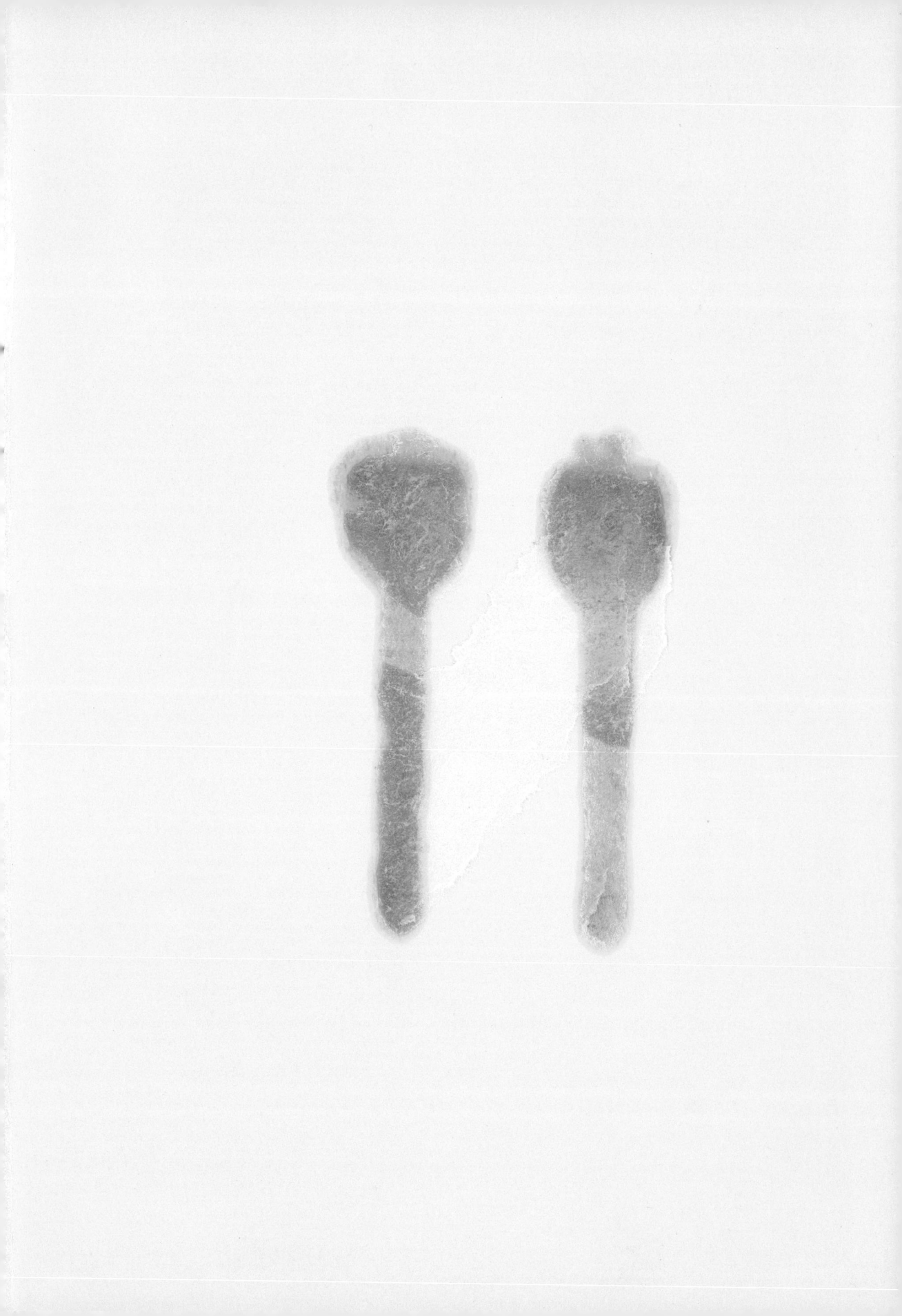

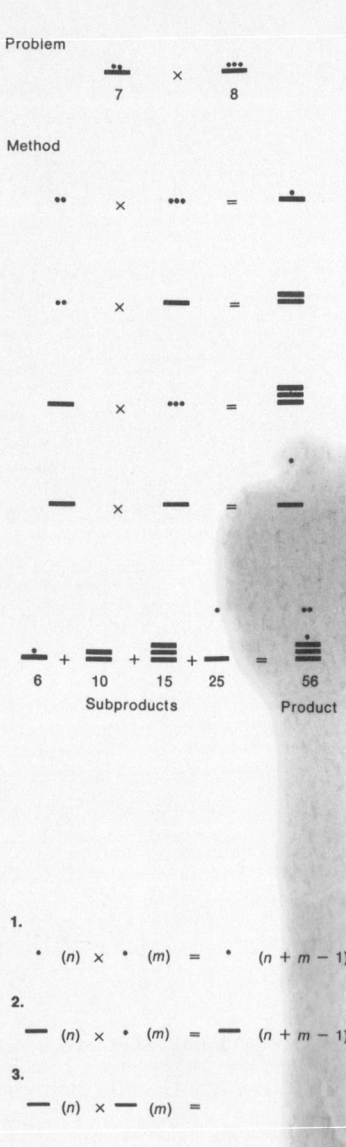

The procedures for multiplication and division described here are his, amplified by our codification of rules.

The single-digit (0–9) Arabic multiplication table contains one hundred entries, but there are actually only thirty-eight distinct operations when regularities (0 times anything is 0; anything times 1 is itself) and commutations ($2 \times 3 = 3 \times 2$) are discounted. Since Maya notation contains only the dot and the bar, only three rules are needed to describe their interactions. These rules can be derived from the Arabic counterparts: (1) since $1 \times 1 = 1$, it follows that dot $\times$ dot = dot; (2) since $1 \times 5 = 5$, it follows that dot $\times$ bar = bar; and (3) since $5 \times 5 = 25$, then bar $\times$ bar = (bar + dot-in-the-next-higher-register)—i.e. 5 + 20, or 1.5 vigesimally. These three rules are equivalent to the thirty-eight operations of Arabic multiplication.

We must also have procedures for determining the "place," or order of magnitude of the product, which in Maya notation corresponds to the register into which the product is placed. To determine the order of magnitude, or highest place of the product, in Arabic notation, in multiplication problems that do not involve carrying, we can take the number of the highest place of the multiplicand and add 1 less than the highest place of the multiplier. Thus any number multiplied by a 1-digit multiplier will give a product with the same number of places as the multiplicand, since we add $1 - 1$, or 0—e.g. $2 \times 43 = 86$; 43 and 86 each have 2 places. (We arbitrarily place the multiplier first and the multiplicand second here.) Multiplication by a 2-digit multiplier adds 1 to the number of places in the multiplicand—e.g. $16 \times 324 = 5,184$—and so on.

Maya multiplication is exactly analogous. A first-register multiplier gives a product in the same register as the multiplicand (1s times any order of magnitude gives the same order of magnitude). A second-register multiplier gives a product in the next register above that of the multiplicand (a multiplier in the 20s register gives a product 1 order of magnitude above the multiplicand). A third-register multiplier gives a product in the second register above that of the multiplicand, and so on. This procedure for determining registers is analogous to collecting exponents in base-10 multiplication. In all these examples, carrying may promote the product to a higher register whenever 4 bars collect in any given register.

Let us consider the simple problem 7×8. Each dot in the multiplicand (8) when multiplied by each dot in the multiplier (7) gives a dot, for a total of 6 (the first subproduct). The multiplicand bar when multiplied by the multiplier dots give 2 bars. The multiplicand dots when multiplied by the multiplier bar give 3 bars. Finally, the multiplicand bar when multiplied by the multiplier bar gives a bar in the same register and a dot in the next higher register. Collection of these four terms gives the product, 56 (2.16 vigesimally). The actual process is far less cumbersome than just outlined. The operation should be done by continuously summing the intermediate figures, as would be done on an abacus or with beans and pebbles on a flat surface. Each subproduct is added directly into the developing sum, and thus when the final operation is completed, the product results automatically.

The problem just given included only first-register numbers. For problems involving higher registers, the order of magnitude of the subproducts and products must be determined. This calculation can be made directly with three rules for the interaction of the symbols. In these rules, n and m refer to the register number: n and m are 1 for the units register, 2 for the 20s, etc. The rules are general for all registers. Thus a dot in any register n times a dot in any register m yields as the product a *dot* that goes into the register just below the one that corresponds to the sum of n and m ($n + m - 1$). A bar times a dot yields a *bar* that goes into the ($n + m - 1$) register. Finally, a bar times a bar gives a *dot* in the ($n + m$) register and a *bar* in the ($n + m - 1$) register.

Take, for instance, the problem 6×106. The dot in the multiplier (6) times the dot in the multiplicand (106) (both in the first register) gives a dot for the product in the first register ($1 + 1 - 1 = 1$). The multiplier dot times the multiplicand bar in the first register gives a bar in the first

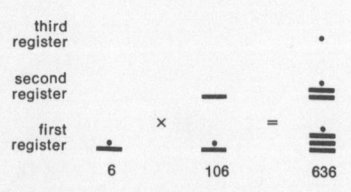

Pages from a modern scientific article dealing with arithmetic operations with the Maya number system. Paper continues as the principal medium for acquisition, storage, and retrieval of human knowledge. Reprinted with permission from "Maya Arithmetic" by Joseph B. Lambert, Barbara Ownbey-McLaughlin, and Charles D. McLaughlin, *American Scientist* 68(3):249-255 (1980).